POULTRY: FEEDS AND NUTRITION

SECOND EDITION

POULTRY: FEEDS AND NUTRITION

Second Edition

Homer Patrick, Ph.D.

Professor Emeritus
West Virginia University
Morgantown, West Virginia

and

Philip J. Schaible, Ph.D. (Deceased)

Formerly, Professor, Poultry Science
Michigan State University
East Lansing, Michigan

avi

AVI PUBLISHING COMPANY, INC.
Westport, Connecticut

Library of Congress Cataloging in Publication Data

Patrick, Homer.
  Poultry, feeds and nutrition.

  Bibliography: p.
  Includes index.
  1. Poultry—Feeding and feeds. I. Schaible, Philip John, 1899-   joint author. II. Title.
SF494.P38 1980   636.5′085   80-17202
ISBN 0-87055-353-4
ISBN 0-87055-360-7 (pbk.)

Printed in the United States of America

Preface to the Second Edition

This new edition presented to the author both an opportunity and a challenge. It provided an opportunity to include new fundamental and applied material, as well as a challenge to present the material as simply as possible.

To accomplish this task without enlarging the book necessitated rewriting most of the text. In presenting the new material, heavy reliance was placed upon nutrition conference publications where both fundamental academic and commercial data were evaluated. General discussions were limited and often replaced with more positive findings.

The general discussions on feed ingredients have been reduced. The Association of Feed Control Officials' tentative and official definitions of feed ingredients along with the permanent National Research Council numbers are presented for most of the ingredients used in poultry feeds.

To simplify reading the text, most of the principal references have been placed in the Appendix under 40 divisions. This type of organization allows the student to study without interruption, yet have a source of references.

Poultry are biological machines operating on chemical energy. They are endothermic and maintain a body temperature of approximately 41.7°C. The chemicals required by poultry are the nutrients which must be provided by the ration. These nutrients must be supplied in amounts dictated by genetic makeup and environmental conditions. The nutritionist must always be aware of the influence of genetic makeup on poultry and provide a ration adequate in the nutrients for each strain as set forth by breeding.

Although rations are composed chiefly of natural feedstuffs, the nutritionist must keep abreast of newer knowledge dealing with the fundamental biochemistry of poultry.

Today, rations are of a simple corn-soybean type structure. As we use more byproducts, a knowledge of nutritional biochemistry will aid greatly in formulating the rations of tomorrow.

The field of poultry nutrition is a rapidly changing field and appears to be continually undergoing adjustments.

Poultry production has displayed more progress than any other field of agriculture, requiring the efforts of many applied and professional helpers. Biochemists, nutritionists, geneticists, pathologists, veterinarians, engineers, food processors, industrialists, feed and equipment manufacturers, drug sanitarians, statisticians, farmers, as well as financial houses and departments of government have made contributions, along with numerous others.

The desire to raise chicks has always been strong in man, and it has been said that he has a genetic factor guiding him into the poultry business.

Poultry is often called a fluid or unstable business because of the short time required to develop and close out the operation. It is frequently entered during an economic or sociological transition. For example, unemployed ocean-going men became poultrymen in New Jersey. Timber harvesters and former cotton producers in Georgia, Arkansas and southeastern Missouri attempted to obtain a new livelihood through poultry during the depression in the 1930s. When crop production became so poor that a normal family life could not be maintained, grain farmers became poultry producers.

Nutrition research resulted in formulated rations which were manufactured by feed mills. The manufacture of poultry feeds today accounts for most animal feed tonnage. As feeds improved, undesirable characteristics in poultry became evident, and through breeding, the industry was provided with more desirable birds. Slow feathering in broilers became noticeable with increases in barebacks. This was corrected by geneticists upon the discovery of a rapid-feathering gene. Specialists in genetics soon developed birds capable of producing more and better eggs, as well as more efficient broilers and turkeys which grew and developed at a rapid rate. A concerted effort by scientists and industrialists is necessary to produce the poultry of today.

Poultry nutrition has been a very active field, supported by all those associated with poultry as well as allied industries. Today, no field of animal or plant nutrition is better informed in fundamental and applied nutrition or the manufacture and distribution of feeds, than is the field of poultry production. Discoveries in poultry nutrition have supported human and farm animal nutrition. Vitamin K, pyridoxine and vitamin B_{12}, all discovered through chick nutrition research, found immediate and direct application in human nutrition and medicine.

Today, many eggs are used in the production of vaccines and in the isolation and characterization of pathogens affecting man. Day-old chicks, which are readily available and cheap throughout the year, are one of the most desirable laboratory tools.

The purpose of this nutrition and feed production treatise is to aid students, scientists, farmers and industrialists in finding better uses for poultry in the poultry field and in using poultry products, such as eggs and chicks, to solve problems.

The author gratefully acknowledges the assistance of the city of Denver, Colorado, and the Arapahoe County libraries. Without their collection of reference books, the revision would have been most difficult. Appreciation is also expressed to the National Science Foundation and the Association of American Feed Control Officials for use of their published material. He wishes to thank Barbara Cowan for her artwork, and Dinette Clark and Pam Bergin for typing the manuscript.

This author would also like to express his appreciation to Dr. D.K. Tressler, Dr. N.W. Desrosier, Mrs. Elizabeth Gerteiny, Ms. Barbara Flouton and Ms. Lisa Melilli of the AVI Publishing Company for their encouragement and assistance in bringing this book into being.

HOMER PATRICK, Ph.D.
Denver, Colorado

January 1980

Preface to the First Edition

From a source of the farmer's wife's pin money to a mass-produced, supermarket loss leader, poultry has made tremendous strides in the past half century. Paralleling its progress, and depending on it to a great extent, has been the feed manufacturing industry, two-thirds of whose volume is poultry feeds. It took a lot of doing, mainly by a team of scientists who ranged from nutritionists, geneticists, and biologists to pathologists, engineers and statisticians, to bring the U.S. poultry industry to its present efficiency. In the middle has been the poultryman trying to manage an enterprise that was growing nationally by leaps and bounds.

From expensive, New York-dressed, "chicken in the pot every Sunday," poultry has modernized to inexpensive, fully-dressed products that are marketed every day in the year, frozen or ready-to-eat. It has shown more progess than any other branch of animal agriculture.

In 1920, an average farm flock consisted of 200 hens or less; whereas today with advances in automation, the commercial poultryman may have 25,000 hens. Previous chores like mixing feed and housekeeping are done now by machinery so he can spend more time checking his chickens and trouble-shooting. Innovations in equipment and labor-saving devices developed by the poultry industry are being adopted by the livestock industry to a great extent.

Nevertheless, it takes more know-how and business acumen now to keep a poultry enterprise profitable. A poultryman must be more aware of other phases of the business. The more knowledgeable he is, the more likely he will survive in a very competitive business. Feed manufacturers must also be aware of the advances being made by poultry scientists so that their feeds measure up to the potential bred into modern birds. The amazing progress that has been made in recent years attests to the fact that communication between the groups is needed.

For many years there has been a need for greater application of the principles of nutrition to feeding practices as well as in preventing diseases of poultry. This treatise attempts to encompass the fields of poultry feeds and feeding in health and in disease in a concise and authoritative manner. It is hoped that it will help the nutritionist, veterinarian, and feeder to become more knowledgeable in the general area of poultry feeds and feeding and become a teaching aid for teachers and students of nutrition and feed technology, as well as for general poultrymen.

The solution of problems of nutritional improvement in poultry feeds and feeding is vital to long-range betterment of standards for raising poultry and production of human food. Experienced feed formulators, feed manufacturers, county agents, vocational agricultural teachers recognize the necessity of using great care in the selection of feed ingredients for poultry. Nutrition principles explain the reasons why many established practices are very important in the formulation and mixing of feeds and provide the ways of assuring quality, uniformity, and the true value of manufactured feeds.

With nearly 80% of the world population needing an improved diet, any contribution toward producing more and better foodstuffs at lower cost is both worthwhile and timely. Feed and poultry production are not exact sciences but are based on knowledge, the base from which all progress is made. Critical phases of feed and poultry production, and current research are explored, and practical recommmendations for meeting a variety of circumstances are made.

The U.S. Congress recently passed legislation authorizing the Bureau of Standards the possible adoption of the metric system. Half the world's population are now on the metric system. Canada and the United States are the only big countries still clinging to the fading imperial system. Britain hopes to have metrification completed by 1975. The simplicity and efficiency of the metric system is self-evident, its basis being the decimal point and the length of the meter. The latter is based on the wave-length of orange-red light given off by the element Krypton-86. The United States has had a legal metric system since 1866 which it does not use, while it uses the foot-pound system that has never been legalized. The benefits of a single measuring system are so great that conversion to the metric system is inevitable. Because of the international aspects of the book, measurements in the text are often given in both systems.

PHILIP J. SCHAIBLE

February 1969

Contents

1

The Poultry Industry

The poultry industry started as a home industry. The first record of poultry production in the United States was around 1607, following the establishment of permanent homes in Jamestown, Virginia. As towns were formed and developed, the demand for poultry also grew. Live poultry in the eastern and southeastern United States were shipped to New York to be sold on the live poultry market. Live marketing grew until after World War I, when refrigerator cars and poultry processing plants provided a more manageable product. The development of the commercial flock and feed mills accompanied by better management and breeding programs brought rapid growth to the poultry industry. Commercial enterprises began to develop goods, such as hatcheries, feeders, waterers, building supplies and drugs, to help produce poultry. During this period, there were also changes occurring in the location of the poultry industry. The Georgia and Arkansas broiler businesses began developing in areas where tree harvesting and cotton production were not providing a desirable standard of living for farmers. In Maryland, poultry production served to provide a living for unemployed ocean-going men. On the other hand, turkey production began in Minnesota because Dr. A.B. Billings had initiated a management program to control the disease, Blackhead, that had been making turkey production unprofitable in Kentucky and Virginia. The entire industry was on the move.

By 1976, United States broiler growers were producing 3.28 billion broilers and turkey growers were producing 139.75 million birds. This represented an increase of 710 million broilers and 24.25 million turkeys over the year 1966. Egg production dropped from the 1966 production of 66,450 million to 64,821 million in 1976. These data are presented in Tables 1.1, 1.2 and 1.3.

Hatchery production has grown in birds hatched from 2.445 billion in 1960 to 3.98 billion in 1976 (Tables 1.4 and 1.5). However, the number of hatcheries has decreased. At one time, there was a hatchery in nearly every town in the corn belt. Now, there are large ones in the centers of

TABLE 1.1. THE 10 LEADING EGG-PRODUCING STATES

	1966		1976	
Rank	State	Production Millions	State	Production Millions
1	California	7,664	California	8,953
2	Georgia	4,501	Georgia	5,591
3	Iowa	3,616	Arkansas	3,734
4	Pennsylvania	3,205	Indiana	3,091
5	Texas	2,734	Alabama	2,919
6	North Carolina	2,717	Pennsylvania	2,701
7	Arkansas	2,699	Florida	2,846
8	Indiana	2,443	North Carolina	2,756
9	Minnesota	2,379	Texas	2,357
10	Alabama	2,358	Minnesota	2,289
Total United States Production		66,450		64,821
Average Egg Production Per Hen		218		235

Source: U.S. Dep. Agric. (1977).

TABLE 1.2. THE 10 LEADING BROILER-PRODUCING STATES

	1966		1976	
Rank	State	Production Thousands	State	Production Thousands
1	Georgia	456,192	Arkansas	540,428
2	Arkansas	361,753	Georgia	451,531
3	Alabama	324,124	Alabama	430,225
4	North Carolina	264,959	North Carolina	315,589
5	Mississippi	183,734	Mississippi	257,442
6	Maryland	154,615	Maryland	199,008
7	Texas	154,918	Texas	190,703
8	Delaware	117,963	Delaware	160,046
9	Massachusetts	72,458	California	104,950
10	California	67,034	Massachusetts	86,659
Total United States Production		2,570,516		3,280,122

Source: U.S. Dep. Agric. (1977).

TABLE 1.3. THE 10 LEADING TURKEY-PRODUCING STATES

	1966		1976	
Rank	State	Production Thousands	State	Production Thousands
1	California	16,972	Minnesota	24,370
2	Minnesota	16,439	California	17,506
3	Missouri	9,990	North Carolina	16,700
4	Iowa	7,223	Arkansas	10,138
5	Arkansas	6,470	Missouri	9,755
6	Texas	6,427	Texas	9,300
7	Virginia	6,234	Virginia	7,297
8	Wisconsin	5,527	Iowa	6,347
9	North Carolina	5,281	Indiana	5,188
10	Indiana	4,200	Wisconsin	5,120
Total United States Production		115,507		139,740

Source: U.S. Dep. Agric. (1977).

TABLE 1.4. CHICKS HATCHED BY COMMERCIAL HATCHERIES: NUMBER, AVERAGE PRICE AND VALUE, UNITED STATES IN 1960–1976

	Chicks Hatched				
Year	Broiler-type (Thousands)	Egg-type (Thousands)	All (Thousands)	Average Price of Baby Chicks per 100 (Dollars)	Value of Chick Production (1000 dollars)
1960	1,965,504	479,666	2,445,170	15.5	379,197
1961	2,119,845	525,876	2,645,721	14.4	381,701
1962	2,197,276	501,973	2,699,249	13.9	376,306
1963	2,254,150	514,780	2,768,930	14.2	392,067
1964	2,305,959	528,371	2,834,330	13.5	381,867
1965	2,513,881	488,965	3,002,846	13.2	395,770
1966	2,747,592	569,333	3,316,925	12.6	384,151
1967	2,758,542	534,818	3,293,360	11.5	350,900
1968	2,799,089	514,301	3,313,390	11.7	358,944
1969	3,025,040	529,888	3,544,928	11.7	386,723
1970	3,189,169	571,142	3,695,425	11.2	385,643
1971	3,167,952	527,473	3,760,311	11.6	405,314
1972	3,267,457	491,354	3,758,811	11.1	389,593
1973	3,206,986	534,256	3,741,242	12.7	441,968
1974	3,104,430	473,403	3,577,833	14.1	473,754
1975	3,169,071	453,804	3,622,875	14.6	493,198
1976	3,488,326	492,180	3,980,506	15.4	568,957

Source: Agric. Statistics (1977).

TABLE 1.5. CHICKS HATCHED BY COMMERCIAL HATCHERIES: NUMBER AND AVERAGE PRICE, BY 10 LEADING STATES IN 1976

	Broiler-type Chicks			Egg-type Chicks	
State	Hatched (Thousands)	Average Price per 100 (Dollars)	State	Hatched (Thousands)	Average Price per 100 (Dollars)
Arkansas	591,275	12.20	Georgia	47,380	36.70
Georgia	484,368	11.30	California	45,831	39.10
Alabama	450,658	10.80	Florida	40,058	30.00
North Carolina	362,449	11.20	Indiana	33,162	36.50
Mississippi	251,000	12.80	Arkansas	27,574	32.30
Maryland	217,306	13.00	Texas	25,137	39.20
Texas	205,700	22.10	Minnesota	22,496	49.30
Delaware	168,460	13.00	Iowa	22,470	49.40
California	113,266	21.80	Pennsylvania	21,674	37.50
Pennsylvania	97,748	13.60	Maine	14,849	40.20
United States	3,488,326	13.50	United States	492,180	37.90

Source: U.S. Dep. Agric. (1977).

poultry production. The production of broiler- and egg-type chicks is under the full control of large, well-developed and well-managed breeding operations. In 1930, Clinton, Missouri, was the hatchery capitol of the world. Today, Arkansas leads in broiler-type chicks and Georgia in egg-type chicks hatched. The 3 states leading in broiler-type chicks hatched are Arkansas, Georgia and Alabama (Table 1.5). These were relatively unknown poultry producers in 1930.

The per capita consumption of poultry from 1960 to 1976 has increased (Table 1.6), and the consumption of eggs has dropped from 312 in 1963 to 276 in 1976. Poultry meats per capita consumption increased from 34.4 lb in 1960 to 52.5 lb in 1976. Turkey consumption during the same period has increased from 6.6 lb to 9.2 lb.

The decrease in egg consumption has not been due chiefly to the nutritional value of eggs (Table 1.7), but rather to a change in eating habits and to the premature publicity given to the dietary effects of cholesterol. Egg substitutes have had little effect on egg consumption; many Americans today do not eat a complete breakfast. However, eggs on the menu at breakfast account for a high percentage of the eggs consumed.

The consumption of broilers has increased in the home, and the introduction of rapid-eating places serving or selling fried chicken has further increased consumption. Fried chicken can be purchased from rapid-eating places throughout the United States and in many foreign nations.

TABLE 1.6. FOOD CONSUMPTION: INDEX OF PER CAPITA CONSUMPTION IN UNITED STATES 1960 TO 1976

	1960	1965	1970	1976
Poultry meat	75.3	90.6	107.1	116.1
Eggs	104.2	97.8	97.1	86.0
Vegetable fat (oils)	80.5	91.5	115.3	136.0
Animal fat	122.7	111.6	90.5	68.5

Source: Agric. Statistics (1977).

TABLE 1.7. PERCENTAGE OF TOTAL NUTRIENTS CONTRIBUTED BY EGGS TO THE DIET OF HUMANS

	Average 1957-1959	1976
Food energy	2.6	1.9
Protein	6.8	5.0
Fat	4.0	2.9
Carbohydrates	0.1	0.1
Calcium	2.8	2.3
Phosphorus	6.7	5.2
Iron	7.1	4.8
Magnesium	1.6	1.2
Vitamin A	7.3	5.7
Thiamin	2.9	2.1
Riboflavin	6.5	4.7
Niacin	0.2	0.2
Vitamin B_6	2.7	1.8
Vitamin B_{12}	11.3	8.1

Source: Agric. Statistics (1977).

As the consumption of poultry increases, production of feed must also increase. The human population throughout the world is increasing faster than feed ingredients can maintain a meat diet. In 1977, the population of the United States was 216 million with an expected population of 262.5 million in the year 2000. This represented a yearly increase of about 0.7%, as compared to a worldwide increase of 1.8%. A projected world population yearly increase is expected to be 1.93% in 1980, 1.84% in 1990 and 1.61% in 2000. Can the world's agricultural industries feed this estimated population in 2000? What will the poultry industries be and will the rations of the year 2000 be similar to those of 1978?

American agriculture has made many changes, one being the size and number of farms. In 1954, there were 4,782,000 farms with a population of 19,019,000. In 1976, there were 2,786,000 farms with a population of 8,253,000. These figures reflect the trend to larger and more highly mechanized farms.

Poultry has become more efficient in egg production, meat production and labor management. Egg production has improved from 122 eggs in 1925 to 235 eggs in 1976 per hen, housed with a lowering of feed efficiency of 7.5 lb to 4.0 in 1976. A 3 lb broiler can be produced today in 8 wk or less than 2 lb feed per lb gain, whereas in 1950, it required about 11 wk and 3 lb feed per lb gain.

The great improvement in poultry is attributable to the united efforts of nutritionists, geneticists, pathologists, feed and equipment manufacturers and state and federal experiment stations. Although most costs for production have increased, the price of poultry for the consumer has remained relatively the same.

Poultry are also used for purposes other than human consumption. Thousands of eggs are used yearly in the development and production of vaccines. The chick is used for research in nutrition, and as a laboratory animal for food evaluation and toxicity studies. The rapid growth of a chicken makes it an excellent animal for use in biological studies, since it increases its weight 3–4 times in 14 days. Thousands of chicks are used each year to test products for fat- and water-soluble vitamin activity. The chick is one of the most effective test animals for the antipernicious anemia factor in humans. The value of antibiotics and trace minerals was first discovered in the chick. Vitamin K was originally discovered to be required by the chick, then found to be effective in human medicine.

The chicken egg has made important contributions to the culture of viruses that cause many human and animal diseases. Eggs stand between man and such universal and fatal contagious diseases as smallpox, yellow fever, encephalomyelitis, polio, influenza, mumps, psittacosis, rabies, canine distemper and fowl pox. The conquest of many diseases has permitted an increase in survival.

PRESENT STATUS OF THE AMERICAN POULTRY INDUSTRY

The United States produces a surplus of poultry and is exporting poultry to, or operating poultry production plants in, many foreign countries. The United States also has a large surplus of feed grains and protein concentrates.

During the period between 1966 and 1976, the leading production states, with the exception of Missouri, maintained their lead (the top ten) as producers of eggs, broilers and turkeys. This stability indicates that the poultry industry has lost much of its fluid character and is becoming more stable. This is due to a great extent to the role played by feed manufacturers in owning or directing complete broiler, turkey or egg businesses. Since 1970, there has been an increase in the use of "brand" names for broilers, as well as for turkeys and eggs (Table 1.8).

It seems certain that the poultry industry will remain strong and active. As food becomes more important in international relations, poultry becomes a more important industry.

TABLE 1.8. CHICKENS: NUMBER AND VALUE, UNITED STATES, DEC. 1, 1969–1976[1]

		Pullets						
Year	Hens (1000s)	Of Laying Age (1000s)	3 Months Old and Older Not of Laying (1000s)	Under 3 Months Old (1000s)	Other Chickens (1000s)	All Chickens (1000s)	Value per Head (1000s)	Total Value (1000 Dollars)
1969	136,874	176,649	50,832	51,083	6,658	422,096	1.28	541,453
1970	134,543	189,302	50,292	53,069	6,697	433,903	1.21	526,163
1971	138,157	182,896	50,874	47,286	6,363	425,576	1.23	524,002
1972	132,206	169,840	48,834	49,155	6,206	406,241	1.28	519,908
1973	127,488	168,974	52,593	57,307	6,141	412,503	1.62	669,564
1974	121,205	163,824	46,718	46,303	5,529	383,579	1.71	654,106
1975	122,855	156,974	48,004	46,040	5,367	379,240	1.74	660,237
1976	117,168	163,328	48,261	44,546	5,638	378,941	1.70	646,052

Source: Agric. Statistics (1977).
[1]Does not include commercial broilers.

SOURCE OF POULTRY AND FEED INFORMATION

Those who work in nutrition should be aware of current trends in poultry and nutrition, at home and abroad. An understanding of population growth, international trade and trade agreements, and crop production statistics is imperative. Scientific and trade journals, federal and state bulletins, reviews, books and college extension news reports should be studied. Many of these publications can be obtained without cost. Attendance at feed or nutrition conferences and seminars is suggested.

2

Economic Structures Developed by the Poultry Industry

Poultry production is similar to dairy production in several ways. Each requires large investments in buildings and machinery, and specific labor practices each day of the week are mandatory. Studies on selection of labor for poultry indicate that those suited for dairy farming also make successful poultry producers.

Economic structures related to the poultry business are many and varied. During the early development of the poultry industry, the farm operation was the principal unit; however, as the size of the farm operation grew, it acquired many structures and needs common to a small business. The term *agribusiness* was created, and today is used when discussing modern poultry production. The following economic structures are active in the poultry business.

The Farm

The farm performs 3 functions:

(1) Housing for the poultry flock
(2) Home for labor and management
(3) Land for crop and waste disposal

Today, many farms must provide a place for disposal of manure, and digestion ponds for purification of liquid wastes. Farms near large population centers are required to provide a specific acreage for utilization of manure. The costs of fuel for operation of dehydrators are so great that direct land application is desirable.

A poultry farm designed to handle 20,000 hens and providing all the above functions can easily cost over $200,000. Many broiler- and egg-producing farms produce 40,000 to 60,000 birds. A broiler house providing for housing and equipment storage for 60,000 birds must contain

60,000 sq ft of floor space, requiring an investment of approximately $120,000. This indicates the expense involved in a large broiler operation.

Financial Houses

Farm land and buildings are generally financed by local banks, credit unions, life insurance companies or federal land banks. Equipment, feed, medication and chickens are often financed through local feed dealers. The largest investment is in feed, which accounts for nearly 75% of the cost of producing a broiler.

Feed Manufacturing Companies

The feed manufacturer generally provides the feed and the chicks, supervises the operation and arranges for the sale of the broilers, eggs or turkeys.

The Hatchery

The hatchery provides the chicks or egg-laying stock and often vaccinates the poultry.

Breeding or Hatchery Supply Operations

Poultry breeding operations are large corporations. Their investments are often in the millions. The breeding stock is selected by computer, and several breeding lines are maintained. The breeding operations furnish the hatchery supply stock to those poultrymen involved in providing hatchery eggs.

Food Processing Plants and Distribution

Eggs, broilers, roasters, turkeys, ducks, geese and game birds are made available by processing plants to food markets and eating places throughout the year. Prepared meals are often sold as frozen food. Rapid-food eating establishments offer chicken daily in many countries. Some provide only fried chicken.

Insurance Companies

Many insurance companies provide liability insurance for feed manufacturers and poultry producers.

Poultry By-product and Dried Poultry Waste Processors

Poultry by-products consist of non-rendered clean parts of the carcasses of slaughtered poultry, such as heads, feet, viscera (free from fecal content and foreign matter, except in such trace amounts as might occur unavoidably in good practices NRC–5–03–800). This product supplies a high percentage of animal protein used in canned pet foods.

Dried poultry manure is used as a fertilizer. It is one of the best organic fertilizers available.

Feathers and poultry by-products are cooked and dried, then sold as a protein concentrate for feeds. This concentrate has high biological value and is preferred over meat scraps and tankage.

Experimental Laboratories

Many vaccines for man involve the use of eggs as a culture medium; eggs are often used in the isolation and identification of new diseases.

The Feed Industry

The total commercial feeds produced in the United States in 1976 was 28,538,000 tons, valued at more than 4 billion dollars. The feed consumed by broilers in 1977, approximately 27,199,680,000 lb, was valued at over 2 billion dollars. These figures stress the size of the poultry industry, which uses many by-products of other industries, such as distillers' solubles and yeasts. The poultry and livestock industries together used over 36 million tons of by-products in 1977.

3

The Feed Industry

The feed industry differs significantly from other major farm supply industries in that (1) most of its raw materials are farm produced and (2) many of the poultry farm businesses are wholly or in part financed, owned, supervised or developed with the cooperation and guidance of the feed manufacturers.

The feed industry has always been extremely self-disciplined. The high quality of the poultry produced is a reflection of the quality of feed and feeding programs provided by the feed industry, whose motto is, "the best is not too good."

The American feed industry has grown from a scoop-shovel mixing operation into today's fully automated mills. In many operations, computer formulation is used. Chemical analysis is performed by multiple automated analytical equipment.

The consumption of feed by livestock and poultry in the United States has increased about 37% from 1940 to 1966. The increase in total feeds, as based on disappearance of ingredients (Table 3.1), increased from 25,877,000 tons in 1965 to 30,573,000 tons in 1975. This represents approximately a 2% yearly growth. During this 10-year period, a decrease or no increase in the use of linseed meal, cottonseed meal, all animal protein ingredients and alfalfa meal was observed. Feed concentrates used from 1966 to 1976 (Table 3.2) showed a decrease in the use of sorghums, oats, barley and rye.

Decrease in use of some feed ingredients may have resulted from inavailability; however, the application of new information in nutrition and the economics of using minor ingredients, are the primary reasons for the reduction. The new synthetic vitamins and amino acids and the readily available corn and soybean ingredients are combined in the formulated ration using the soybean-corn structure most often employed today. Use of automated mill equipment encourages fewer concentrates because of the cost of automated ingredient, storage and weighing units.

TABLE 3.1. COMMERCIAL FEEDS: DISAPPEARANCE OF INGREDIENTS IN 1965 AND 1975 (EXPRESSED AS 1000 TONS)

	1965	1975
Oil cake and meal		
soybean	10,274	15,613
cottonseed	1,755	1,266
linseed	248	87
peanut	115	313
copra	98	—
total	13,028	17,279
Animal proteins		
tankage and meat meal	2,068	2,001
fish meal	827	504
dried milk	255	162
total	3,150	2,667
Mill products		
wheat mill feeds	4,612	4,667
gluten feed and meals	1,135	1,490
brewes dried grains	304	321
brewes dried grains	426	400
rice mill feeds	395	547
dried molasses and beet pulp	1,153	1,650
alfalfa meal	1,652	1,552
total	9,677	10,627
Total commercial feeds	25,877	30,573

Source: U.S. Dep. Agric. (1977).

TABLE 3.2. FEED CONCENTRATES FED TO LIVESTOCK AND POULTRY 1966 AND 1976 (EXPRESSED IN MILLIONS OF TONS)

	1966	1976
Feed grains		
corn	93.3	104.3
sorghums	16.8	14.0
oats and barley	17.0	13.2
total	127.0	131.5
wheat	3.6	5.5
rye	0.3	0.2
by-products[1]	31.2	35.8
total concentrates	162.1	173.0

Source: U.S. Dep. Agric. (1977).
[1]Oilseed meals, animal protein feeds and mill products.

The feed industry in the future will face many ingredient changes. The production of feed grains will probably not continue to increase at a rate sufficient to meet the food needs brought about by world population growth. As the human population increases, the demand for grains also increases. A crop disaster on 1 continent can influence the overall feed program for poultry and livestock. For example, only twice during the last 28 years have the 4 leading wheat-producing nations (China, India, the United States and the U.S.S.R.) obtained an above-average production of wheat. In 1970 and 1978, each of these nations produced an above-average crop of wheat. The United States generally produces an above-average crop, whereas India and the U.S.S.R. often have crop failures. The United States always produces a surplus of wheat; in fact, we must export over 40% of our wheat crop if our storage facilities are not to be overburdened.

Total world production of the major grains in 1976 is presented in Table 3.3. The yearly increase in food production is not maintaining a growth rate in the production of protein and energy to offset the decline in decreases caused by the reduced production of grains, such as oats and barley, or to meet the increased demands of rapid human population growth.

TABLE 3.3. PRODUCTION OF GRAINS AND SOYBEANS IN 1976 (EXPRESSED IN 1000 TONS)

	Wheat	Corn	Rice	Soybeans
North America	85,058	173,466	6,927	34,972
United States	58,432	157,895	5,307	34,425
South America	15,882	30,124	11,608	12,306
Europe	85,376	44,389	1,812	295
U.S.S.R.	96,882	10,260	2,015	495
Africa	10,684	24,743	7,908	18
Asia	106,419	51,855	313,021	21,072
Oceania	12,103	408	490	365
World total	412,402	335,245	348,781	69,118

Source: Agric. Statistics (1977).

CONSOLIDATION OF FEED MANUFACTURERS

Many feed manufacturers have consolidated to form stronger companies and to develop a research program for quality control and product development. Some companies sponsor research with private investigators and agricultural experiment stations. For many years, nutrition conferences and poultry association meetings have been jointly financed or sponsored by industries associated with the poultry business.

FARM COOPERATIVES

Farm cooperatives have made important contributions to the feed industry by sponsoring research at agricultural experiment stations and conducting studies on cooperative-owned facilities.

Feed manufacturers' research has played a very prominent role in the development of the poultry industry. Their use of large populations with management and with facilities similar to those recommended to the poultrymen, gives a valid on-the-farm evaluation before offering it to the poultrymen. The recommendations resulting from this type of evaluation are very important to poultrymen in light of feed-price ratios today. The change in feed-price ratios from 1960 to 1976 is summarized in Table 3.4. There has been a decline of 0.8 lb over a 16-year period for broilers, 3.1 lb for eggs and 4 lb for turkeys. For poultrymen to operate economically in spite of this decline in buying power requires sound management. Since the poultry industry has continued to grow and to adjust production to meet demands, a sound production program has been provided. During the early development of the poultry industry, extension workers, provided by educational institutions, guided the business. Today, the feed industry has assumed most of the responsibility formerly assumed by the educational institutions' extension personnel.

TABLE 3.4. POULTRY: FEED-PRICE RATIOS AND POULTRY RATION COST UNITED STATES IN 1960–1976

	Ratios[1]		
Year	Egg-feed lb	Broiler-feed lb	Turkey-feed lb
1960	10.9	3.6	7.7
1961	10.6	3.0	6.1
1962	9.9	3.3	6.1
1963	8.1	3.2	4.9
1964	8.1	3.1	4.8
1965	8.1	3.3	5.1
1966	9.1	3.3	5.1
1967	7.3	2.9	4.4
1968	8.5	3.2	4.7
1969	10.0	3.4	5.0
1970	9.3	2.9	5.0
1971	7.2	2.8	4.6
1972	7.2	2.9	4.5
1973	7.9	3.3	4.8
1974	7.0	2.6	3.2
1975	7.0	3.2	4.2
1976	7.8	2.8	3.7

Source: Agric. Statistics (1977).

INTEGRATED PROGRAMS

Many feed manufacturers have developed a cooperative poultry program known as an integrated program. The poultry farmer provides the poultry-growing facility (housing, equipment, fuel) and labor. The feed industry provides the feed, medication and vaccination, and markets the finished product: the eggs, pullets, broilers or turkeys. The feed manufacturer may also own, or have a contract with, the hatchery, processor, hatchery flocks, pullet replacement contractor and breeder contractor. Such integration allows the farmer to devote his full interest and finances to growing the chickens and turkeys or producing the eggs, yet with some assurance that he will receive a fair share of the earnings.

Each contract shows the minimum proportion of the sales income each farmer will receive; each segment, too, will receive a bonus for a job well done. Bonuses are generally based on mortality and on feed efficiency.

THE AMERICAN FEED MANUFACTURERS' ASSOCIATION (AFMA)

The feed industry is organized on a national, regional and state level. The American Feed Manufacturers' Association (AFMA), 1701 North Fort Myer Drive, Arlington, Virginia, 22209, is a trade association of great importance and represents the feed industry and allied businesses. Regional branches of the organization—the Northeastern Feed Manufacturers' Association, the Southern Feed Manufacturers' Association, the Midwest Feed Manufacturers' Association and the Northeast Feed Manufacturers' Association—cooperate with the American Feed Control officials and various state agricultural departments in the control of the feed and food business.

Since it was founded, AFMA has expanded its activities to cover the important needs of feed manufacturers, and committees have been organized for different areas of the industry: Feed Control Relations, Industry Advisory, Feed Marketing Survey, Legislative Market Research, Additives, Public Relations, Purchasing Agents, Traffic, Nutrition Council, and Sales Executives. Agricultural economists from the agricultural experiment stations make up still another committee; they determine the annual feed needs for livestock and poultry.

FUTURE OF FEED MANUFACTURING

The United States is a manufacturing nation. The United States Department of Agriculture, United States Foreign Aid, the many commercial foundations, such as the Rockefeller Foundation, have provided research funds and trained scientists for many nations seeking support in meeting their food and clothing needs. Although great achievements have been made in developing new farm crop methods and improving crop yields, the rapid population growth and the lag in education have not provided a worldwide surplus in feeds. Only a surplus can encourage the manufacturing of feeds in underdeveloped nations.

In the United States, change is in progress in all phases of life. The world's largest feed mill is now idle, the heretofore largest feed manufacturer has discontinued many of its feed mills, and since 1970, many new small automated feed mills have been constructed. Feed manufacturing will continue to change from large centers to small plants close to the center of poultry production, designed for bulk delivery.

The world population by the year 2000 will consume our present surplus. The scarcity and high cost of energy, needed to provide many of the required nutrients—vitamins, minerals and proteins, as well as drugs—will automatically reduce the production of meat. This, in turn, will bring about a reorganization of the feed industry. The rations will be more complicated and by-products will make up most of the ingredients. Poultry will compete with people and swine for high-protein/high-energy feeds. A shift in dairy farming will probably develop, using improved strains of dairy cattle that will produce both meat and milk. Since World War II, we have observed this in Europe. The Flo Holstein breed in America is moving in this direction. In the future, more poultry "offals" will be used in pet foods. This will reduce the availability of animal protein ingredients for poultry and livestock.

The possibility of producing leaf proteins in tropical areas is being studied. The final results will probably see soybean protein being directed toward human foods, and leaf meals and leaf protein concentrate being used in poultry and livestock feeds.

4

Poultry and Feed Supplies Around the World

World population reached a total of 3.89 billion in mid 1975. At this rate of growth, 2.1% annually, the world's population will have doubled by the year 2006. The average annual rate of population growth between 1965 and 1975 was:

	%
Africa	2.7
Asia	2.1
Europe	0.6
Latin America	2.7
North America	1.0
South America	1.0

Studies have shown that over 60% of the world population is underfed; however, the expected lifespan has increased to an average of over 60 years for males born in 1978. Those nations having a life expectancy of less than 55 years are:

Burma	40.8
Colombia	44.2
Egypt	51.6
India	41.9
Jordan	52.6
Kenya	46.9
Syria	54.5

The highest life expectancy in Sweden is 72.1 years. In the United States, it is 68.2 years. Females have a life expectancy of nearly 7 years longer than males.

Consumption of food in the United States for the periods 1960, 1965, 1970 and 1976 shows a large decline in eggs and a large increase in poultry meats (Table 4.1).

TABLE 4.1. PER CAPITA CONSUMPTION OF POULTRY AND EGGS IN THE UNITED STATES IN 1960 TO 1976 (LB)

	1960	1965	1970	1976
Poultry meats	34.4	41.3	48.9	52.5
Eggs	42.2	39.8	39.5	35.0
Turkeys	6.6	7.4	8.0	9.2

Source: Agric. Statistics (1977).

Food production in most countries is on the increase. Dramatic new high-yielding varieties of food grains are being disseminated around the world and yields are proving better than was first thought possible. Nevertheless, food production in the underdeveloped countries has scarcely kept pace with population growth. Food is no longer a free commodity, but is being exchanged for certain commitments, primarily to improve agriculture. Feed production figures in various countries are given in Tables 4.2, 4.3, 4.4, 4.5, 4.6, 4.7. Production of all grains has increased sharply during the past 10 years, except of oats and barley, which declined in 1977. Exciting new varieties of rice, wheat, grain, sorghum and corn, developed at the International Rice Institute in the Philippines, the International Maize Improvement Center in Mexico, and in the United States, are now available. These out-yield traditional varieties and are much more responsive to fertilizer. However, they are generally lower in protein than the traditional varieties and the cereal grains are deficient in lysine. High-lysine varieties of corn (Opaque senis) provide more crude protein and lysine, but the yield is not equal to the regular hybrids.

The addition of amino acids to feed for improvement of protein quality is simple in the United States, because we are a manufacturing nation. However, in many nations, the addition of supplements is difficult because of a lack of milling machinery. Food quality is also being improved through the development and utilization of low-cost sources of protein.

FEEDING EXPANDING POPULATIONS

A significant percentage of the population in developed countries eats well, indeed overeats, but this is an exception to the general pattern. In the underdeveloped countries, food production gains have been offset by population growth.

Experts in poultry science and feed manufacturing can be of considerable assistance in this regard, particularly, as there is a need for

TABLE 4.2. WORLD MAIZE (CORN) PRODUCTION (1000 METRIC TONS)

	1965	1975
World	216,189	324,721
Africa	16,039	25,018
North America	105,825	161,630
South America	17,826	27,551
Asia	37,487	55,127
Europe	25,696	47,763
Oceania	196	344
U.S.S.R.	13,122	7,328
United States	95,561	147,251

Source: Statistical Yearbook, U.N. (1976).

TABLE 4.3. WORLD PRODUCTION OF WHEAT (1000 METRIC TONS)

	1965	1975
World	254,534	355,895
Africa	6,375	9,765
North America	50,109	78,024
South America	10,053	12,160
Asia	55,968	100,421
Europe	59,352	77,118
Oceania	8,470	12,183
U.S.S.R.	64,207	66,224
United States	33,040	58,102

Source: Statistical Yearbook, U.N. (1976).

TABLE 4.4. WORLD PRODUCTION OF RICE (1000 METRIC TONS)

	1965	1975
World	253,223	348,374
Africa	5,493	7,764
North America	4,062	6,761
South America	8,039	11,638
Asia	233,564	316,950
Europe	1,517	1,932
Oceania	157	409
U.S.S.R.	390	2,009
United States	3,084	5,805

Source: Statistical Yearbook, U.N. (1976).

agriculture based on animals to supply protein for the diet. Since poultry production is a relatively easy field to break into and provides a continuous market for feed manufacturers, the combination of these skills should help considerably. Perhaps all that is needed, in some cases, is a plant to manufacture foolproof feeds.

THE IMPORTANCE OF POULTRY TO FEEDING THE WORLD

How will we feed the expanding world population? Poultry is perhaps more important to human welfare today than ever before in history.

TABLE 4.5. WORLD SOYBEAN PRODUCTION (1000 METRIC TONS)

	1965	1975
World	32,474	68,927
Africa	65	96
North America	19,791	43,145
South America	408	10,740
Asia	11,800	13,729
Europe	15	362
Oceania	1	74
U.S.S.R.	394	780
United States	19,560	42,079
Brazil	354	9,892
Romania	3	213

Source: Statistical Yearbook, U.N. (1976).

TABLE 4.6. WORLD PRODUCTION OF OATS (1000 METRIC TONS)

	1965	1975
World	47,811	49,005
Africa	175	238
North America	19,998	14,100
South America	860	670
Asia	2,404	3,549
Europe	17,116	16,748
Oceania	1,206	1,206
U.S.S.R.	6,052	12,495
United States	13,848	9,546

Source: Statistical Yearbook, U.N. (1976).

TABLE 4.7. WORLD BARLEY PRODUCTION (1000 METRIC TONS)

	1965	1975
World	99,682	156,577
Africa	3,544	3,921
North America	12,711	18,321
South America	1,249	1,148
Asia	27,004	34,390
Europe	33,780	59,477
Oceania	1,076	3,512
U.S.S.R.	20,318	35,808
United States	8,676	8,368

Source: Statistical Yearbook, U.N. (1976).

Poultry products are dependable sources of life-supporting nutrients. With the trend in animal agriculture toward bigger and fewer production units, feeding operations are being located nearer processing installations and consumers. Larger feeding operations and use of automation make complete manufactured feeds desirable. Feed manufacturers are in a more favorable position to serve large feeders. Many factors other than ingredient cost make up the cost of a manufactured feed. Qualified nutritionists are needed to develop feed formulas. Nutritional judgment must direct electronic calculations—mere arithmetic is not enough. Formula feeds are made largely from by-products of other industries, from grain products and a wide variety of feed additives. Chemical, microbial or enzyme ingredients are used to increase nutritional value.

Rice, corn, wheat, soybeans, chick peas and certain oil seeds which make up a large share of the diet of the world's population are inexpensive but lack certain essential amino acids. These foods can be balanced with proteins to improve the diet of the world. However, for religious and esthetic reasons, a massive program will be necessary to promote animal foods. Thirty grams of animal protein per day will provide all the protein required to maintain the human body in vigorous health. Highly-productive though it is, the North American continent cannot continue to feed the world. The answer lies in strengthening the economy of the underdeveloped nations so they can feed themselves.

Two-thirds of the world's population now live in countries that are nutritionally deficient, where the daily diet averages 300 Cal less than the 2300 Cal considered desirable, and the daily consumption of protein is less than two-thirds of that in diet-adequate countries.

The United States feed and poultry industries can help to close the food gap. Their contribution of vitamins, drugs, hormones, antibiotics, etc., can accelerate the conversion of feed to food. United States scientists can make the most important contributions from the vast reservoir of knowledge and technical skills developed over the years. The application of this special knowledge and technology in the underdeveloped nations of the world should prove beneficial to all concerned.

THE IMPORTANCE OF RICE

More than half the world relies on the tropical rice plant for its basic food supply, which literally falls down on the job. If the rice plant is to feed more than half of the 5 billion people who will be on the earth in 10 years' time, a dramatic change in its structure is essential. The development of short-stemmed rice varieties that will not fall over when fertilized, yet will give high yields and resist disease, has been partially achieved by the International Rice Institute in the Philippines. Such new varieties are making a substantial impact on rice production. The need for research on problems of this kind has long been recognized in the United States. A network of agricultural research stations is a major component of a highly-developed farming industry. An unfortunate aspect of these high-yielding varieties of rice is that they are much lower in protein than the conventional varieties. Many populations who depend upon rice eat measured proportions daily, and although the newer rice strains would provide similar amounts of energy, the lower protein content, 6% as compared to 9 or 10% for the conventional varieties, obviously would worsen the apparent protein deficiency. American Foundations work beyond the nation's border to advance human welfare. The Ford Foundation, which conducts the largest foundation program overseas, spends $50 million annually, enabling skilled consultants to train personnel that many countries need for their development. The United States Agency for International Development spends annually more than $2 billion abroad.

FERTILIZERS

As we increase the yield capabilities of grain-producing crops through breeding, we also increase the need for fertilizers. For example, a 50% increase in corn (maize) production would bring about a proportionate increase in the need for nitrogen, phosphorus and potassium fertilizers. Tables 4.8 and 4.9 show the change in nitrogen, phosphate and potash fertilizers used throughout the continents, as well as in some selected nations, from 1965 to 1975. An increase in fertilizer use is an index to increased food production. The 2 continents, South America and Africa, have had a very small increase in their use of fertilizer, and their overall production of grains has also shown only a relatively small increase. Asia, with an approximate 100% increase in fertilizer use, has shown a large increase in food production.

During this period of increased production, the United States has increased its exports of grain. The United States exported over 60% of the total grains and oil seeds shipped in 1977. The exports and imports of grains and oil seeds in 1973, 1974 and 1975 have shown an increase in wheat, soybeans, soybean meal and peanuts (Tables 4.10 and 4.11). A decrease in exports of corn, sorghum grain and barley occurred during this same period. Although the United States imported only a small amount of oil seeds, there was a large increase in sesame seeds (Table 4.12). Sesame seed is used in bakery bread, candy and oil meal production. Between 1966 and 1975, the production of fish meal increased (Table 4.13); however, reports of "overfishing" in many oceans has been observed. Modern sophisticated fishing gear and failure to comply with "catch-size" limits will soon bring about a noticeable decrease in fish harvested.

The imports of fish meal into the United States have decreased over 60% between 1972 and 1976. The overall world catch has dropped only about 5% (Table 4.14).

The economic factors bringing about the decline in fish meal imports have been:

TABLE 4.8. WORLD USAGE OF NITROGENOUS FERTILIZERS (1000 METRIC TONS)

	1965	1975
World	14,987	43,301
Africa	450	1,247
North America	4,434	11,174
Mexico	199	833
United States	3,922	9,385
South America	236	798
Brazil	58	410
Colombia	40	131
Asia	2,951	10,282
China	1,001	4,555
India	429	2,031
Japan	736	638
Korea	302	732
Europe	5,385	12,256
Germany	1,071	1,906
France	766	1,708
Italy	393	724
Netherlands	286	453
Poland	358	1,224
Romania	82	788
Spain	763	351
United Kingdom	581	1,045
Oceania	66	186
Australia	57	165
New Zealand	5	10
U.S.S.R.	1,466	7,357

Source: Statistical Yearbook, U.N. (1976).

(1) The discovery and commercial availability of Vitamin B_{12}. This reduced the need for animal protein as a source. Fish meal and fish solubles were the best sources of the vitamin before its synthesis.

(2) The increased production of high quality soybean meal.

(3) The improvement in poultry through breeding has resulted in strains which are often called "corn-soybean strains."

(4) The synthesis of economically useable methionine for usage in correcting the methionine deficiency in soybean meal.

(5) The cost of a unit of protein from vegetable sources is less than from fish or meat scraps.

WORLD POULTRY AND EGG PRODUCTION

The average yearly United States chicken population decreased from 366,000 million in 1974 to 347,000 million in 1976 (Table 4.15). However, world egg production has increased from 16,337 million metric tons in 1965 to about 23,538 million in 1975 (Table 4.16). This increase has

TABLE 4.9. PHOSPHATE AND POTASH (K_2O) FERTILIZERS (1000 METRIC TONS)

	1965		1975	
	P_2O_5	K_2O	P_2O_5	K_2O
World	12,622	10,280	24,071	21,335
Africa	313	128	819	340
North America	3,403	2,672	5,639	5,200
Canada	252	118	526	215
United States	3,008	2,460	4,731	4,724
Asia	1,342	868	3,808	1,720
China	331	120	1,253	401
Japan	495	568	624	517
India	114	56	453	270
Korea	151	30.5	361	161
Europe	5,227	5,175	7,989	8,196
France	1,135	920	1,618	1,314
Germany	999	1,667	1,221	1,800
Italy	398	139	490	276
United Kingdom	442	440	391	399
Oceania	999	135	854	195
Australia	715	54	490	73
New Zealand	283	81	360	120
U.S.S.R.	1,084	1,148	3,883	4,997
South America	250	153	1,129	688
Brazil	86	80	914	588
Colombia	48	34	46	39

Source: Statistical Yearbok, U.N. (1976).

TABLE 4.10. UNITED STATES EXPORTS OF GRAINS AND SOYBEANS

	1973	1974	1975
Soybeans 1000 bu	539,129	420,704	555,104
Soybean meal 1000 t	5,548	4,299	5,145
Sorghum grain 1000 bu	244,314	191,569	237,997
Barley 1000 bu	87,559	39,098	22,728
Oats 1000 bu	466,678	10,857	12,349
Corn 1000 bu	1,372,064	1,131,030	1,555,448
Wheat 1000 bu	1,106,684	1,001,100	1,129,179

Source: Statistical Yearbook, U.N. (1976).

TABLE 4.11. UNITED STATES EXPORTS OF OILSEEDS AND NUTS

	1965 1000 lb	1975 1000 lb
Cottonseeds	18,040	110,622
Flaxseeds	295,104	53,592
Peanuts (shelled)	177,820	338,714
Soybeans	15,395,030	33,962,040
Soybean meal	2,524,447	4,894,725

Source: Statistical Yearbook, U.N. (1976).

TABLE 4.12. OILSEEDS IMPORTED INTO THE UNITED STATES

	1965 1000 lb	1975 1000 lb
Castor beans	2,940	10
Copra	580,323	0
Flaxseed	0	9,163
Shelled peanuts	617	411
Poppy seed	6,573	16,287
Rapeseed	1,074	51,963
Sesame seed	28,293	51,963
Soybean	28	110

Source: Statistical Yearbook, U.N. (1976).

TABLE 4.13. PRODUCTION OF FISH MEAL AND SOLUBLES

	1966 1000 Metric Tons	1975 1000 Metric Tons
World production	4108	4354
Chile	220	169
Denmark	125	339
Ireland	176	107
Japan	361	790
Norway	422	337
[1]Peru	1471	687
South Africa	270	250
[2]U.S.S.R.	239	638
United States	278	369

Source: Agric. Statistics (1977).
[1]1970 production—2,253,000 metric tons.
1971 production—1,935,000 metric tons.
[2]U.S.S.R. in 1975 produced 17.1 million metric tons whale meal. World production was 39 million tons. World production of whale meal dropped from 62 million metric tons in 1966 to 39 million in 1975.

TABLE 4.14. UNITED STATES PRODUCTION AND IMPORTS OF FISH MEAL

	1972		1976	
	1000 Metric Tons	Value $1000	1000 Metric Tons	Value $1000
Fish meal imports	392	$51,627	140	$32,872
U.S. production	286	$48,289	309	$96,729

Source: Agric. Statistics (1977).

TABLE 4.15. CHICKEN POPULATION THROUGHOUT THE WORLD

	1974 Millions	1975 Millions	1976 Millions
North America	366	354	347
United States	297	286	281
South America	76	76	74
Europe	617	591	570
Total E.E.C.	290	278	273
Total Western Europe	379	373	372
Total Eastern Europe	238	218	198
U.S.S.R.	450	455	450
Africa	12	13	13
Asia and Oceania	191	208	207

Source: Agric. Statistics (1977).

TABLE 4.16. WORLD EGG PRODUCTION

	1965 1000 Metric Tons	1975 1000 Metric Tons
World	16,337	23,538
Africa	427	700
North America[1]	4,435	4,792
South America	703	1,116
Asia	4,306	6,960
Europe	4,698	6,542
Oceania	193	252
U.S.S.R.	1,575	3,176
[1]United States	3,823	3,797

Source: Statistical Yearbook, U.N. (1976).

been accompanied by a large increase in the number of eggs per hen, and also by an improvement in feed efficiency.

UNITED STATES POULTRY EXPORTS

The United States has been able to increase its exports of poultry produced (Table 4.17) from $80,015,000 in 1971 to $207,129,000 in

TABLE 4.17. EXPORTS OF POULTRY AND POULTRY PRODUCTS

Product	1971 1000 Dollars	1976 1000 Dollars
Poultry and poultry products	80,015	207,129
Live poultry		
breeding chicks	15,314	16,114
baby chicks	3,180	5,003
other	1,940	8,797
Poultry meat, fresh, frozen		
chickens	24,434	81,509
turkeys	11,843	42,215
other	3,184	7,960
Poultry canned	2,543	2,396
Eggs, hatching	12,099	20,996
Eggs, table	367	6,075
Eggs, dried	712	6,222
Other	4,399	9,842

Source: Agric. Statistics (1977).

1976. The development of poultry businesses in the common market countries—the United Kingdom and Japan—has provided an outlet for our surplus poultry. Major exports have been fresh, frozen broilers and turkeys, and also hatching eggs.

The United States has always imported some poultry (Table 4.18). Before and after World War I, we imported dried eggs from China, and some shell and dried eggs from the United Kingdom. Today, poultry production has increased in Mexico, and some poultry is exported into the United States near the border. Mexican production is modern and of high quality, and the breeding stock is probably of United States' origin.

TABLE 4.18. IMPORTS OF POULTRY PRODUCTS

Product	1971 1000 Dollars	1976 1000 Dollars
Poultry products	6153	3493
eggs, dried	1491	71
eggs, shelled	3426	1976
poultry meat	1236	1446

Source: Agric. Statistics (1977).

FUTURE FEED SUPPLIES

The future production of feed available for poultry will depend on whether or not there is a surplus above that required by man. The present rate of population growth strongly suggests that by the year 2006, there will not be a surplus, unless food production is increased by 100%.

Increasing feed production requires:

(1) Increased production of fertilizers

The production of nitrogen requires natural gas or electrical energy. Today, there is limited production of both. The production of phosphates and potash will be difficult because new sources have not been found and the energy costs accompanied by energy shortages will make production more difficult in future years.

(2) Land needs and supply

The amount of land devoted to field crop production cannot be increased without making several decisions:

(a) Water supply—Irrigation water is scarce and much future land development will come from arid or semiarid land areas.

(b) Fertilizers—Arid and semiarid land generally requires heavy fertilization.

(c) Financial needs—Land development, crop planting, machinery and crop harvesting require capital unavailable to many nations.

(d) Education—People must be educated for new jobs in new types of food production. This cannot be accomplished overnight.

(e) Government stability—Governments may not desire improvement or may not be stable enough to encourage foreign investors.

(f) Religion or spiritual considerations—The development of agricultural production may not have the guidance and support of all social units; religion could prevent the development of areas of land, lakes and streams. Poultry, however, is consumed by nearly all religious groups.

(g) People—The people may not be adaptable to improvement of agricultural methods. Population growth may not be controlled. The present rate of population growth and food production, as well as political unrest, will retard economic stability.

The improvement in feed production will be necessary, yet difficult, in many instances.

Poultry will, nevertheless, continue to play an important role in food production. The relatively short period required to become active in poultry production makes it an ideal field to enter, particularly during times of surplus.

5

Poultry: A Biological Machine

A functional knowledge of poultry anatomy and physiology is desirable before attempting to understand their nutritional requirements.

The chicken is a vertebrate whose anatomical structure is adapted to flight. Chickens are warmblooded, having an average body temperature of 105° to 107°F (40.6°–41.7°C). Their heart rate is rapid, sometimes reaching 300 pulsations a minute; small birds have the most rapid rate.

Metabotically, the chicken is a well-developed animal, and is often called a "fast-living" animal. It has been domesticated and improved through breeding and management for many years. Today, the statement is often made that we have a corn-soybean strain of chicken because the geneticist has been using a corn-soybean type ration to feed breeding stock for more than 25 generations.

The crowing of the cock has traditionally awakened and informed and often involved man in superstition. However, the chick has had only limited religious connotations, and therefore, is virtually a universally accepted food for man.

Bird feathers have been prominent in the North American Indian's tribal customs—decoration for headgear, as well as their use as primitive writing tools. Today, the "down" of water-fowl is used in cold-weather jackets and sleeping bags.

BEHAVIOR PATTERNS

Adult chickens and pheasants are known to have social patterns called "peck-orders." These have great nutritional influence on their eating habits. The birds at the top of the peck-order have complete control of eating facilities, and those remaining have graduated control, based on their rankings below the top of the peck-order. This peck-order is extremely well-developed in pheasants. If adequate feeder space is not provided, those birds in the lower peck-order will be deprived of food, hence becoming undernourished.

Poultry do not have selective taste but prefer some colors over others. This is very noticeable in turkey poults regarding color of drinking fountains and color of feed particles. Colored feed particles or bright light spots focused on feed may stimulate turkey poults to eat.

Turkeys will not eat when hungry, even if feed is present, often for as long as 36 hr. The gizzard lining of starved poults generally becomes loose and sloughs off. Accompanying this physiological change is the loss of the desire to eat and drink. Starters or tuitors, such as White Leghorn chicks, are often used to stimulate poults to start eating.

Sudden noises or light intensity changes can cause birds to panic. Ducks may stampede in the dark. Airplanes can cause turkeys to stampede; a songbird flying over a range roost at night can frighten turkeys into a destructive stampede.

Imprinting, a specialized form of learning, is seen clearly in many kinds of birds. Ducks, geese and turkeys when first hatched will follow after their mother or humans. This following behavior is the result of imprinting and is used by nutritionists in determining feeding patterns of wild birds, because the young will follow man through the day as though he were their mother.

Birds are thought to be stupid, yet they have always served man. They have limited learning power, yet can be trained to perform. Hens have been taught to operate a feeding device; the hungry hen jumps on a lever which strikes a golf club that drives a ball. The ball then upsets a feed cup, spilling a small helping of pellets.

THE SKELETON

The bones in the skeleton are adapted for flight, body frame support and egg production. The skeleton has 2 types of bones: (1) the skeleton frame (Fig. 5.1) and (2) the medullary bone, formed in the medulla of the long bones of mature hens beginning about 10 days before egg production starts.

The function of the medullary bone is to provide a readily available source of calcium for egg production. The medullary bone is most clearly seen in the shaft of the femur; however, analysis of the complete skeleton shows the distribution to be widespread (Table 5.1). The formation of the medullary bone is controlled by the reproductive hormones. About 25% of the calcium used in egg shell formation and egg yolk fat transfer comes from the medullary bone. The remaining 75% comes from the daily dietary calcium.

The bones of the metatarsus are fused, and their joining with the tibia forms what is commonly called the tibia-metatarsal or hock joint. This joint often becomes distorted or abnormal when the diet is deficient in

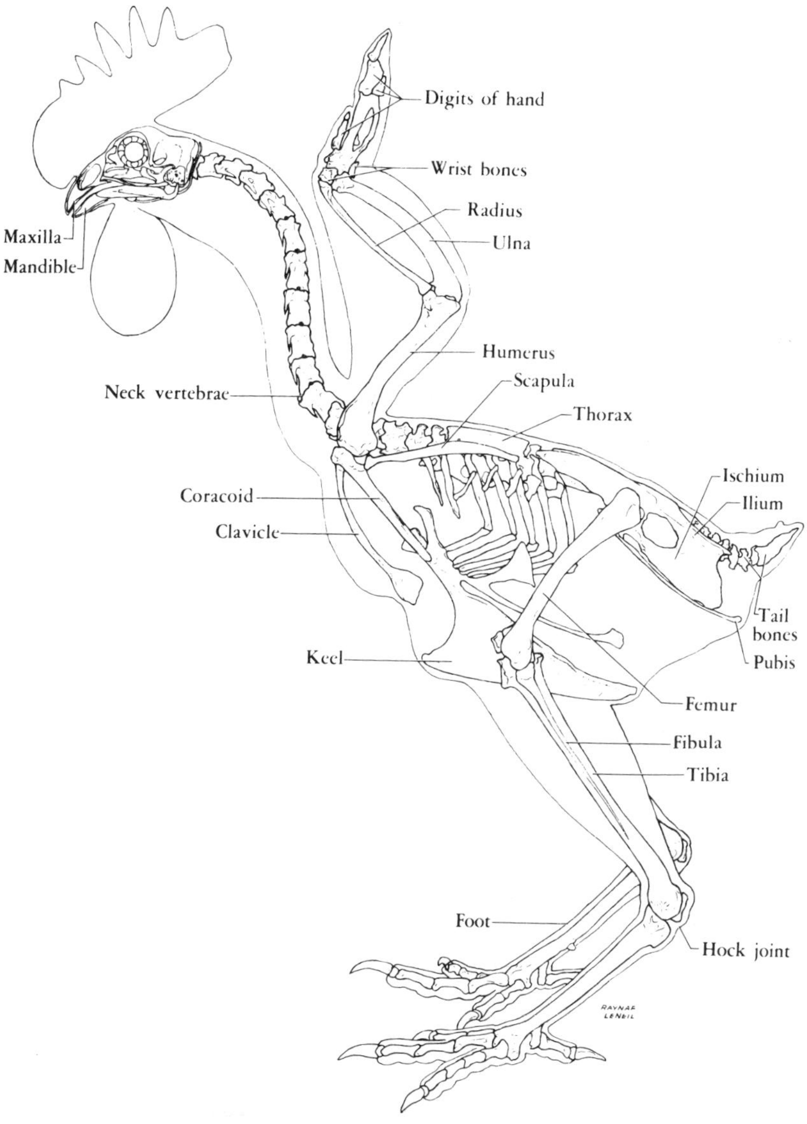

FIG. 5.1. VARIOUS SPECIES OF DIATOMS CONTAINING SILICEOUS SKELETAL MATERIAL FOUND IN DEPOSITS IN MANY PARTS OF THE WORLD

TABLE 5.1. DISTRIBUTION OF MEDULLARY BONE IN THE SKELETON OF THE DOMESTIC FOWL

Bone	% of Total Bone Present as Medullary Bone
Tibia	12.4
Femora	21.8
Ilia, Ischia, Pubis	20.8
Sternum	15.5
Ribs	29.2
Ulnae	4.0
Toes	1.3
Scapula	16.1
Total skeleton	11.7

Source: Simkiss (1961).

certain trace minerals or vitamins, or contains excess calcium or magnesium. The deformity of the hock joint is generally called Perosis. Young, rapidly growing birds readily develop this nutritional disease. The developing bone cells in perotic chicks are abnormally shaped but the ash content is normal. Since this is probably an abnormal protein structure, it cannot be cured; therefore, it must be prevented through adequate diet.

ANATOMY AND FUNCTION

The Head

The design of the head places the eyes nearly on the forward side of the skull. This makes it necessary for the bird to turn his head sidewise to identify and select a particle of food before he pecks forward. After pecking, he again turns his head sidewise to evaluate future selection,and by so doing, with his beak, throws the feed sidewise, sometimes out of the feed trough. This sidewise-forward motion causes heavy feed waste in overly filled or improperly placed feed troughs.

Breastbone

The breastbone adapts the bird for flying and also gives adaptation for incubation. The keel or sternum (breastbone) may become deformed by bruising, nutritional deficiencies of Vitamin D, calcium, or some anti-perosis minerals or water-soluble vitamins. Wet, contaminated litter may cause the feathers protecting the breastbone to be lost and allow the bone to be bruised, forming breast blisters. Also, some diseases cause accumulation of fluid under the skin which covers the breastbone.

The geneticist selects breeding stock with wide breasts. Most meat birds have wide breasts. Some birds have overly-wide breasts which are not adequately covered with feathers and muscle tissue. When this occurs, the incidence of breast blisters increases.

Embryonic Bone

In the chick embryo, mesenchyme that eventually forms bone contains Vitamin C distributed in the cells in the same way as in precartilage. When the differentiation into osteoblasts begins, the cells lose their Vitamin C in well-developed bone such as is seen in the femur 2 days after hatching. The early osteoblasts generally contain few granules of Vitamin C, but rapidly accumulate this vitamin as they mature. Then a few cells become heavily impregnated. The chicken does not require dietary Vitamin C, yet in bone development, we find a concentration of this vitamin near the site of the osteoblasts.

The Vitamin C found in the chick is synthesized within the body. It is a sugar derivative that is also a fairly strong organic acid and stable in the crystalline state, but readily oxidized in body solutions. Besides bone formation, Vitamin C is active in reducing the valence of hexavalent (+6) chromium to trivalent (+3) chromium. The chick readily excretes trivolent chromium.

The Digestive System

Poultry must, as a matter of efficiency, select energy-rich foods which are also high in protein, minerals and vitamins. They must digest them quickly and completely because of an excessively short digestive system which must be provided useable nutrients, as poultry are among the "fastest-living" of vertebrates. They also require food that allows procurement by the beak (Fig. 5.2 and 5.3).

The digestive system in a mature chicken is about 60 in. long and has both primary and supplementary digestive organs. The pH of the digestive system is shown in Table 5.2.

The digestive system of poultry performs 7 major functions related to nutrition:

(1) It collects and makes small pieces from large particles or pieces.
(2) It softens food so that enzymes may readily react.
(3) It provides a desirable environment for microorganisms.
(4) It promotes or brings about synthesis in the gut.
(5) It allows for absorption and water balance.
(6) It absorbs, excretes and recycles substances.
(7) It produces and disposes of feces.

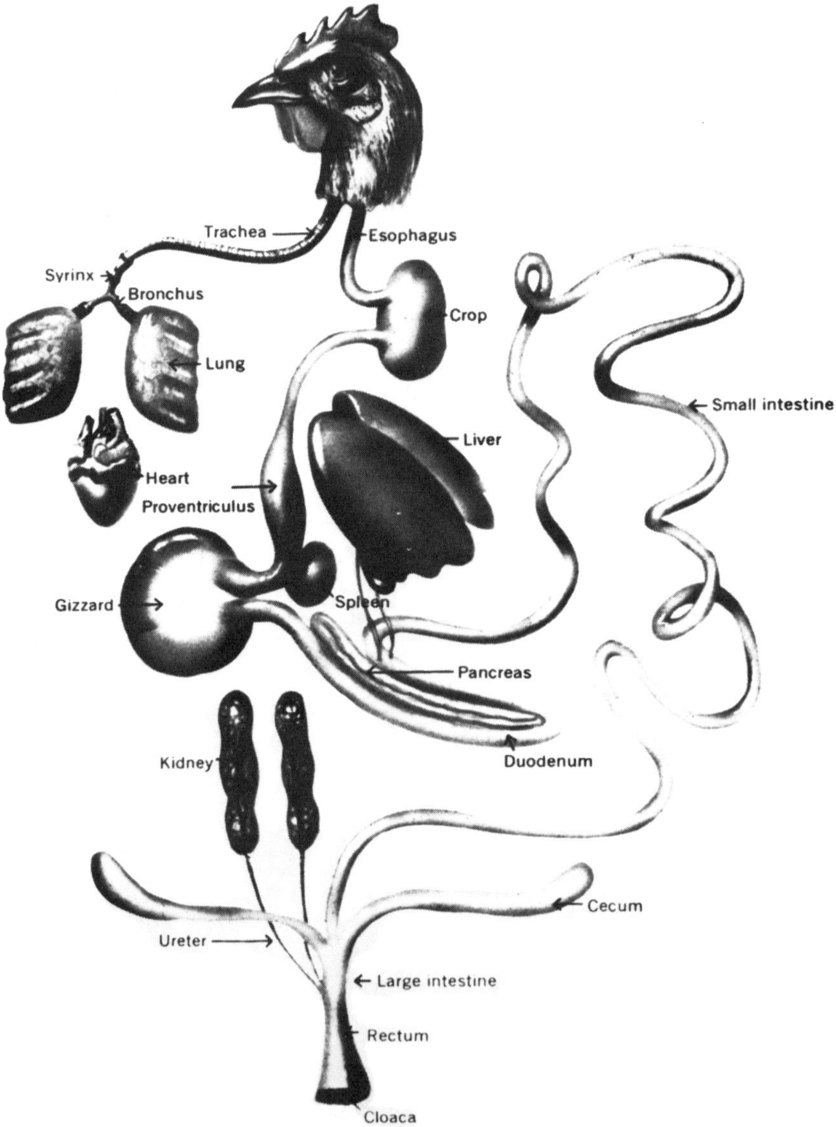

FIG. 5.2. NORMAL ANATOMY OF DOMESTIC FOWL

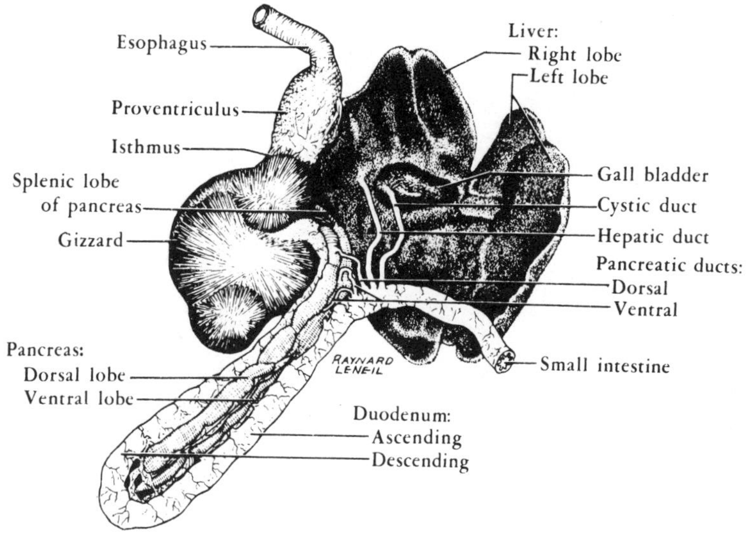

FIG. 5.3. DIGESTIVE GLANDS OF CHICKEN SHOWING DUCTS FROM LIVER AND FROM PANCREAS TO THE DUODENUM

TABLE 5.2. THE pH MEAN AND RANGE OF THE DIGESTIVE TRACT OF CHICKENS

Region	Mean	Range	
Crop	4.67	4.39	4.82
Proventriculus	4.48	4.30	4.60
Gizzard	2.94	2.83	3.01
Duodenum	6.13	5.95	6.40
Jejunum	6.29	6.03	6.62
Ileum	6.58	6.36	6.81
Ceca	6.14	5.75	6.50
Rectum	6.82	6.62	7.21
Bile (gall bladder)	6.51	6.12	6.64
Pancreas	6.58	6.38	6.80

Mouth

The mouth consists of an upper and lower mandible which together forms the beak. There is a limited supply of salivary gland secretions in the mouth of the chicken; therefore, salivary digestion is nearly absent. A complete summary of digestive juices and enzymes is shown in Table 5.3.

The beaks on yellow-skin birds become colored due to the deposit of the pigment, xanthophyll. This pigment is fat-soluble and is accumulated in the beak when the birds are not laying. Laying birds divert this pigment to the egg yolks, while that in the beaks and skin bleaches out.

The beak becomes hard when adequate calcium and Vitamin D are fed. Rickets is suspect if the beak area around the nasal openings is soft and pliable.

TABLE 5.3. POULTRY ENZYMES AND DIGESTIVE JUICES

Enzyme	Enzyme Found in Which Digestive Juice	Digestive Juice Secreted By	Digestive Juice Secreted into What Part of Digestive Tract	Enzyme Acts on What?	Enzyme Active in Acid or Alkaline Medium	Action Produces
Salivary amylase	Saliva	Salivary glands	Mouth	Starch	Neutral to slightly alkaline	Maltose
Salivary maltase	Saliva	Salivary glands	Mouth	Maltose	Neutral to slightly alkaline	Glucose
Pepsin	Gastric juice	Wall of proventriculus	Proventriculus	Proteins	Acid	Proteoses, polypeptides, peptides
Gastric lipase	Gastric juice	Wall of proventriculus	Proventriculus	Fats	Acid	Fatty acids, glycerol
Pancreatic amylase	Pancreatic juice	Pancreas	Upper small intestine (duodenum)	Starch	Alkaline[1]	Maltose
Trypsin, etc.	Pancreatic juice	Pancreas	Upper small intestine (duodenum)	Proteins, proteoses, polypeptides peptides	Alkaline[1]	Intermediate protein breakdown products, amino acids
Pancreatic lipase	Pancreatic juice	Pancreas	Upper small intestine (duodenum)	Fats	Alkaline[1]	Fatty acids, glycerol monoglycerides
Intestinal peptidases (erepsin)	Intestinal juice	Wall of small intestine	Small intestine	Intermediate protein breakdown products	Alkaline[1]	Amino acids
Intestinal maltase	Intestinal juice	Wall of small intestine	Small intestine	Maltose	Alkaline[1]	Glucose
Sucrase	Intestinal juice	Wall of small intestine	Small intestine	Sucrose	Alkaline[1]	Glucose
Lactose	Intestinal juice	Wall of small intestine	Small intestine	Lactose	Alkaline[1]	Glucose fructose
	Bile	Liver	Upper small intestine (duodenum)	Bile reacts with fats	Alkaline[1]	Soap, glycerol

[1]pH ranges from 3 to 6.8 in poultry.

The shape of the beak indicates feeding habits. Short beaks generally indicate seed eaters and long beaks are found on many birds that collect food from stream bottoms. Flat (bill) beaks indicate water feeders.

Esophagus

The esophagus or gullet is the tube which connects the mouth to the crop. A mucous-type material secreted in the gullet sometimes coats the food to aid in swallowing.

Crop

The crop is a modification of the esophagus and its major function in poultry is to store food when the bird is eating rapidly. In the pigeon, we find the crop secreting "pigeon milk," the food given young squabs. Both male and female pigeons produce pigeon milk.

The nerve or muscles of the crop may become injured by mechanical means, pathological or nutritional reactions. A bird may fly or strike the crop on hard objects, such as limbs or stones, destroying the nerves or tearing an opening in the crop membrane. A disease affecting the nerves, commonly known as "Range Paralysis," can destroy the nerves controlling the tonus of the crop muscles. The crop will become enlarged and food will accumulate, producing a pendulous crop.

Food particles may block the lower opening to the crop by swelling or by coarseness. When blocked, the ingested food ferments and a "sour" crop develops.

Proventriculus

This is the glandular part of the stomach. Its purpose is to secrete pepsin and hydrochloric acid for protein digestions. The rapid movement of food through the proventriculus does not provide an opportunity for digestion until it enters the gizzard.

Gizzard

The gizzard is the muscular stomach or ventriculus whose function is to crush feeds into fine particles and mix with water to form a paste-like consistency called chyme. Peptic digestion occurs after this mixture is produced. The mixture then passes into the upper portion, the duodenal loop, of the small intestine. The size and strength of the gizzard is related to the hardness of the food particles and the presence of crushed insoluble stones (grit). A well-developed gizzard may exert pressures up to 500 lb per sq in. Recording instruments have shown that the crushing motion is rhythmical. When the gizzard becomes well-developed, the remaining portion of the digestive system is stimulated to grow. Thus, we have a stronger, healthier and larger digestive system brought about by feeding the chicken grit or hard feed particles. If the grit or hard feed particles are removed, the gizzard atrophies becomes soft and weak. The gizzard is not an essential organ, nutritionally, provided the feed is well-pulverized; economically, however, it is essential in mash-grain feeding programs or where coarse ground ingredients are fed. This use of a coarse grind or whole grain lowers the feed cost and also seems to aid in controlling feather-picking.

The gizzard has been used in the production of bogus antique jewelery. Egyptian dealers have been accused of placing rings and other small pieces in the gizzard of a chicken and allowing the organ to scratch and change the appearance to resemble an antique.

Ulcers are often found in the gizzards of newly-hatched chicks. So far, no cause has been established. The ulcers heal within the first 2 wk of life.

When an organ changes in size or function, there is generally a change in biochemical activity, as well. The feeding of granite grit has increased the size and the strength of the gizzard. The feeding of granite grit resulted in approximately a 50% increase in the size of the gizzard and a 50% increase in alkaline phosphatase activity in the duodenal loop, tibia and blood plasma, as shown below:

	Chick Starter Only	Chick Starter and Granite Grit
Alkaline Phosphatase Content		
Duodenal loop	2780 phosphatase units	4760 phosphatase units
Tibia	470 phosphatase units	720 phosphatase units
Blood serum	3.6 phosphatate units	5.1 phosphatase units
ORGAN SIZE		
Gizzard weight	5.88 g	7.47 g
Small intestine length	26.5 in. (67.3 cm)	27.8 in. (70.6 cm)

The Small Intestine

The small intestine of the chicken is a sensitive organ, and some of its digestion and absorption qualities are embryonic in nature. The sensitive nature of the digestive system continues until the chick is 21 to 28 days of age, when the system becomes similar to that of the mature bird. It is well-established that many of the early nutritional requirements of the chick are similar to those of the breeding hen, and remain so until the chick is 21 to 28 days of age. A hen can utilize high-calcium uptake, 2 to 4%, without bringing about a trace element deficiency (manganese or zinc), whereas a chick of 1 to 21 days of age readily fails to absorb or utilize manganese or zinc when over 1.6% calcium is fed. This sensitive nature of the small intestine dictates the upper level of dietary calcium fed baby chicks and poults.

Since most digestion is completed in the duodenum, the upper portion of the small intestine completes the absorption and serves as a type of reprocessing organ where the phenomenon of recycling becomes very pronounced.

Intestinal recycling is a phenomenon seldom recognized. In essence, the duodenum, with the upper small intestine, serves as both an absorptive and excretory organ. Minerals and amino acids which are not used or impounded when absorbed after digestion, are recycled back into the small intestine where they can be reabsorbed. When proteins are being digested in the duodenum and small intestine, there are proteins being digested in body cells to the amino acid stage, which are then recycled into the small intestine. Amino acids from intestinal digestion and recycled amino acids are blended, and absorption/reabsorption occurs. This recycling phenomenon causes rapid amino acid depletion. A more efficient use of minerals occurs because amino acids form soluble chilates or complexes with many minerals to facilitate absorption or to form a truer stable solution. The rapid amino acid depletion and loss is speeded up because each time the acids pass through the liver, some are oxidized and deaminized.

Duodenum

The duodenum is a loop of the upper small intestine where the digestive processes concerned with the hydrolysis of crude nutrients, such as starches, fats and proteins, is completed. Absorption of the end products of digestion takes place primarily in the duodenum.

Pancreas

An organ, the pancreas, lies in the fold of the duodenal loop. Its function is to secrete pancreatic juices which complete the digestion of starches, fats and proteins.

Liver

The liver secretes bile into the lower portion of the duodenum by way of 2 ducts. It performs 2 functions in digestion:

(1) It neutralizes the hydrochloric acid.
(2) It forms soluble soaps with the free fatty acids.

This helps in absorption and translocation of fatty acids.

One of the bile ducts is enlarged to form a bile storage compartment, the gall bladder. When a chicken has not eaten for several hours, the gall bladder is enlarged and filled with bile. Two important bile acids found in the bile fluid are taurocholic and glycocholic acid. The most important functions of the bile acids are:

(1) They facilitate digestion of fats by forming emulsions.
(2) They activate pancreatic lipase.
(3) They aid in absorption of fatty acids, cholesterol and fat-soluble vitamins.
(4) They stimulate hepatic bile flow.
(5) They keep cholesterol in solution in the bile.

Cecum

There are 2 ceca, or blind guts, but they serve little purpose. Several diseases, such as Blackhead in turkeys and Coccidiosis in chickens, can develop in the cecum. Some undigested nutrients are decomposed by the cecae microorganisms, but very little, if any, absorption occurs.

Large Intestine

Microorganisms decompose many of the undigested food particles and form feces. Ureters from the kidneys run into the large intestines, where the feces are collected, and the water content of feces and urine are reduced to approximately 72 to 75%.

The composition of the urine is presented in Table 5.4, and shows the influence of diet on urine content.

Poultry excrete ammonia, mostly in the form of uric acid. In general, animals which excrete uric acid lay eggs enclosed in water-resistant coverings. Such eggs are called "cleidoic" eggs, a term that means literally, "enclosed," and refers to the fact that everything necessary for embryonic development and growth has been enclosed with the egg covering. Excretory products of the embryo must be retained inside the egg until hatched. Uric acid, the end product of protein metabolism in poultry, is relatively insoluble and nontoxic, whereas urea and free

TABLE 5.4. DISTRIBUTION OF THE MAJOR NITROGENOUS COMPOUNDS IN CHICK URINE[1]

Source of Dietary Nitrogen	Total Urinary N mg	Uric Acid N %	NH_3N %	Urea N %	Creatine N %
Corn-soya diet 22.3% crude protein	814	80.7	10.5	4.5	0.9
35% Casein	1038	79.6	15.0	2.0	0.4
35% Casein + 1.2% Arginine	1208	75.9	15.4	5.6	2.9
25% Casein + 10% Gelatin	961	80.4	13.4	1.9	0.6
35% Liver protein	865	75.1	16.0	3.8	0.7

Source: Data obtained from O'Dell *et al.* (1960).
[1]Calculation on daily basis.

ammonia are toxic and soluble. The formulation of uric acid requires more glycine than to form urea, hence, poultry require more glycine than other farm animals.

FOOD PASSAGE

Food moves rapidly through the digestive tract; however, several factors influence this movement:

(1) An increase in the amount of food consumed speeds up the passage rate.
(2) Finely-ground food moves faster than coarse particles.
(3) Soluble food passes faster than coarse particles.

Table 5.5 summarizes the time required for food to pass through the alimentary tract of chickens and turkeys.

EGG PRODUCTION

Ten days prior to laying her first egg, the pullet has hundreds of small yolks about the size of sorghum seeds in her ovary. When 100 days old, the pullet senses the circadian rhythm of her environment. A major factor is the length of daily periods of darkness. She is stimulated by a succession of days during which the dark period becomes shorter each day, such as occurs from March to June in the United States. Egg production starts by stimulus to a hen's brain by nerves to the hypothalamus, and neural secretory cells to produce hormone(s) known as releasing factors which seep through the capillaries to the portal system of the pituitary gland. The nerves are telephones; the hormones are mes-

TABLE 5.5. THE TIME REQUIRED FOR FEED TO PASS THROUGH THE ALIMENTARY TRACT OF CHICKENS AND TURKEYS

	Age (Years)	Av. Food Passage Time (Hr)
Young turkey hens	4 to 5 months	2.27
Old turkey hens	1½ to 2½	3.52
Chicken layers	1 to 2	3.42
Chicken nonlayers	1 to 2	3.50
High temp (90°F)	1 to 2	3.46
Low temp (60°F)	1 to 2	3.51
Chicken hens	1 to 2	3.40
Turkey hens	1½ to 2½	3.45
Pullets	6 to 8 months	3.12
Cocks	1 to 2	3.20

sengers. This stimulates specific cells of the anterior lobe of the pituitary. Here, the follicle-stimulating (FSH) and leuteinizing hormones (LH) are produced which also seep into the bloodstream.

When these hormones reach a certain level in the ovary, they stimulate one of the hundreds of yolks to start a 10-day period of rapid growth; the next day, a second yolk starts enlarging. Consequently, a new egg starts to grow rapidly each day for 3 to 5 days followed by a day missed before another yolk starts to develop. The number of days of the clutch is determined by both genetic and environmental effects. When the yolk reaches its normal size, it is ready for ovulation. This occurs about 7 hr before it is destined to start down the long path of packaging so it can break out of the follicle. Ovulation also is programmed in a similar manner as that required for rapid yolk growth. The hormone which sets the time fuse is leuteinizing hormone (LH) from the pituitary. The laying of the egg and ovulation of the next yolk are programmed in a linear sequence about 15 min apart, even though one is not dependent on the other. During this period, the bird changes in physical appearance—she eats voraciously and gains weight—from one-half to 1 lb in 2 to 4 weeks—and her comb enlarges and reddens. Some of these changes are masculine, although predominantly feminine. She is off the roost at the crack of dawn, scratching in her yard, even though she does not need to do so. The interstitial cells of the cortex of the ovary in close proximity to the developing yolk produce a chemical known as alpha estradiol, an estrogen or feminizing hormone. At the same time, in the medulla of the ovary, a second path is programmed to interstitial cells which secrete testosterone, a masculinizing hormone. Both hormones seep into the bloodstream and make remarkable internal changes. The pubic bones spread so that when the bird starts to lay, the egg can slip between her bones readily. The oviduct enlarges rapidly in preparation for production. Metabolic rate, adrenal function and calcium level in the blood are all programmed together, so that in 10 days, egg production is in full swing.

At ovulation, the egg drops free into the body cavity, where it is immediately picked up by the funnel of the oviduct. The mechanism of production is such that no hen can produce 2 eggs less than 22 to 23 hr apart. Egg-type hens require 24 to 25 hr to place the membranes and shell on each egg. Ovulation for the first egg of a clutch occurs just after lights are turned on. Thus, the hen is not a self-starter—she must have a rhythm or beat on which she can rely, and this comes from her environment. Stresses like crowding, feed restriction, social rank, fighting, cold, heat, for example, can interrupt the hen's egg production.

HORMONES

The term "hormone" was created by two British physiologists, Davis and Starling, in 1903, from the Greek word, "hormao," which means "I excite." Hormones are excitatory in their action and are often referred to as chemical messengers. Sex behavior and instincts depend upon specific hormone secretions elaborated by the respective reproductive glands. The most obvious external effects are seen in the comb, wattles and plumage of male and female birds. Curiously, there exists a dual potentiality of the sexes—all males and females combine potentialities of both sexes. In the hen, a preponderance of female hormone is secreted by the ovary; in the cock, the male hormone predominates. In certain conditions, as in disease or castration, balance between the male and female hormones becomes disturbed, or one or the other is completely repressed. When castration is performed upon young cockerels, secondary sex characteristics and sex instincts are suppressed, while body weight is greater than that of uncastrated males of the same age. Old males are fattened and their muscles made tender by administration of estrogen.

Since estrogen has a feminizing effect, it could be undesirable for persons eating chicken should estrogen remain in the dressed bird. Regulatory agencies are responsible for seeing that residual estrogen in the dressed carcass is of a sufficiently low level to be harmless to humans. Otherwise, it may not be sold.

Hormones give poultrymen control over the timing of egg production. A large dose of progesterone makes hens quit laying on the day of injection, but production is back to normal 4 weeks afterwards. Most of the birds molt completely; the rest lose about half their feathers. This injection forces birds out of production when egg prices are low and allows them to come into heavy production when egg prices rise. Progesterone is an effective and harmless way of molting chickens because they maintain their body weight during the molt better than chickens force-molted by other methods.

Heritability of thyroid weight has been shown through breeding and selection. Calcitonin or thyrocalcitonin, a hormone in the ultimobranchial rather than thyroid gland, has an important function in regulating the level of calcium in body fluids. The anterior pituitary controls growth; too little pituitary hormone produces dwarfs, and too much, giants.

Drugs, such as sulfocyanide, thiourea, and thiouracil inhibit the formation and secretion of thryoid hormone and thus are powerful goitrogenic compounds. In chickens, addition of thiouracil to chicken rations at levels of up to 0.2% of the ration consistently improves carcass quality in broilers. In most cases, however, thiouracil decreases the rate of growth.

Hormonizing by estrogens was first introduced to the poultry industry in 1947 by injection of pellets of hormone. Pellets are absorbed very slowly, and harmful amounts of the hormone are not found in edible portions of the bird. Birds are implanted by means of a simple device—an implanter which holds a number of pellets in clip form. One pellet is injected under the loose skin of the neck next to the head. An experienced operator with 4 people to catch birds can implant as many as 1500 birds per hr. If the litter becomes too wet, which often happens with hormonized birds, a low protein diet is one method of cutting down excess dampness. A 16 to 17% protein feed is used by most growers to minimize the problem, although some start on a 14% protein diet from 4 to 5 days before implanting and carry the same diet through to marketing time. Plenty of water is made available after caponizing.

Cabbage seed, rapeseed and steamed black or white mustard seed all produce pronounced thyroid enlargement. These plants belong to the mustard family, Brassicaceae Cruciferae. Plants other than Cruciferae also cause development of goiter—for example, raw soybeans and a variety of chemical materials. Thiocyanate and related monovalent anions inhibit the concentration of inorganic iodide ions in the thyroid and thus decrease the formation of the thyroid hormone. Certain plants which contain antithyroid compounds in their edible parts have goitrogenic materials as inactive thioglucoside precursors. These thioglucosides are hydrolyzed enzymatically to liberate the active aglycone. Other antithyroid substances, such as the thiocyanates, are present in certain feeds, but their goitrogenicity is overcome by the addition of iodine to the diet.

All the hormones discussed are classed technically as drugs, and as such, come under the provision of the food and drug laws. Consequently, before any of them can be obtained for general use, United States Government officials must be sure that their use will not cause a health hazard to consumers of the poultry. The current status of the acceptability of hormones and related chemical substances should be cleared through state and federal feed control officials before being used.

In the case of thyroprotein, approval has been given for its use in poultry feeds. No evidence, so far, has been found that the hormone residues will be found in the flesh or eggs of birds that would ultimately affect the consumer. The goitrogenic drugs, thiouracil, etc., are still restricted, and so cannot be used for fattening, except under carefully supervised conditions.

The female sex hormones were formerly released for use conditionally, providing the pellets were removed before the carcass of the bird was marketed. There still may be some question as to the advisability of using these compounds, however, because recent research has shown that traces of estrogenic compounds are retained in the fat storage depots of the body for considerable periods of time. It is evident that in any circumstance, hormone therapy must be controlled very carefully in order to obtain uniformly good results and to protect consumers.

Endocrine glands act as a complicated set of delicate controls over many vital body processes. The endocrine glands include: pituitary, located at the base of the brain; pineal, also in the brain; thyroid, parathyroid, ultimobranchial body, and thymus (not considered by some to be endocrine), located in the neck; adrenals, in the abdominal cavity near each kidney; pancreas, also in the abdominal cavity; and testes, in the male, and ovaries, in the female. The secretions of these glands are emptied directly into the blood vessels within themselves. That is why they are know as ductless glands or glands of internal secretions. Each gland secretes one or more substances, called hormones, in very minute quantities and normally formed as needed through nervous impulse. Hormones control sexual growth and the various stages of reproduction.

The thyroid gland, through its hormones, thyroxine and triiodothyronine, controls the rate of activity of most metabolic functions in birds. It is concerned with body growth, feather development and egg production—in short, practically all functions that are of economic importance in the poultry industry. Chemically, thyroxine is a derivative of the amino acid, tyrosine, formed in the thyroid gland by reaction between tyrosine and iodine. The production of thyroxine is under the influence of the pituitary gland which, in turn, is influenced by many environmental factors, such as light, temperature and stress. The hormone is distributed through the blood to all cells of the body.

Thyroxine-like compounds produced commercially by the addition of iodine to casein have many of the benefits of thyroxine. A product, "Protamone," is effective in improving growth, feathers and carcass composition in market ducks, especially during the hot summer months. In summer, feathering is poor, and ducks tend to lay on excessive amounts of fat. Protamone also improves the feathering of growing turkeys, but not growth rate.

Hormones regulate the metabolism of protein, fat and carbohydrate, and thus it is important to understand these substances in the study of nutrition. It is logical to assume that many, if not most, of the actions of nutrition on reproduction are mediated through altering hormone production by the endocrine glands. Endocrine imbalances also increase requirements for dietary factors necessary for growth and reproduction. The only type of malnutrition which has been conclusively proven to impair the secretion of gonadotropic hormones by the pituitary is underfeeding or any state leading to inanition, including inadequacies of the B-vitamins. Thyroid function also is decreased by undernutrition via the pituitary; but in addition, it can be depressed directly by lack of iodine or the presence of goitrogenic substances in the diet. There is considerable basis for the belief that the gonads and sexual activities are more easily impaired during inanition than other body functions.

Hormonal imbalances may impair reproductive activities by several means. Hypo- or hyperthyroidism reduces the secretion of gonadotropic hormones by the pituitary, alters gonadal response by gonadotropins and changes dietary requirements. Hyperthyroidism prevents conversion of carotenes to Vitamin A.

The pituitary gland is unquestionably the master gland of the body. It consists of 2 parts—a larger anterior lobe and a smaller posterior lobe. It is located in the bony cavity at the base of the brain surrounded by a rich supply of blood vessels. The development of such sex characteristics as body conformation and, in the female, the ripening of the ovum in the ovary, is directly or indirectly the result of stimulation by hormones secreted by the anterior lobe of the pituitary gland.

Carbohydrate metabolism is influenced by a hormone secreted by the pituitary. Digestion and assimilation of fats and proteins are influenced by anterior pituitary hormones, as is secretion of insulin by the pancreas. Some forms of obesity are attributed to glandular disfunctions.

The posterior lobe of the pituitary stores and releases the hormone, pituitrin. The kidneys are influenced directly by the posterior lobe through the stimulation of the filtering mechanism. The testis has 2 major functions—production of spermatozoa and secretion of the male hormone, testosterone. It is antagonistic to the female hormone, estrogen, which induces female characteristics. It even produces male characteristics in females. Testosterone has been isolated, analyzed and synthesized so it is available in pure form.

The ovary is a reproductive gland of the mature female and is located on the left side of the body in the abdominal cavity, connected to the uterus by a tube which serves for passage of the egg. Its functions are the production of the egg, as well as the follicle, which secretes the hormone, estrogen, and the hormone, progesterone. Estrogen has been analyzed chemically and prepared synthetically. It induces the development of secondary sex characteristics in the female, such as distinctive plumage, and inhibits the pituitary from secreting the follicle-stimulating principle.

The parathyroids are 2 to 4 ductless glands located under each lobe of the thyroid gland. Parathormone, produced by the parathyroids, maintains equilibrium between the soluble calcium and phosphorus compounds in the skeleton and the blood.

The adrenal glands are located in front of each kidney and consist of 2 parts: the medullary (inner) and cortical (outer) portions. The hormone, adrenin, is secreted by the medullary part, and the hormone, cortin, is contained in the cortical portion of the gland. The latter is similar to testosterone in chemical makeup and its antagonistic reaction to estrogen.

The thyroid gland resembles a tiny shield and is located on either side of the upper part of the windpipe. Chemical analysis shows that thyroxine contains large amounts of iodine. An increased production of thyroxine gives individuals an excessive amount of energy (metabolic rate may be doubled), whereas a decrease makes individuals sluggish. A deficiency of iodine in the diet results in a decreased functioning of the thyroid gland.

In poultry husbandry, it has been common practice to castrate males to improve the tenderness and tastiness of the meat. As previously stated, an estrogen is a substance or hormone normally produced by the ovary of a mature fowl. Substances that have the same effect as the natural hormone have been made synthetically. They are used in commercial hormonization of poultry, a logical successor to caponization, which it resembles in some respects. Caponization is the surgical removal of the sex organs, whereas, hormonization is the addition of a hormone which produces extensive depression of the size and function of the testes—an effect that has given rise to the common misnomer for treated birds: "chemical capons" or "caponettes."

Estrogen produces numerous effects on birds; some have considerable significance from a commercial viewpoint. Hormonization rapidly increases the cholesterol, phospholipid, triglyceride and protein contents in the blood. The percentage volume of red cells is decreased. Estrogen causes hyperossification of endosteal bone, which causes fragile bones, but which have a normal percentage of ash. The birds' combs shrink and become pale, wrinkled and nonwaxy. The masculine sex drive is lost, aggressiveness diminishes, fighting and crowing cease. Males squat when approached and spermatogenesis ceases.

Turkeys respond similarly to chickens in regression of head furnishings and diminished fighting instinct, but in this species, there is heightened sexual activity. The ovary is quite resistant to estrogen treatment and there is no reduction in egg production. Estrogen produces an enlargement of the vent and spreading of the pubic bones such as normally occurs in laying chickens. A single injection of estrogen interrupts broodiness in hens and returns the birds to egg production. Feather growth is stimulated but the development of new pin feathers inhibited by estrogens. This effect is more marked in turkeys than in chickens. Improved skin texture is a consequence of estrogen treatment, presumably due to hypertrophy of the epidermal strata.

Diethylstilbestrol was the first compound used commercially, subcutaneously implanted as pellets in the necks of the birds. Effects are noticeable within a week, but require weeks for satisfactory results. A paste suspension containing diethylstilbestrol is also used for implantation. Attempts were made early to feed diethylstilbestrol but the results were unsatisfactory. This led to a search for new compounds of greater potency by the oral route. Dienestrol diacetate in the feed has given good results and is used commercially. A moderately increased rate of fat deposition is obtained by feeding this compound. Turkeys fatten less dramatically than chickens in response to estrogens. Oral administration of estrogens to turkey has led to some unfavorable results, although the average market grade of turkey broilers is improved.

Estrogen treatment has at least some influence on each of the components of carcass quality, and for the most part, the influence is favorable. Fatness is the best advantage; extra subcutaneous fat improves the finish of the dressed carcass. Muscle and leg fat increases and improves the "juiciness" of the meat, lightens flesh color and increases tenderness. Flavor evaluation is inconclusive and inconsistent. Stability of meat in storage is unaffected. Blemishes due to fighting are decreased in chickens but not in turkeys. A smooth, silky texture is imparted to dressed carcasses. Skin tenderness is increased and, therefore, extra care is necessary during scalding and plucking. Extra care in processing is also necessary because of increased bone fragility.

Extent of effects on carcass quality depends on the length of the treatment with estrogen, its route of administration and the age of bird at the time of treatment. Estrogen mixed in the feed at moderate dosages increases feed intake, but has no effect on rate of gain of growing birds. There may be a slight growth depression with turkeys.

Kafir, wheat and corn are superior to oats and barley in fattening rations containing estrogen. High-density mashes also improve rate of gain, feed efficiency and finish of the dressed carcasses, and average grades are as good as those observed in birds treated with estrogen, regardless of the ration.

Certain unfavorable features of estrogen treatment sometimes appear. Repeated implantations of pellets produce severe atheromatous lesions in cocks.

Treated birds sometimes go "off feed," make subnormal gains during hot weather, and react with more than usual severity to disease. Exceptionally moist droppings may result. Estrogen treatment of poultry is not permitted in Canada but pellets were permitted in the United States for a number of years, although now the practice is not approved by the Food and Drug Administration (Table 5.6).

TABLE 5.6. FEEDING PROGRAM DURING TREATMENT OF BIRDS WITH ESTROGEN PELLETS

Type of Birds	4 to 5 Days Before Treatment	After Treatment
Small broilers	Small amt. of corn	10 to 25% corn to market
Broilers		
ration 20 to 22%	Feed up to 50% corn	Up to 50% corn to market
ration 17 to 18%	Feed up to 40% corn	Up to 40% corn to market
broiler	4½ parts concentrate, 38% to 15½ parts corn	Same schedule to market
Caponette		
roasters	Up to 50% corn	To market (these birds are shot again 2½ weeks after first shot; increase corn if desired after 2nd shot)
Stags	Free choice	To market (2nd shot 2½ weeks after first); free choice
Cull hens	Free choice	Market in 3 weeks

To recapitulate, finishing quality in all classes of meat chickens, except mature hens, is improved by estrogen treatment. Slight increases in weight gains are obtained in chickens, along with an increase in feed consumption and feed efficiency. Estrogen treatments for turkeys are less useful than for chickens, being most advantageous for finishing immature birds. Turkeys fatten less dramatically than chickens, but other effects, such as feathering, are improved as much or more. Diethylstilbestrol was usually implanted as pellets or paste in the neck at the base of the skull, but is no longer permitted by the Food and Drug Administration. Offal from estrogen-treated birds should not be used for animal feeds. Dienestrol diacetate is used commercially in the feed for low-cost superior quality broiler and roaster production (Table 5.7).

ENZYMES

Enzymes are organic substances produced by living cells possessing the ability to catalyze specific chemical reactions. They synthesize or break down chemical compounds or transform them from 1 type to another.

TABLE 5.7. DIENESTROL DIACETATE TREATMENT PERIODS: WITHDRAW 48 HR BEFORE SLAUGHTER

Type of Birds	Age at Start of Treatment (Weeks)	Lipamone lb (Dienestrol Diacetate) per Ton Feed	Feeding Period (Weeks)
Broilers	5 or 6	1/3	4 to 6
Broilers	3	1	3
Roasters	5 or 6	1/3	6 to 10
Roasters	8 or 9	1/2	5 to 7
Turkey broilers	8 to 10	1	3

They are highly specific and catalyze only 1 reaction or even act upon only 1 isomer of a particular compound. There may be many different enzymes in a single cell. The number of known enzymes is around 750, of which 100 have been isolated in crystalline form. All enzymes are conjugated proteins. Many have traces of Zn, Co and Cu which stay with the protein even in crystalline form.

Enzymes go back to the earliest written documents of human activites. Housewives used leaven for making bread, calf stomachs for curdling milk to make cheese, fermentation with microorganisms to produce wine, beer and medicinals, tanned leather, etc. There is large-scale production of enzymes called industrial enzymes for specific uses in the food and chemical industry.

The most general way to name enzymes is to add the suffix, "-ase," to the name of the medium on which the enzymes work. For example, "maltase" hydrolyzes maltose. Another way is to add the suffix, "-in," to the name of the source from which it is produced. Papain is produced from the plant, papaya, ficin from the plant, ficus, and pancreatin from the pancreas. A third manner of naming enzymes is to add the suffix, "-ase," to the reaction the enzyme catalyzes. For example, "invertase" inverts sucrose to glucose and fructose.

The animal's own enzymes in its digestive tract digest or break down proteins, carbohydrates and fats in feeds to simpler materials before they are used by the animal for energy and growth. Supplemental enzymes in the feed are sometimes used to augment the animal's own supply of digestive enzymes to enable it to utilize feeds more rapidly and efficiently. This is particularly true of grains with poor feeding value. For example, Washington State University researchers in 1957 showed that the feeding value of Pacific Northwest barley for chicks and poults could be upgraded to equal approximately that of corn by either water treatment or enzyme supplementation. Both bacterial and fungal enzyme preparations are effective, although the bacterial products are more economical.

Enzyme-containing materials as feed additives are not new, and early reports indicated that they were of doubtful value. The general conclusion was that the digestive tract of a normal animal is not deficient in the enzyme required for the digestion of rations usually fed to those animals. The recent advent of high-energy feeds with increased emphasis on feeding for faster growth and improved feed efficiency of poorly digested grains has stimulated interest in taking another look. Recent work indicates that in certain conditions enzymes may have some value as feed additives. Their future would seem to depend on products standardized with respect to the content of the specific enzyme for increasing the availability of a specific substance in the ration. So, it remains to be seen whether or not such products can be supplied at a price permitting possible use in feeds.

Beneficial effects of enzyme treatment have been: improvement in growth, feed utilization and litter condition. Northwest barley substituted for corn in an ordinary chick or turkey ration depresses growth and feed conversion by 20%. Wet litter problems occur when barley is used in place of corn, due to increased water consumption. The addition of enzymes to feed lowers water consumption and the water content of the manure appreciably. This is of practical importance to the poultryman. Barleys grown in the Midwest and East do not respond to enzymes as do those grown in the Rocky Mountain and Pacific coastal states and western Canada. Rye and oats are also markedly improved by enzymes; wheat is improved to some extent. Some benefits to corn and milo have been obtained.

Results with enzyme treatment of feeds for laying hens are less clear-cut; some improvement in egg production, feed efficiency and hatchability are claimed. In general, there is a greater response from both feed and growth efficiency by water treatment than by enzyme preparation.

Enzymes are organic catalysts that by their mere presence in trace amounts initiate or accelerate the speed of reactions occurring in organic matter that would not otherwise proceed at an appreciable rate. Enzymes catalyze all the chemical changes occurring in living plants and animals. Enzymes are very sensitive to the temperature and pH of the medium in which they operate. Temperatures of maximum activity usually fall within the range of 86° to 122°F. Above the optimum temperature, they are permanently inactivated; lower than optimum, they become inactive but regain their activity when warmed. Optimum pH varies from neutrality to decidedly acid.

Chemical changes induced by enzymes are gradual step-by-step processes, not marked by severe energy changes. Conversion of starch into glucose is the end result of several intermediate changes, each requiring a different enzyme. Some enzymes are secreted in both active and inactive forms. The inactive form becomes active only in the presence of some other substance, such as a trace metal, a vitamin or a complex protein compound. Living cells are not essential for enzymatic action; enzymes can be separated from the cells without loss of catalytic activity.

6

Metabolic Responses in Chicks

The patterns of growth, body weight and organ weight have been studied in many breeds of both chickens and turkeys. There is a relationship of growth, body and organ weight to feed efficiency. Early studies on growth demonstrated that the relationship between weight gain and age is not linear, and obvious changes in the growth rate are apparent at 11, 13 and 16 wk of age. The increase in growth weight versus age of White Leghorn chickens shows 3 phases of postnatal growth:

(1) A period of relatively slow increase in weight
(2) A period of rapid growth
(3) A slow increase after the time of sexual maturity

This general type of sigmoidal curve is characteristic of the chicken, not only for whole body growth, but specifically, for muscles, skeleton, digestive tract, lungs, heart and kidneys.

The growth of the cockerel testes is an index to feed efficiency, because when the testes reaches maturity, the feed efficiency is very poor. During the period from hatching to 8 wk of age, the growth rate of White Leghorn chicks in equal numbers of both sexes was slow until 2 wk of age, after which the growth rate accelerated and continued almost linearly until termination at 8 wk. A similar study with New Hampshire cockerels showed that whereas it took 8 wk in the White Leghorn, it took between 16 and 18 wk with this meat breed. Early signs of growth in the testes of the White Plymouth Rock cockerel appeared at 10 to 12 wk of age. The difference in sexual growth between the White Leghorn and the White Plymouth Rock or New Hampshire Red is expressed in total body growth and hence, feed efficiency. The White Leghorn has poor feed efficiency after 6 wk of life and the heavy breeds are efficient up to 12 wk of life.

It is well-established that growth and sexual development are under hormonal control. A broiler or turkey should be marketed before the growth slows as a result of the interaction of the sex hormones and the growth hormones. Some of the major hormonal effects are summarized in Table 6.1.

Several hormones of the anterior pituitary are well-known for their role in controlling growth and development of the reproductive system. One of the important feeding programs is to feed heavy broiler-strain pullets a limited amount of feed so as to reduce sexual maturity, yet provide for growth.

Egyptians obtained data which correlated the development of certain endocrine glands in Fayomi cockerels. The largest average adrenal weight increase occurred between 12 and 16 wk, which preceded the largest average testes weight increase.

The environment influences the growth of the thyroid and also survival of the chicks. Thyroid hormone secretion in both male and female chicks during the first 2 wk after hatching may show a 3-fold variation from a maximum in winter months to minimum values in summer. Day-old White Plymouth Rock cockerels raised at a constant ambient temperature to 4 wk of age had the highest feed efficiency, body weight and livability when reared for the next 4 wk at 19°C. The highest mortality was in birds raised at 8°C, and the lowest body weight, feed efficiency and hemotocrat values were observed at 30°C. This demonstrated that birds must have more fortified diets in warm weather because of the lowered feed consumption.

Accompanying growth is skeletal (bone) development. Also during bone growth there is bone resorption, which is influenced by the parathyroid hormone. The parathyroid hormone, which stimulates bone mobilization, increases citrate production by bone. Bone citrate and calcium content in bone vary about the same percentage throughout growth. Well-calcified bones have a high level of citrate.

Testosterone increases the synthesis of proteins and decreases amino acid catabolism. The total carcass protein is increased by testosterone, and there is synergistic action between testosterone and growth hormone in increasing protein synthesis. The growth hormone brings about a reduction in plasma amino acids because of the greater tissue uptake.

A reversible steady-state relationship has been shown to exist between the tissue, and its environment which influences the uptake of plasma amino acids, either in the intact animal or in isolated tissues. For best interpretation, plasma levels and tissue levels of amino acids should be considered together in order to demonstrate a net change in the utilization of an amino acid. A limited interpretation can be based upon plasma amino acid levels. For example, in very young birds, a large proportion of sulfur amino acids was diverted to the growth of feathers containing about 10% cystine. The ratio of feather synthesis to tissue-protein synthesis may determine the sulfur amino acid requirement. The White

Leghorn, a rapidly-feathering breed, showed a marked growth and feathering response to higher dietary levels of arginine and glycine, while a more slowly-feathering breed displayed little of the same response. Further studies on young chicks up to 3 wk of age demonstrated a high arginine requirement accentuated by the requirement for this amino acid in muscle creatine synthesis. Since creatine increased rapidly during early life and reached an optimal level at approximately 26 days, the subsequent requirement of arginine for creatine synthesis declined. It is also known that testosterone can increase the rate of synthesis of creatine. Thus, by 4 wk, avian muscle has attained a composition approximately comparable to the adult at 27 wk with regard to proportions of amino acids and electrolytes. An early investigator speculated that the requirement would be higher for both fast-growing and younger birds of any breed, since muscle and organ tissue contain high amounts of lysine. In various dietary treatments, the plasma level was always higher at 2 wk when compared with the level at 8 wk. Determination of the dietary lysine requirement likewise confirmed that chicks up to 4 wk have a 1.1% lysine requirement in diets with 20% good quality protein and 974 Cal. of productive energy per lb feed, a requirement which declined at the rate of 0.05% per wk to a minimum 0.8% lysine requirement at 10 wk of age. The investigators hypothesized that the decreasing lysine requirement was due to a shift in the ratio of amino acid to energy expenditure. In a growing bird, the rate of gain constantly decreases in proportion to the body weight, while the enlarging body mass continuously increases the energy requirement. The above evidence partially explains the decreasing lysine requirement with age. It is evident that a high degree of correlation exists between proportions of essential amino acids contained in muscle tissue and the requirements for these amino acids by either rats or chicks. Results on the accumulation of DL-lysine-3H in rat tissues revealed that the highest *in vitro* rate occurred at 1 wk and declined continuously through 6 wk of age. The most rapid uptake was found in the descending order of: motor neurons, liver, plasma, muscle, and spinal gray matter. Other recent reports illustrated a progressive decline in the rate of net intestinal absorption of L-valine.

The great influence of diet, environmental temperature and hormones on body growth and sexual development is recognized. The influence of these factors on organ growth and blood plasma content is needed in developing better rations. The data presented in the tables is from the dissertation of R.J. Carlotti (1971). Manifestations of "Metabolic Bursts" during development of the cockerel, Graduate school, West Virginia University, Morgantown, West Virginia (98 pages). White Leghorn cockerels were use to establish these values. The normal growth of White Leghorn cockerels is presented in Table 6.2. Also, growth of the

TABLE 6.1. SUMMARY OF SELECTED HORMONES AND THEIR ACTIONS

Hormones	Principal Actions Pituitary Hormones	Observed Changes
Somatotrophin (STH growth hormone)	Causes growth of bone and muscle; anabolic effect on nitrogen metabolism; carbohydrate and lipid metabolism; elevates glycogen stores of skeletal and cardiac muscle.	Increases nitrogen retention tissue uptake of amino acids; increases glycogenesis. Decreases rate of nitrogen excretion, plasma amino acid levels. Enhances effect of tropic hormones for specific organs.
Gonadotrophins (a) luteinizing or	Testis: stimulates the intestitial cells of Lydig, thus promoting production of androgen. Acts synergistically with FSH in stimulating Lydig cells to produce testosterone.	Increases synthesis of protein, calcium retention and levels of RNA in accessory sex organs. Decreases rate of nitrogen excretion, rate of amino acid catabolism.
(b) follicle-stimulating hormone (FSH)	Testis: has possible action on seminiferous tubules to promote spermatogenesis.	
Adrenocorticotrophin (ACTH)	Stimulates the secretion of adrenal cortical steroids by the adrenal cortex.	
Thyrotrophin (TSH)	Stimulates the thyroid gland to form and release thyroid hormones.	

Source: Turner (1962); Hunter (1968); White *et al.* (1968).

TABLE 6.1. *(Continued)*

Hormones	Principal Actions Hormes Other than Those from the Pituitary Gland	Observed Changes
Testis		
testosterone	Causes maturation and functional maintenance of male reproductive tract; development of secondary sex characters; direct stimulation of testis itself; influences sex behavior.	Extratesticular effects are the same as those described for LH.
Parathyroid		
parathormone	Conditions the metabolism of calcium and phosphorus; acts on the skeleton and possibly the kidney.	Increases blood calcium and phosphorus levels. Accelerates the transfer of calcium and phosphorus from bone to blood.
Ultimobranchial body (in birds)		
calcitonin	Antagonizes parathormone actions on calcium and phosphorus metabolism.	Decreases blood calcium and phosphorus levels. Accelerates transfer of calcium from blood to bone.
Thyroid		
thyroxine and triiodothyronine	Accelerates metabolic rate and oxygen consumption by tissues.	Accelerates growth of whole organism by stimulating protein and nucleic acid synthesis.
Pancreatic islets		
insulin	Regulates the storage and utilization of carbohydrate.	Reduces blood sugar.
glucagon	Promotes liver glycogenolysis.	Elevates blood.
Adrenal cortex		
aldosterone other cortical steriod hormones	Is a steroid hormone of special importance in electrolyte and water metabolism. Metabolizes carbohydrates, fats and proteins. Has effects on fluid shifts; gives immunity and resistance to infection, inflammation and hypersensitivity.	Increases serum sodium. Decreases serum potassium. Diminishes water loss through kidney. Maintains blood sugar in fasting state by sparing carbohydrate utilization and increasing gluconeogensis.[1]

Source: Carlotti (1971).

[1]In chickens, the normal blood glucose levels of 175 to 225 mg per 100 ml are approximately double those found in most mammals (Sturkie 1963). During a 24 hr fast, the blood declines to the lower limits of the normal range. However, the fasting period must be extended to 3 to 6 days or longer to initiate gluconeogenesis indicated by the return of blood glucose levels.

TABLE 6.2. WHOLE BODY WEIGHTS AND RATE OF GAIN

Cockerel Age	Body Weight in g		Daily Rate of Gain in g	
	Mean[1,2]	S.E.	Mean	S.E.
Day 1	36.4	± 1.41	—	—
Week 1	51.2	± 1.00	2.46	±0.39
Week 2	108.6	± 1.41	8.21	±0.071
Week 3	191.9	± 1.73	11.9	±0.16
Week 4	229.2	± 6.16	5.32	±1.14
Week 5	335.9	± 23.7	15.3	±1.94
Week 6	434.9	± 47.3	14.2	±1.05
Week 7	540.2	± 32.5	15.0	±3.59
Week 8	699.9	± 2.62	22.8	±0.26
Week 10	911.7	± 4.52	15.1	±0.89
Week 12	1143.3	0	16.5	±0.29
Week 14	1256.3	± 5.00	8.07	±0.35
Week 16	1451.9	± 40.8	14.0	±2.55
Week 18	1647.5	± 2.73	14.0	±2.70
Week 20	1773.5	± 34.4	8.98	±2.63
Week 22	1759.8	±138.3	−9.50	−0.97 ± 9.7

Source: Carlotti (1971).
[1]The mean ± the standard error of the mean represents the number of groups of cockerels, organs, tissues or other analyses.
[2]Four groups of cockerels.

testes, liver and femur are shown in Fig. 6.1. The femur and body growth show a similar percentage rate, whereas the liver and testes show rapid gains during the 18- and 20-wk periods. A rapid testis growth is shown at the age of 8 to 12 wk. During this 8- to 12-wk period, the White Leghorn cockerels become sexually active.

Bone development in birds has demanded much consideration. Hock disease caused heavy losses before the vitamins and trace minerals associated with this nutritional disease were recognized. The major substances or elements in bone are water, protein, calcium, phosphorus, citrate and crude fat (ether extract). The trace minerals associated with bone formation (protein structure) are zinc and manganese, and water-soluble vitamins are choline and biotin. Vitamin B_{12} (cobone) is very active in bone structure during the first 5 days of embryo development (often called congenital malformations or defects). After this period, it exerts the greatest effect in red blood cell (erythrocytes) formation which occurs in the bone marrow.

The weight and percentage of water in the femur of developing White Leghorn cockerels are presented in Fig. 6.2 and Table 6.3. The decline in water content is accompanied by a similar rate of bone growth. Bone growth is greatest during the tenth- to twelfth-week period, whereas the decline in water is gradual. The increase in dry matter is also greatest during the tenth- to twelfth-week period.

Bone ash, total calcium/phosphorus ratio and citrate/calcium ratio are shown in Table 6.4 and Fig. 6.3.

The bone ash, calcium, phosphorus and citrate continued to increase throughout the growing period of 22 wk. The ratio of calcium to phosphorus changed from 1.71 to 1.86, whereas the percentage of increase in calcium was 8.l, compared to 2.1 for phosphorus (Table 6.5). The rate of calcification and phosphorlation displays 4 peaks, each at the third, eighth, twelfth and eighteenth wk of life. There is a change in the nutritional response to calcium by the twenty-first day of life. This is expressed by the digestive system changing from a very sensitive to a less sensitive organ regarding pH. When the pH becomes over 6.5–7 before 3 wk of age, the zinc and manganese become unavailable and a bone deformity (Perosis) develops. The increase in calcium at 18 wk may be related to the development of the medullary bone in the pullet, although this has not been investigated. In birds, there seems to be, at times, some common relationship between the male and female. An excellent example of this is the simultaneous development of both the male and female crop in the pigeon to produce "pigeon-milk" for feeding the young.

The testes makes a rather rapid growth on the twelfth wk and the calcium uptake is also increased. Also, at this period, there often appears a bone deformity. The eighth-week peak coincides with some peaks for

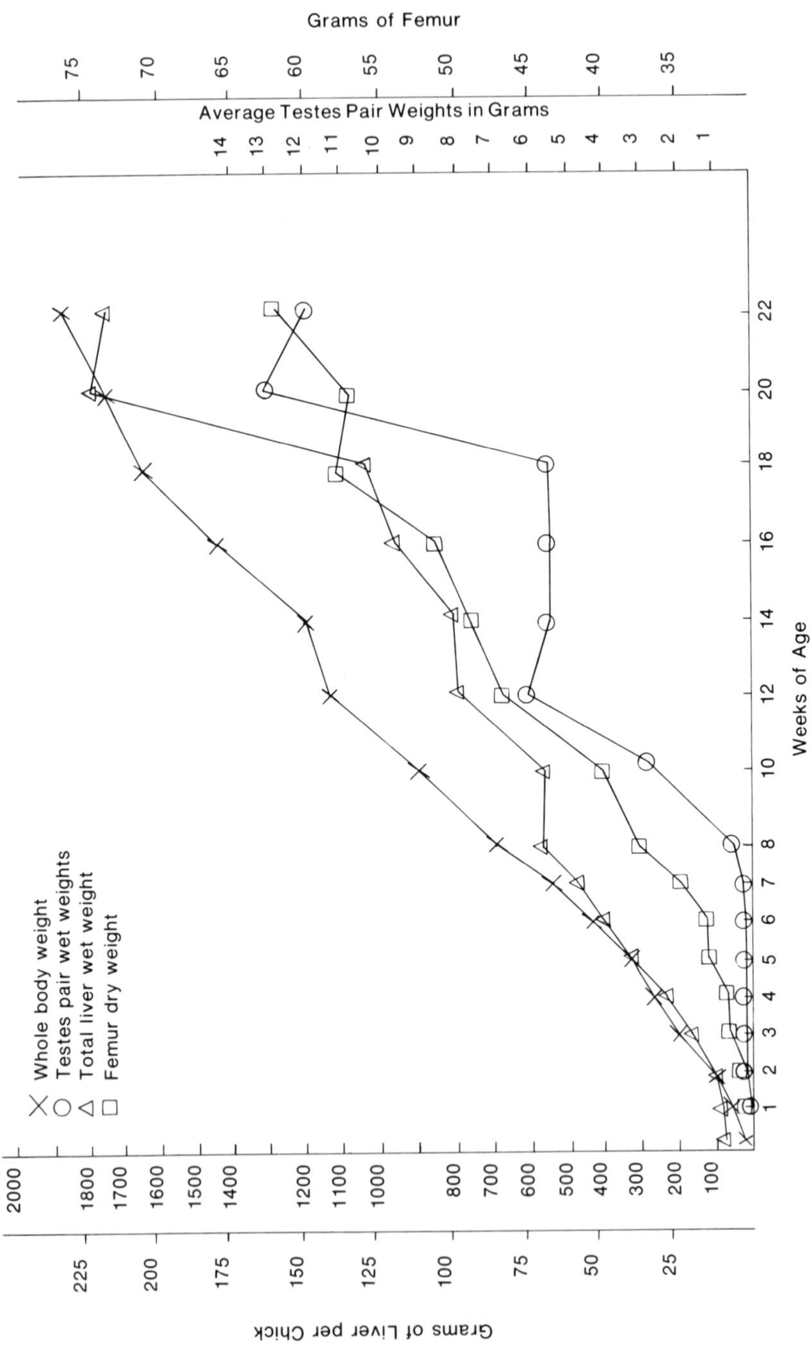

FIG. 6.1. WHOLE BODY, TESTES, LIVER WET AND FEMUR DRY WEIGHTS

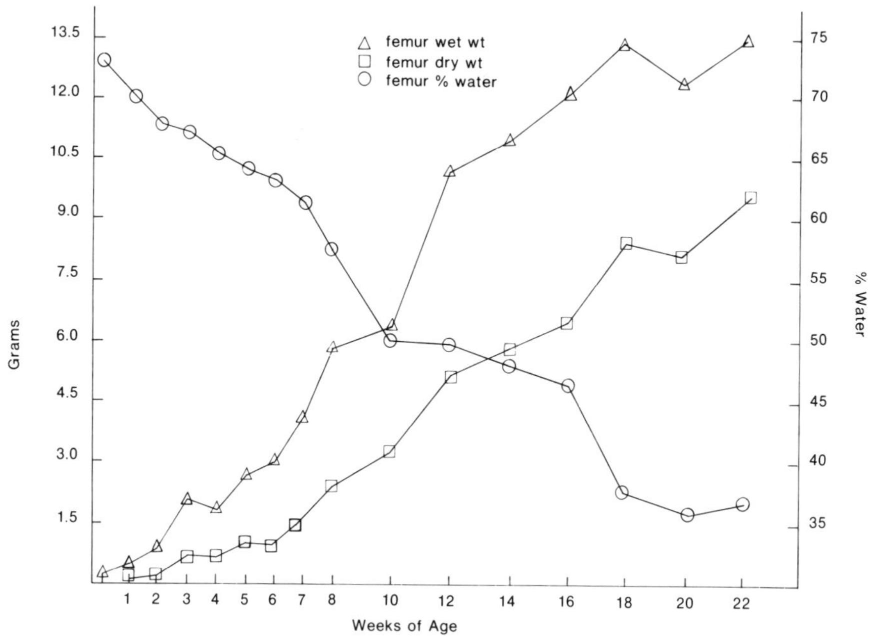

FIG. 6.2. WET AND DRY FEMUR WEIGHTS AND PERCENTAGE OF WATER

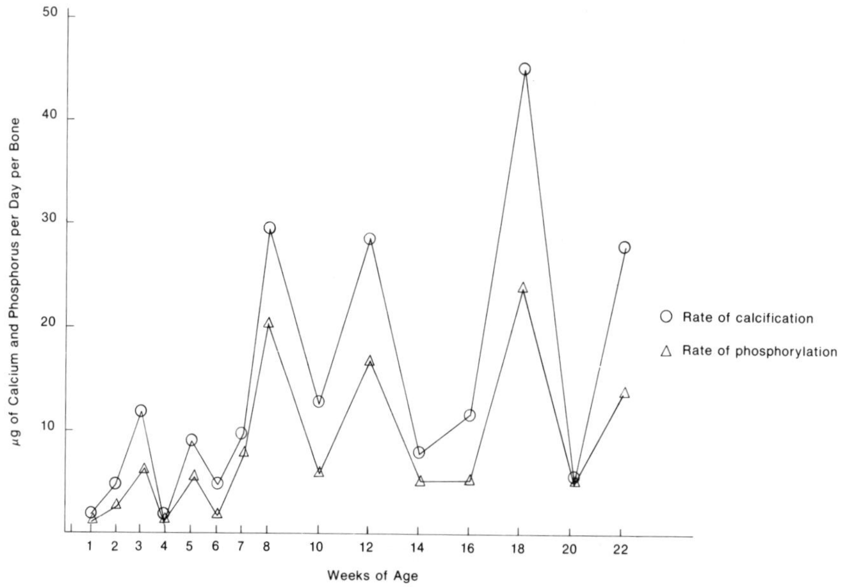

FIG. 6.3. RATE OF CALCIFICATION AND PHOSPHORYLATION IN FEMUR

TABLE 6.3. FEMUR WEIGHTS AND PERCENTAGE WATER

Cockerel Age	Wet Weight in g		Dry Weight in g		% Water	
	Mean[1,2]	S.E.	Mean	S.E.	Mean	S.E.
Day 1	0.189	±0.007	0.051	0	72.95	±1.58
Week 1	0.385	±0.010	0.110	0	71.43	±2.00
Week 2	0.817	±0.14	0.262	±0.012	67.93	0
Week 3	2.030	±0.13	0.663	±0.012	67.25	±1.87
Week 4	1.790	±0.13	0.614	0	65.58	±2.12
Week 5	2.635	0	0.946	0	64.06	0
Week 6	2.936	±0.25	1.075	±0.13	63.30	0
Week 7	3.956	±0.27	1.523	±0.14	61.55	±0.17
Week 8	5.766	±0.26	2.453	±0.10	57.48	0
Week 10	6.207	±0.88	4.709	±0.66	50.09	±1.00
Week 12	10.386	±1.16	5.186	±0.48	49.90	±1.41
Week 14	10.983	±0.07	5.760	0	48.01	±0.71
Week 16	12.075	±0.46	6.452	±0.23	46.58	0
Week 18	13.403	±0.29	8.395	±0.26	37.42	0
Week 20	12.483	±0.07	8.002	±0.20	35.90	0
Week 22	13.620	±2.11	9.330	±0.44	36.50	±6.08

Source: Carlotti (1971).
[1]The mean ± the standard error of the mean represents the number of groups of cockerels, organs, tissues or other analyses.
[2]Two groups of femurs.

TABLE 6.4. FEMUR CITRATE

Cockerel Age	mg Citrate per g Bone		mg Citrate per mg Calcium		mg Citrate per Bone	
	Mean	S.E.	Mean	S.E.	Mean	S.E.
Day 1	5.25	±0.16	42.69	0	0.268	±0.0087
Week 1	6.94	±0.064	46.28	±0.77	0.760	±0.011
Week 2	7.43	±0.16	39.85	±1.63	1.95	±0.10
Week 3	8.17	±0.39	42.11	±1.79	5.42	±0.33
Week 4	8.02	±0.48	37.80	±0.15	4.92	±0.33
Week 5	7.47	0	37.07	±0.089	7.07	±0.02
Week 6	8.46	±0.045	40.86	±0.81	9.11	±0.65
Week 7	8.27	±0.28	43.18	±2.77	12.4	±0.95
Week 8	9.20	±0.34	45.29	±0.24	22.6	±0.65
Week 10	8.85	±0.24	41.69	±0.55	28.3	±6.55
Week 12	9.10	±0.42	44.60	±1.83	47.0	±2.10
Week 14	9.23	±0.15	42.19	±1.21	53.1	±0.90
Week 16	8.77	±0.11	39.73	±2.33	56.5	±0.20
Week 18	8.97	±0.67	36.62	±0.055	75.3	±4.05
Week 20	9.58	±0.14	37.05	±0.88	76.6	±0.10
Week 22	9.29	±0.30	35.27	±0.62	86.7	±4.35

Source: Carlotti (1971).

TABLE 6.5. MINERAL COMPOSITION OF FEMUR

Cockerel Age	% Ash		% Calcium		% Phosphorus		Calcium: Phosphorus Ratio	
	Mean	S.E.	Mean	S.E.	Mean	S.E.	Mean	S.E.
Day 1	28.53	±0.75	42.42	±0.42	24.95	±0.17	1.71	0
Week 1	34.40	±0.68	44.03	±0.51	26.92	±0.21	1.64	±0.040
Week 2	39.49	±0.95	47.24	±0.18	26.44	0	1.80	0
Week 3	40.57	±0.12	47.77	±0.16	26.36	±0.086	1.83	±0.025
Week 4	42.37	±0.71	47.78	±1.41	26.39	±0.12	1.82	±0.16
Week 5	42.58	±0.17	47.35	0	26.46	±0.12	1.79	0
Week 6	41.61	±0.071	49.68	±0.86	26.65	±0.20	1.87	±0.055
Week 7	42.76	±2.10	44.99	±0.57	26.40	±0.24	1.71	±0.035
Week 8	44.09	±1.20	46.13	±0.66	27.20	±0.0070	1.70	±0.030
Week 10	43.24	±2.76	49.33	±0.58	27.41	±0.28	1.800	0
Week 12	42.96	±1.26	47.53	±1.16	27.41	±0.065	1.74	±0.045
Week 14	44.56	±1.12	49.14	±0.46	27.96	±0.28	1.76	±0.012
Week 16	44.54	±0.85	49.57	0	27.55	±0.21	1.80	0
Week 18	49.26	±0.71	49.64	±0.68	27.16	±0.24	1.83	±0.012
Week 20	51.73	±0.36	50.00	±0.16	27.59	±0.24	1.82	±0.020
Week 22	52.40	±1.68	50.32	±0.53	27.09	±0.37	1.86	±0.020

Source: Carlotti (1971).

plasma amino acids that may be related to calcium-amino acid chelates. At this age, there is a heavy demand for some amino acids for feather development.

The amino acid level in the blood of chickens does not remain as constant as in other animals. The plasma level of most amino acids drops sharply after the eighth wk of life. Also, whereas the plasma level of most amino acids lowers during fasting, the chick's plasma lysine and threonine may increase from 3 to 5 times.

Alanine and tyrosine are the only plasma amino acids which increase in concentration with age. The level of the lysine will increase if the dietary level is increased above the requirements. Leucine, tyrosine, lysine, theonine, serene, alanine, cystine and phenlyalanine make pronounced increases in plasma level between 4 and 8 wk of age (Tables 6.6 and 6.7). The feed efficiency drops rapidly from 6 to 8 wk of age in White Leghorn cockerels.

TABLE 6.6. PLASMA FREE ESSENTIAL AMINO ACIDS

Cockerel Age	Leucine[1]		Phenylalanine		Lysine		Histidine		Arginine	
	Mean	S.E.	Mean	S.E.	Mean	S.E.	Mean	S.E.	Mean	S.E.
Day 1	4.13	±0.12	2.23	±0.09	12.19	±0.87	2.84	±0.27	6.49	±0.59
Week 1	3.96	±0.12	1.80	±0.13	11.99	±0.26	2.09	±0.05	3.03	±0.28
Week 2	4.12	±0.03	2.19	±0.10	11.01	±1.62	2.24	±0.09	3.26	±0.04
Week 3	3.99	±0.38	2.20	±0.15	15.30	±1.99	2.32	±0.04	3.83	±0.38
Week 4	4.41	±0.32	2.55	±0.04	12.90	±2.44	2.96	±0.36	4.95	±0.48
Week 5	4.36	±0.29	1.99	±0.11	11.90	—	2.54	±0.31	3.97	±0.43
Week 6	4.43	±0.05	2.08	—	11.07	±1.31	2.37	±0.15	5.06	—
Week 7	5.30	±0.76	2.03	±0.53	15.05	±2.42	2.63	±0.32	6.18	±0.78
Week 8	5.73	±0.20	2.03	±0.31	10.28	±0.71	2.65	±0.24	6.10	±0.48
Week 10	3.74	±0.66	2.36	±2.07	8.26	±3.28	2.62	±0.13	8.42	±0.62
Week 12	2.20	±0.83	1.63	±0.81	9.00	±0.37	2.37	±0.14	—	—
Week 14	2.59	±0.86	2.02	±0.14	10.58	±0.21	2.15	±0.20	6.42	±0.27
Week 16	3.29	±0.15	1.75	±0.01	9.73	±2.41	2.15	±0.01	8.84	±0.07
Week 18	2.38	—	2.62	—	6.38	—	2.59	—	5.67	—
Week 20	3.28	±0.10	1.95	±0.02	9.52	±1.75	2.49	±0.09	6.57	±0.09
Week 22	3.30	±0.21	2.00	±0.10	6.78	±1.75	2.47	±0.11	6.61	±0.40

Source: Carlotti (1971).
[1]All amino acid values are in mg per 100 ml.

TABLE 6.6. ***(Continued)***

Cockerel Age	Threonine		Glycine		Valine		Methionine		Isoleucine	
	Mean	S.E.	Mean	S.E.	Mean	S.E.	Mean	S.E.	Mean	S.E.
Day 1	2.10	±0.16	4.11	±0.22	7.22	±0.69	1.47	±0.24	3.25	±0.26
Week 1	1.74	±0.23	5.65	±0.06	4.90	±0.15	0.82	±0.02	2.74	±0.17
Week 2	2.17	±0.25	6.61	±0.20	4.23	±0.46	1.01	±0.03	2.94	±0.35
Week 3	3.07	±0.11	6.76	±0.48	4.28	±0.58	1.21	±0.04	2.91	±0.42
Week 4	4.60	±1.36	5.65	±0.24	4.26	±0.02	1.26	±0.09	3.07	±0.26
Week 5	3.20	±0.78	5.55	±1.55	3.86	±0.22	1.18	±0.12	3.06	±0.19
Week 6	2.13	±0.02	5.26	±0.26	4.19	±0.69	1.27	±0.01	3.23	±0.08
Week 7	1.83	±0.23	4.57	±0.50	3.75	±0.17	1.35	±0.06	2.45	±0.35
Week 8	1.84	±0.13	4.87	±0.21	3.81	±0.39	1.23	±0.02	2.65	±0.34
Week 10	1.39	±0.04	5.28	±0.10	3.36	±0.35	1.20	±0.07	2.24	±0.37
Week 12	1.24	±0.10	6.07	±0.25	3.39	0	0.75	±0.23	1.85	±0.14
Week 14	1.17	0	5.35	±0.12	3.78	±0.21	1.17	±0.19	2.23	±0.13
Week 16	0.91	±0.54	6.15	±0.20	3.42	±0.04	0.93	±0.08	2.14	±0.53
Week 18	1.20	—	3.88	—	2.89	—	0.77	—	1.50	—
Week 20	0.88	±0.31	4.13	—	4.34	±0.06	0.96	±0.14	2.09	±0.09
Week 22	1.07	±0.10	4.99	±0.03	3.78	±0.22	0.91	±0.16	1.75	±0.07

Source: Carlotti (1971).

TABLE 6.7. PLASMA FREE NONESSENTIAL AMINO ACIDS

Cockerel Age	Aspartic[1] Acid		Serine		Alanine		Cystine		Tyrosine	
	Mean[1]	S.E.	Mean	S.E.	Mean	S.E.	Mean	S.E.	Mean	S.E.
Day 1	0.41	±0.09	7.62	±0.79	3.92	±0.27	0.99	±0.21	1.93	±0.09
Week 1	1.97	±0.24	9.75	±0.62	3.91	±0.30	1.36	±0.05	2.19	±0.14
Week 2	1.61	±0.09	10.41	±0.25	3.87	±0.12	1.51	±0.04	2.62	±0.13
Week 3	1.19	±0.02	10.31	±0.20	4.47	±0.17	1.85	±0.38	3.32	±0.36
Week 4	1.56	±0.19	10.78	±0.16	5.29	±0.33	1.93	±0.73	3.73	±0.19
Week 5	1.26	±0.22	10.63	±1.91	5.31	±0.36	1.18	±0.07	3.42	±0.39
Week 6	1.89	±0	10.52	±0.03	6.42	±0.47	1.58	±0.22	4.31	±0.36
Week 7	1.88	±0.35	10.07	±0.86	6.27	±0.63	1.50	±0.13	3.80	±0.50
Week 8	1.19	±0.13	8.34	±0.03	5.36	±0.15	1.66	±0.30	2.99	±0.31
Week 10	1.62	±0.13	8.92	±1.15	5.63	±0.29	1.36	±0.13	2.69	±0.30
Week 12	2.24	±0.03	8.94	±1.06	6.67	±0.62	1.19	±0.23	2.73	±0.07
Week 14	1.86	±0.08	6.95	±0.41	5.40	±0.38	0.96	±0.03	2.25	±0.01
Week 16	2.29	±0.66	7.91	±0.98	5.76	±0.37	1.23	±0.02	2.61	±0.22
Week 18	0.90	—	7.18	—	4.78	—	0.93	—	2.08	—
Week 20	1.80	±0.11	5.46	±0.52	4.35	—	0.716	±0.02	2.37	±0.33
Week 22	1.94	±0.25	5.51	±0.15	4.26	±0.18	0.240	±0.07	2.56	±0.19

Source: Carlotti (1971).
[1]All amino-acid values are in mg per 100 ml plasma.

7

Energy Nutrition

Energy is defined as the capacity to do work; therefore, one of the major intentions of a poultry nutritionist is to provide the nutrients essential for maximum work capacity. Work for poultry is expressed as egg production, meat, feathers and game bird sports.

The first energy is heat energy, which serves to maintain the body temperature. The second is free energy, which is available for work. Both types of energy can be expressed as calories, a small calorie being defined as the amount of heat required to raise the temperature of 1 g water 1 degree centigrade. (A large calorie or kilocalorie equals 1000 small calories and if often abbreviated as Cal. or kcal.)

THE ORIGIN OF ENERGY

The only source of energy biologically useful to the earth is sunlight. The energy from the sun is a nuclear fusion reaction and is called solar energy. Green plants capture and utilize solar energy by means of a process called photosynthesis. This process takes place in the chlorophyll of the plant leaves, and is able to synthesize carbohydrates (starches) from simple molecules of carbon dioxide (CO_2) and water (H_2O). By-products of this reaction are oxygen and a large amount of energy stored in the bonds of the carbohydrates.

Poultry consumes carbohydrates, and if digestible, they yield simple sugars which are absorbed, and then the enzymes in the animal reverse the reaction of the green plant. The energy stored is released in the carbohydrates. The chief function of carbohydrates in poultry is that of a fuel. Protein and fats are the other major energy sources. The major end products of bioenergetics are energy, water and carbon dioxide.

Today, poultry rations consist of corn, soybean meal and fat, sources of protein and energy. The rations in the future will use many by-products, and research must be conducted to form mixtures of feed ingredients which will give the best energy yield, as well as perform the other functions accompanying energy metabolism.

The energy value of feeds may be determined by several methods. The simplest one results in gross energy which is determined directly by burning a sample. The results are in agreement with those obtained by calculation. Organic acids, such as malic, citric, oxalic, lactic, succinic, acetic and malonic, are distributed widely in feeds. In some feeds, these acids are sufficiently abundant to be determined separately for estimations of energy values, since their heat of combustion is lower than for carbohydrates. Many items of animal origin contain small amounts of carbohydrates. Fats of plant origin are more important today than a number of years ago because of the processing of vegetable proteins. General factors of 4 Cal. for protein, 9 Cal. for fat and 4 Cal. for carbohydrate provide a quick means of calculating the physiological fuel value from composition data of the total feed supply. These factors are used for estimating values of average diets from the total quantity of protein, fat and carbohydrate. More specific factors should be used for experimental rations, individual foods and food supplies of a totally different character, as well as for areas of the world where the food supplies consist largely of unrefined feedstuffs.

THE RELATIONSHIP OF ENERGY AND FOOD COMPOSITION

Recent investigations show the importance of relationships of energy to various nutrients. Scott *et al.* (1947) demonstrated that high-energy diets permitted broilers to grow more rapidly and utilize their feed more efficiently than low-energy diets. Later, Cornell University investigators found that the protein requirement for maximum growth appeared to be a relatively constant absolute quantity, but as a percentage of the ration, was related to productive energy level through its influence on feed intake. Feed intake of poultry is, in turn, determined largely by energy concentration, providing the diet is adequate and bulkiness does not limit intake. Since protein and energy are the most expensive nutrients in a practical ration, their efficient use is critical for economical production of meat and eggs. Consequently, it is desirable to relate the level of protein (amino acids) to the energy level of the ration. In order to satisfy energy requirements when moderate amounts of cellulose replace glucose, birds consume more feed. This means more protein is available for growth. At high levels of cellulose, the volume of food consumed is high, but growth rate is limited by an energy deficiency caused by the low-energy level of the diet. The calorie-protein (C/P) ratio is defined as the productive energy (kcal) per lb divided by the percentage of crude protein in the ration. Best results are obtained with a broiler starter ration having a C/P ratio of 41 and a finisher ration with a C/P ratio of 49 (Table 7.1). Poor feathering occurs in birds receiving a calorie-protein ratio of 54.

TABLE 7.1. SUGGESTED CAL.-PROTEIN RATIOS[1] FOR POULTRY

	Metabolizable	Productive
Broiler starter (0–5 weeks)	60–65	42–45
Broiler finisher (after 5 weeks)	69–75	48–53
Chick starter (0–8 weeks)	63–66	44–46
Chick grower (8–18 weeks)	77–86	54–60
Layer, 50% production	91–94	64–66
Layer, 70% production	84–87	59–61
Layer, 90% production	80–83	56–58
Turkey starter (0–6 weeks)	43–46	30–32
Turkey grower (6–12 weeks)	57–60	40–42
Turkey grower (12–18 weeks)	71–74	50–52
Turkey grower (18–24 weeks)	85–100	60–70
Turkey finisher	100–105	70–72
Turkey breeders	86–91	6–64

[1]C/P ratio is defined as $\frac{\text{kcal per lb}}{\text{\% protein}}$

There is a linear relationship between energy value and efficiency of egg production in Leghorns (Table 7.2). The beneficial effect obtained from fat supplementation can be attributed to increased density and improved palatability. If the ration is pelleted, supplementary fat does not increase growth rate. Body composition is influenced by altering the C/P ratio, primarily due to its effect on appetite. Laying hens fed low-protein diets overconsume on feed energy while withdrawing body protein in order to sustain good egg production.

TABLE 7.2. EFFECT OF ENERGY LEVEL ON FEED REQUIREMENTS AND BODY WEIGHT OF LAYERS

Productive Energy Cal. per lb	lb Feed per Dozen Eggs	% Egg Production	Body Wt of Layers (lb)	
			at 5 Months	at 14 Months
740	5.74	64.9	—	—
840	5.19	66.9	—	—
945	4.62	66.5	4.36	4.61
985	4.35	69.5	4.24	4.63
1025	4.16	70.9	4.34	4.71

There are interrelationships of energy with other nutrients. The level and type of dietary fat has an influence on calcium utilization by the chick as measured by growth rate and tibia bone ash. Although there is some controversy concerning this effect, the greatest divergence of opinion occurs when extremely high, impractical fat levels and low calcium are employed. With the addition of fats, the requirements of chicks for choline, riboflavin, folic acid and methionine are increased. When sufficient energy is available, the methionine requirement expressed as a percentage of the diet increases as the protein level increases. The best correlation between growth performance and calculated requirement is obtained when requirement is considered as a function of the calorie-protein ratio. An increased requirement of sulfur amino acids in growing chickens occurs on high-energy rations. The methionine requirement varies with the energy level because when rations contain more energy, birds simply eat less. Therefore, the intake of methionine is less. The requirement for methionine can be expected to vary considerably when expressed as a percentage of the ration with birds of different sizes, the season of the year, production level, egg size and energy content of the feed. A similar set of relationships holds for the lysine requirement. The proportion of energy to protein content appears to fix amino acid requirements for broilers. If the requirement for energy is the primary factor controlling feed intake, the need for other essential nutrients is related directly to the energy content of the ration.

Fraps of Texas showed that feed ingredients with higher fiber content contained less energy than feedstuffs with lower fiber composition. The amount of nutrients a chicken consumes per day depends on the density of the diet rather than on its total weight. The weight of feed required to produce 1 lb of gain, therefore, does not give a true picture of the amount of nutrients consumed by chickens on high-fiber diets. The food intake of birds on a low-energy diet is greater than those on a high-energy diet. Decreasing the caloric content of a 16% protein breeder diet reduces hatchability. The chief limiting effect of fibrous materials for both gain and feed efficiency is their effect upon the calorie-protein ratio. Reports on the level of energy to specific vitamins are rather limited. Studies with turkey poults showed that the riboflavin requirement is more highly correlated with the energy content of the diet or with the daily intake of riboflavin than with the amount of riboflavin per lb of feed. Increasing the energy content of the chick diet also increases the need for potassium. Thus, the use of C/P ratios has helped in the practical formulation of poultry feeds.

From the standpoint of economy, the energy content of feed is the most important single consideration in reducing feed costs of meat and egg production. This is true, because birds consume more low-energy than high-energy feed to produce the same number of eggs or pounds of meat. Feeding light-weight oats instead of corn to high-producing light-breed layers reduces egg production. The increased energy content of the corn improves feed efficiency. Since both oats and corn are palatable to birds, difference in energy content is the major economic consideration. Oats average 700 to 800 Cal. per lb, whereas corn contains 1150 Cal. per lb. Substitution of 300 lb oats for corn in a ton of feed reduces total energy by 60 Cal. per lb. Energy levels in feed for egg production, growth and fattening are, in many instances, the limiting nutrient for maximum efficiency of performance.

ENERGY REQUIREMENT

The energy requirement of birds is met entirely by the chemical energy contained in the feed. Energy retained by the body is metabolizable energy (M.E.), accounting for 70 to 90% of the gross energy, depending on a number of factors. Absorption of energy-producing nutrients from the gastrointestinal tract is followed soon after by an increase in heat production, referred to as the heat increment, calorgenic effect, or the specific dynamic action (S.D.A.). The amount of heat produced by a bird in a given time is measured by direct and indirect means. The direct method involves enclosure of the animal in some form of calorimeter and the measurement of the temperature change of the air circulating through the walls of the calorimeter. This measures the total amount of heat lost from the animal; if the body temperature does not change during the course of the measurements, the heat loss will be an index of the heat production.

A recent development is the gradient layer calorimeter, which measures the nonevaporative heat loss from the bird, giving a temperature difference across a gradient layer incorporated in the walls of the calorimeter. From the temperature difference, the heat flow can be obtained. This gives a fast response and a continuous record of heat loss.

Calorimeters are expensive, complex instruments. Heat production can be measured more simply by indirect calorimetry. The amount of oxygen consumed by the bird in a given time is measured and the heat production is calculated from the oxygen consumption. The respiratory quotient (R.Q.) is calculated; this is the ratio of carbon dioxide produced to the volume of oxygen consumed in a given time. In the case of carbohydrate, the R.Q. is 1. With pure fat, it is 0.71, and with protein is 0.73. Values for fat and protein are low because their molecules contain less oxygen, and therefore require more oxygen for combustion than do carbohydrates. Measurement of oxygen consumption is usually made by enclosing a bird in a chamber through which air flows at a constant rate. The air leaving the chamber is passed through an automatic carbon dioxide and oxygen gas analyzer. From the rate of flow and the difference in oxygen, carbon dioxide content of the air entering and leaving the chamber, the rate of oxygen consumption, and carbon dioxide production can be calculated. The amount of food consumed and feces produced are then measured and their energy content determined. The amount of energy absorbed from the intestinal tract can then be measured.

Energy retention of a group of animals over a period of time can be obtained simply by killing half of them at the start of the experiment and the other half of them at the end of the experiment. The caloric values of the carcasses of both groups are determined. This is known as the comparative slaughter method and is particularly suitable for small animals. Different breeds of chickens have different rates of heat production. White Leghorns have a higher heat production than Rhode Island Reds which, in turn, produce more heat than New Hampshire and Cornish breeds. When birds are deprived of feed, their heat production diminishes. The R.Q. diminishes because fat is preferentially metabolized during starvation. When chickens are fed a sufficient amount of food to maintain constant body weight on a so-called maintenance diet, their heat production is approximately 50% greater than their standard metabolic rate.

When birds are exposed to environmental temperatures below their critical temperature, their heat production may be 3 or 4 times greater than normal. The heat production is brought about largely by shivering. Growth rate of chicks is adversely affected by cold wind because heat loss is accelerated and energy that would otherwise be available for growth is used to maintain body temperature.

Energy is one of the most important single factors required in poultry feeds (Table 7.3). Life itself is dependent on many catabolic and anabolic energy interchanges involving oxidation and hydrolysis of various reversible and irreversible types. A major factor in poultry production is the extent to which energy or its equivalent is upgraded from feedstuffs into materials useful to poultry and on an economically competitive basis.

As has been indicated previously, energy is not a nutrient, but is an important function of some nutrients. It is fuel for the maintenance of body temperature, movements of the body, digestion and utilization of feeds. All life processes depend on the energy from the feed that is eaten.

SOURCE OF ENERGY

Carbohydrates, fats and proteins are the 3 recognized sources of energy for poultry. Each of these sources is manufactured by plants which derive and store energy from the sun. Whereas carbohydrates and fats are fed primarily as a source of energy, protein is fed as a source of amino acids for building body protein. The energy obtained from amino acid metabolism comes after it has served its role in protein tissue usage. When these energy-containing products are consumed, digested and absorbed by poultry, they can be either deposited in the body or broken down into CO_2 and water with a subsequent release of energy. Both carbohydrates and protein contain 4 Cal. per g (Tables 7.4 and 7.5) and fats 9 Cal. The only energy of value in a ration is that which is utilizable by poultry.

Corn and milo are the major sources of energy for poultry feeds. Wheat is also considered a high-energy ingredient, but its cost makes it undesirable economically. Some by-products from corn and milo processing,

TABLE 7.3. METABOLIZABLE ENERGY VALUES OF COMMON FEEDSTUFFS FOR CHICKENS

Feedstuff	Metabolizable Energy Average (Cal. per lb)	Metabolizable Energy Range (Cal. per lb)
Corn	1600	1550–1670
Hominy feed	1320	1190–1530
Milo	1500	—
Wheat	1520	1480–1570
Wheat red dog	1270	—
Wheat flour middlings	1220	—
Wheat stand middlings	840	—
Wheat bran	540	—
Oats	1210	1130–1290
Barley	1310	1260–1400
Rye (*Tetrapetkus*)	1330	—
Molasses (blackstrap)	910	—
Soybean oil meal, 44% protein	1030	970–1110
Soybean oil meal, 50% protein	1110	990–1160
Meat scraps (50 to 55%)	910	860–1190
Corn distillers solubles	1300	—
Dried whey	850	810–860
Alfalfa meal	640	530–880
Stabilized tallow	3300	—
Lard	3960	—
Corn gluten meal	1475	—
Corn oil	3975	—

TABLE 7.4. METABOLIZABLE ENERGY VALUE OF DIGESTIBLE NITROGEN-FREE EXTRACT

Class of Feedstuff	Cal. per G of Digestible Nitrogen-free Extract
Grains, and most other seeds	4.2
Legume seeds (e.g., soybean) & rice by-products	4.0
Meat and fish by-products	3.9
Legume laves and stems (e.g., alfalfa)	3.8
Milk by-products	3.7
Total energy value of carbohydrates	
Carbohydrate	Cal. per g
pentoses and pentosans	3.72–4.38
hexoses	3.75–3.86
sucrose	3.96
dextrin	4.11
glycogen	4.19
starch	4.20
cellulose	4.20
Total energy value fats and oils	
Meat, fish and eggs	9.50
Dairy products	9.25
Cereals, vegetables and fruits	9.30

TABLE 7.5. ENERGY VALUES OF FEEDSTUFFS

Class of Feedstuff	Cal. per G of Digestible "Conventional" Protein (6.25 × N)
Animal Products	
casein	4.35
eggs	4.35
fish	4.25
gelatin	3.30
meat	4.25
milk	4.40
Cereals and cereal products	
barley	4.00
corn	4.40
millet	4.00
oats	4.00
rice	4.10
rye	4.00
sorghum grains	4.40
wheat	4.00
wheat bran	4.20
Legumes	
alfalfa (leaf and stem)	3.65
beans	4.30
peanuts	3.60
soybeans	3.90
Other	
sunflower seed	3.40

such as starch, sugar and hominy feeds, are high in energy. Heavy oats and barley are medium-energy ingredients and light oats, low-energy.

Molasses and fat are used to provide energy. The amount of molasses added must be neither greater than that which will mix well in the ration nor great enough to cause a laxative condition. Fat must be used only when it is of good quality, stabilized, and needed to provide additional energy.

RELATION OF ENERGY UPTAKE TO GROWTH AND COMPOSITION OF BIRDS

Energy which is not required for maintenance is used for growth and fattening. The feed requirement for growth and fattening increases as the fiber content of the diet increases and the dietary energy level decreases (Table 7.6). Finishing rations should be higher in energy than growing rations.

One of the uses of high-fiber, low-protein and low-energy rations is to reduce cannibalism in confined birds. Oats are rated as excellent in reducing feather-picking and cannibalism.

TABLE 7.6. THE EFFECT OF DIETARY ENERGY LEVEL ON GROWTH, FOOD INTAKE, CALORIC INTAKE, AND FAT CONTENT OF CHICKS

	Basal Diet				
	High Calorie	Plus 10% Hulls	Plus 20% Hulls	Plus 30% Hulls	Plus 40% Hulls
Energy level (Cal per lb)	975	858	741	623	505
Calorie-protein ratio	49	43	37	31	25
Avg weights 11 wk (lb)	3.28	3.26	3.35	3.21	3.20
Avg feed consumption 11 wk (lb)	9.6	9.7	10.8	11.2	12.0
Avg calorie intake 11 wk (Cal)	9470	8400	8100	7090	6160
Fat content of males, dry basis 11 wk (%)	26.8	23.2	21.1	18.1	16.1

ENERGY VALUE AND FEED FORMULATION

The energy value of a feed must be determined by the use of poultry. The largest single dietary need of poultry is for a source of energy, and only that portion of the total energy value of a feedstuff that can be utilized by the chicken is useful.

The physiological processes the animal requires energy for are:

Work (movement)
Respiration
Circulation
Temperature regulation
Absorption
Excretion
Nerves
Reproduction

Since the energy requirement for maintenance functions must be first met before growth or reproduction needs are supported, it becomes of major importance that optimum energy needs are provided. A considerable portion of the feed consumed by all classes of poultry is used for maintenance. Therefore, poultry and all other homiotherms must be fed to maintain life, whether they are producing eggs or growing.

The maintenance requirement for energy includes the need for normal activity and basal metabolism. Basal metabolism (BM) is recognized as the chemical change which takes place in the cells of an animal in the fasting and resting state, when it uses only enough energy to maintain vital cellular activity, respiration and circulation. This is a constant for poultry maintaining a normal status. It can be computed because it is a function of surface area or body weight. The equation generally used is kcal/24 hr = (W kg. .75).

Most poultry are allowed to consume as much feed as they desire, and the major factor controlling feed uptake when other nutrients are present in adequate amounts is energy content. Table 7.7 summarizes the relation of energy and environment temperature to egg production, feed intake and efficiency; Table 7.2 demonstrates the relation of egg production to feed efficiency. Attempts have been made by some producers to feed pelleted high-fiber feeds to growing pullets, only to observe a great increase in feed consumption. Had the poultryman chosen a high-fiber/low-density ration (non-pelleted), the birds would have tired of eating or the crop would have filled before the energy needs were met.

TABLE 7.7. THE INFLUENCE OF ENERGY AND ENVIRONMENTAL TEMPERATURE ON EGG PRODUCTION, FEED CONSUMPTION, BODY WEIGHT AND MORTALITY

Cal. M.E. per lb Ration	Average % Production	Feed Consumed per 100 Hens Daily lb	Feed per Doz Eggs lb	Change in Body Weight %	% Mortality %
		Entire Experiment (280 days)			
1150	50.7	27.2	6.4	L 12.2	10
1250	58.6	23.9	4.9	L 9.4	12
1350	57.5	20.8	4.4	L 8.3	17
1450	58.6	18.7	4.0	L 8.0	17
1550	58.9	18.7	3.8	L 6.4	23
		112 Consecutive Coldest Days (66° to 54°F)			
1150	58.0	31.5	6.2	L 6.9	4
1250	69.6	28.0	4.8	L 3.4	2
1350	70.5	24.9	4.2	L 0.6	2
1450	66.0	22.9	4.1	G 0.7	3
1550	68.7	22.1	3.8	G 2.2	1
		112 Consecutive Hottest Days (100° to 88°F)			
1150	42.9	22.7	6.3	L 5.5	4
1250	47.3	19.3	4.9	L 6.1	9
1350	44.6	16.5	4.4	L 7.7	13
1450	48.2	15.7	3.9	L 8.7	13
1550	49.1	15.0	3.7	L 8.3	21

Source: Values obtained from Heywang and Vavich (1962).

BIOLOGICAL PARTITION OF ENERGY

Energy of feeds is generally divided in 4 parts: Grass Energy (GE), Digested Energy (DE), Metabolizable Energy (ME) and Net Energy (NE). Grass energy (GE) is determined by burning the feed sample in a colorimeter (Fig. 7.1). The digested energy (DE) is obtained by subtracting the fecal energy loss (obtained by burning the dry feces in a colorimeter) from the grass energy (GE). The metabolizable energy (ME) is obtained by subtracting the urinary energy loss from the digested energy (DE). Since the urine and feces are excreted together, a surgical step is necessary for separation of the urine from the feces. If only the metabolizable energy (ME) is needed, the feces and urine as excreted can be dried and burned in the colorimeter. The net energy (NE) is the metabolizable energy less the energy loss as heat.

The metabolizable energy (ME) is the value used in calculations for the poultry rations.

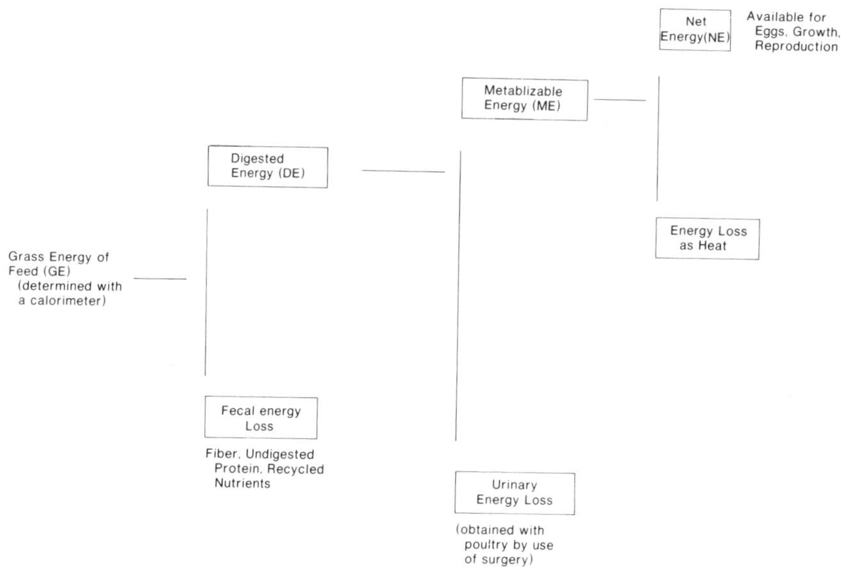

FIG. 7.1. BIOLOGICAL PARTITION OF ENERGY

8

Nutrition of Fats

Fats and fatty acids are ether-soluble substances which are the highest energy sources in feedstuffs. Fats in plant materials are generally found in small micelles, particulates or glands, whereas in poultry, most of the fat is in adipose tissue. Fat deposits in the poultry other than in adipose tissue contain largely phospholipids, cholesterol and glycerol.

Fat deposits in feeds and poultry contain mainly neutral fat (triglyceride), by which is meant esters of glycerol with 3 fatty acids.

Fats are generally divided into 3 groups, the simple fats being found in adipose tissue and the compound lipids in nerve and bone marrow tissue.

(1) Simple Fats
- (a) Fatty acids
- (b) Neutral fats (mono-, di-, triglycerides)
- (c) Waxes (esters of fatty acids with higher alcohols)
  - (i) Sterol esters (i.e., cholesterol esters with fatty acids)
  - (ii) Nonsterol esters (i.e., Vitamin A esters, etc.)

(2) Compound Fats
- (a) Phospholipids
  - (i) Phosphatidic acids (phosphoglycerides), lecithins, cephalins, etc.
  - (ii) Plasmalogens
  - (iii) Sphingomyelins
- (b) Glycolipids (carbohydrate-containing)
  - (i) Cerebrosides
  - (ii) Gangliosides
- (c) Lipoproteins (fats in combination with protein)

(3) Derived Lipids (alcohols, including sterols, hydrocarbons)

The distribution of fatty acids in glycerides is not completely random; however, most of the fatty acids contain 16 to 18 carbons. These fatty acids are oleic (18C), palymitic (16C) and stearic (18C). The fatty acids in fats and oils and in feeds are presented in Tables 8.1 and 8.2. The fatty acid composition of poultry meats, offals, feathers and 4 commonly used feed ingredients are summarized in Table 8.3. The fatty acid makeup content of natural plant material and poultry products is genetically controlled. However, it is possible to change or alter adipose tissue by feeding sex hormones, and feeding excess amounts of fat which will not allow the body's metabolic system to reconstruct the fat characteristic of the animal. The feeding of some carbohydrate sources also produces a different fat than others. When poultry are fed oats and barley, they place more fat under the skin than when fed corn. The fat is also firmer. Corn-fed poultry tend to have more and softer fat around the viscera than either oats-, barley- or wheat-fed animals.

The implanting of diethylstilbesterol (DES) under the skin of poultry causes a very heavy accumulation of fat in the viscerol with oil being present. The melting point is influenced by the fatty acids present; shorter chain or more unsaturated fatty acids tend to impart lower melting points. (Fats which have melting points at room temperature are called oils.)

Poultry require one or more unsaturated fatty acids (Table 8.4). The polyunsaturated acids—linolenic, linoleic and arachidonic—have often been designated the essential fatty acids, although current practice is to consider only linoleic acid as significant in poultry nutrition. Linoleic is an unsaturated fatty acid having 2 double bonds and 18 carbons, which must be supplied by dietary sources. An essential fatty-acid deficiency is characterized by poor growth, fatty livers, reduced egg size and poor hatchability. The best sources of unsaturated fatty acids are oils from corn, soybeans or safflowers.

The animal cell is capable of forming some unsaturated fatty acids, but the types of acid that can be formed are limited. Any new double bonds must be introduced between the carboxyl group and the first double bond of the fatty acid molecule. Thus, linoleic acid may be converted to arachidonic acid in animal cells, but apparently little or no conversion of other acids to linoleic acid occurs. However, plant cells are able to introduce a second double bond between carbon atoms 12 and 13 of oleic acid to form linoleic acid. (See Fig. 8.1.)

Because of the double bonds in the molecule, unsaturated fatty acids may exist as isomers. This type of isomerism is known as cis-trans isomerism and involves changes in the geometrical configuration of the molecule. Cis isomerism results in a doubling-back of the molecule in a horseshoe-like configuration, whereas trans isomerism has the effect of

TABLE 8.1. COMPOSITION AND CONSTANTS OF FATS AND OILS

Fatty Acid	Chemical Formula	Double Bonds	Corn	Cotton-seed	Lin-seed	Olive	Peanut	Rape-seed	Saf-flower	Sesame	Soy-bean	Lard	Tal-low	Her-ring	Menha-den
Saturated															
caproic	$C_6H_{12}O_6$	0	—	—	—	—	—	—	—	—	—	—	—	—	—
caprylic	$C_8H_{16}O_2$	0	—	—	—	—	—	—	—	—	—	—	—	—	—
capric	$C_{10}H_{20}O_2$	0	—	—	—	—	—	—	—	—	—	—	—	—	—
lauric	$C_{12}H_{23}O_2$	0	—	—	—	—	—	—	—	—	—	—	0.1	—	0.1
myristic	$C_{14}H_{28}O_2$	0		0.5		0.5		0.5	0.1		0.1	1	3	7	7
palmitic	$C_{16}H_{32}O_2$	0	8	21	5.5	9.5	7	2	4.5	8	8	28	29	14	16
stearic	$C_{18}H_{36}O_2$	0	3.5	2	4	2.0	4	0.9	2	4	4	13	20	0.5	2
arachidic	$C_{20}H_{40}O_2$	0	0.5	0.2	0.3		3	0.5	0.4	0.5	0.6		0.8		
behenic	$C_{22}H_{44}O_2$	0					2	1		0.1					
lignoceric	$C_{24}H_{48}O_2$	0	0.2	0.3	0.2		2	1		0.1					
Unsaturated															
myristoleic	$C_{14}H_{28}O_2$	1	—	—	—	—	—				0.1	0.2	0.5	0.5	0.1
palmitic	$C_{15}H_{30}O_2$	1	—	—	—	—	—	0.1			0.2	3	2	6	16
oleic	$C_{18}H_{34}O_2$	1	46	29	22	81	60	18	20	45	28	46	42	21	15
erucic	$C_{22}H_{42}O_2$	1	—	—	—	—	—	51							
linoleic	$C_{18}H_{32}O_2$	2	42	45	17	7	22	22	70	42	54	6	2		7
linolenic	$C_{18}H_{30}O_2$	2	—	2	51	—	—	3	3		5	0.7	0.5		3
arachidonic	$C_{20}H_{32}O_2$	3	—	—	—	—	—					2	0.1	28	17
clupanodonic	$C_{22}H_{36}O_2$	4	—	—	—	—	—							23	11
	$C_{24}H_{38}O_2$	5	—	—	—	—	—								4
	C_{26}		—	—	—	—	—								1
	Over C_{26}	—	—	—	—	—									2

TABLE 8.2. FATTY ACIDS OF FEED FATS

Acetic Series ($C_nH_{2n}O_2$)
- lignoceric, $C_{24}H_{48}O_2$—rape, peanut, soy, sesame, sunflower, maize
- behenic, $C_{22}H_{44}O_2$—rape and other cruciferous seeds
- arachidic, $C_{20}H_{40}O_2$—peanut, soy, cottonseed, olive, sesame, sunflower, maize
- stearic, $C_{18}H_{36}O_2$—solid fats and most oils
- palmitic, $C_{16}H_{32}O_2$—most fats and many oils
- myristic, $C_{14}H_{28}O_2$—palm kernel, coconut, walnut, rape, cottonseed, olive
- lauric, $C_{12}H_{24}O_2$—pine nut, palm kernel, coconut
- capric, $C_{10}H_{20}O_2$; caprylic, $C_8H_{16}O_2$; caproic, $C_6H_{12}O_2$ —butter, palm kernel, coconut
- butyric, $C_4H_8O_2$—butter

Oleic series ($C_nH_{2n-2}O_2$)
- erucic, $C_{22}H_{42}O_2$—rape, cod liver oil
- oleic, $C_{18}H_{34}O_2$—oils and most fats

Linoleic series ($C_nH_{2n-4}O_2$)
- linoleic, $C_{18}H_{32}O_2$—hemp, poppy, rape, almond, peanut, soy, linseed, cottonseed, olive, sesame, sunflower, maize

Linolenic series ($C_nH_{2n-6}O_2$)
- linolenic, $C_{18}H_{30}O_2$—beechnut, hemp, poppy, rape, soy, linseed
- isolinolenic—hemp, linseed, madia

Clupanodonic series ($C_nH_{2n-8}O_2$)
- clupanodonic, $C_{18}H_{28}O_2$—marine oils

extending the molecule. As an illustration, the cis and trans forms for the C_{18} monounsaturated acid are shown later.

Fatty acids can be synthesized in the liver by converting the 2 carbon fragments acetyl CoA to malonyl CoA. By condensing several units of malonyl CoA, the fatty acid of the desired length can be obtained. Also, the chick's liver can oxidize a long-chain fatty acid by reducing it by 2 carbons as acetyl CoA each time or it can elongate the carbon chain condensing acetyl CoA to an already-existing fatty acid.

Oleic acid
($C_{18}H_{34}O_2$)

$CH_3(CH_2)_7CH{=}CH(CH_2)_7COOH$

Linolenic acid
($C_{18}H_{30}O_2$)
$CH_3CH_2{-}CH{=}CH{-}CH_2{-}CH{=}CH{-}CH_2CH{=}CH(CH_2)_7COOH$

Linoleic acid
($C_{18}H_{32}O_2$)
$CH_3(CH_2)_4CH{=}CH{-}CH_2{-}CH{=}CH(CH_2)_7COOH$

Arachidonic acid
($C_{20}H_{30}O_2$)
$CH_3(CH_2)_4CH{=}CH{-}CH_2{-}CH{=}CH{-}CH_2{-}CH{=}CHCH_2CH{=}CH(CH_2)_3COOH$

FIG. 8.1. THE STRUCTURE OF THE UNSATURATED FATTY ACIDS

TABLE 8.3. FATTY ACID COMPOSITION OF POULTRY PRODUCTS AND INGREDIENTS[1]

	Lauric 12:0	Myristic 14:0	Palmitic 16:0	Stearic 18:0	Oleic 18:1	Linoleic 18:2	Linolenic 18:3
Chicken breast	0.2	0.6	19.4	7.5	1.9	29.9	1.0
Heart		0.4	22.1	7.6	5.0	21.4	
Liver		0.2	25.9	14.8	3.0	13.5	
Skin		0.8	16.1	4.6	3.1	44.0	1.6
Thigh	0.8	0.5	16.2	6.0	2.4	38.7	1.8
Eggs		0.1	25.5	8.9	47.6	12.8	
Offal oil	0.2	1.4	21.4	5.9	39.5	23.5	1.0
Products meal		1.9	21.9	12.3	35.9	16.8	2.0
Feather meal	0.2	1.9	29.9	14.6	29.7	13.1	
Corn			16.3	2.6	30.9	47.9	2.3
Oats		0.2	23.3	2.0	32.5	38.2	3.1
Soybean meal 50		0.3	23.8	4.7	16.2	46.6	6.9
Barley	0.5		27.6	1.5	20.5	43.3	4.3

[1]Values are percentage by weight of total fatty acids in the lipid extract.

TABLE 8.4. UNSATURATED FATTY ACIDS FOR POULTRY

Name	Designation[1]	Melting Point °F	Melting Point °C	Comments
Palmitoleic	16:1	34.7	1.5	
Oleic	18:1	57.2	14.0	Most common unsaturated fatty acid in animal tissue
Linoleic	18:2	23.0	5.0	Essential fatty acid
Linolenic	18:3	6.8	14.0	Essential fatty acid
Arachidonic	20.4	57.1	49.5	Essential fatty acid—can be synthesized from linoleic acid

[1]Designation: First number refers to the number of carbons and second number refers to the number of double bonds or unsaturation.

When fats are oxidized in the liver, it is by 2 carbon fragments, and is called beta oxidation. If it is oxidized in the feed bag, the oxidization is by use of surface energy catalyzed by metal ions. This pattern of oxidation may involve one or more carbon fragment formations, as well as leaving the carbon fragments as free radicals. This type of fat destruction when fat is added to poultry feeds has caused many problems. The fat-soluble vitamins can be destroyed by peroxides, and this requires an antioxidant for dietary control of peroxides. Vitamin E (alpha-tocopherol) is essential for protection of the chick's body tissue and membranes and selenium functions in the elimination of free radical damage to tissues and membranes within the chick's body.

Fat is perhaps the most challenging nutrient added to poultry rations. It can provide energy and improve palatability. When extracted from its natural state, it readily becomes rancid, and then the peroxides formed are active in the destruction of Vitamin A and bring about the destruction of all membranes. Fats introduce a high requirement for Vitamin E and a need for an antioxidant, as well as extra effort in reducing the action of free radicals. They also influence the flavor of meat and change its storage qualities.

Fats were first added to poultry feeds as an energy-supplying ingredient during the early period of development of the corn-soybean type ration. The establishment of the theory that poultry eat to supply their energy needs led to the addition of high-energy ingredients, such as fat, to rations for prevention of protein waste.

Before the addition of fat, most of the fat deposited in the adipose tissue was derived from ingested digestible carbohydrates. When a combination of corn and oats or barley was fed, turkeys and broilers appeared uniformly fleshed under the skin and viscera. Corn alone tended to increase visceral fat. When cockerels were given stilbesterol implants and only corn as the grain, there was not only heavy deposition of fat under the skin and around the viscera, but also a heavy accumulation of oil in the visceral cavity.

The addition of high levels of fat to broiler rations has resulted in excessive visceral fat. Some broilers have an excess of oils or fats which remain fluid in chilled broilers to the extent that the term "greasy" is applied. These "greasy" broilers are difficult to keep coated with batter while cooking.

The excess visceral fat is removed, and represents a financial loss to the consumer. Many educational programs are strongly recommending a reduction in dietary fat. Although there appears very little connection between dietary fat and cholesterol formation, excessive fat is not needed in the average American diet.

Today, pork producers feed hogs diets containing less than 70% TDP and not less than 15% protein to develop carcasses with low back-fat. In the early 50's, consumers complained of excessive visceral fat in poultry. We must discontinue developing rations based on efficiency alone and look toward "effective" food value. Tests must be developed to estimate the fat content of live broilers so that they can be produced with only enough fat to place them in the minimum prime grade.

Besides the unneeded dietary fat in man's diet, we must be mindful that fats added to poultry rations have caused problems in the loss of Vitamins A and E by oxidation, have brought about body membrane destruction and changes in the flavor of poultry meats.

9

Nutrition of Carbohydrates

Carbohydrates are polyhydroxylic aldehydes and ketones whose chief functions in poultry are as major sources of energy. The end products of carbohydrate digestion are simple sugars and the end products of metabolism are water, carbon dioxide and energy. Some biochemical products resulting from the metabolism of carbohydrates can act as catalysts, promoters of oxidation and starting material for the biological synthesis of other types of compounds in the body, such as fatty acids and certain amino acids. Quantitatively, the 3 polysaccharides of greatest importance in poultry nutrition are starch, cellulose and glycogen. All of these are polymers of glucose, generally classified as homopolysaccharides, since they are pure glucose polymers.

Starches are the most common digestible polysaccharides in plants, present primarily in the cells of grain, fruits, and tubers in the form of granules, which under microscopic examination, appear to be typical of each starch. The composition of starches differs somewhat, but all types contain both amylose, a straight-chain polymer, and amylo-pectin, a branched chain. When digested, or upon hydrolysis, starch yields dextrins and maltose, and eventually glucose. The glucose units in starch are linked together by the -D-1,4- or -D-1,6-glucosidic linkage. The chick's digestive system secretes an enzyme capable of hydrolyzing both the -D-1,4- and the -D-1,6- linkages. The digestibility of starch is approximately 95%.

Glycogen is the only homopolysaccharide of importance in poultry metabolism. It is often called animal starch, and its role in the chick's body is primarily that of a storage carbohydrate similar to the role of starch in plant cells. It is found in small amounts in the liver and muscle cells, where it serves in the homostatic mechanism regulating blood glucose levels, and as a source of energy for the muscles. Glycogen is not digested, but broken down by an enzyme, phosphorylase, forming glucose-6-phosphate.

Cellulose is a straight-chain polymer of glucose and is a constituent of the cell wall of plants. The glucose units are joined through carbon 1 and

4 by a B-glucosidic linkage. Where starch is a stored form of nutritionally-available carbohydrates in plants, cellulose is not an available form of carbohydrates for poultry. Chickens do not produce enzymes which can digest cellulose.

Cellulose supports plants and may provide a protective structure for grains. The chicken's proventriculus (gizzard) grinds the hard particles of feed, resulting in a well-developed digestive tract. Since the digestive tract of the chicken is short, the cellulose structures aid absorption by helping control the movement of the food through the intestinal tract. If not controlled, the food can move through in about 2 hr.

Although all homopolysaccharides contain glucose as the building unit, there is a difference in the body distribution of fat caused by feeding different grains.

Poultry convert excess glucose from plant starch into body fat. This fat appears to differ from 1 plant carbohydrate source to another. Whereas corn produces fat which concentrates around the viscera and under the skin, oats, wheat and barley produce fat which concentrate chiefly under the skin with limited fat accumulation around the viscera. Turkeys will eat heavy oats but not corn in warm weather. The feeding of heavy oats has allowed turkeys to be finished for market during warm weather. Visceral fat generally becomes a loss to the processor.

CARBOHYDRATE SYNTHESIS

Green plants synthesize simple carbohydrates from carbon dioxide obtained from the air with water from the soil, through the action of chlorophyll in the leaves. The energy to synthesize carbohydrates comes from sunlight and is commonly called solar energy. This chemical reaction, called photosynthesis, is as follows:

$$\underset{\text{Carbon Dioxide}}{6\,CO_2} + \underset{\text{Water}}{12\,H_2O} \xrightarrow[\text{chlorophyll}]{\text{sunlight}} \underset{\text{Sugar}}{C_6H_{12}O_6} + \underset{\text{Oxygen}}{6\,O_2} + \underset{\text{Water}}{6\,H_2O}$$

The carbohydrates found in feed (Table 9.1) are many, but most important are polymers of glucose. The chief carbohydrate in plants is cellulose, but starch is the principal digestible carbohydrate. Although there are many complex carbohydrates, the 2 most significant in poultry nutrition are (1) starch as an energy source and (2) cellulose as a source of bulk.

CRUDE FIBER

Crude fiber is the portion of the total carbohydrates of an ingredient that is resistant to the acid and alkali treatment as described in the

TABLE 9.1. CARBOHYDRATES IN FEEDS

Monosaccharides
dimethyldioses ($C_4H_8O_2$)
dimethylglycolose, $CH_2 \cdot CHOH \cdot CO \cdot CH_3$—occurs in vinegar
trioses ($C_3H_6O_3$)
dioxyacetone, $CH_2OH \cdot CO \cdot CH_2OH$—fermentation product
oxymethyltetroses ($C_5H_{10}O_3$)
apoise or beta-oxymethyltetrose $(CH_2OH)_2 \cdot COH \cdot CHOH \cdot CHO$—a constituent of the glucoside apiin of parsley
pentoses ($C_3H_{10}O_3$)
L-arabinose, $CH_2OH \cdot HOCH{:}HCOH \cdot HOCH \cdot CHO$—formed by hydrolysis of araban
L-xylose, $CH_2OH \cdot HOCH \cdot HCOH \cdot HOCH\text{-}CHO$—formed by hydrolysis of xylan
D-ribose, $CH_2OH \cdot HOCH \cdot HOCH \cdot CHO$—constituent of nucleic acids
methylpentoses ($C_6H_{12}O_3$)
rhamnose, $CH_3 \cdot CHOH \cdot HCOH \cdot HOCH \cdot HOCH \cdot CHO$—constituent of glucosides
fucose, $CH_3 \cdot CHOH \cdot HOCH \cdot HOCH \cdot HCOH \cdot CHO$—formed by hydrolysis of fucosan of seaweeds
hexoses
D-glucose, $CH_2OH \cdot HOCH \cdot HOCH \cdot HCOH \cdot HOCH \cdot CHO$
D-mannose, $CH_2OH \cdot HCOH \cdot HOCH \cdot HCOH \cdot HCOH \cdot CHO$—in plant juices formed by hydrolysis of mannan
D-gallactose, $CH_2OH \cdot HOCH \cdot HCOH \cdot HCOH \cdot HCOH \cdot CHO$—constituent of glucosides and formed by hydrolysis of lactose
D-fructose, $CH_2OH \cdot HOCH \cdot HOCH \cdot HCOH \cdot CO \cdot CH_2OH$
D-sorbase, $CH_2OH \cdot HCOH \cdot HOCH \cdot HCOH \cdot CO \cdot CH_2OH$—in fermented juice of *Sorbus domestica*

Disaccharides ($C_{12}H_{22}O_{11}$)
sucrose, $C_6H_{11}O_5 \cdot O \cdot C_6H_{11}O_5$—in sugar cane, sugar beets, and many natural products
maltose, $C_6H_{11}O_5 \cdot O \cdot C_6H_{11}O_5 + H_2O$—formed by the action of malt extract on starch
trehalose, $C_6H_{11}O_5 \cdot O \cdot C_6H_{11}O_5 + 2\ H_2O$
lactose, $C_6H_{11}O_6 \cdot O \cdot C_6H_{11}O_3 + H_2O$—the carbohydrate of milk

Trisaccharides ($C_{18}H_{32}O_{16}$)
raffinose (melitriose or gossypose), $C_6H_{11}O_5 \cdot O \cdot C_6H_{10}O_4 \cdot O \cdot C_6H_{11}O_6 + 5\ H_2O$—occurs in beet juice and cottonseed
Polysaccharides—To this group belong xylan and araban, both assigned tentatively the formula ($C_6H_8O_4$) which are widely distributed in nature and are the mother substances of the pentoses xylose and arabinose, respectively, also starch, dextrin, mannan, the mother substance of mannose, glycogen, and cellulose with the general formula ($C_6H_{10}O_5$). The pectins derived from the middle lamellae of many fruits are closely related substances.

separation of fractions by proximate analysis. Approximately 95% of crude fiber is cellulose, the remaining portion being lignin and hemicellulose. (Lignin is not considered a carbohydrate.)

Cellulose is composed of glucose units joined by the beta-D-glucopyranoside linkage. Some investigators have reported that cellulose is digested in the cecum of poultry; however, the cecum does not secrete digestive juices and any change in cellulose content would be due to bacterial decomposition. So far as established, poultry does not secrete enzymes which digest cellulose.

Cellulose and hemicellulose play important roles in digestion, and also an important physical role. The influence of fiber in the development of the physical structure of the digestive system has been documented. The indigestible residue, largely crude fiber, is highly responsible for physiologically active bulk. The normal peristaltic movements of the intestinal tract are dependent, in part, on internal distention developed by indigested residues. The hydrophylic residue, such as hemicellulose, appears to have a high water-holding capacity which helps maintain a moist and soft condition of the fecal mass. Also, the decomposition of cellulose in the cloaca and large intestine by bacteria, produces free fatty acids that may stimulate peristalsis.

The bulk produced by some feed ingredients fills the digestive tract before the nutrients required for maximum growth are obtained. This results in slower sexual development of breeding stock. Some breeding stock mature sexually before physiological maturity, thus are not able to sustain high egg production.

FOOD MOVEMENT

The presence of cellulose in the digestive tract tends to reduce the movement of food through the tract. This improves absorption and aids in digestion. Cellulose also improves the physical structure of the feces and develops a more solid mass with water.

10

Nutrition of Proteins

Proteins are complex compounds, macromolecules or polymers of amino acids joined in peptide linkages.

They contain approximately 16% nitrogen and sometimes sulfur, iron or phosphorus. Nearly half the weight of body cells is protein. The enzymes in the cells, hormones, antibodies and membranes all have protein structures. Protein is so important that no life can exist without it.

There are several classifications of proteins; however, a useable one for poultry nutrition is based on solubility and composition: (1) simple proteins, (2) conjugated proteins and (3) derived proteins (Table 10.1). The simple proteins upon hydrolysis yield only amino acids and the conjugated proteins yield some substances other than amino acids. These substances may be iron, phosphorus, fats or glycogen. The derived proteins are the native proteins which have undergone complex changes.

Simple and conjugated proteins are important in nutrition. The terms, vegetable proteins and animal proteins, are collective terms in that both simple and conjugated proteins are included. The plant glutelins are generally deficient in one or more amino acids or possess some substance or molecular configuration that reduces availability of the amino acids.

TABLE 10.1. CLASSIFICATION OF PROTEINS

Simple proteins:	*Conjugated proteins (contd)*
albumins	lipoproteins
globulins	flavoproteins
glutelins	miscellaneous metalloproteins
prolamines	
scleroproteins	*Derived proteins:*
	split products of conjugated proteins
Conjugated proteins:	protamines
nucleoproteins	histones
phosphoproteins	others
porphyrinoproteins	denatured proteins
hemoproteins	hydrolytic products
chlorophylloproteins	proteoses and peptones
glycoproteins	peptides

The animal proteins classified as albuminoids or sceleroproteins are generally poorly digested and deficient in one or more amino acids. This group includes keratin from feathers, collagen from tendons, skin and bone, and elastin from ligament. Egg albumen is one of the highest-quality proteins. A phosphoprotein of relatively good biological value is casein from milk. Muscle proteins have high biological value.

Natural feeds contain many different types of protein. Each type of protein has a different organization, molecular configuration and amounts of each of the amino acids. Whereas a simple protein type in a feed may be deficient in one or more amino acids, it is possible for another type of protein to be present which is adequate in the amino acids deficient in the other. One protein would supplement another to produce a good quality protein.

Amino acids are fundamental units of proteins and are generally alpha organic acid which, in most cases, has a nitrogen attached to the carbon next to the carboxyl group. The basic structure of an amino acid is:

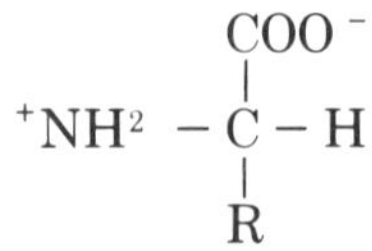

The peptide linkage is:

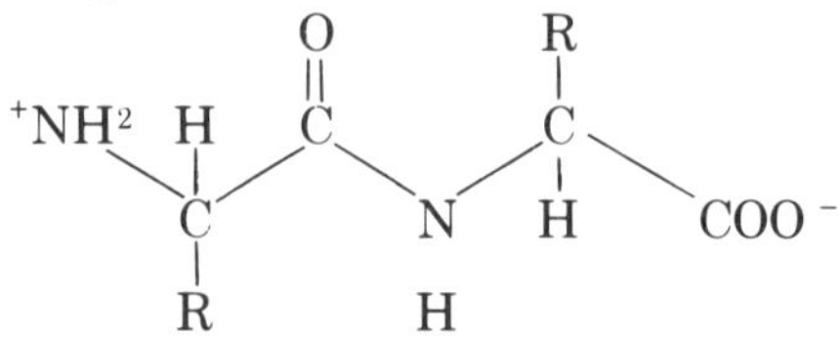

A total of 20 amino acids is known to exist in most natural proteins (Table 10.2) of which 13 are known to be nutritionally essential for poultry. The essential and nonessential amino acids are:

	Essential	Nonessential
(1)	phenylalanine	alanine
(2)	isoleucine	aspantic acid
(3)	lysine	cystine
(4)	threonine	cysteine
(5)	histidine	hydroxyproline
(6)	arginine	proline[1]
(7)	tryptophan	serine
(8)	methionine	tyrosine
(9)	valine	glutamic acid
(10)	leucine	
(11)	glycine	

[1]Required in some conditions.

TABLE 10.2. AMINO ACIDS IN FEEDS

Monoamino monocarboxylic	
[1]glycine	$H-CH(NH_2)-COOH$
alanine	$CH_3-CH(NH_2)-COOH$
[1]valine	$(CH_3)_2CH-CH(NH_2)-COOH$
[1]leucine	$(CH_3)_2CH-CH_2-CH(NH_2)-COOH$
[1]isoleucine	$CH_3-CH_2-CH(CH_3)-CH(NH_2)-COOH$
Hydroxyl-containing	
serine	$HO-CH_2-CH(NH_2)-COOH$
[1]threonine	$CH_3-CH(OH)-CH(NH_2)-COOH$
Sulfur-containing	
cystine (and cysteine)	$S-CH_2-CH(NH_2)-COOH$ / $S-CH_2-CH(NH_2)-COOH$ (S—S linked); $HS-CH_2-CH(NH_2)-COOH$
[1]methionine	$CH_3-S-CH_2-CH_2-CH(NH_2)-COOH$
Aromatic	
tyrosine	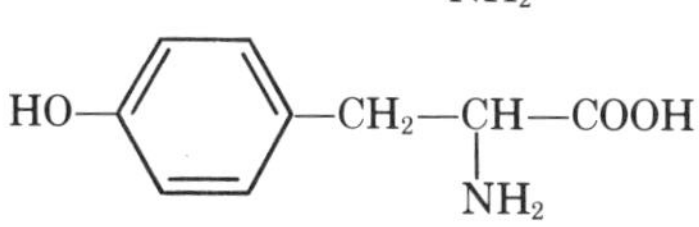
[1]phenylalanine	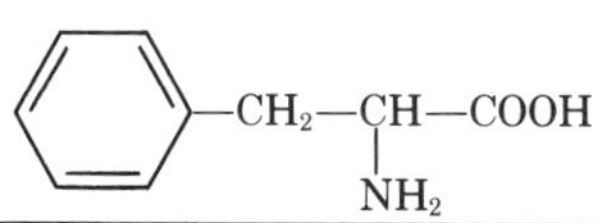

TABLE 10.2. *(Continued)*

Heterocyclic	
[2]proline	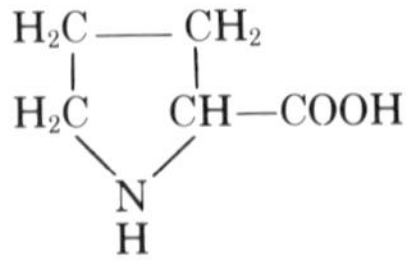
hydroxyproline	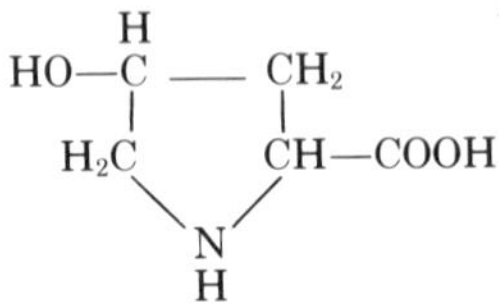
[1]histidine	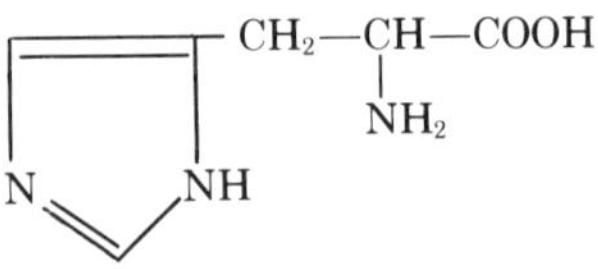
[1]tryptophan	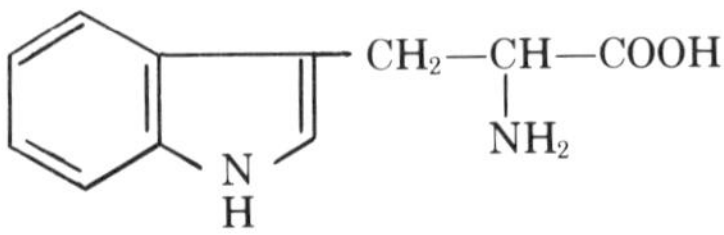
Diamino Monocarboxylic (Basic)	
[1]arginine	$H_2N-C(=NH)-NH-CH_2-CH_2-CH_2-CH(NH_2)-COOH$
[1]lysine	$H_2N-CH_2-CH_2-CH_2-CH_2-CH(NH_2)-COOH$
Monoamino Dicarboxylic (Acidic)	
glutamic acid	$HOOC-CH_2-CH_2-CH(NH_2)-COOH$
aspartic acid	$HOOC-CH_2-CH(NH_2)-COOH$

Source: Patrick (1965).
[1]Essential amino acids.
[2]Essential under some conditions.

Economically, all amino acids are required because when the nonessential amino acids are not present in amounts needed, the chick's body will generally convert an essential amino acid into a nonessential one.

A protein which contains all the essential amino acids in an available state and in amounts required by poultry is said to have high biological value.

An understanding of protein nutrition generally centers on the development of a protein-quality program designed to establish biological value. Many methods are used to determine protein quality in isolated or crude proteins. There are several factors which influence the biological value of a protein. These should be recognized and given careful evaluation before the biological value of proteins is considered. These factors are:

Protein Level.—When determining the biological value or requirement of any nutrient, the level incorporated into the assay ration should be less than the minimum optimum level. A 20% deficiency is often used as a desirable test level.

Environmental Temperature.—Chickens require less energy and more protein as the environmental temperature increases. For example, poultry production in warm weather or in the southern states requires more protein and less energy than in cold weather or in the northern states. Since the chicken's body temperature is approximately 107°F (41.7°C), the energy requirement for maintenance is much higher than for animals having a lower body temperature. The release of energy needed to maintain homeostasis is higher because water evaporation by respiration is the major method employed.

Age of Poultry.—Young growing chickens have a higher total protein, and also a higher amino acid, requirement than mature fowl. Therefore, when determining the biological value of a protein for formulation of poultry rations, the young chick from 1 to 42 days of age is the most desirable. The chick during this period requires 13 amino acids, whereas the mature fowl may not require over 9 to 10 amino acids. From 1 to 28 days of age, the lysine content of the plasma from fasted chicks is approximately 4 times as high as from fed chicks. It requires approximately 24 hr after feeding for the lysine content of the plasma of a fasted chick to return to normal. The decline of both lysine and theonine from the large accumulation status by 8 wk of age suggests that from 1 to 42 days of life is the desirable period for protein evaluation studies, and from 1 to 28 days is the more sensitive period for evaluation.

Species, Breed and Strain.— The protein requirement for different species of poultry varies markedly when expressed as a percentage of the ration, in g of protein per unit of metabolizable energy or in amount per day. The protein requirement per unit of gain is similar in different species of poultry. Ducks and squabs make better gain per unit of protein than chickens or turkeys. A strain difference in protein requirement of White Leghorn hens became apparent when the protein uptake was near the minimum protein requirement for this breed.

A measurable difference in arginine and lysine requirements has been established. A breed difference for methionine has been reported between White Leghorn and Australorp chicks. White Leghorn chicks apparently have the higher requirement.

Amino Acid Content.—Proteins which provide only the desirable level of the essential amino acids are preferred to those which have a high excess of some essential with a minimum level of another essential. This is because a large surplus of 1 amino acid can be antagonistic to another. A very common example of this is the antagonism between arginine and lysine. The feeding of excess lysine to chicks causes a marked increase in kidney arginase activity, which in turn increases the need for arginine. Casein, a good protein, has nearly twice as much lysine as arginine, and when fed to chicks, an arginine-lysine antagonism develops which can be corrected by supplementing the casein with arginine.

Digestibility.—Poultry have a very short digestive tract and also one that is particularly sensitive to pH changes. Some natural crude proteins are slowly digested, hence the available amino acids could be absorbed and deaminized before those slow to be released are available for absorption. The liver cannot store amino acids, so if the amino acids are not absorbed when needed, they cannot be used for protein synthesis.

There are many factors affecting digestibility of proteins. Whereas chemical analysis and amino acid content may show a complete essential amino acid picture, factors such as solubility, structure of protein macromolecule and type of protein can influence digestibility. Studies on the relation of breeding to digestion and absorption have shown that these functions are not readily influenced by breeding.

Availability.—Many proteins contain amino acids that are linked so that the enzymes cannot free them for absorption. Good examples of this are methionine in raw soybean meal and lysine in cooked cottonseed meal. Heating inactivates an undenatured protein which prevents trypsin from hydrolyzing the soybean protein to free the methionine for absorption. Heating cottonseed meal causes the lysine to become unavailable and the proteolytic enzymes to be inactivated.

USE OF PROTEIN AND AMINO ACIDS

Protein or amino acids have many uses in the body. When excess protein is fed, the unneeded amino acids are converted to carbohydrates, fats, or directly to energy.

The following reactions use protein or amino acids in poultry:

(1) Building tissue protein
(2) Building enzyme systems
(3) Energy

(4) Fat, fat translocation and storage
(5) Mineral chelation, transportation and storage
(6) Buffering and osmotic pressure
(7) Reproduction, sperm, ovum (eggs are 55% moisture and 30% protein)
(8) Colloidal structure
(9) Oxygen transport

Figure 10.1 summarizes the fate of ration proteins and amino acids in poultry.

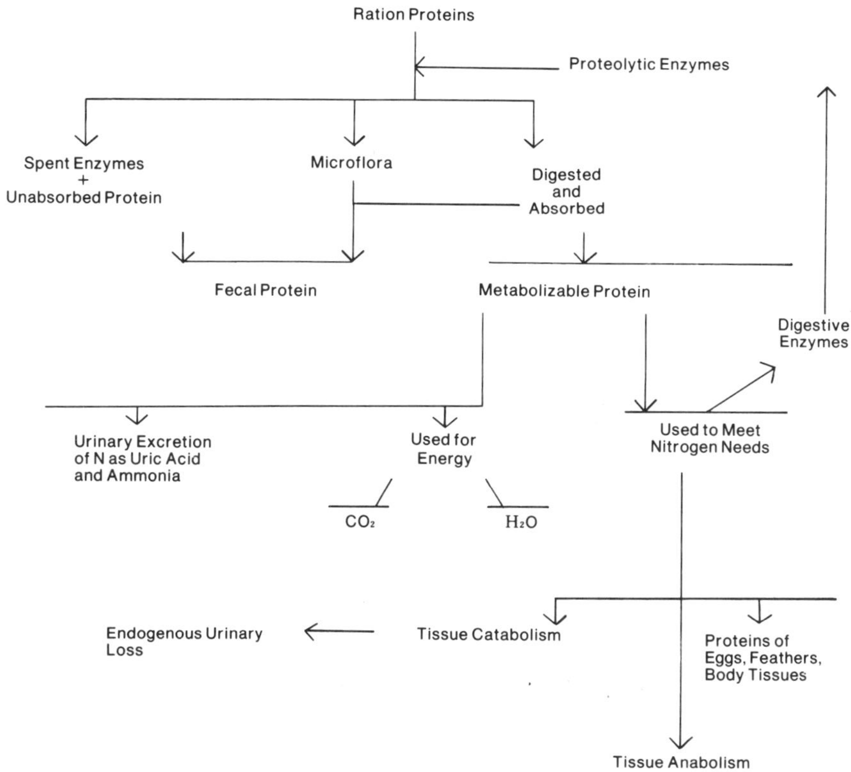

FIG. 10.1. PROTEIN METABOLIC FLOW IN POULTRY

FACTORS INFLUENCING THE PROTEIN NEEDS OF POULTRY

There are many factors which influence the needs for protein and amino acids by poultry.

(1) Age.—The young starting birds have the highest protein requirement because their bodies are forming new cells which are mostly protein and water. As the chickens mature, their protein requirements become reduced.

(2) Growth Rate.—Growth is accompanied by cell division, hence an increase in protein requirement. When growth-stimulating factors, such as antibiotics, toxic substances (arsenicals) and hormones are fed or administered at growth-promoting levels, an increase in protein requirements can occur.

(3) Reproduction.—When mature birds begin laying, there is an increase in protein, vitamin and mineral requirements per unit of feed consumed and an increase in energy needs. If the energy level of reproductive rations is not lowered, or the protein, vitamin and mineral level increased, poor egg production and hatchability may occur. This is because energy uptake by poultry influences feed consumption.

(4) Climate.—Poultry in cold climates or cold seasons require more energy than in warm weather. If the energy level is not adequate to meet the maintenance, growth or reproductive needs, the chicken will continue to consume feed until its needs for energy are met. Reproductive rations for hens in cold climates, such as exist in the northern states, may be as low as 12% protein, whereas in warm climates, they may be 15 to 17% protein.

(5) Energy Level.—Since the chicken eats only until his energy needs are met, it becomes necessary that as the energy level increases, the protein level must increase; otherwise, the desired physiological functions—growth, maintenance or reproduction— cannot be properly performed. A calorie-protein ratio for each ration must be developed.

(6) Diseases.—Diseases which influence feed uptake, digestion or absorption also influence protein needs. When a disease damages either the absorptive membrane in the small intestine or the secretion of digestive juices, a change in protein requirements becomes evident. Coccidiosis, a disease of the small intestine, may increase the requirements for some minerals and other nutrients.

(7) Breed and Strain.—The methinine requirement of White Leghorns is higher than that of Australorps. Some strains of White Leghorn chickens and Japanese quail have higher protein requirements than others. Studies on the influence of genetic factors on arginine requirements demonstrated that chicks having a high arginine requirement also have a high level of arginase in the kidneys. Chicks which have a high kidney arginase level possess a higher level of plasma lysine. This strain of chicks is very sensitive to increase in dietary lysine.

INFLUENCE OF A PROTEIN OR AMINO ACID DEFICIENCY ON THE CHICK

When a ration is deficient in protein or one or more essential amino acids, several physiological conditions may occur. These are:

(1) Increased Protein Synthesis in the Liver.—When a diet is deficient in an essential amino acid, the chick's body will cycle amino acids from within the body to the digestive tract for reabsorption, along with the present dietary amino acids. This results in a negative amino acid balance in the body in which the liver attempts to increase its rate of protein synthesis. Since the use of the recycled amino acids and the dietary amino acids is not 100% utilized for liver protein syntheses, a negative recovery is observed in the inter metabolism of body protein. If the recycled amino acids are greater than the utilizable resynthesized protein, an increase in protein synthesis in the liver occurs.

(2) Increased Fat Synthesis.—When deficiency of an amino acid exists to the extent that one or more of the essential amino acids needed are not available, the chick's body will deaminize the amino acids and use the carbon portion for energy needs, if necessary. If the body does not immediately need the energy, it will synthesize fat. Fattening generally occurs when low-protein or amino-acid deficient rations are fed.

(3) Increased Sensitivity to Aflatoxins.—Aflatoxins are fat-soluble substances; therefore, any biochemical reaction which increases fat synthesis will provide better chances for these to be transported in the chylomicrons, since most of the fat is transported in these particulates. The aflatoxins are not readily metabolized in the liver but are deposited in the liver fat cells.

(4) Increased Sensitivity to Chlorinated Hydrocarbons.—Chlorinated hydrocarbons are absorbed as fat-soluble substances and transported to the liver, where they are deposited in, or transferred to, the fat cells by way of the chylomicrons. The chicken cannot readily metabolize the chlorinated hydrocarbons; therefore, their half-life in the chicken is very long, and related to the fat removal process or life span of the fat cells.

(5) Poor Growth.—The first symptoms of a protein or amino acid deficiency is poor growth because of the reduced new-cell formation and decrease in nitrogen retention. Amino acids perform many functions in the body besides formation of protein structures. Some of these are hormonal and enzymatic. Without these substances, the chicken cannot grow.

(6) Reduced Feed Efficiency.—An amino-acid or protein deficiency results in increased fat synthesis and greater oxidation of the amino acids. Since feed efficiency is an expression of feed requirement per unit gain, the failure to grow results in poor feed efficiency.

(7) Poor Reproduction.—Eggs are high in protein, and if the diet is deficient in protein or one or more essential amino acids, the hen cannot produce a protein-deficient egg because the reproductive process forms only eggs complete in protein. This also means that the protein content of the ration does not readily influence egg composition.

(8) General Appearance.—Some amino acids influence the growth and pigmentation of feathers. Feathers may grow very slowly and appear out-of-shape and broken. Dark-feathered birds may show lack of pigment in the feathers when lysine is deficient.

DETERMINATION OF THE BIOLOGICAL VALUE OF PROTEINS FOR POULTRY

There are 6 methods of determining the biological value of proteins. The first determination made with proteins is the nitrogen content which establishes the crude protein values. This determination is generally performed by the procedure outlined in the official publication of Agricultural Chemists, commonly known as the Kjeldahl determination. Although this method determines nitrogen, the results in feeds are converted to crude protein by multiplying N × 6.25. The factor, 6.25, is derived from the average analysis of protein, being 16%. The nitrogen content of proteins varies from different sources; however, the 16% value provides a relatively good evaluation, unless the ingredient has been incorporated with materials containing urea, ammonium salts, or other non-protein nitrogen compounds.

(1) Amino Acid Content.— Some control laboratories use selected conversion factors for proteins. These conversion factors are shown in Table 10.3. Unless the ingredient has a non-protein nitrogen, such as urea or some ammonium salts, the factor, 6.25, for mixed poultry feeds and natural ingredients is a reliable working value. The amino acid content of ingredients provides a good index to protein quality. Several methods are used to determine the amino acid content. The microbiological and chemical methods are the most commonly used. These methods have been perfected so that the results are reasonably accurate. The amino acid content of most of the ingredients is summarized in Table 10.4. These results are calculated averages of published data obtained by all methods used for amino acid determinations.

TABLE 10.3. SELECTED CONVERSION FACTORS FOR PROTEINS

Ingredient Originated	% N in Protein	Conversion Factor
Oil seed protein	18.5	5.40
Cereal protein	17.0	5.90
Plant leaf	15.0	6.60
Animal or fish	16.0	6.25
Eggs	16.0	6.25
Milk	15.8	6.38
Gelatin	18.0	5.55
Corn, grain	16.0	6.25
Barley, grain	17.2	5.83

Source: Jones (1931).

(2) Nitrogen Balance.— Nitrogen balance is one of the oldest and most reliable methods used to establish protein quality. When a chicken excretes nitrogen, it demonstrates body oxidation of protein and amino acids. The poorer the protein quality, the greater the nitrogen excretion. When the quality is poor, the amino acids are deaminized, the nitrogen is converted to uric acid for excretion, and the carbon chain is converted to fatty acids, carbohydrates, or used immediately for energy needs. Nitrogen balance studies are generally made with animals which cast the feces and urine separately. The chicken can be used, but requires surgical procedures to obtain separation of urine and feces. Some protein concentrates prepared from tissues, such as feathers, beaks and toes, are poorly attacked by the digestive enzymes. When dried, the enzymes cannot react with the insoluble protein to free the amino acids. Digestion cannot be performed unless the protein is in a solution or in a water-protein colloidal state.

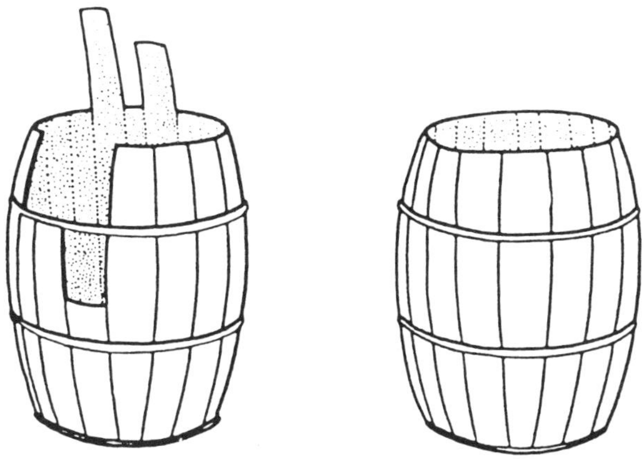

FIG. 10.2. BARREL STAVE ANALOGY OF LIMITING AMINO ACID CONCEPT
An absence or deficiency of any 1 amino acid prohibits or limits protein synthesis.

TABLE 10.4. AMINO ACID CONTENT OF FARM FEEDS

Product	Crude Protein g per 100 g Product	Arginine, % Sample	Arginine, % Crude Protein	Histidine, % Sample	Histidine, % Crude Protein	Isoleucine, % Sample	Isoleucine, % Crude Protein	Leucine, % Sample	Leucine, % Crude Protein
Algae									
Ascomeal	5.75	0.23	4.00	0.064	1.11	0.25	4.35	0.35	6.09
Cladophora rupestris	27.22	1.44	5.29	0.29	1.07	0.94	3.45	1.41	5.18
Laminaria cloustoni	10.91	0.25	2.31	0.14	1.28	0.28	2.57	0.39	3.57
Neptune's Bounty	6.01	0.15	2.50	0.052	0.87	0.19	3.16	0.27	4.49
Rhodymenia palmata	20.69	1.08	5.22	0.33	1.59	0.88	4.25	1.12	5.41
Animal By-products									
Blood meal	80.50	3.37	4.18	4.25	5.28	0.99	1.23	11.10	13.80
Bone meal	23.42	1.77	7.56	0.19	0.81	0.46	1.96	0.97	4.14
Dried buttermilk	30.73	1.06	3.45	0.75	2.44	1.73	5.63	3.01	9.79
Dried cheese whey	12.78	0.32	2.50	0.20	1.56	0.74	5.79	1.14	8.92
Dried skim milk	33.44	1.08	3.23	0.78	2.33	2.00	5.98	3.09	9.23
Dried whey	12.10	0.24	1.98	0.15	1.24	0.74	6.11	1.10	9.09
Meat and bone scraps	52.37	3.79	7.24	0.91	1.74	1.76	3.36	3.27	6.24
Meat scraps									
Sample 1	45.34	3.38	7.46	0.81	1.79	1.48	3.26	2.96	6.53
Sample 2	48.41	3.36	6.94	0.74	1.53	1.50	3.10	3.04	6.28
Tankage	61.02	1.40	2.29	2.05	3.36	1.40	2.29	5.74	9.40
Fermentation Feeds									
Dried brewer's and distiller's grains									
Brewer's dried grains	32.63	1.59	4.87	0.81	2.48	1.66	5.09	3.12	9.56
Distiller's dried grains	26.09	1.09	4.18	0.66	2.53	1.14	4.37	3.14	12.03
Distiller's dried grains with solubles	29.50	1.04	3.53	0.73	2.47	1.19	4.03	2.91	9.86

Corn distiller's dried grains, Distiller's Feed Research Council, composite sample	25.77	1.08	4.19	0.61	2.37	0.94	3.65	2.29	8.89
Corn distiller's dried grains with solubles, Distiller's Feed Research Council, composite sample	27.82	1.12	4.03	0.75	2.70	1.17	4.21	2.95	10.60
Dried fermentation solubles									
Distiller's solubles	27.03	0.96	3.55	0.64	2.37	0.97	3.59	1.61	5.96
Corn distiller's solubles, Distiller's Feed Research Council, composite sample	27.72	1.19	4.29	0.67	2.42	1.13	4.08	2.42	8.73
Corn butyl fermentation solubles	32.13	0.92	2.86	0.52	1.62	1.34	4.17	2.72	8.47
Cane syrup butyl fermentation solubles	25.13	0.47	1.87	0.26	1.03	1.13	4.50	1.30	5.17
Molasses ethyl fermentation solubles (Vacatone 40)	7.85	0.041	0.52	0.037	0.47	0.12	1.53	0.15	1.91
Malt sprouts	29.01	1.25	4.31	0.53	1.83	1.09	3.76	1.64	5.65
Distatic barley malt	12.28	0.65	5.29	0.29	2.36	0.52	4.23	0.81	6.60
Yeast, Fleishmann's, irradiated, dry	47.38	2.46	5.19	1.00	2.11	2.94	6.20	3.56	7.51
Fish By-products									
Crab bran	40.05	2.21	5.52	0.80	2.00	1.75	4.37	2.59	6.47
Fish press water	59.72	2.17	3.63	2.02	3.38	0.99	1.66	1.87	3.13
Herring meal	72.25	4.13	5.72	1.71	2.37	4.15	5.74	5.89	8.15
Perch meal	58.42	3.64	6.23	1.14	1.95	2.69	4.60	4.08	6.98
Sardine meal	68.25	4.02	5.89	1.80	2.64	3.48	5.10	5.33	7.81
Shrimp bran	33.44	1.79	5.35	0.67	2.00	1.23	3.68	1.97	5.89
Shrimp meal	42.28	2.25	5.32	0.70	1.66	1.58	3.74	2.27	5.37
Whale meal	53.02	2.50	4.72	1.19	2.24	2.73	5.15	4.29	8.09
White fish meal	67.38	4.38	6.50	1.24	1.84	2.78	4.12	4.48	6.65

TABLE 10.4. *(Continued)*

Lysine, %		Methionine, %		Phenylalanine, %		Threonine, %		Tryptophan, %		Valine, %	
Sample	Crude Protein	Sample	Crude Protein	Sample	Crude Protein	Sample	Crude Protein	Sample	Crude Protein	Sample	Crude Protein
Algae											
0.29	5.04	0.10	1.74	0.28	4.87	0.26	4.52	0.074	1.29	0.31	5.46
1.55	5.69	0.43	1.58	0.96	3.53	1.17	4.30	0.31	1.14	1.23	4.52
0.39	3.57	0.15	1.37	0.29	2.66	0.44	4.03	0.09	0.83	0.47	4.30
0.178	2.96	0.075	1.25	0.21	3.49	0.20	3.33	0.063	1.05	0.24	4.06
1.49	7.20	0.37	1.79	0.81	3.91	0.89	4.30	0.27	1.30	1.25	6.04
Animal By-products											
3.87	4.81	1.28	1.59	5.80	7.22	4.34	5.39	1.17	1.45	7.76	9.64
0.88	3.76	0.18	0.77	0.56	2.39	0.58	2.48	0.05	0.21	0.72	3.07
2.11	6.36	0.76	2.47	1.44	4.68	1.37	4.46	0.47	1.53	2.11	6.86
1.07	8.37	0.25	1.96	0.36	2.82	0.83	6.50	0.22	1.72	0.73	5.71
2.11	6.31	0.91	2.72	1.73	5.17	1.63	4.87	0.48	1.43	2.29	6.84
0.62	5.12	0.18	1.49	0.29	2.40	0.73	6.03	0.25	2.07	0.67	5.54
2.87	5.48	0.70	1.34	1.79	3.42	1.72	3.28	0.36	0.69	2.50	4.77
2.49	5.49	0.64	1.41	1.73	3.82	1.64	3.62	0.33	0.73	2.38	5.25
2.68	5.54	0.66	1.36	1.64	3.39	1.53	3.16	0.32	0.66	2.17	4.48
4.40	7.21	0.76	1.24	0.78	1.28	2.00	3.28	0.61	1.00	4.18	6.85
Fermentation Feeds											
1.33	4.08	0.65	1.99	1.75	5.36	1.29	3.95	0.37	1.12	1.94	5.95

0.83	3.18	0.56	2.15	1.14	4.37	0.99	3.79	0.23	0.88	1.52	5.83
0.98	3.31	0.64	2.17	1.33	4.51	1.02	3.46	0.24	0.81	1.51	5.12
0.80	3.11	0.49	1.90	1.03	4.00	0.88	3.41	0.22	0.85	1.19	4.62
0.75	2.70	0.60	2.16	1.26	4.53	1.08	3.88	0.23	0.83	1.47	5.28
0.67	2.48	0.48	1.78	1.00	3.70	0.90	3.33	0.22	0.81	1.18	4.37
0.90	3.25	0.55	1.98	1.26	4.55	1.09	3.93	0.25	0.90	1.46	5.27
1.51	4.70	0.49	1.53	1.12	3.49	1.23	3.83	0.24	0.75	1.35	4.20
0.95	3.76	0.41	1.63	0.72	2.87	0.97	3.86	0.19	0.77	1.21	4.81
0.07	0.94	0.057	0.73	0.11	1.40	0.14	1.79	0.03	0.38	0.18	2.29
1.36	4.69	0.39	1.34	0.91	3.14	1.01	3.48	0.41	1.41	1.46	5.03
0.49	3.99	0.18	1.46	0.62	5.05	0.43	3.50	0.22	1.79	0.70	5.70
3.70	7.81	1.00	2.11	2.77	5.85	2.41	5.21	0.73	1.54	3.06	6.46
					Fish By-products						
1.	4.62	0.73	1.82	1.70	4.25	1.54	3.84	0.45	1.12	2.00	4.99
2.	4.02	0.80	1.34	0.96	1.61	0.10	0.167	0.17	0.28		
5.	8.19	2.11	2.92	2.92	4.04	3.26	4.51	1.12	1.55	4.37	6.05
4.	7.52	1.61	2.76	2.22	3.80	2.46	4.21	0.70	1.20	2.56	4.38
5.	8.58	2.04	2.99	2.92	4.28	3.04	4.45	0.89	1.30	4.07	5.96
1.	4.42	0.62	1.85	1.33	3.98	1.34	4.01	0.37	1.11	1.62	4.84
2.	4.78	0.72	1.70	1.55	3.66	1.47	3.48	0.39	0.92	1.98	4.68
3.	6.56	1.01	1.90	2.07	3.90	1.64	3.09	0.82	1.55	2.82	5.32
4.	6.78	1.75	2.60	2.26	3.35	2.76	4.10	0.68	1.01	3.03	4.50

TABLE 10.4. *(Continued)*

Product	Crude Protein g per 100 g Product	Arginine, % Sample	Arginine, % Crude Protein	Histidine, % Sample	Histidine, % Crude Protein	Isoleucine, % Sample	Isoleucine, % Crude Protein	Leucine, % Sample	Leucine, % Crude Protein
			Grains						
Barley, Cordona	13.93	0.59	4.24	0.33	2.37	0.57	4.09	0.95	6.82
Barley, Goliad	13.97	0.67	4.80	0.35	2.51	0.52	3.72	0.86	6.16
Buckwheat	11.86	1.13	9.53	0.26	2.19	0.41	3.46	0.62	5.23
Corn, Country Gentleman	11.08	0.45	4.06	0.29	2.62	0.48	4.32	1.55	13.99
Corn, yellow	8.05	0.37	4.60	0.22	2.73	0.33	4.10	0.95	11.80
Darso	8.90	0.36	4.05	0.17	1.91	0.46	5.18	1.30	14.60
Feterita	12.05	0.47	3.90	0.25	2.07	0.62	5.14	1.86	15.40
Hegari	9.94	0.32	3.22	0.19	1.91	0.49	4.94	1.45	14.60
Kafir	11.59	0.44	3.80	0.26	2.24	0.59	5.09	1.75	15.10
Millet, Hog (Proso)	12.14	0.35	2.89	0.23	1.90	0.49	4.04	1.23	10.13
Milo, Martin	10.25	0.45	4.39	0.22	2.15	0.46	4.49	1.22	11.90
Oats, Alamo	12.15	0.87	7.16	0.26	2.14	0.55	4.53	0.85	7.00
Oats, Fultex	8.98	0.64	7.10	0.19	2.11	0.39	4.33	0.63	7.01
Oats, Mustang	11.60	0.57	4.91	0.20	1.72	0.45	3.88	0.74	6.38
Oats, Ranger	10.45	0.74	7.08	0.20	1.91	0.46	4.40	0.72	6.89
Rice, Early prolific	6.39	0.54	8.44	0.14	2.19	0.28	4.38	0.51	7.98
Rice, Texas Patna	7.84	0.65	8.29	0.18	2.30	0.36	4.59	0.62	7.91
Rice, Texas Patna, polished	7.88	0.65	8.25	0.19	2.41	0.35	4.44	0.63	7.99
Rye	11.19	0.59	5.27	0.25	2.23	0.43	3.84	0.67	5.99
Shallu	8.23	0.34	4.13	0.19	2.31	0.40	4.87	1.04	12.70
Speltz	9.56	0.46	4.81	0.20	2.09	0.42	4.39	0.67	7.01
Wheat, Austin	11.15	0.55	4.93	0.26	2.33	0.44	6.91	0.80	7.17
Wheat, Frisco	15.28	0.81	5.30	0.38	2.49	0.61	3.99	0.96	6.28

Wheat, Quanah	18.16	0.86	4.74	0.44	2.42	0.77	4.24	1.21	6.66
Wheat, Seabreeze	18.50	1.01	5.46	0.50	2.70	0.71	3.84	1.13	6.11
Wheat, Selkirk	17.00	0.83	4.88	0.48	2.82	0.60	3.53	1.05	6.18
	Grain By-products from Milling and Processing								
Corn by-products									
corn meal, yellow	9.34	0.45	4.82	0.24	2.57	0.36	3.85	1.12	12.00
Corn gluten meal	45.72	1.50	3.28	1.00	2.19	2.03	4.44	7.53	16.47
Corn gluten feed	22.86	0.98	4.29	0.69	3.02	0.81	3.54	2.34	10.23
Corn oil cake meal (germ meal)									
Sample 1	22.16	1.39	6.27	0.67	3.02	0.86	3.88	1.82	8.21
Sample 2	22.69	1.51	6.65	0.78	3.44	0.96	4.23	1.97	8.68
Hominy, yellow	10.56	0.61	5.78	0.29	2.75	0.43	4.07	1.17	11.08
Grain sorghum by-products									
Grain sorghum oil cake meal (germ meal)									
Sample 1	21.62	1.48	6.84	0.73	3.38	0.98	4.53	1.88	8.70
Sample 2	18.50	1.05	5.68	0.64	3.46	0.78	4.22	1.45	7.84
Grain sorghum gluten feed	22.26	0.95	4.27	0.61	2.74	0.98	4.40	2.53	11.37
Grain sorghum gluten meal									
Sample1	44.54	1.19	2.67	0.81	1.82	2.26	5.07	7.30	16.39
Sample 2	64.69	1.38	2.13	1.09	1.68	3.28	5.07	11.11	17.17
Milo heavy steep water 54.7% solids	17.82	1.58	8.87	0.64	3.61	0.33	1.85	0.62	3.46
Oatmeal	14.76	0.99	6.71	0.31	2.10	0.63	4.27	1.14	7.72
Rice by-products									
Rice bran	12.54	1.07	8.53	0.35	2.79	0.55	4.38	0.89	7.10
Rice, polish	12.78	1.04	8.14	0.32	2.50	0.44	3.44	0.81	6.34
Wheat by-products									
Wheat bran	14.58	1.09	7.48	0.44	3.02	0.54	3.70	0.92	6.31
Wheat germ meal	28.08	2.08	7.41	0.64	2.28	0.97	3.45	1.67	5.95
Wheat grey shorts	15.88	0.96	6.04	0.36	2.27	0.58	3.65	0.93	5.86
Wheat, Red Dog	14.22	0.55	3.87	0.29	2.04	0.59	4.15	1.02	7.17
Wheat shorts	16.85	1.07	6.35	0.48	2.85	0.63	3.74	1.03	6.11

TABLE 10.4. *(Continued)*

Lysine, %		Methionine, %		Phenylalanine, %		Threonine, %		Tryptophan, %		Valine, %	
Sample	Crude Protein	Sample	Crude Protein	Sample	Crude Protein	Sample	Crude Protein	Sample	Crude Protein	Sample	Crude Protein
Grains											
0.48	3.45	0.20	1.44	0.77	5.53	0.48	3.45	0.178	1.28	0.76	5.45
0.51	3.65	0.18	1.29	0.67	4.80	0.47	3.36	0.177	1.27	0.70	5.01
0.66	5.56	0.22	1.85	0.45	3.79	0.50	4.22	0.200	1.68	0.56	4.72
0.33	2.98	0.21	1.90	0.55	4.97	0.39	3.52	0.089	0.80	0.59	5.32
0.27	3.35	0.18	2.24	0.42	5.22	0.31	3.85	0.087	1.17	0.43	5.34
0.19	2.14	0.13	1.46	0.51	5.73	0.34	3.82	0.122	1.37	0.51	5.73
0.23	1.91	0.19	1.58	0.72	5.98	0.47	3.90	0.159	1.32	0.71	5.89
0.17	1.71	0.14	1.41	0.55	5.53	0.37	3.72	0.122	1.23	0.56	5.63
0.25	2.16	0.21	1.81	0.66	5.70	0.46	3.97	0.150	1.30	0.66	5.69
0.25	2.06	0.30	2.47	0.59	4.86	0.44	3.63	0.168	1.38	0.62	5.12
0.24	2.34	0.18	1.75	0.50	4.88	0.41	4.00	0.140	1.36	0.58	5.66
0.53	4.36	0.16	1.32	0.59	4.86	0.47	3.87	0.173	1.42	0.70	5.76
0.43	4.78	0.14	1.56	0.44	4.90	0.36	4.00	0.103	1.15	0.53	5.89
0.38	3.28	0.10	0.83	0.51	4.40	0.38	3.28	0.156	1.34	0.60	5.17
0.41	3.92	0.12	1.15	0.53	5.07	0.39	3.72	0.155	1.48	0.61	5.84
0.28	4.38	0.14	2.19	0.31	4.85	0.22	3.44	0.098	1.53	0.40	6.26
0.31	3.96	0.18	2.30	0.38	4.87	0.28	3.57	0.125	1.59	0.50	6.38
0.26	3.30	0.20	2.54	0.36	4.57	0.27	3.43	0.125	1.59	0.50	6.34
0.45	4.02	0.18	1.61	0.47	4.20	0.37	3.31	0.140	1.25	0.56	5.00
0.20	2.43	0.15	1.82	0.43	5.23	0.33	4.01	0.101	1.23	0.46	5.59
0.29	3.03	0.16	1.67	0.46	4.81	0.38	3.97	0.122	1.28	0.47	4.92
0.32	2.87	0.17	1.52	0.57	5.11	0.42	3.77	0.180	1.61	0.52	4.66

0.41	2.68	0.19	1.24	0.72	4.71	0.45	2.95	0.241	1.58	0.69	4.52
0.50	2.75	0.24	1.32	0.88	4.85	0.46	2.53	0.230	1.27	0.71	3.91
0.50	2.70	0.23	1.24	0.90	4.86	0.53	2.86	0.224	1.21	0.85	4.61
0.43	2.53	0.22	1.29	0.86	5.06	0.52	3.06	0.249	1.46	0.78	4.59
Grain By-products from Milling and Processing											
0.29	3.10	0.21	2.25	0.41	4.39	0.34	3.64	0.082	0.90	0.50	5.36
0.96	2.10	1.22	2.67	2.80	6.12	1.63	3.56	0.24	0.52	2.37	5.18
0.74	3.24	0.48	2.10	0.87	3.80	0.81	3.54	0.16	0.70	1.26	5.51
1.05	4.74	0.44	1.98	0.91	4.11	0.86	3.88	0.23	1.04	1.31	5.91
1.19	5.24	0.43	1.90	0.95	4.19	0.94	4.14	0.29	1.28	1.50	6.61
0.46	4.36	0.20	1.89	0.48	4.54	0.43	4.07	0.12	1.14	0.60	5.68
0.86	3.98	0.35	1.62	1.00	4.62	0.74	3.42	0.22	1.02	1.44	6.66
0.71	3.82	0.29	1.57	0.75	4.03	0.77	4.15	0.21	1.16	1.15	6.22
0.67	3.01	0.37	1.66	1.02	4.58	0.80	3.59	0.19	0.85	1.29	5.80
0.59	1.32	0.72	1.62	2.58	5.79	1.34	3.01	0.45	1.01	2.52	5.66
0.70	1.08	0.97	1.50	3.64	5.63	1.97	3.05	0.76	1.17	2.45	3.79
0.45	2.53	0.42	2.35	0.31	1.72	0.95	5.33	0.124	0.70	1.08	6.06
0.62	4.20	0.25	1.69	0.78	5.28	0.48	3.25	0.23	1.56	0.86	5.83
0.67	5.34	0.24	1.91	0.58	4.62	0.49	3.91	0.24	1.91	0.75	5.98
0.64	5.01	0.21	1.64	0.48	3.76	0.45	3.52	0.19	1.49	0.75	5.87
0.64	4.39	0.24	1.65	0.55	3.77	0.52	3.57	0.41	2.81	0.81	5.56
1.84	6.55	0.46	1.64	1.11	3.96	1.11	3.95	0.30	1.07	1.41	5.02
0.66	4.16	0.24	1.51	0.57	3.59	0.49	3.08	0.30	1.89	0.78	4.91
0.31	2.18	0.22	1.55	0.66	4.64	0.40	2.81	0.22	1.55	0.63	4.43
0.68	4.04	0.30	1.78	0.65	3.86	0.59	3.50	0.27	1.60	0.87	5.16

TABLE 10.4. *(Continued)*

Product	Crude Protein, g per 100 g Product	Arginine, % Sample	Arginine, % Crude Protein	Histidine, % Sample	Histidine, % Crude Protein	Isoleucine, % Sample	Isoleucine, % Crude Protein	Leucine, % Sample	Leucine, % Crude Protein
Oil Seed Residues									
Babassu meal	22.72	3.19	14.04	0.41	1.81	0.88	3.87	1.40	6.17
Castor flour	65.03	8.39	12.90	1.36	2.09	3.46	5.32	4.16	6.40
Castor pomace	39.79	3.99	10.03	0.67	1.68	1.85	4.65	2.25	5.65
Copra meal	22.22	2.54	11.43	0.38	1.71	0.89	4.00	1.40	6.30
Cottonseed meal	39.61	4.38	11.02	1.07	2.70	1.59	4.01	2.46	6.20
Cottonseed flour (Proflo)	54.07	6.11	11.30	1.39	2.57	2.13	3.94	3.37	6.23
Linseed meal	36.73	3.41	9.28	0.66	1.80	1.68	4.57	2.19	5.96
Palm kernel meal	19.16	2.54	13.26	0.31	1.62	0.76	3.97	1.23	6.42
Peanut meal	38.95	4.02	10.33	0.84	2.16	1.69	4.33	2.60	6.68
Peanut flour	61.00	6.87	11.26	1.32	2.16	2.64	4.33	4.11	6.74
Safflower meal	22.10	1.72	7.78	0.44	1.99	0.85	3.85	1.22	5.52
Sesamine meal	46.08	5.49	11.91	1.02	2.21	1.97	4.27	3.19	6.92
Soybean meal	45.29	3.38	7.46	1.12	2.49	2.49	5.50	3.48	7.69
Soybean protein (commercial)	79.50	5.72	7.20	1.85	2.33	4.65	5.85	6.18	7.77
Sunflower seed meal	21.02	1.63	7.76	0.46	2.19	0.95	4.52	1.25	5.95
Tung meal	20.88	2.27	10.87	0.48	2.30	1.03	4.93	1.58	7.57

Walnut oil meal	13.03	0.97	7.44	0.21	1.61	0.42	3.22	0.68	5.22
			Peas and Beans						
Black-eye peas	22.31	1.43	6.41	0.63	2.82	1.03	4.62	1.61	7.22
Cow peas, mixed	24.14	1.27	5.27	0.73	3.03	1.14	4.72	1.73	7.17
Mexican Pinto beans	22.58	1.55	6.85	0.64	2.83	1.14	5.05	1.11	4.92
Soybeans, Laredo	35.02	2.40	6.84	0.80	2.29	1.88	4.68	2.03	5.80
Soybeans, Arksay	39.77	2.95	7.42	0.90	2.26	2.13	5.36	3.05	7.68
Soybeans, Red Tanner	42.08	3.27	7.78	1.00	2.39	2.21	5.26	3.24	7.71
			Miscellaneous						
Alfalfa leaf meal	19.47	0.96	4.93	0.42	2.18	1.00	5.14	1.54	7.91
Animal protein factor[1]	52.94	1.98	3.74	0.49	0.93	1.58	2.98	1.79	3.38
Antibiotic feed supplement[2]	46.13	1.27	2.75	0.98	2.12	1.98	4.29	1.80	3.90
Citrous pulp	5.82	0.28	4.81	0.09	1.55	0.18	3.10	0.31	5.33
Coffee pulp	10.59	0.38	3.68	0.21	1.98	0.40	3.78	0.58	5.48
Cow pea foliage, dry	26.01	1.11	4.27	0.45	1.73	1.26	4.84	2.00	7.69
Gourd seed meal *C. foetidissima*	63.50	9.05	14.25	1.18	1.86	2.78	4.38	4.22	6.65
Pumpkin seed	30.81	3.39	11.00	0.65	2.10	1.12	3.64	1.65	5.37
Sweet potato stock feed	3.88	0.094	2.42	0.061	1.57	0.20	5.15	0.27	6.96

[1]Dried fermentation product prepared as source of Vitamin B_{12}. Amino acid composition of products is due to the carrier substance. Other animal protein factor preparations may be quite different in amino acid content according to the process used.

[2]Dried fermentation by-product from bacitracin production.

TABLE 10.4. *(Continued)*

Lysine, %		Methionine, %		Phenylalanine, %		Threonine, %		Tryptophan, %		Valine, %	
Sample	Crude Protein	Sample	Crude Protein	Sample	Crude Protein	Sample	Crude Protein	Sample	Crude Protein	Sample	Crude Protein
					Oil Seed Residues						
0.98	4.31	0.53	2.33	1.35	5.94	0.71	3.13	0.24	1.06	1.19	5.24
2.20	3.38	1.16	1.78	3.38	5.19	2.50	3.84	0.93	1.43	4.38	6.74
0.12	3.02	0.58	1.46	1.86	4.68	1.29	3.24	0.44	1.11	2.15	5.40
0.68	3.06	0.35	1.59	0.94	4.23	0.84	3.78	0.21	0.94	1.22	5.49
1.67	4.20	0.59	1.49	2.09	5.25	1.38	3.47	0.63	1.59	1.98	4.98
2.25	4.16	0.87	1.61	2.81	5.20	1.82	3.36	0.84	1.55	2.62	4.84
1.33	3.62	0.61	1.66	1.65	4.49	1.39	3.78	0.64	1.74	2.04	5.55
0.66	3.44	0.41	2.14	0.82	4.28	0.60	3.13	0.20	1.04	1.03	5.38
1.38	3.53	0.41	1.04	1.94	4.97	1.16	2.98	0.47	1.22	1.88	4.82
1.95	3.20	0.62	1.02	3.10	5.08	1.63	2.67	0.78	1.28	2.83	4.64
0.60	2.71	0.34	1.54	1.16	5.25	0.65	2.94	0.26	1.18	1.09	4.93
1.27	2.76	1.22	2.65	2.18	4.73	1.68	3.64	0.88	1.91	2.33	5.06
2.79	6.17	0.63	1.39	2.20	4.86	1.82	4.03	0.76	1.69	2.45	5.40
4.59	5.77	0.93	1.17	4.40	5.53	2.83	3.56	1.16	1.46	4.09	5.14
0.80	3.81	0.46	2.19	1.76	5.12	0.72	3.43	0.29	1.38	1.03	4.90
0.87	4.17	0.47	2.24	1.52	7.28	0.96	4.60	0.35	1.68	1.74	8.33
0.28	2.15	0.16	1.23	0.43	3.30	0.34	2.61	0.20	1.53	0.50	3.82

					Peas and Beans						
1.47	6.59	0.34	1.52	1.20	5.38	0.80	3.58	0.23	1.03	1.18	5.29
1.52	6.30	0.28	1.16	1.25	5.18	1.04	4.31	0.32	1.33	1.34	5.56
1.60	7.08	0.26	1.15	1.20	5.31	1.09	4.82	0.32	1.42	1.23	5.54
2.26	6.47	0.48	1.37	1.67	4.77	1.29	3.68	0.56	1.60	1.89	5.39
2.68	6.73	0.62	1.57	2.07	5.20	1.57	3.96	0.56	1.42	2.06	5.18
2.64	6.28	0.54	1.28	2.14	5.08	1.57	3.73	0.56	1.33	2.30	5.46
					Miscellaneous						
1.08	5.55	0.30	1.54	0.99	5.08	0.90	4.62	0.41	2.11	1.10	5.65
1.63	3.08	0.40	0.76	1.04	1.96	1.49	2.81	0.31	0.59	1.67	3.15
2.08	4.51	0.45	0.98	1.67	3.62	0.95	2.06	0.40	0.87	1.20	2.60
0.20	3.44	0.08	1.37	0.18	3.09	0.18	3.09	0.06	1.03	0.25	4.30
0.36	3.40	0.28	2.64	0.33	3.12	0.33	3.12	0.10	0.94	0.50	4.72
1.08	4.15	0.51	1.96	1.25	4.81	1.06	4.08	0.52	2.00	1.43	5.50
2.07	3.26	1.23	1.94	2.91	4.58	1.65	2.60	1.02	1.61	3.05	4.80
1.56	5.06	0.63	2.06	1.32	4.28	0.72	2.34	0.46	1.49	1.30	4.21
0.076	1.96	0.08	2.06	0.20	5.15	0.18	4.64	0.069	1.78	0.25	6.44

(3) Plasma Amino Acid Method.—Plasma amino acid methods were developed following the automation of amino acid analysis. This method involves intubation feeding of a slurry of the diet to either fasted chicks or rats. If properly conducted, the results are accurate when compared to the amino acid content, growth rate and nitrogen balance. When chicks are fasted, the lysine content of the plasma increases to approximately 4 to 5 times the normal (Table 10.5). When the fasted chicks are refed, the time required for the plasma lysine to return to normal is approximately 24 hr; however, the level at 6 hr is a good index of protein quality, because the percentage decrease is greatest for the sixth period and it also allows the refeeding test to be started and completed within 1 day. The plasma lysine content is generally determined by a microbiological method. Chemical tests can be performed for lysine but the microbiological method is rapid and accurate.

TABLE 10.5. LYSINE CONCENTRATION IN PLASMA OF 21-DAY-OLD CHICKS SUBJECT TO DIFFERENT DIETARY TREATMENTS

	mg/100 ml Plasma
Full fed	4.86
Fasted 24 hr	20.12
Refed 1 hr	18.32
Refed 2 hr	16.04
Refed 3 hr	15.19
Refed 4 hr	13.55
Refed 5 hr	11.78
Refed 6 hr	9.56
Refed 8 hr	7.09
Refed 24 hr	4.35

Source: Ponton (1972).

(4) Rate of Amino Acid Availability.—The chick's body does not store amino acids; therefore, the amino acids required for protein synthesis must be available as needed. If the release of some amino acids from the proteins being digested does not occur in proper schedule for body protein synthesis, the amino-acid oxidation enzyme system will deaminize the amino acids. An example of this is the reduced release of methionine and lysine in uncooked soybean meal. These amino acids are released so slowly from the raw soybeans that they are oxidized, and hence, actually unavailable to the body for protein synthesis.

Rate of amino acid availability can be determined by selective-enzymes digestion for only a limited period of time—30 to 60 min.

(5) Growth Response.—The chick growth response is a rapid and desirable method for protein evaluation of all protein sources used in poultry rations. A semipurified ration or a corn-soybean structure can be used. In the semipurified ration, corn sugar or dextrin is often used as a source of carbohydrates. A complete vitamin and mineral mixture is also used. The proteins tested can be soybean meal or isolated soybean protein or protein replacements for these. Since the corn-soybean structure is used in formulated rations, this basal ration is desirable for development of commercial evaluation because it is simple, yet reasonably accurate. These rations are presented in Table 10.6.

TABLE 10.6. BASAL CHICK RATIONS FOR PROTEIN EVALUATION

	Natural Basal	Purified Basal
Ground yellow corn	68	—
Corn sugar or dextrin	—	70
Soybean meal (50)	20	—
Isolated soybean protein[1]	—	18
Wesson oil or soybean oil	5	5
Mineral premix	5–	5–
Vitamin premix	2	2

Source: Patrick (1960).
[1]Isolated soybean protein should be tested for growth values.

(6) Digestibility.—Digestibility of proteins is sometimes used for evaluation of purified proteins. The results do not appear as applicable for starting rations as for growing and breeding rations. The food appears to move faster through the digestive tract in the young chicken or poult (approximately 3 hr); therefore, the digestibility tests which generally use a 12-hr exposure to the enzymes is not desirable. The older birds appear to be more adaptable than the younger birds to the use of proteins which are slowly digested.

NATURE OF PROTEIN IMBALANCE

Some of the amino acids are not properly balanced as far as nutritional requirements are concerned. The imbalance may bring about a change in niacin requirement because of a tryptophan content, a poor response to glycine because of a choline deficiency, and a good response to the level of methionine because of the presence of betaine. Any factor which can exert a sparing action or become antagonistic to an amino acid will play a very definite role in protein imbalance.

A high biological-value protein contains a balance of available amino acids in proportions required by the body, and as the balance of amino acids in a protein deviates from these figures, the biological value declines. The concept of amino acid imbalance is observed in adverse effects beyond the expected decrease in protein utilization resulting from feeding low-protein diets supplemented with amino acids or unbalanced protein. The growth depression can be reversed by supplying an additional quantity of the most limiting amino acid in the diet. Often, a slight increase in the uptake of a certain amino acid can increase the requirements of others, to maintain a set growth rate when total protein intake is low. The arginine-lysine imbalance is frequently observed.

Feeding a large excess of individual amino acids can be toxic in some conditions. Toxicity of this type can be reduced or corrected by feeding the first limiting amino acid and sometimes by feeding an excess of certain water-soluble vitamins. The feeding of a toxic level of glycine or methionine can often be corrected by feeding an excess of Vitamin B_{12}.

Poultry rations use chiefly a corn-soybean mixture in which their proteins are complementary (Fig. 10.3). The 4 critical amino acids in corn and soybean meal are:

	Ground Corn[2]	Soybean Meal[2]
Methionine	2.24	1.39
Lysine	3.50	6.17
Tryptophan	1.17	1.69
Arginine	4.60	7.46

Corn and soybean meal supplement each other except for methionine. A slight deficiency of methionine could occur during rapid growth, feather development and egg production. This deficiency is corrected by use of synthetic methionine or methionine hydroxy analog. The decision to add a synthetic amino acid to a ration involves an understanding of the relationships of the added amino acid to the limiting amino acids. An excess of free arginine can produce a retardation of growth. However, this can often be corrected by the addition of lysine or methionine.

LEVEL OF ENERGY

A chicken will eat until its energy needs are fulfilled. Therefore, if the diet is high in fiber and low in energy and protein, the chicken will continue eating, trying to meet its energy needs. In this situation, it probably will obtain enough protein. If, however, the diet is high in energy and low in protein, the energy needs of the chicks will soon be met and they will stop eating before consuming enough protein.

The protein requirement is often expressed as calories per unit of protein (Table 10.7). A standard of this type is necessary because of the development and use of the high-energy ration which recognizes that poultry eat until their energy needs are satisfied.

GROWTH RATE

A young chicken in the most rapid growth stage will have the highest total protein requirement and also the highest essential amino acid

---

[2]Expressed as percent of crude protein.

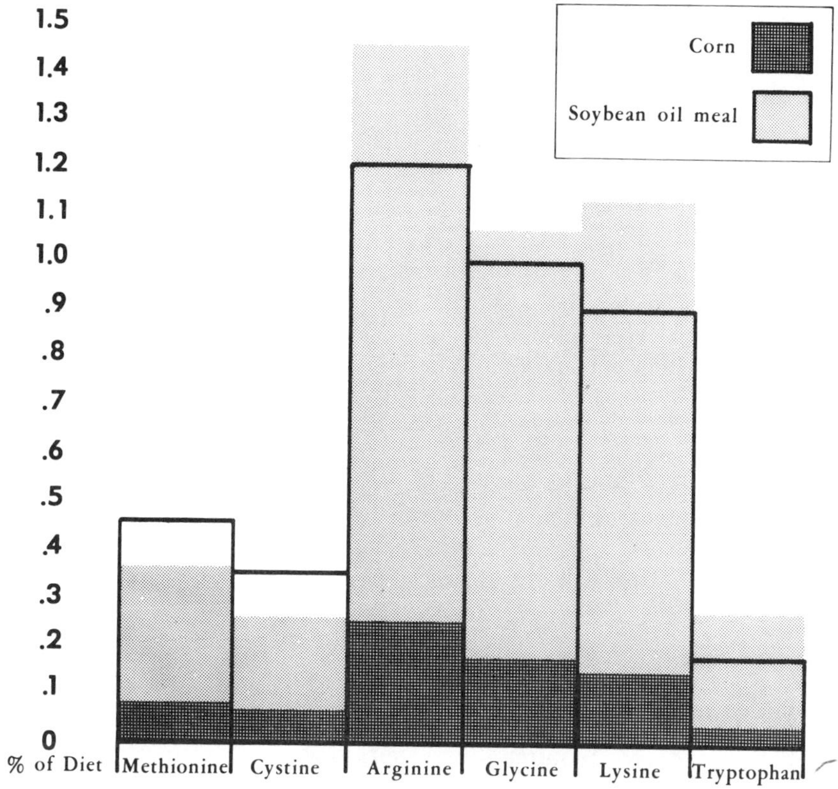

FIG. 10.3. CORN AND SOYBEAN OIL MEAL: PERCENTAGE OF DIET

TABLE 10.7. SUGGESTED CAL.–PROTEIN RATIOS[1] FOR POULTRY

	Metabolizable	Productive
Broiler starter (0–5 weeks)	60–65	42–45
Broiler finisher (after 5 weeks)	69–75	48–53
Chick starter (0–8 weeks)	63–66	44–46
Chick grower (8–18 weeks)	77–86	54–60
Layer, 50% production	91–94	64–66
Layer, 70% production	84–87	59–61
Layer, 90% production	80–83	56–58
Turkey starter (0–6 weeks)	43–46	30–32
Turkey grower (6–12 weeks)	57–60	40–42
Turkey grower (12–18 weeks)	71–74	50–52
Turkey grower (18–24 weeks)	85–100	60–70
Turkey finisher	100–105	70–72
Turkey breeders	86–91	56–60

[1]C/P ratio is defined as $\frac{\text{kcal per lb}}{\text{\% protein}}$.

requirement. If a protein is found to have a high biological value for the young chick, it will also have a high biological value for the growing, laying and breeding birds.

SPARING ACTION FACTOR

There are many dietary factors which are related to amino acid metabolism and amino acid requirements; therefore, their presence or absence can influence the biological value of the protein. Tryptophan is converted into niacin when needed; therefore, if the diet were low in both tryptophan and niacin, then the conversion to niacin would produce a deficiency of tryptophan. Choline and methionine are labile methy donors. Betaine can also provide methyl groups. This relationship can influence the demand upon methionine.

DISEASES

Diseases, such as Coccidiosis, can influence absorption of the amino acids, vitamins and minerals from the small intestine. This impairment in absorption will decrease the biological value. Diseases which elevate body temperature will also decrease protein values.

PROBABLE CRITICAL ESSENTIAL AMINO ACIDS

Methionine

Methionine is the required sulfur-containing amino acid. Two other sulfur-containing amino acids are cystine and cysteine in which the animal can convert methionine into cysteine. The chick also converts methionine hydroxy analogue to methionine.

The sulfur-amino acid requirements of poultry are high because:

(1) Feathers are high in sulfur-amino acids.
(2) Methionine provides active methyl groups.
(3) The inorganic sulfur requirement is provided chiefly by sulfur-amino acid metabolism.
(4) Sulfur-amino acids may help remove toxins.
(5) Methionine often is present in an unavailable form; therefore, the requirement from natural sources is increased.
(6) Methionine can be converted to taurine which is an essential compound in the body.

Forms of Methionine and Hydroxy Methionine Metabolic Use

Methionine is present in both the D and L forms, but only the L form is of great importance in intercellular proteins. There are few substances with such universal diversity in its use as methionine. It is necessary for growth, repair and metabolism of all tissues, and for reproduction, and is active in taurine synthesis. The general metabolic use of methionine is presented in Fig. 10.4.

Methionine can be formed in the chicken from homocysteine or from methionine hydroxy analogue. The chemical structures of each are:

$H - S - CH_2 - CH_2 - CH - NH_2 - COOH$ Homocysteine

$CH_3 - S - CH_2 - CH_2 - CH - OH - COOH$ Hydroxy methionine analogue

$CH_3 - S - CH_2 - CH_2 - CH - NH_2 - COOH$ Methionine

Methionine hydroxy analog is used in poultry rations instead of methionine because it is often cheaper, and the steric configuration does not prevent poultry from converting it into the L-configuration of L-methionine.

The labile methyl group and the carbon-sulfur linkage account for much of the versatility of methionine. These, along with the use of carbons 2-3-4 and sulfur in synthesis of taurine, place methionine as one of the most important amino acids.

Methionine is the only amino acid added in the synthetic form to poultry rations.

Lysine

The second synthetic amino acid to be added to some poultry rations is lysine. Lysine has many uses in the body, and it accumulates in the blood plasma when the ration is deficient in protein. It also accumulates in the plasma when the diet is adequate or fortified in this amino acid.

The Barrel Stave Analogy

All amino acids required for tissue synthesis must be present at the same time and in the proper amount at the site of the biological transformation into protein. A deficiency of any one essential amino acid means that the others remaining are wasted. Because of this, one can no longer consider simply the protein level of the diet in the determination of the nutritional adequacy. This is illustrated by the use of a barrel stave analogy. On the left there is a most unusual barrel with staves of irregular lengths. Some stick out above the rim; others are shorter than the rim of the barrel. It is readily apparent that the amount of water that can be held by the barrel is limited by the shortest stave, and that the rest of the staves serve no purpose. Suppose each barrel stave represented a particular amino acid and the rim or top of the barrel the amino acid requirements of the animal. Some proteins have less or more of a particular amino acid to satisfy requirements. Somewhat the same situation holds in the feeding of unbalanced proteins as with the water-

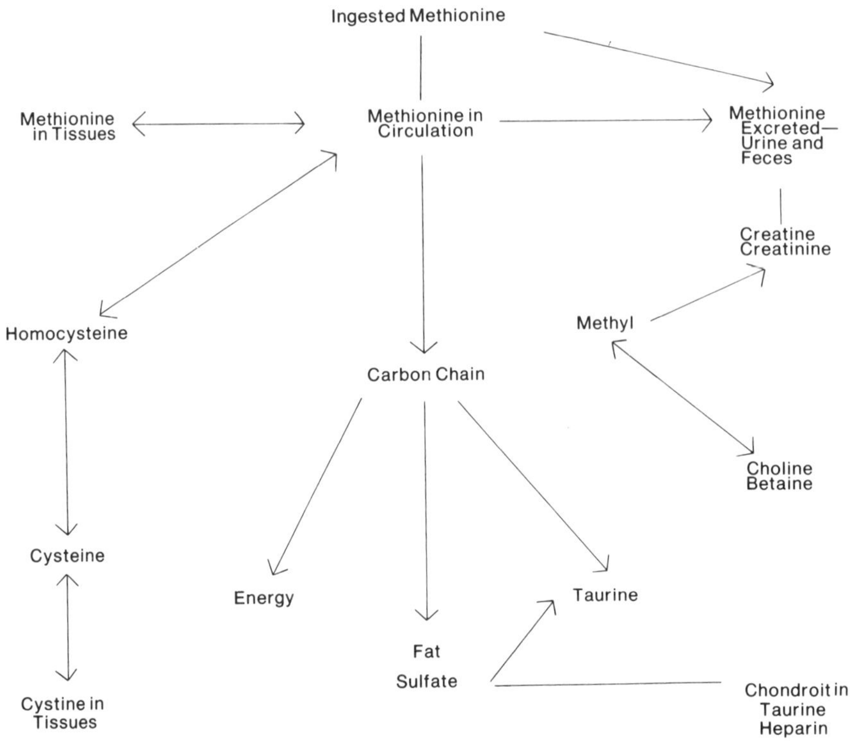

FIG. 10.4. METHIONINE METABOLISM

holding capacity of the barrel. With all good analogies, there must be a moral and the moral of our barrel story, therefore, would be that the barrel is no larger than its smallest stave. The amount of protein that can be utilized by the animal for the tissue synthesis is determined by the first limiting amino acid with the remaining amino acids being lost to the animal. A complete deficiency of an essential amino acid would be represented by a missing stave and a barrel in which no water could be held. Such protein, if fed as the sole source of amino acid, is of no value to the animal because no protein synthesis can occur. The ideal barrel has all the staves of uniform length with no deficiency of staves. The ultimate in amino acid supplementation is when all amino acids present in the feedstuff are utilized with no surpluses for the animal to use up. An excess of amino acids in the diet is not only wasteful of the amino acids, but also of energy in that such energy could be more efficiently and economically obtained from fats or carbohydrates (Fig. 10.2).

Methionine hydroxy analogue (MHA) is often used as a supplement to high-soybean meal rations because of lower costs. Methionine gives greater improvement in both growth and feed efficiency to a soybean meal ration than does MHA. However, when urea or diammonium citrate is combined with MHA, the growth response and feed efficiency is similar to DL-methionine as a supplement to a corn-soybean ration. A summary showing the comparative value of nitrogen sources as supplements to methionine and MHA is presented in Tables 10.8 and 10.9. Amino acids, except lysine and threonine, become involved in the chemical reaction known as transamination. (Transamination is the transfer of the amino group from 1 amino acid to what is originally an alpha-keto acid, forming a new amino acid and keto acid, without the appearance of ammonia in the free state.) Vitamin B_6 (pyridoxine) plays a very important role in transamination. Lysine is the amino acid forming the linkage in connective tissue giving elasticity to the arteries.

TABLE 10.8. RELATIVE EFFICIENCIES OF METHIONINE SOURCES[1]

	Relative Efficiency
L-Methionine	100%
DL-Methionine	97%
D-Methionine	82%
Hydroxy Analog	98%

[1]The results on the comparative efficiency of the isomers of methionine demonstrate that using a commercial-type ration makes only a slight difference.

TABLE 10.9. FEED EFFICIENCY AND WEIGHT GAINS AS AFFECTED BY THE ADDITION OF UREA AND DIAMMONIUM CITRATE TO METHIONINE HYDROXY ANALOGUE AND DL-METHIONINE SUPPLEMENTED 12% PROTEIN FEEDS

Modification of Basal Ration	Average Gain in 5 Weeks (g)	g Feed per g Gain (g)
DL-Methionine, 0.288%	123	3.25
DL-Methionine, 0.288% + urea, 0.06%	110	3.27
DL-Methionine, 0.228% + diammonium citrate, 0.226%	131	3.34
MHA, 0.347%	90	4.25
MHA, 0.347% + urea, 0.06%	120	3.20
MHA, 0.347% + diammonium citrate, 0.226%	122	3.37

Source: Summary of data from Sullivan and Bird (1957).

The proteins from corn (corn gluten) or wheat (wheat gluten) are deficient in lysine. An improved corn, recognized as Opaque 2, is a high-lysine variety. This variety of corn is being grown in Mexico and may become important in the United States when the yield is improved.

Lysine is unstable to heat. The amino group on the sixth carbon is readily removed by heat or acid treatment.

Cereal grains used in breakfast foods and bread are often supplemented with lysine. Soybeans are an excellent source of lysine.

Tryptophan

The third amino acid that may be deficient in poultry rations is tryptophan. It is deficient in corn gluten, peanut meal, and some productions of meat scraps or tankage.

Tryptophan is a precursor of nicotinic acid. Vitamin B_6 is a cofactor in this reaction. Poultry rations are supplemented with nicotinic acid, and this spares tryptophan.

PROVIDING PROTEIN AND ESSENTIAL AMINO ACIDS FOR POULTRY

Protein supplements can be divided into groups based on their origin and method of production. These groups are:

(1) Animal Proteins
(2) Vegetable Proteins
(3) Microbiological Ingredients
(4) Synthetic Amino Acids
(5) Purified Proteins

Animal proteins are generally more nutritionally complete in the required amino acids and minerals than are vegetable proteins. In general, all vegetable proteins are deficient in one or more essential amino acids and are also inadequate in minerals.

Only the amino acids—methionine, lysine, tryptophan and arginine—need particular attention in practical rations. Most poultry starting rations are supplemented with synthetic methionine or methionine hydroxy analog.

CORRECTING AMINO ACID DEFICIENCIES

There are 3 ways of correcting an amino acid deficiency:

(1) Add synthetic amino acids.
(2) Feed a higher percentage of protein in the diet.
(3) Use a combination of proteins so that the shortcomings of 1 protein is compensated by an abundance of a certain required amino acid in the other proteins.

11

Nutrition of Water–Soluble Vitamins

A great many feeding experiments with chickens pointed to the fact that substances other than minerals, water, carbohydrates, protein and fat were required for adequate nutrition. This led to the development of basic approaches to nutrition. Fundamental studies in nutrition using extracts from yeasts, liver meal and milk resulted in the establishment of a class of substances called "accessary food factors."

Rice polishings were found to cure polyneurities, a primary symptom of Beriberi, in chickens raised on a polished-rice diet. When polyneuritic chickens were fed an extract prepared from rice polishings, the deficiency symptoms were dramatically alleviated. This observation was reported by the Dutch physician, Eijkman, in Java, as early as 1897. Funk, in 1912, isolated the active substance from rice polishings and observed that small amounts of the crystalline substance cured Beriberi in pigeons. A similar substance was isolated from yeast. Since the substance was apparently essential for life, and chemical tests indicated that it was an amine, the name "vitamine" was coined. The term, without the "e," is used today to designate a large group of "accessory food factors."

Beginning in 1912, the experimental approach to nutritional problems was becoming significant, and fundamental nutrition became important. A great deal of effort was expended toward isolating purified factors and assigning names using the word, vitamin. For those unidentified growth factors that were not purified, the term U.G.F. was often used. Vitamins were divided into 2 major groups: (1) Fat-soluble and (2) Water-soluble Vitamins.

Today, although few attempts are directed toward defining vitamins, it is recognized that they are organic compounds, occurring in natural foods, either as such, or as utilizable precursors which are required in minute quantities for normal life processes, such as growth, maintenance and reproduction (egg production and hatchability).

With the exception of Vitamin C (ascorbic acid), poultry require all the known vitamins. Today, all known vitamins have been characterized and synthesized. After more than 60 years of vitamin research, many functions of some of these substances in the chain of metabolic events is understood yet we cannot unequivocally state that the complete function of any one of the vitamins in the body of the chicken is known. Many of the vitamins have been shown to serve as coenzymes in fundamental metabolic reactions, and the fundamental functional activity of all vitamins may ultimately be found to be due to their influence on, or participation in, enzyme systems.

For our purposes, vitamins will be divided into fat- and water-soluble groups. Unrecognized growth factors will be identified as functionary substances.

Many vitamins have close synergism with other vitamins. These are outlined in Table 11.1. Water-soluble vitamins demonstrate more synergism than fat-soluble vitamins.

A vitamin can be known by several names. A glossary of the B-Vitamins is shown in Table 11.2. In general, after a vitamin is iolated and synthesized, the chemical name engenders a renaming of the factor using some chemical identification.

VITAMIN DEFICIENCY SYMPTOMS

Investigators have often observed that deficiency symptoms reported vary from 1 experiment station to another. A number of factors contribute to this variation with the following being the most important.

Breeder Rations: Breeder rations fortified in the vitamins will produce hatching eggs that are fortified in those vitamins which are readily stored in the egg.

The chick's depletion time and deficiency symptoms will vary with the fortification of the egg. If the hatching egg contains only enough of the vitamins for hatchability, the newly-hatched chicks, if placed on a vitamin-deficient ration, will probably die before developing or showing the expected deficiency symptoms.

The degree of deficiency of the starting ration will influence the chicks' depletion time and the deficiency symptoms. An example of this is observed with a riboflavin deficiency. If the starting ration is free of riboflavin, the chicks will not grow and die. If the starting ration contains about 50% of the requirements, the legs will be paralyzed and some growth will be observed. A 30% deficiency will result in curled toes, a small amount of paralysis and slow growth. If growth cannot be produced on the deficient diet, the expected deficiency symptoms will probably not develop because of death.

Newly-hatched turkey poults from eggs produced by hens whose riboflavin intake was enough to provide the minimum requirements for

TABLE 11.1. SYNERGISM (+) AMONG THE VITAMINS

Vitamins	A	B_1	Ribo[1]	B_6	B_{12}	D	E	K	Biotin	F.A.	Niacin	P.A.
A	—	—	+	—	+	—	±	—	—	—	—	—
B_1	—	—	+	+	+	—	—	—	—	—	+	+
Ribo	+	+	—	+	+	—	—	—	+	+	+	+
B_6	—	+	+	—	—	—	+	—	+	+	+	—
B_{12}	+	+	+	—	—	—	+	—	+	+	+	+
D	—	—	—	—	—	—	—	—	—	—	+	—
E	—	—	—	+	+	—	—	+	—	+	—	—
K	—	—	—	—	—	—	+	—	—	—	—	—
Biotin	—	—	+	+	+	—	—	—	—	+	—	+
F.A.[2]	—	—	+	+	+	—	+	—	+	—	+	+
Niacin	—	+	+	+	+	+	—	—	—	+	—	+
P.A.[3]	—	+	+	—	+	—	—	—	+	+	+	—

Source: With modification, from Kutsky (1973).
[1]Riboflavin.
[2]Folic acid.
[3]Pantothenic acid.

TABLE 11.2. GLOSSARY OF THE B VITAMINS

Name	Synonym or Chemical Name
Adenylic acid	Adenine nucleotide Vitamin B_6
Adermin	Vitamin B_6
Aneurin	Vitamin B_1
Antiacrodynia rat factor	Vitamin B_6
Antialopecia mouse factor	Inositol
Antiberiberi vitamin	Vitamin B_1
Anticanitic factor	*p*-Aminobenzoic acid
Antidermatitis chick factor	Pantothenic acid
Antidermatitis rat factor	Vitamin B_6, pyridoxine
Antidermatosis vitamin	Pantothenic acid
Anti-egg white injury factor	Biotin
Anti-gizzard erosion factor	Possibly Vitamin B_3
Anti-gray-hair factor	*p*-Aminobenzoic acid
Antineuritic vitamin	Vitamin B_1
Antipellagra (chick) factor	Pantothenic acid
Antipellagra factor	Nicotinic acid Nicotinamide
Anti-spectacled-eye factor	May be inositol, biotin or pantothenic acid
Bios 1	Inositol
Bios 2	Biotin
Bios 2*b*	Biotin
Biotin	Vitamin H
Chick A-P factor	Choline
Choline	Beta-hydroxyethyltrimethyl ammonium hydroxide
Chromotrichia factor	*p*-Aminobenzoic acid
Coenzyme-R	Biotin
Complimentary factor	Vitamin B_6
Factor 1	Vitamin B_6
Factor 2 from liver	Pantothenic acid
Factor B_x	*p*-Aminobenzoic acid
Factor S	Biotin
Factor U	Stokstad-Manning factor
Factor W	Biotin
Factor X	Biotin
Factor Y	Vitamin B_6
Filtrate factor	Pantothenic acid
Filtrate factor 1	Possibly Vitamin B_6, nicotinic acid Peters factor
Filtrate factor 11	Pantothenic acid
Folic acid	May be Vitamin B_c
Inosite	Inositol
Inositol	Hexahydroxycyclohexane
Lactoflavin	Vitamin B_2
Liver filtrate factor	Pantothenic acid
Niacin	Nicotinic acid
Niacin amide	Nicotinamide
Niamide	Nicotinamide
Oryzamin	Vitamin B_1
Ovaflavin	Vitamin B_2
Panothen	Pantothenic acid
Pellagra-preventative factor	Nicotinic acid Nicotinamide
Polyneuramin	Vitamin B_1
P-P factor	Nicotinic acid Nicotinamide
Pyridoxine	Vitamin B_6
Pellagramin	Nicotinic acid Nicotinamide (+)-alpha, y-dihydroxy,-B,B-dimethyl buryryl-B-alanide
Riboflavin	Vitamin B_2

H-riboflavin	Vitamin B_2
Skin factor	Biotin
Thiamin	Vitamin B_1
Torulin	Vitamin B_1
Trichochromogenic factor	*p*-Aminobenzoic acid
Vitamin B_1	4-Methyl-5-B-hydroxyethyl-N- 2-methyl-4-aminopyrimidyl-(5)-methyl-thiazolium chloride hydrochloride, Thiamin
Vitamin B_2	6, 7-Dimethyl-9-(d-l-ribityl)-isoalloxazine, Riboflavin
Vitamin B_3	Probably pantothenic acid Williams-Waterman factor
Vitamin B_4	Reader factor; may be arginine or glycine
Vitamin B_5	Peters factor May be Vitamin B_4 or nicotinic acid
Vitamin B_6	3-Hydroxy-4, 5-bix(hydroxymethyl)-2-methylpyridine; pyridoxine
Vitamin B_7	Vitamin I
Vitamin B_8	Adenylic acid
Vitamin B_c	May be folic acid
Vitamin B_p	Antiperosis factor
Vitamin B_w	Biotin
Vitamin B_x	Pantothenic acid
Vitamin B_x	*p*-aminobenzoic acid
Vitamin F	Vitamin B_1
Vitamin G	Vitamin B_2
Vitamin G	Pantothenic acid
Vitamin H	Biotin
Vitamin H	Vitamin B_6
Vitamin I	Vitamin B_7
Vitamin L_1) Vitamin L_2)	Lactation vitamins
Vitamin M	Day's factor
Yeast eluate factor	Vitamin B_4
Yeast filtrate factor	Pantothenic acid

hatchability, had high mortality the first 24 to 36 hr of life. Hens receiving diets border-line deficient in manganese can produce eggs which hatch, yet the chicks become perotic within 3 to 7 days.

STARTING RATION

A well-fortified breeding ration can provide the newly-hatched chick with enough vitamins for several days' survival when placed on deficient rations. Sometimes, a deficiency in the breeding ration may result in an abnormal digestive tract as well as other organ changes which prevent the young chick from utilizing the essential elements in the starting ration.

Protein level and amino acid content of the starting rations can promote a deficiency of several vitamins. The Vitamin B_{12} requirements are increased when high levels of soybean meal is used in the starting feeds. Methionine intake can influence choline and Vitamin B_{12} requirements. The feeding of low-tryptophan diets increases the niacin requirement.

Environmental temperature and fat content of the ration affects the Vitamin B_{12} requirements. The pantothenic acid requirement increases when chicks are subjected to low or sudden changes in temperatures. Some investigators have called Vitamin B_{12} and pantothenic acid, thermal factors.

Diseases, such as Coccidiosis, increase the vitamin requirements and reduce zinc absorption.

To summarize, any factor which influences depletion time, absorption or requirements will also cause a difference in the overall deficiency syndrome.

WATER-SOLUBLE VITAMINS

The Vitamin B complex belongs to a group of vitamins that are water-soluble. This is important from a nutritional standpoint because the water-soluble vitamins are not stored to any extent in the body. Large excesses over and above what is required by the chicken each day are excreted. This is in contrast to the fat-soluble vitamins (A, D, E, and K) which can be stored in the body. Because the water-soluble vitamins are not stored, a constant supply must be provided in the daily diet.

DISCOVERY OF THE B VITAMINS

For many generations, certain human population groups of the world, generally in areas where rice is the major food, have suffered from a disease called beriberi. This disease is especially common in areas where polished rice is the staple food.

Takaki, a Japanese medical officer, was the first to suggest that beriberi was related to diet. He observed, between 1878 and 1883, that the instance of beriberi among Japanese naval men was greatly reduced when the diet contained meat, milk or vegetables. However, Takaki thought that beriberi was related to the protein content of the diet.

Some 14 years later, in 1897, a disease similar to beriberi was observed in poultry that consumed hospital food scraps. Research showed that the disease was nutritional and that the outer layers of rice contained a protective material missing from polished rice. Other investigators interpreted the results of the poultry experiment to mean that some essential nutrient was absent from the polished rice diet. In 1911, the term "Vitamin" was coined for the biologically active substance in the rice's outer layers (polishings) which could prevent beriberi. Later, the term, "water-soluble B" was used until it was established that this was not a single compound, but rather, a group of compounds. The group of compounds was referred to as the Vitamin B complex. Vitamin B_1 (thiamin) was found to be responsible for preventing or curing beriberi. Today, nutritionists feel that all the members of the Vitamin B complex have been identified. Poultry has played a very prominent role in identifying and characterizing the nutritional properties of each vitamin. The water-soluble vitamins required by poultry are:

Thiamin (Vitamin B_1)
Riboflavin
Pyridoxin (Vitamin B_6)
Cyanocobalamin (Vitamin B_{12})
Niacin
Pantothenic acid
Folic acid
Choline
Biotin
Inosital

Vitamin premixes often contain more of the vitamins than are required. The water-soluble vitamins generally added to premixes are:

Riboflavin
Niacin
Pantothenic acid
Choline
Vitamin B_{12}

Vitamin B_1 (Thiamin)

Vitamin B_1 or thiamin was the first water-soluble vitamin found to be required by poultry. It is a rather simple chemical compound, being composed of a pyrimidine and a thiazole ring. It is very soluble in water and fairly soluble in 95% alcohol. Thiamin is unstable in alkaline solutions and stable in acid solutions.

Thiamin is phophorylated to form a coenzyme, thiamin pyrophosphate. This coenzyme is also know an cocarboxylase and has 3 components: (1) thiamin, (2) Mg^{++} and (3) the apoenzyme protein. Thiamin pyrophosphate functions in several critical metabolic reactions in the chick.

Nutrition.—Thiamin-deficient chicks have an accumulation of pyruvates in their tissues. A nervous disorder and poor appetite are observed because of the toxicity of pyruvic acid.

Sources.—Natural feedstuffs are good sources of thiamin. Synthetic thiamin is available.

Assay.—Thiamin can be easily determined by the microbiological or the chemical method. Most control laboratories use the chemical method.

Requirements.—The natural feedstuffs have adequate thiamin. A deficiency can only be produced by feeding a highly purified diet. The requirements are about 2 mg per kg of ration.

Feed Formulation.—Today's feed ingredients are adequate in thiamin; therefore, there is no special need to provide supplements of this factor. When the rations use a large number of by-products obtained from solvent extraction processes, the thiamin content should be investigated and placed under surveillance.

Thiamin is easily destroyed by some sulfur salts that are used in protein preparations. The sulfate should be removed from the protein preparations by dialysis if present in large amounts.

Riboflavin

Riboflavin is the compound in milk and egg albumen which imparts the yellow fluorescent color. It is of major importance in poultry nutrition because it is generally inadequate in rations formulated of common natural feedstuffs.

The flavins are widely distributed in nature and a number of analogues exist which have vitamin activity. However, only riboflavin is synthesized and used in poultry rations.

The enzymes containing riboflavin are called flavoproteins. Two coenzymes are known: (1) flavin monucleotide (FMN) and (2) flavin adenine dinucleotide (FAD). These coenzymes serve in the electron transport system leading to the formation of the high-energy compound, ATP.

Nutrition.—A riboflavin deficiency results in retarded growth, curled toes, leg paralysis, poor hatchability and high mortality.

The time required to deplete or replete a breeding flock of chickens or turkeys is approximately 14 days. Some riboflavin is stored in the liver but the daily uptake should be adequate and not dependent upon body storage.

The requirement for riboflavin is well-established and the general ration formulation program adds enough supplementary riboflavin above that present in the natural feedstuffs to provide for the needs of poultry. The riboflavin requirement of chicks is genetically controlled. A specific problem in riboflavin utilization by chickens involves a riboflavin protein transport. Mutant strains of hens had no riboflavin-binding protein in the blood, egg yolk or egg albumen, while in normal hens, the protein was present. Fertile eggs from the mutant strain did not hatch unless riboflavin was injected into the egg. The commonly used natural feedstuffs, such as ground corn and soybean meal, supply about 30% of the requirements. Tables 35.1 to 35.8 present the recommended levels to use in sample rations.

Synthetic riboflavin, yeasts, milk, fermentation products and liver meal are the best sources of this factor.

Feed Formulation.—Riboflavin is generally secured by using vitamin premixes, fermentation products, yeasts or milk. It is stable in the mixed feed, but will be destroyed if exposed to strong light. No ration should be formulated without adding a riboflavin concentrate.

Vitamin B_{12} (Cobalamin)

Vitamin B_{12} (Cobalamin) is the active factor isolated from the Animal Protein Factor (APF) concentrates. This was the last vitamin discovered and synthesized.

Vitamin B_{12} is red-colored, due to the cobalt. The cobalt accounts for 4.35% of the molecular weight. It has a coordination number of 6 and the vitamin has a net charge of 1.

Vitamin B_{12}, although the largest molecule of the vitamins, mol. wt. 1355, is stable. There are several B_{12} Vitamins, but only cobalamin has been developed commercially.

Vitamin B_{12} forms several coenzymes, and functions in metabolic processes resulting in methylation, hydrogen transfer and hemoglobin formation.

It is widely used in human medicine and is used in all rations for simple-stomach animals, including poultry.

Nutrition.—Vitamin B_{12} was first introduced to the field of Poultry Nutrition as the Animal Protein Factor (APF), the cow-manure factor and the fish-soluble factor. It became evident that a factor was needed when the corn-soybean meal ration was used and when fish solubles were observed to enhance the growth in broilers. Natural Vitamin B_{12} is found only in animal tissues, animal feces and rumen fluids.

The presence of Vitamin B_{12} in cow manure and poultry feces led to the discovery that the intestinal or rumen microorganisms synthesize Vitamin B_{12}, provided that cobalt is present. Also, a chicken can have a Vitamin B_{12} deficiency, yet when cobalt is present, the intestinal microorganisms will synthesize this vitamin and the chicken will excrete it in its feces. The chick can not utilize this vitamin which is produced in its lower intestines because it is synthesized in the part of the digestive tract where absorption does not occur. In the case of the ruminant, some of the Vitamin B_{12} is absorbed from the duodenum and some is excreted. Vitamin B_{12} is poorly absorbed, yet retained for long periods of time.

Chickens, rabbits, rats and hogs obtain Vitamin B_{12} by eating their feces.

One of the largest commercial Vitamin B_{12} sources is from feces. Fish solubles was one of the first sources and the A.P.F. or F.S.F. was introduced because of its presence in this feed ingredient.

The growth-promoting effect of Vitamin B_{12} on chicks receiving a corn-soybean meal ration led nutritionists to the association of methionine to this vitamin because of a methionine deficiency in soybean meal. Since methionine has been identified with choline, and choline with glycine, 4 nutrients—methionine, choline, glycine and Vitamin B_{12}—have been placed as factors requiring consideration when corn-soybean meal type rations are formulated for poultry.

Anemia, poor growth and hatchability, fatty livers and poor feathering are common deficiency symptoms. (See Fig. 11.1.)

FIG. 11.1. RHODE ISLAND RED CHICKS HATCHED FROM EGGS LAID BY HENS FED AN ALL-PLANT PROTEIN LAYER FEED DEVOID OF VITAMIN B_{12}
The larger bird was hatched from an egg injected with one-millionth of a gram of Vitamin B_{12} at 10 days of incubation. The smaller bird came from an untreated egg. Both chicks were fed a Vitamin B_{12} deficient diet from hatch through 5 weeks of age.

The requirement for Vitamin B_{12} increases with the level of fat and the use of soybean protein. The only sources of this vitamin are animal products, feces and microbiological preparations. The levels required are shown in Tables 35.1 to 35.8. The requirement for starting and broiler chicks is about 3 times that of laying hens or breeders. Turkeys have about the same requirements as layers, which is 0.003 mg per kg of complete ration. Vitamin B_{12} requirements of chicks increase when the chicks are subjected to sudden temperature changes. This vitamin is a "thermal factor," required to help young birds withstand sudden temperature changes.

Feed Formulation.—Feed manufacturers add 8 to 12 mg of Vitamin B_{12} per ton poultry feeds. They use either fish solubles or a microbiologically-derived product. (Feces is a microbiologically-derived product.)

Niacin (Nicotinic Acid)

A niacin deficiency in man was probably the most harmful vitamin deficiency in the United States. Niacin was found to cure canine blacktongue and human pellegra many years before it was found to be required by poultry.

It was discovered that niacin is an essential constituent of a coenzyme involved in hydrogen transpart. Today, coenzyme I and coenzyme II have been found to have niacin as their prosthetic group.

Its name was changed from nicotinic acid to niacin because it was frequently and erroneously associated with nicotine, the toxic substance in tobacco.

Niacin is a very stable compound in natural feedstuffs. It is present as free niacin, bound niacin, or nicotinamide in the coenzyme form. The natural feedstuffs are relatively low in this vitamin, but many of the proteins contain adequate tryptophan which can be converted to niacin.

Nutrition.—A niacin deficiency in chicks is characterized by slow growth, inflamed mouth (dark red lining), ragged feathers, scaly skin and enlarged hocks.

A niacin deficiency was first observed in laboratory experimental animals when degerminated corn meal or hominy grits along with corn gluten were used in the diets. Corn is not only deficient in niacin but also in tryptophan. Pellegra developed in Southerners because corn grits and corn meal as their staple foods did not provide adequate niacin or tryptophan.

Yeast, fish meal, fish solubles, distillers' solubles, meal from poultry by-products, rice bran, rice polishings, wheat bran and sunflower meal are good sources of available niacin. Meat scraps are a poor source when chick assays are made. Niacin is available commercially in the synthetic form.

Niacin is added to all commercial starting and growing poultry rations. Duck, pheasant, turkey and quail rations require more niacin than chick rations, but due to the low cost, an excess is added to the chick or broiler rations. A deficiency in chicks or broilers is seldom observed except under laboratory conditions.

Feed Manufacturing.—Most commercial vitamin preparations are formulated to provide 8 to 18 g niacin per ton of mixed complete starting, finishing or growing rations.

Pantothenic Acid

Pantothenic acid, also known as the chick antidermatitis factor, is widely distributed in cereals, animal proteins, vegetable proteins, waste and leafy plants. It has been synthesized and is available commercially. It is highly water-soluble.

Pantothenic acid is a part of coenzyme A. This coenzyme enters into metabolic pathways involving most classes of compounds. Its relation to the formation of Acetyl CoA and the tricarboxylic acid cycle links it to the metabolism of carbohydrates, lipids and protein. It is also very active in the synthesis of fats.

Nutrition.—Chicks deficient in this factor develop dermatitis around the beak and eyelids and between the toes. Growth, egg production, hatchability and livability are reduced when the diet is deficient. Pantothenic acid plays a role in survival to low or sudden changes in temperature. This vitamin and Vitamin B_{12} can be called thermofactors. Rations which are adequate for chick growth under good management will not support the need for withstanding sudden, chilling temperature changes. Young chicks under heat lamps for brooding are subjected to a great temperature change when the thermostat turns off the electricity. A change of as much as 20°F (11°C) often results.

Feed Manufacturing.—The requirements are presented in Tables 35.1 to 35.8. The panthothenic acid content of soybean meal, ground corn, and alfalfa meal will meet the minimum needs of the chick; therefore, an additional 6 to 8 g per ton feed is recommended for stress conditions. The synthetic material is desirable. Yeasts, molasses and grains are good sources. It is water-soluble and stable.

Biotin

Biotin was the first vitamin found to have an antivitamin. A protein in egg white called avidin was found to unite chemically with the boitin in the digestive tract rendering it unavailable to the chick or poult. Cooking or heating of the egg white denatured the avidin, thus eliminating its chemical ability to bind the biotin.

Biotin is a stable compound and a rather simple monocarboxylic acid. It is involved in metabolism of fats, carbohydrates and protein.

Nutrition.—Natural feedstuffs supply adequate biotin to the ration. Perosis, poor growth, and dermatitis around the beak, foot pads and the skin covering the toes are deficiency symptoms. It differs from a pantothenic acid deficiency in that the dermatitis that forms around the toes can result in a loss of the appendage. The skin covering the shanks of the legs becomes "scaly."

Synthetic biotin is available. Yeast and liver meal are good sources; however, it is present in all plant and animal products. Eggs are rich in biotin. Its primary function is as a coenzyme for enzymatic reactions in the addition of carbon dioxide to lengthen the carbon chain. This process is often called carbon dioxide fixation and is a very important reaction in the synthesis of fatty acids.

Feed Manufacturing.—Today, feeds do not require supplementary biotin; however, as more by-products are used, close observation must be made for a possible deficiency. Biotin is being included in some swine rations. Since it is highly involved in the synthesis of fatty acids, and poultry rations are often high in fat, it is advisable to observe the biotin content of the natural feeds used in the ration.

Folic Acid

Folic acid, an antianemia factor, was also known as Vitamin B_c, L. casei factor, and pteroylmonoglutamic acid. It was isolated from liver and yeast and found to be required by poultry.

Sources.—Liver, kidney, fish meal, yeasts, intestinal bacteria and most animal proteins are good sources. It is present in plant leaves and seeds in an unavailable form.

Function.—Folic acid is involved in the structure of several enzymes, one of which is folicin acid (citrovorum factor). These enzyme systems are found in the liver in which reductases, transferases and synthetases are among those involved in purine, pyrimidine, choline and methionine synthesis. Folic acid plays an important role in cell mitosis and red blood cell growth. When absent in the diet, red blood cell maturity is reduced and the mature cells grow into larger cells than are normal in chicks. This results in the development of macrocytic anemia.

Natural feed ingredients generally provide enough folic acid, but when sulfa drugs are fed, a deficiency can occur. Sulfa drugs can reduce intestinal synthesis of this vitamin.

Pyridoxine (Vitamin B_6)

Vitamin B_6 (Pyridoxine) is a vitamin involved in protein, carbohydrate and lipid metabolism. It functions as a coenzyme in many phases of amino acid metabolism. One of the most common reactions with amino acids is transamination and deamination. A pyridoxine deficiency in chicks results in a nervous disorder and sudden death.

Most grains and protein concentrates are good sources of pyridoxine. Cereal grains, soybean meal, yeast and animal proteins are good sources. The average ration is adequate in this vitamin. So far, deficiencies occur only in laboratory studies.

Three closely-related compounds—pyridoxal, pyridoxol and pyridoxamine—form the group known as Vitamin B_6. They are stable and water-soluble.

Choline

Many investigators do not consider choline a vitamin because of the large amount required—approximately one-tenth to fifteen-hundredth %. Choline is required by poultry, and unlike 4-footed animals, can only be synthesized in very limited amounts in the body cells of chicks.

Choline is a constituent of phospholipids. Its lipotropic action is due to the formation of phosphatidal cholines (lecithins) and the subsequent transport of fatty acids from liver cells. Choline is also a source of labile methyl and a constituent of acetyl-choline, a neurohormone. Poultry have rather high requirements for methyl groups which apparently are not met by their regular metabolic reactions.

Requirements.—Poultry nutrition evaluates choline as an antiperosis factor. Although the diet is adequate in glycine, chicks will readily develop perosis on choline-deficient rations. Choline also plays a role in egg production and growth.

Poultry requires between 0.1 and 0.15% choline. Turkeys, ducks and pheasants have a higher requirement than chickens.

Sources.—Choline is found in soybean meal and animal protein sources. Commercially-synthetic choline is used to supplement feeds. Most commercial choline preparations contain 50 to 70% choline carried on distillers' grains. As a pure chemical compound, choline is a strong alkaline, hydroscopic substance. It is relatively stable in feeds. Some feed manufacturers add betaine, which exerts a sparing action on choline. Beet sugar and molasses are good sources of betaine.

Inositol

Inositols are cyclic alcohols (cyclohexanehexols) occurring naturally in both plants and animals. Because of the presence of several hydroxyl groups in the molecule, they are related chemically to the sugars.

The phosphoric acid esters of inositol are the most common inositol derivatives. The hexaphosphate ester, called phytic acid, is most heavily concentrated in seeds and grains where it accounts for 86% of the total phosphorus. Inositol in soybeans represents 40% of the total phosphorus. Phytin phosphorus is considered relatively unavailable to poultry. Also, phytic acid can form chelates with calcium and zinc, rendering them unavailable to poultry. Sesame seed and soybean meal are rich in phytic acid. Therefore, rations high in these ingredients plus high dietary calcium, bring about the formation of an unavailable zinc chelate in both poultry and swine.

Inositol is found in many of the vital organs of the animal body, being richest in the heart, brain, stomach, kidney, spleen and liver, chiefly in the cytoplasm of the cell. It is found in most biological materials (Table 11.3).

Although inositol has not been thoroughly investigated in poultry, several fundamental observations have been reported. In animals, it is considered a lipotropic agent reducing the fat in the liver. Its physiological action is apparently directly related to the function of phospholipids.

Myo-inositol has been found essential for the survival and growth of cultured human cells.

An inositol deficiency in poultry is difficult to achieve because of its widespread nature and because the whole animal can synthesize inositol. In chickens, this substance has been shown in some dietary conditions to possess growth-promoting properties.

Inositol is a factor of primary importance to the growth of yeast, and is generally accepted as a member of the Vitamin B complex.

Ascorbic Acid (Vitamin C)

Nutritionally, poultry does not require dietary Vitamin C. However, biochemically, Vitamin C is synthesized in the embryo and in the individual. It is found concentrated in embryonic or new bone and apparently plays a very important role in bone development.

Vitamin C was not found to influence egg production or chick growth. Although it is unstable and easily oxidized, it plays a definite role in converting chromium from a valence of 3 to a valence of 6 in the blood.

Chromium 6 is readily excreted. It may also be found to function in stability of fat in the animal.

TABLE 11.3. THE INOSITOL CONTENT OF BIOLOGICAL MATERIALS

Material	Inositol Content, Gamma per g of Fresh Tissue
Cereals	
Wheat germ	6000
Whole wheat	1700
Flour, white	830
Flour, whole wheat	1105
Oats	1000
Corn	500
Vegetables	
Peas, English, green	1620
Cabbage	950
Potatoes, sweet	660
Potatoes, white	290
Lettuce	550
Carrots	480
Spinach	270
Miscellaneous	
Peanuts, roasted	1800
Molasses	1500
Tea leaves, dry	10,000
Yeast, *Torulopsis utilis*	2700
Yeast, brewers'	500
Meats	
Beef, muscle	115
Beef, heart	2600 16,000
Beef, brain	6000 2000
Beef, liver	3400 510
Pork, loin	360-450
Veal chop	320-350
Fowl and Fish	
Chicken, breast	480
Oyster	440
Halibut	170
Salmon	170
Egg	220
Milk products	
Whole milk, cow's	500
Cheese	250

12

Nutrition of Fat–Soluble Vitamins

Fat-soluble vitamins contain only carbon, oxygen and hydrogen, whereas the water-soluble vitamins contain these elements plus either sulfur, cobalt or nitrogen. This group of factors is readily stored in the body wherever there are fat deposits. The fat-soluble vitamins are excreted wholly in the feces, whereas many of the water-soluble vitamins are synthesized in the intestinal tract by microorganisms. Only Vitamin K of the fat-soluble group is synthesized in this organ.

The water-soluble vitamins are generally associated with the transfer of energy and the fat-soluble vitamins are required for the regulation of the metabolism of structural units, each vitamin having a separate and specific role. The synergistic scope of the vitamins summarized in Fig. 12.1 illustrates this interrelationship.

The fat-soluble vitamins differ from the water-soluble vitamins in that they may be present in plants in the form of a provitamin.

VITAMIN A

Vitamin A was the first of the accessory food factors to be identified as a component of specific foods. True Vitamin A exists only in the animal kingdom. There are several A vitamins, namely, Vitamin A_1, A_2, A_3, etc. Today, the names retinol, retinal and retionic acid are accepted chemical nomenclatures for different forms of Vitamin A.

Vitamin A activity in plants resides in the carotenoid pigments which are precursors to the Vitamin A in the chicken's body and are, therefore, designated the Provitamins A. Conversion of the provitamins to Vitamin A occurs in cells of the intestinal mucosa or liver.

The B-ionine ring is essential for Vitamin A activity. The B-carotene molecule contains 2 B-ionine rings; however, during digestion and metabolism of B-carotene, there is destruction of the molecule to the extent that an average of less than 1 B-ionine ring is converted to Vitamin A. The 3 carotenes possessing Vitamin A activity are alpha, beta and gamma.

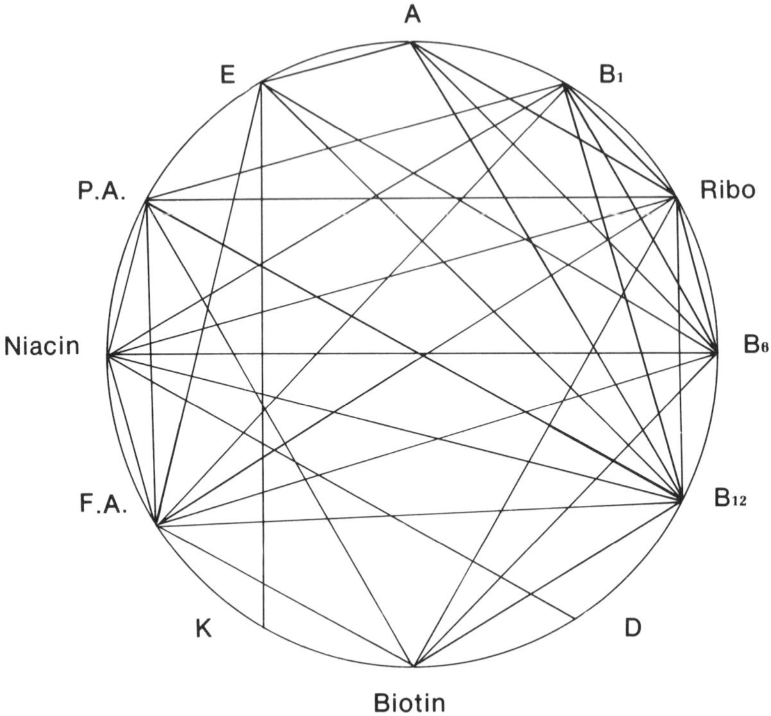

FIG. 12.1. SYNERGISM AMONG THE VITAMINS

Utilization of Crystalline Carotene and Vitamin A by Chicks and Turkey Poults

Carotene provided most of the Vitamin A activity for poultry rations during the development of the industry. Cod liver oil was used as a source of Vitamin A, Vitamin D and fat in early poultry rations. The cod liver oils were of low potency, and 1% level was a common recommendation. The 2 forms of Vitamin A activity, carotene and Vitamin A, did not give uniform response, and there was a variation between and within species on efficiency of utilization of carotene as compared to Vitamin A for growth, blood plasma concentration and liver content.

The Vitamin A content of the blood plasma and liver in chickens as influenced by the level of Vitamin A activity as provided by feeding crystalline carotene or crystalline Vitamin A acetate is summarized in Table 12.1. Both crystalline carotene and crystalline Vitamin A acetate, when ingested in sufficient amounts to raise the concentration in the

TABLE 12.1. THE INFLUENCE OF AMOUNT AND FORM OF VITAMIN A ACTIVITY ON GAIN BY CHICKS IN BODY WEIGHT AND VITAMIN A CONCENTRATION IN LIVER AND PLASMA

Vitamin A per lb Diet I.U.	Average Gain in 42 Days g	Vitamin A in Blood Plasma[1] 8th Week I.U./100 ml Plasma	Vitamin A in Liver[1] 8th Week I.U./ g Liver
		Crystalline Carotene	
500	578	61	2
6,000	640	234	74
12,000	637	253	243
18,000	614	258	277
24,000	598	271	446
		Crystalline Vitamin A Acetate	
500	542	54	2
6,000	648	255	316
12,000	640	283	728
18,000	614	288	1163
24,000	624	356	2291
		Black Cod Liver Oil	
500	558	48	3
6,000	606	220	219
12,000	615	255	685
18,000	596	258	1065
24,000	588	355	1645

Source: Castano *et al.* (1951).
[1]The Vitamin A content of the blood plasma at the beginning of this feeding trial was: 105 I.U. per 100 ml plasma and 9 I.U. per g liver.

blood to a certain level, were equally effective in sustaining a satisfactory rate of growth. The amount of Vitamin A in the blood plasma should be approximately 50 I.U. per 100 ml of blood plasma if liver storage is to be maintained. The plasma Vitamin A level tends to remain relatively static at the expense of liver storage of this vitamin.

4000 I.U. of Vitamin A activity from crystalline carotene is required to equal 2000 I.U. of activity from crystalline Vitamin A acetate or about 3000 I.U. of Vitamin A from "black cod" liver oil. When fed a level of 6000 I.U. of Vitamin A activity per lb feed, all 3 sources gave similar growth response (Table 12.1). The greatest liver storage occurred with crystalline Vitamin A acetate and the lowest storage with crystalline carotene. When fed at levels above 12,000 I.U. per lb feed, a slight retardation in growth was observed.

Similar studies with White Holland poults demonstrated that 1000 I.U. of crystalline Vitamin A acetate supported growth equal to 3800 I.U. of pure crystalline carotene. Turkey poults stored over 20 times as much Vitamin A in the livers from crystalline Vitamin A acetate as from crystalline carotene during their 8 wk feeding treatment (Table 12.2).

TABLE 12.2. AVERAGE BODY WEIGHTS AND VITAMIN A CONTENT OF LIVERS OF TURKEY POULTS FED VARYING AMOUNTS OF CRYSTALLINE CAROTENE AND VITAMIN A

Vitamin A per lb of Diet I.U.	No. Poults at 8th Week	5th Week Weight g	8th Week Weight g	8th Week Liver Vit. A I.U.
		Crystalline Carotene		
238	0	[1]	[1]	[2]
475	2	371	674	[2]
950	2	352	768	[2]
1,900	7	405	927	[2]
3,800	11	402	1,004	99
7,600	10	409	977	423
15,200	8	406	1,046	1,217
30,400	5	404	1,090	2,377
		Crystalline Vitamin A Acetate		
250	3	391	665	[2]
500	6	373	884	[2]
1,000	11	412	1,016	47
2,000	15	427	1,065	985
4,000	11	420	1,043	2,189
8,000	9	386	879	10,179
16,000	9	388	935	34,877
32,000	7	426	1,024	78,731

Source: Reorganized data from Gurçay *et al.* (1950).
[1]All poults were dead by the 5th week.
[2]No measurable amount was found.

The results from feeding carotene and Vitamin A to both poults and chickens demonstrated that very high levels of carotene must be fed to assure liver storage. It is very important to develop liver storage in breeding birds and also in any bird likely to be subjected to stress conditions which will result in reduced feed consumption.

A specific coenzyme function has not yet been demonstrated for Vitamin A. It has been well-defined as a constituent of the visual pigment, rhodopsin, present in the rod cells of the retina.

Nutrition of Vitamin A

Vitamin A is very important to epithelial tissue. Keratinization of epithelial tissue is an early sign of a Vitamin A deficiency and so this vitamin probably plays a prominent role in muco-polysaccharide formation. Young chicks having a deficiency secrete an over production of mucous-type connective tissue. Vitamin A may be involved in activation of sulfate prior to its incorporation into mucopolysaccharides. Some research evidence suggests that Vitamin A is active in the synthesis of certain steriod hormones of the adrenal cortex, as well in the maintenance of the integrity of the adrenal cortex.

Poor growth, cheesy exudates in the eyes and respiratory tract, rough scales on legs, poor egg production and hatchability, and poor feathering are often observed in Vitamin A-deficient chickens. There is also a lowered resistance to certain diseases, such as coccidiosis.

Sources

Liver oils are the best natural sources of Vitamin A (retinol) and synthetic Vitamin A acetate is the most important and most stable source of this nutrient. We must never forget that Vitamin A and also the carotenes are the most unstable fat-soluble vitamin-active compounds. Alfalfa meal, green pastures, corn gluten and yellow corn are the most important Provitamin A sources.

Vitamin A acetate concentrates are prepared from gelatin-coated beads of Vitamin A, plus an antioxidant. A desirable carrier for Vitamin A concentrates from liver or synthetic retinal is soybean meal. This vegetable protein is an excellent carrier for all fat-soluble vitamins. It retards oxidation and does not readily absorb moisture. Moisture, air and peroxides must be excluded from fat-soluble substances if they are to be stabilized.

Alfalfa meal, an excellent source of Provitamin A, is treated with an antioxidant, plus a pelleting aid, and is then pelleted, cooled, dried and stored in a cool place or under an inert gas. Well-preserved alfalfa meal, yellow corn and corn gluten can supply ample Vitamin A activity for growing chickens. The addition of extra Vitamin A acetate for turkey and chicken breeders and egg producers is highly recommended. The additional amount is desirable because it takes more Vitamin A activity from carotene to produce body storage than from pure Vitamin A (retinal). The Vitamin A requirements of different poultry are summarized in Table 12.3.

Assay

The Carr-Price colorimetric method of assay was the chemical method used by early investigators. This method is based on a blue color formed with antimony trichloride in chloroform. Its results are accurate and compare well with biological methods.

Vitamin A in solution shows a maximum absorption in the ultraviolet region of the spectrum of light at a wavelength of 325 to 328 m. With a spectrophotameter, it is possible to determine Vitamin A quantitatively by comparing the amount of light of 325 m wavelength transmitted by the Vitamin A solution.

Chick bioassay methods are used for both Vitamin A and Provitamin A (carotene). The chick method is preferred in natural crude preparations,

TABLE 12.3. FAT-SOLUBLE VITAMINS REQUIRED BY POULTRY PER KG OF RATION

Poultry	Vitamin A I.U.	Vitamin D_3 I.C.U.	Vitamin E I.U.	Vitamin K mg
Chickens				
starting	1500	200	10	0.5
growing	1500	200	5	0.5
laying	4000	500	5	0.5
breeding	4000	500	12	0.5
Turkeys				
starting	400	900	12	1.0
growing	4000	900	10	0.8
breeding	4000	900	25	1.0
Pheasants and quail				
starting	3000	1200	12	1.0
growing	3000	900	10	1.0
breeding	4000	1200	18	1.0
Waterfowl—ducks and geese				
starting and growing	4000	300	10	0.5
breeding	4000	300	10	0.5

Source: Patrick (1965).

whereas the spectrophotametric method is rapid and reliable for synthetic Vitamin A. Growth rate and liver storage are the criteria for chicks.

Provitamin A (carotene) can be determined by colormetric methods; however, chick growth and liver storage are more desirable. Some studies have shown that the Provitamin A (carotene) molecules show higher chemical values than the chick assay will substantiate.

VITAMIN D (ERGOCALCIFEROL, CHOLECALCIFEROL)

Vitamin D, called the antirachitic or sunshine vitamin, was found to heal or prevent rickets in poultry. The beneficial effects of light in the prevention of rickets were recognized long before Vitamin D was postulated. Early fed a Vitamin D-deficient ration showed a slight deficiency in 4 days and a complete deficiency in 14 days. When deficient chicks were given supplements of Vitamin D_3 and radioactive calcium (Ca^{45}), some response was evident in 4 hr and normal calcium absorption was manifested by 24 hr.

Young chickens do not readily store Vitamin D in their livers. When day-old chicks were given 300 ICU of Vitamin D_3 orally, the depletion time was 10 days, as compared to 4 days for the unsupplemented group. This gain of only 6 days in depletion time shows how poorly this vitamin is stored (Table 12.4).

Vitamin D_3 is stable (Table 12.5), and carriers, such as soybean meal, corn meal and distillers' grains, are often used. Regardless of its stability, it should be stored in a cool, dry and unlighted room. Strong light will destroy this factor; however, it can be incorporated in premixes containing minerals.

TABLE 12.4. VITAMIN D DEPLETION IN CHICKS USING Ca^{45}

	Bone Mineralization[1]					
	0 Days	4 Days	7 Days	10 Days	14 Days	18 Days
Basal ration	13.4	12.2	9.4	9.0	7.2	7.8
Basal + Vitamin D_3	—	13.8	13.0	12.4	13.0	13.4
Basal (fortified chicks)[2]	—	14.1	14.0	12.1	8.9	8.1

Source: Patrick and Schwietzer (1955).
[1]Percentage of a dose of radioactive calcium deposited in both tibia within 2 hr after oral supplementation.
[2]Each chick was given orally 300 I.C. units of Vitamin D_3 at 1 day of age.

TABLE 12.5. EFFECT OF STORAGE ON VITAMIN D ADDED AS FISH OIL TO POULTRY FEEDS

	A.O.A.C. Units Vit. D per 100 g Feed		
Feed Number	Fresh	3 Months	Difference (%)
A.O.A.C. basal	25	25	0
B-79	78	72	7.7
B-80	26	21	19.2
B-94	60	53	11.7
B-107	32	29	9.4
B-128	90	86	4.4
B-130	57	42	26.3
B-131	82	88	+7.3
B-132	49	44	10.2
B-133	83	60	27.7
B-143	99	98	1.0
B-194	80	50	37.5
B-195	80	52	35.0
B-199	127	135	+6.3
B-212	133	133	0
B-213	80	11	13.7
B-214	117	107	8.5
B-240	90	45	50.0
B-241	135	130	3.7
B-393	43	12	72.1
B-394	40	24	40.0

Deficiency Symptoms

Rickets can be readily detected by bone or toe ash. If the attendant will observe the ribs, small beads will be found. The beak area around the nasal openings, as well as the upper beak, will display a rubbery condition. Ricketic chicks sit and walk with difficulty. Poultrymen allow the chickens to have "runs" to provide exercise and sunshine. When it was discovered that cod liver oil would prevent rickets, as well as provide Vitamin A, the term Vitamin D was established.

Chemistry

There are several Vitamins D (Vitamin D, D_2, D_3 and D_4); however, Vitamins D_2 and D_3 are the most important. Vitamin D_2 (calciferol) is formed in plants by sunlight activating ergosterol to form ergocalciferol. This Vitamin D is of value to man and 4-footed animals, demonstrating very few antirichetic properties in poultry. Vitamin D_3 (cholecalciferol) is utilized by poultry, man and 4-footed animals. Vitamin D_3 is formed in animals when sunshine activates 7-dehydrocholesterol in the animal's skin.

Vitamin D_3 (cholecalciferol) is the only Vitamin D used in poultry rations. It is relatively stable and not easily oxidized in feeds.

Vitamins D_2 and D_3 are produced synthetically by irradiation of plant sources of ergosterol and purified sources of 7-dehydrocholesterol.

Nutrition

Vitamin D_3 is the antiriachitic vitamin in poultry, and its primary function is to control the absorption of calcium from the duodenum and the small intestine.

Laying hens which receive rations deficient in Vitamin D will lay soft shelled eggs or calcium crystals will appear as encrustations on the shell surface.

Turkey eggs from layers deficient in Vitamin D will have very light brown markings as well as visible calcium encrustations on the shell surface. The laying turkey hens will also develop crooked breastbones with a reduced calcium content as the laying season progresses. Many broken bones will be observed when transporting Vitamin D deficient hens.

Sources of Vitamin D and D_3

Vitamin D can be obtained from fish liver oils; Vitamin D_3 is obtained from chemical synthesis and is the principal form used. Poultrymen should never rely upon sunshine to provide enough vitamin D_3 because very little sunshine is provided under today's management programs.

Vitamin D Assay

A chick A.O.A.C. method is recommended for assay of all Vitamin D concentrates used in poultry rations. This is an official method and published in the A.O.A.C.

Requirements

Vitamin D_3 requirements vary with age and species of poultry. The requirements for Vitamin D_3 by chickens, turkeys, quail, pheasants, ducks and geese are summarized in Table 12.1. Breeders have the highest requirement and their rations should contain enough Vitamin D to provide body storage in the starting birds.

A regular ration should not contain over 5 times the recommended requirement, since toxic levels can cause excess excretion of zinc, calcium and magnesium.

Ration

Vitamin D is generally secured in a vitamin premix; however, regardless of the source, the proper time to add it to a ration is with or after the soybean meal and before the minerals.

Vitamin D should be stored in the dark and away from moisture. Only synthetic Vitamin D_3 should be used, unless the cod liver oil Vitamin D preparation is not rancid and if not over 0.2% of the oil used. Rancid fish oils destroy Vitamin A and alpha tocopherol.

VITAMIN E (ALPHA-TOCOPHEROL)

Early investigators observed that a fat-soluble factor found in plants was related to reproduction in rats. It was named Vitamin E, and today, 7 Vitamins E are known. The term, antisterility or fertility vitamin, was also designated. An alcohol was prepared from the nonsapomfiable fraction of wheat germ oil which had high Vitamin E activity. This alcohol was called -tocopherol (tokas, childbirth; phero, to bear). The tocopherols are oils, fat-soluble, and antioxidants functioning as protectors from destructive oxidation of Vitamin A and polyunsaturated fatty acids. The tocopherols are easily oxidized.

Chicks fed a ration which had been soaked in an iron solution developed nutritional emphalomalacia. The iron catalyzed the oxidation of the fat, tocopherols and Vitamin A.

Nutrition

The Vitamin E-deficient rations fed chickens used either (1) iron-treated diets (2) high levels of fish oils (3) torula yeasts and/or (4) lard (Vitamin E-free). Also, Vitamin E was introduced in commercial feeds during the time that fat was added to increase the energy content. Many discussions have developed on the true role of Vitamin E; some scientists postulate that it is an antioxidant protecting Vitamin A and the quality of fat and phospholipids available within the cell membranes.

The accumulation of body fluids in the cavity of Vitamin E-deficient chicks probably results from the disruption of disintegration of cell membranes. When fats are added to feedstuffs, they coat the surface of the feed particles where surface oxidation occurs. Surface oxidation can break the carbon chain in different places, whereas beta oxidation removes 2 carbon fragments each time. Normal oxidation of fats is by beta oxidation. The surface-oxidized fatty acids produce free radicals which can disrupt normal formation of the cell membranes. Located in cells which perform digestive functions are organelles, called lysosomes. These organelles are rather large or abundant and are membranes enclosing complex enzymes capable of digestion or of breaking down complex nutrients. As long as the membrane remains unaltered or preserved, the lysosomes can perform necessary digestive functions. Upon destruction of the membrane, the enzymes are freed and the cell digests itself.

The disruption of the normal function of the membrane probably occurs when the cell has an accumulation of peroxides, free radicals, abnormal phospholipids in the membrane, or the formation of free radicals or peroxides within the phospholipid layer of the membrane.

Vitamin E and Selenium in Peroxides and Free Radicals

After the introduction of the high-energy ration, cheaper and more concentrated energy sources were secured. The most common is fat from animal rendering and oils from vegetable-protein processing. The 2 major feedstuffs in the high-energy ration are corn and soybean meal. When fat is added to rations, it is in contact with minerals which can catalyze oxidation, and surfaces which can provide the environment necessary for surface oxidation.

Selenium, which is reported to have a relation to Vitamin E, is also very active in controlling free radicals.

Several attempts have been directed toward establishing an enzyme relationship of Vitamin E. So far, this reaction has not been established. One common proposal for the role of Vitamin E in metabolism is largely that of a nonspecific antioxidant. This is based on data showing that several unrelated antioxidants perform functions similar to the vitamin. This proposal agrees with the role of tocopherols as nonspecific antioxidants but suggests that a more specific role exists.

Sources

Tocopherols are found in the oil glands or globules of most plants or seeds. Wheat germ is one of the oldest known sources. Vitamin E (tocopherol) concentrates are obtained by molecular distillation of vegetable oils. Another source of tocopherols is the synthetic product.

Standards or Assay Methods

A standard procedure is to determine the value of the tocopherol in prevention of fetal resorption in depleted rats.

A liver storage method is used with chicks; however, the oxidation of Vitamin E during the assay often gives erroneous results.

Requirements

Since Vitamin E functions as an antioxidant and can be easily oxidized, the requirement is influenced by the presence of peroxides, free radicals and minerals which catalyze oxidation. The NRC has provided recommended levels (Table 12.3).

Feed Formulation

When developing a ration, the stability and source of fat should be carefully executed. Also, the means of incorporating the preparation into the feeds and the source must be evaluated.

Only clean, stabilized fat should be used, and if possible, should be added to the soybean meal and yellow corn before the minerals or animal proteins are incorporated. The vitamin, though an antioxidant, must be protected in the feed before it enters the digestive system.

Summary of Vitamin E

Birds require exogenous antioxidants for the stabilization of tissue lipids in general, and for the lipoprotein components of cellular and subcellular membranes in particular.

The tocopherols are the prime natural lipid-soluble antioxidants and their role at the molecular level appears to be nonspecific. The chemical antioxidants are effective in stabilizing dietary fat but often are ineffective in biological systems.

The requirement for alpha-tocopherol (Vitamin E) is primarily related to the proportion of polyunsaturated fatty acids consumed in the dietary lipids.

Dietary Vitamin E increases the stability of fat in processed chickens and turkeys. There is also an improvement in meat flavor of both broilers and turkeys upon extended freezer storage.

VITAMIN K

Vitamin K, the antihemmorrhagic vitamin, is required by chicks to maintain normal blood-clotting tissue. Several K vitamins, such as Vitamin K_1 (found in alfalfa), Vitamin K_2 (found in fish meal) and the synthetic form of Vitamin K_3, registered under the name of menadione, are known. The chemical name of synthetic Vitamin K_3 is 2-methyl-1, 4-

naphthoquinone, and it is soluble in hot water, whereas the natural forms of K_1 and K_2 are fat-soluble. Another synthetic form is phythiocol, which also is a water-soluble derivative, and like menadione, has more Vitamin K activity than the natural forms.

The active chemical structure is the naphthoquinone. The natural Vitamin K_1 compound contains the phytyl radical, whereas the synthetic compound, menadione, does not contain the phytyl radical. Apparently, the phytyl radical reduces the blood-clotting ability.

Vitamin K is involved in electron transport and oxidative phosphorylation in bacteria. Since Vitamin K is synthesized in the lower intestines by bacteria where absorption is poor, a secondary dietary need may arise.

Requirements

Natural feedstuffs generally provide adequate Vitamin K for healthy chicks; however, synthetic Vitamin K_3 is generally added as a precaution at the rate of 0.5 to 1 mg per lb feed. This vitamin is generally added to broiler rations to protect the birds from mold toxins (Table 12.3).

Sources

Alfalfa, fish meal, soybean meal and feces are good sources of Vitamin K. Synthetic Vitamin K compounds—menadione and phythiocol—are the most active and available sources, being approximately twice as active as natural Vitamins K_1 and K_2.

Vitamin K activity is determined by a very accurate chick assay procedure. It depends upon the degree of regeneration of prothrombin as a result of feeding Vitamin K-active substance to depleted chicks. Vitamin K is added to all chicken and turkey broiler and starting rations. It is added as a disease preventive factor rather than one to correct a nutritional deficiency.

Nutrition Vitamin K Summary

The Vitamin K requirement of the chick is increased when a toxic level of some molds is present. For this reason, supplementary Vitamin K is added, realizing that most diets contain enough naturally-occurring Vitamin K. Spoiled sweet clover caused cattle and hogs to have prolonged blood clotting. The blood prothrombin was found to be lowered, and a compound, dicumarol, was found to be the toxic substance. Vitamin K helps prevent or heal this condition.

Vitamin K is required by chickens because they do not have blood platelets to aid in the release of prothrombin, which in the presence of

Ca^{++}, forms thrombin. The thrombin brings about chemical change in the fibrous water-soluble protein, fibrinogen, to form an insoluble protein, fibrin.

Vitamin K is essential to the synthesis of prothrombin in liver cells. Without prothrombin or Vitamin K in the chick, the clotting time of the blood is increased.

13

Nutrition of Minerals

Minerals comprise the ash of the biological materials remaining after the organic materials have been completely oxidized and the water removed. Although most of the minerals are located within the skeleton, a very significant amount is found in the soft tissues.

Minerals can be divided into 4 groups: (1) required microminerals, (2) required macrominerals, (3) functionary minerals, and (4) contaminants or toxic minerals.

(1) Microminerals.—Micromineral requirements are expressed in parts per million and generally are associated in activation or integrated parts of enzymes. These include:

(a) Cobalt
(b) Copper
(c) Iodine
(d) Iron
(e) Manganese
(f) Selenium
(g) Zinc

(2) Macrominerals.—Macrominerals are required in large amounts, generally expressed in percentage, and are often structured parts or acid-base elements. The macrominerals are:

(a) Calcium
(b) Phosphorus
(c) Potassium
(d) Magnesium
(e) Sulfur
(f) Salt (NaCl)

(3) Functionary Minerals.—Functionary minerals are those that may improve growth, feed efficiency and general appearance, or alleviate toxicity, but are not directly a required enzymatic factor. They are also sometimes referred to as "man-made" requirements. They may include:

(a) Arsenic
(b) Chromium
(c) Fluorine
(d) Molybdenum
(e) Silica
(f) Vanadium

(4) Contaminants.—Contaminants include:

(a) Radioactive elements from nuclear explosions
(b) Selenium—some geographical areas
(c) Molybdenum—some geographical areas
(e) Mercury

Minerals make up 3 to 5% of the chicken's body. Since minerals cannot be synthesized, they must be provided by the diet.

The physiological functions of minerals in the chick's body are:

(1) To maintain proper acid-base balance in body fluids.
(2) As chelates.
(3) For support. Most of the skeleton is made up of minerals.
(4) As active parts of protein structures (hemoglobin, phosphoproteins).
(5) To maintain the proper contractibility of muscles, especially the heart.
(6) As parts of amino acids.
(7) For proper cellular osmotic pressure.
(8) For enzyme activation.
(9) As transport mechanisms.

An understanding of minerals can best be developed by first becoming acquainted with the periodic table of the elements (Table 13.1). The required elements are in bold type and should be identified along with those adjoining, since the functions of the latter may overlap the functions of the former. For example, arsenic may be found replacing phosphorus, cesium replacing potassium and strontium replacing calcium. This replacement possibility is of great importance in reducing the toxicity of minerals. The toxicity of selenium can be reduced by feeding arsenic, and copper toxicity reduced by feeding manganese and sulfur.

TABLE 13.1. PERIODIC CHART OF THE ELEMENTS

IA	IIA	IIIB	IVB	VB	VIB	VIIB	VIII			IB	IIB	IIIA	IVA	VA	VIA	VIIA	INERT GASES	
1 H 1.0080																1 H 1.0080	2 He 4.003	2
3 Li 6.940	4 Be 9.013											5 B 10.82	6 C 12.011	7 N 14.008	8 O 16.000	9 F 19.00	10 Ne 20.183	2 8
11 Na 22.991	12 Mg 24.32											13 Al 26.98	14 Si 28.09	15 P 30.975	16 S 32.066	17 Cl 35.457	18 Ar 39.944	2 8 8
19 K 39.100	20 Ca 40.08	21 Sc 44.96	22 It 47.90	23 V 50.95	24 Cr 52.01	25 Mn 54.94	26 Fe 55.85	27 Co 58.94	28 Ni 58.71	29 Cu 63.54	30 Zn 65.38	31 Ga 69.72	32 Ge 72.60	33 As 74.91	34 Se 78.96	35 Br 79.916	36 Kr 83.80	2 8 18 8
37 Rb 85.48	38 Sr 87.63	39 Y 88.92	40 Zr 91.22	41 Nb 92.91	42 Mo 95.95	43 Tc (97)	44 Ru 101.1	45 Rh 102.91	46 Pd 106.4	47 Ag 107.880	48 Cd 112.41	49 In 114.82	50 Sn 118.70	51 Sb 121.76	52 Te 127.61	53 I 126.91	54 Xe 131.30	2 8 18 18 8
55 Cs 132.91	56 Ba 137.36	57 *La 138.92	72 Hf 178.50	73 Ta 180.95	74 W 183.86	75 Re 186.22	76 Os 190.2	77 Ir 192.2	78 Pt 195.09	79 Au 197.0	80 Hg 200.61	81 Tl 204.39	82 Pb 207.21	83 Bi 209.00	84 Po (210)	85 At (210)	86 Rn (222)	2 8 18 32 18 8
87 Fr (223)	88 Ra (226)	89 +Ac (227)	Note—The Lanthanum and Actinium series have been omitted.															

*The nutrition student should identify the required elements in the Periodic Table, their valence(s), molecular weight, molecular size (orbits) and solubility of salts or compounds formed. Also to be identified are the elements adjoining the required elements, their properties and the possibility of being toxic or of value in detoxification.

THE MOLECULAR EXCHANGE THEORY OF MINERAL REPLACEMENT

The absorption, metabolism and biological transport mechanism and pathways of required minerals are apparently controlled directly or indirectly by one or more enzyme systems. Most minerals in the chicken's body are in an active state of movement in-and-out of the molecular configuration. Some movements occur upon death and digestion of the cell, and others occur by molecular exchange. A mineral which moves by molecular exchange does not involve a need for energy in producing the reaction. An example of this is the movement of fluorine into the hydroxy apatite bone crystal.

$$2Ca_{10}(PO_4)_6(OH)_2 + CaF_2 \rightleftharpoons 2Ca_{10}(PO_4)_6F(OH) + Ca(OH)_2$$

Hydroxyapatite Fluorohydroxyapatite

The molecular exchange of fluorine in the apatite structure requires no energy, and whereas the hydroxyapatite is soluble, the fluoroapatite is insoluble.

Molecular exchange occurs in many molecules. In general, the carbon atom does not exchange position with another carbon atom in a carbon chain, whereas Ca will exchange with another calcium or with the element strontium, located next to it in the periodic table. Also, potassium will exchange with rubidium or cesium.

The phenomenon of molecular exchange is used in removing toxic minerals from the body, or to impound toxic minerals that may accumulate in the plasma if not removed. The impoundment of fluorine in the apatite structure is often considered a detoxification process. Upon removal of the high fluorine diet, the fluorine in the apatite structure exchanges with an (OH) ion in the body fluid which can be excreted in the feces.

MINERAL ABSORPTION

Minerals are either actively or passively absorbed. Many biochemists feel that all required minerals are actively absorbed. By active absorption, we mean that an enzyme system and energy are involved. Passive absorption refers to the movement of a mineral by osmosis, solvent drag or by force of Donnan equilibrium. If 2 solutions are separated by a membrane which has become permeable to the solvent and solutes (minerals), the concentration of the solutes (minerals) will be identical when equilibrium is established. In this reaction, both the chemical composition and the osmotic pressures of the 2 solutions are the same.

Solution 1	Solution 2
Na^+	
	K^+
	I^-
Cl^-	

Membrane

Solution 1	Solution 2
Na^+	Na^+
K^+	K^+
I^-	I^-
Cl^-	Cl^-

Membrane

When the membrane is semipermeable to one ion (can be penetrated by one dissociated electrolyte) and not another, yet be penetrated by both ions from another electrolyte, there will be an equilibrium developed known as the Donnan equilibrium. When biological membranes are semipermeable, the Donnan equilibrium can become effective.

The permeability is often under the control of one or more enzymes for required minerals. However, minerals which have close association in the periodic table may follow the permeability opportunity of the required element. An example of this is when the calcium moves through the walls of the intestinal tract. Then, there will be an opportunity for strontium, a non-required element, yet located close to calcium in the periodic table, to move through the walls of the intestinal tract. Another example is rubidium and cesium, which will follow the body distribution of potassium.

39.1 K $^{+1}$	40.0 Ca $^{+2}$
85.5 Rb $^{+1}$	86.6 Sr $^{+2}$
133.0 Cs $^{+1}$	137.4 Ba $^{+2}$

The efficiency of mineral absorption is difficult to determine because the urine and feces are cast together, and also endogenous calcium and phosphorus are re-excreted into the intestinal tract. The use of a surgical procedure allows the separation of urine and feces of poultry.

COMPLEXING OR CHELATION

Complexing or chelation plays a definite role in mineral absorption from the intestinal tract. Many minerals form simple complexes or chelates.

One of the most common substances forming simple complexes in the digestive system is the cellulose micelle. Weak chelates (soluble chelates) are generally formed with amino acids or other compounds or particulates having active amino or hydroxyl groups. When minerals are complexed or chelated, their absorption into the body is generally highly controlled by enzymes; therefore, the absorption is an active process. This process of controlled absorption helps reduce uptake of toxic elements, unneeded amounts of required elements and helps control the rate of absorption.

Some chelates are highly insoluble. The Ca-Mg-Phytate is an insoluble chelate and only forms when excess calcium is present. It renders zinc unavailable, and is, therefore, perhaps the major cause of a zinc deficiency in poultry.

BODY DILUTION

Minerals which have gained entrance into the body tissues may need to be removed. When the cell dies and is digested, the mineral can be excreted or reabsorbed by another cell. If it is to be removed before death, then the process of body dilution is used. This involves feeding an excess of the mineral whose removal is desired, or if it is a nonrequired mineral, then the feeding of an excess of the required mineral closest to the nonrequired mineral in the periodic table is used. The removal of radioactive cesium is obtained by feeding excess potassium. Radioactive strontium is removed by feeding excess calcium.

MINERAL EXCRETION

Excretion of minerals is a complicated process. The presence of minerals in the feces of poultry does not necessarily mean that they were unavailable or unabsorbed by the bird. Poultry absorb most of the minerals and these are circulated throughout the body to the various organs or tissues. Each organ or tissue will absorb or "take up" enough to satisfy the needs or as much as is available from the plasma. The minerals remaining in the plasma may be secreted into the kidney tubules or returned to the intestinal tract where they are mixed with the contents and then reabsorbed; if not reabsorbed, they will be excreted as part of the feces. Depending upon where the minerals are excreted into the intestinal tract, it is possible for a mineral to be absorbed, excreted or reabsorbed several times.

POOLING OF MINERALS

Some minerals are pooled in the chick's body and destined for a specific purpose:

(1) Iodine is pooled in the thyroid gland for synthesis of thyroxin.
(2) Calcium is pooled in the medallany bone for egg shell formation.
(3) Zinc is pooled in the testes for reproduction.
(4) Phosphorous is pooled in the testes for formation of DNA.
(5) The calcium proteinate concentrated in the blood before egg production is pooled and destined for transport of fat into the egg yolk.

MINERAL FEEDBACK

Minerals leave or enter the blood stream by a feedback mechanism similar to an electrical feedback system. Several forces work together, either to increase or decrease the plasma level of minerals.

A simple feedback mechanism is the one showing the action of the parathyroid and the female sex hormones in maintaining the bound and ionizable calcium in the blood plasma during egg production. The sex hormone signals a need for more calcium in egg production and the parathyroid gives a signal to maintain only a designated calcium blood level (Fig. 13.1).

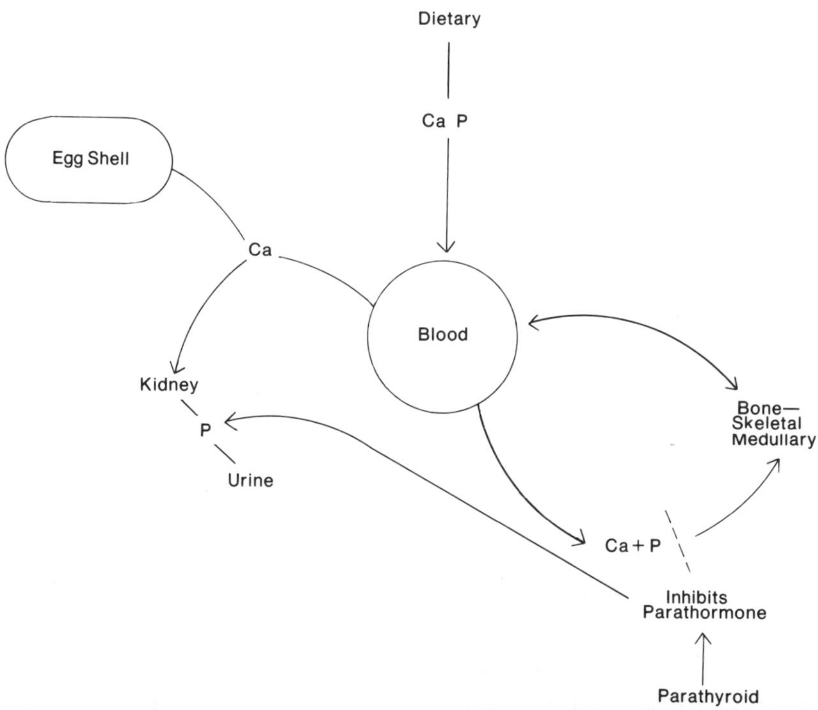

FIG. 13.1. ACTION OF PARATHYROID IN MAINTAINING BLOOD LEVELS OF Ca FOR EGG PRODUCTION

CALCIUM AND PHOSPHORUS

Calcium and phosphorus make up over 70% of the body ash. About 99% of the calcium and 80% of the phosphorus are found in the bones.

The main functions of calcium in poultry nutrition are:

(1) Skeleton formation
(2) Fat translocation from blood to egg yolk
(3) Blood clotting
(4) Egg shell formation
(5) Muscular contraction

A calcium-phosphorus deficiency is characterized by rubbery bones (rickets). In laying hens, the egg shells are weak and thin, and the hens' bones are ricketic with the breast or sternum crooked.

There are many economical and desirable sources of calcium; however, all sources purchased should meet specifications for calcium. These specifications are:

(1) Not less than 95–98% calcium carbonate
(2) Not contain over 2% magnesium carbonate
(3) Free of dirt or foreign material (toxic substances) which are not desirable for poultry

Calcium is available in all calcium carbonates, such as limestone, marble, oyster shell and calcite. The calcium in bone meal and phosphates is also available.

Dolamatic limestones (a mixture of $CaCO_3$ and $MgCO_3$) are often too high in magnesium for laying hens so should not be used as the major source of calcium. Excess magnesium causes wet droppings.

The absorption of calcium takes place principally in the duodenum and the proximal part of the small intestine, while phosphorus is absorbed mainly in the upper half of the small intestine. If the starting chick and poult diet contains approximately 1 and 1.5% calcium, respectively, the pH of the small intestine will be in the range of 5 to 6. If the pH is above 6, insoluble calcium and phosphorus particulates will form which tend to reduce the availability or absorption efficiency of manganese and zinc.

There are 4 factors which greatly influence the absorption of calcium and phosphorus:

(1) Source of Form in Which Ingested.—In general, the inorganic sources of calcium are highly available; however, phosphorus may be anywhere from 40 to 95% available. All sources of phosphorus should be tested for availability by the chick bone-ash method, and also tested for fluorine, since it is often found in rock phosphates at toxic levels.

(2) The pH of the Intestinal Contents.—Phosphorus is poorly absorbed at a pH above 6.5 and best absorbed at a pH below 6.0. The pH can be controlled by not feeding an excess of calcium.

(3) Calcium and Phosphorus Ratios.—Early agricultural chemists reported that bone ash contained approximately 2 parts of calcium to 1 part of phosphorus, a 2:1 ratio. Many livestock rations were formulated to provide this ratio. During the embryonic investigations conducted by Needham, the calcium and phosphorus ratio of 20-day-old chick embryos was approximately 1.4:1. During the early period of embryonic development, the ratio was less than 1:1.

The National Research Council recommends a calcium-phosphorus ratio of 1.67:1 for broilers when based on crude phosphorus in the average ration. When using available phosphorus rather than total inorganic phosphorus, a ratio of 2.2:1 is used. Most nutritionists do not use ratios for calcium phosphorus calculations but rather the percentage available or total (crude) phosphorus. Table 13.2 shows the calcium and total phosphorus requirements expressed as ratios and percentages. The data summarized in this table indicates that ratios of 1.4 or 1.6 to 1 for starting poultry is satisfactory, but layers and breeders have such a high and variable calcium requirement that a ratio cannot be used.

(4) Dietary Vitamin D_3.—The major function of Vitamin D_3 is in calcium absorption. However, when calcium absorption is improved, greater efficiency in utilization of phosphorus is also obtained.

The following data on the influence of Vitamin D_3 as related to Ca^{45} bone mineralization in chicks demonstrates that the major role of Vitamin D_3 is in calcium absorption.

	Method of Administering Vitamin D_3	
	Oral	Injected
Vitamin D_3 deficient ration[1]	6.2	11.1
Vitamin D_3 adequate ration[1]	11.4	12.1

Calcium Requirement

The starting chick and the laying hen must be given special consideration when developing the dietary levels of this element. The chick's digestive system from 1 day to 3 wk of life is extremely sensitive to dietary levels of calcium which would increase the pH of the intestinal tract to pH 6.5 or above. When this pH is reached, the manganese forms an insoluble complex and therefore, is not absorbed. Also, zinc forms a complex with phytic acid at this pH and is unavailable to the young

[1] These results are expressed as percent retention of a single dose of Ca^{45}.

TABLE 13.2. CALCIUM AND PHOSPHORUS REQUIREMENTS AND RATIOS FOR POULTRY

	Calcium	Total Phosphorus	Ratio C:P
Chickens			
starting	0.9	0.7	1.3:1
growing	0.6	0.4	1.5:1
laying	3.25	0.5	6.5:1
breeding	2.75	0.5	5.5:1
Turkeys			
starting	1.2	0.8	1.5:1
growing	0.8	0.7	1.45:1
breeding	2.25	0.7	3.2:1
Pheasants			
starting	1.0	0.8	1.25:1
growing	0.7	0.65	1.1:1
Quails			
starting	0.8	0.65	1.2:1
breeding	2.4	1.0	2.4:1
Ducks and geese			
starting	0.8	0.6	1.3:1
breeding	2.6	0.6	4.3:1

Source: Patrick (1965).

chick. The sensitive nature of the digestive system in the young starting chick persists from hatching to about 21 days of age. After 21 days, high levels can be tolerated but are not recommended for better development of the mineral control mechanisms.

The calcium level in the starting and growing pullet ration must stimulate the development of the glands or organs controlling the calcium feedback mechanism which is essential for calcium mobilizaton in egg production. Providing the starting chick with only adequate calcium, 0.8 to 1%, will initiate the development and functioning of the calcium feedback control mechanism so that when calcium mobilization for eggs is needed, the laying hen's body will respond.

Starting and growing pullet rations containing above 1.4% calcium will not bring about the necessary physiological development of the pullet for calcium mobilization.

When the chick receives an excess of calcium (over 1.4%), the normal body distribution changes, as shown in Table 13.3. Apparently, the chick's body moves a specific amount of calcium through the bones, and when the calcium level is excessive, this bone uptake function is changed.

The level of calcium in the egg-laying rations should start at approximately 2% at 2 wk before the onset of egg production, and gradually increase to 3 to 3.6% when top production is attained. Soft-shelled eggs are not necessarily caused by lack of dietary calcium. A Vitamin D deficiency, the size of the clutch, the age of the hen, genetics and pa-

TABLE 13.3. INFLUENCE OF DIETARY LEVEL OF CALCIUM ON BONE UPTAKE OF A SINGLE DOSE OF Ca^{45}

Percentage Calcium Uptake	Ca^{45}[1]
0.6	4.4
0.8	4.4
1.0	4.3
1.2	4.0
1.4	3.5
1.6	3.1
1.8	2.7
2.0	2.6
2.4	2.4

Source: Patrick (1970).
[1]% of Ca^{45} retained in both tibia of chicks which had received orally 10 microcuries of Ca^{45} at 21 days of age.

thological or drug interference all influence egg shell quality. When the number of eggs per clutch is above 18, an increased number of cracked eggs is produced. When the clutch size reaches 30, as much as 18% cracks have been recorded.

If the hen does not have the ability to absorb and mobilize calcium rapidly, she will deplete the medullary bone and the true skeleton to the point where the egg shells are thin and rickets is present. Some strains or breeding lines of chickens have better calcium mobilization than others. Many brown-egg breeds produce better egg shells than the white-shell breeds because of genetic differences.

Calcium and Hormonal Control

Although the chick is able to grow and have good feed efficiency on a wide range of calcium uptake during the growing period, the future use of the bird for egg production may become impaired. When the dietary calcium level is above 1.4% during the growing period, the demand on the parathyroid gland to maintain homeostasis of calcium in the chicken's body is so insignificant that this gland does not develop normally for calcium homeostasis during egg production. As a result, the young laying hen may develop kidney trouble and die. Feeding radioactive calcium, Ca-45, to young chicks (about 21 days of age) receiving purified rations of varying calcium content indicated that chicks apparently absorb and follow a set metabolic distribution for a high percentage of the dietary calcium below 1.4%; however, at levels of 1.4% and above, the metabolic distribution and body dilution pattern is altered and the percent calcium metabolized by the bones decreases as the calcium level increases above this dietary level. The chick probably follows specific metabolic patterns at the lower calcium levels, but above the 1.4% level, we encounter a "spill-over" or breakdown in the metabolic pattern which also results in the parathyroid becoming less sensitized when overly-excess calcium is ingested. Therefore, when egg production begins, the parathyroid gland is not able to control calcium homeostasis necessary for this reproductive function. An apparent calcium load is then placed on the kidney which becomes damaged as excess calcium is excreted and reabsorbed from the kidney tubules. Also of interest is the observation that the right kidney is the first to become damaged.

PHOSPHORUS

Phosphorus plays a major role in:

(1) Energy transfer (ATP)
(2) Bone structure
(3) Enzyme systems
(4) DNA and RNA linkages
(5) Acid-base balance
(6) Fat translocation

About 20% of the phosphorus in the chicken's body is in the soft tissues. The biological half-life of phosphorus and calcium in the bones of young chicks is approximately 36 days; however, as a phospholipid in the liver, it is about 72 hr. Poultry are the most susceptible of animals to phosphorus deficiency because they use inorganic phosphates, primarily. Most phosphorus in plants is present as a phytate which is highly unavailable to poultry. In poultry, the phosphorus is in the form of bone ash (apatite) or phosphoprotein, which are readily available.

A good biological evaluation program for phosphates should be maintained. Specifications should provide the following information:

(1) Florine Content.—Less than 0.2%.

(2) Vanadium.—Some rock phosphates, bone meal or animal protein sources have a toxic level of vanadium which is very harmful to chicks. It interferes with protein synthesis; therefore, a chick growth test can be used.

(3) Availability.—A chick bone-ash assay should be made to determine the availability of the phosphorus. A good phosphate has at least 90% available phosphate.

POTASSIUM

Potassium is an intracellular mineral, and the chick can be depleted in 3 to 5 days. It is necessary for carbohydrate and protein metabolism. Excess uptake of water tends to cause a greater loss of potassium.

Molasses and potassium chloride are good sources of potassium. There is some evidence to suggest that caged layers require more potassium than floor-managed hens. Data has been released from United States space flights showing that man in space has a heavy loss of potassium and a lowering of specific gravity. The cause appears to be one of confinement, and for this reason, caged birds may have a potassium imbalance.

There is a continuous exchange of intercellular and extracellular potassium in the chick as in all animals. The factors controlling potassium in the chick are not known.

The first symptom observed upon feeding a potassium-deficient diet to 1-day-old or 10-day-old chicks is a rapid reduction in feed consumption and growth rate. Also, a significant reduction in bone ash is observed after feeding the potassium-deficient diet for 7 days. The bone calcium is not reduced but the potassium and phosphorous content are reduced, and bone sodium is increased. The primary effect of potassium on bone calcification is in phosphorus metabolism.

SALT (SODIUM CHLORIDE)

Salt has always been rated as important in poultry nutrition, and most early-formulated rations contained 1%. As early as 1926, investigators observed that chicks receiving high cereal rations grew slowly with no added salt. The addition of 0.5% salt produced satisfactory growth and as much as 1% did not prove detrimental when diets contained ground corn, wheat, wheat by-products, ground oats and alfalfa meal. With the development of the high-energy ration, the usage of 0.5% salt produced wet droppings and also wet litter. The desired amount of salt needed for high-energy rations was found to be approximately 0.25 to 0.35% above that found in the natural ingredient. Studies with purified diets showed that sodium was the most important part of salt in which the sodium requirement was found to be between 0.10 and 0.14% (0.25 to 0.35% salt).

Excess salt readily produces wet droppings and wet litter. Apparently, there also exists an antagonistic relationship between sodium and potassium, potassium being found chiefly in the intercellular plasma and sodium in the extracellular fluids. When either element is present in considerable excess of the other, growth is retarded and mortality increased. Diets high in fiber have higher sodium requirements than the high energy-low fiber diets. Generally, high-fiber ingredients, such as alfalfa meals and wheat bran, contain more potassium than lower-fiber and high-energy ingredients, like corn and milo.

Several feed ingredients, such as fish meals, corn gluten, meat meal, whey, and sunflower meal contain high levels of sodium, and their usage often requires adjustment in the level of supplemental salt. Since excess salts cause an increase in the uptake of water, and the kidneys play a very prominent role in water and sodium excretion, it is extremely important that a diet be fed which does not increase the renal clearance work.

MAGNESIUM

Magnesium is an essential constituent of tissues and body fluids. Approximately 65% of body magnesium is located in a relatively free (loosely absorbed) form, and about 30% is in combination with phosphate in the bone tissue. It ranks next to potassium among the cations of intracellular fluids.

Magnesium ions serve as activators of important enzymes involved in intermediary metabolism. It is required for cellular respiration.

Little is known about the factors involved in regulating the magnesium content of the blood; however, most animals tend to excrete more magnesium in cold weather than in warm weather. A cerebellar disorder is produced when chicks are fed a magnesium-deficient ration. A chick can be depleted in magnesium in about 3 to 5 days. The most common magnesium deficiency symptoms observed are a marked disturbance of equilibrium with ataxia, falling and staggering gait, incoordination of movements, tremors, poor growth and feathering. Natural feedstuffs contain adequate amounts of magnesium.

The most common source of magnesium is dolomitic limestone, but great care must be exercised in the usage of dolomitic limestone in egg rations because excess magnesium is laxative and interferes with calcium usage. The magnesium requirement increases with egg production.

SULFUR

Sulfur is one of the most abundant elements in the earth and has many functions in poultry nutritional biochemistry. Inorganic sulfur, considered for many years to be ineffective in meeting the nutritional requirements of poultry for this mineral, is today recognized as an important required element. Sulfur was not considered by early investigators as a required inorganic element because the amount ingested in the inorganic form was considered insignificant, as well as being chiefly a metabolic waste. Today, because opinions regarding sulfur metabolism have changed, this element is considered a requirement in the inorganic form.

The metabolic processes or reactions involving sulfur in poultry are many and complicated. The reactions can be either biochemical, biophysical, inorganic, physical or organic. Although both anabolic and catabolic sulfur pathways exist in poultry, only the catabolic pathway is highly developed. Most of the inorganic sulfur needs of poultry are met by metabolism of organic compounds. The role of the sulfate ion in specific reactions, such as the formation of sulfated mucopolysaccharides, the synthesis of taurine and the reactions of detoxifications, has been established.

It has been shown that inorganic sulfate S-35 is metabolized by chicks in the production of taurine, and that it increases the absorption and retention of dietary antibiotics, such as the tetracyclines.

Feeding excess inorganic sulfates can be detrimental; destruction of Vitamin D, increased production of feces, and loss of membrane permeability can result in the young chick. Fluids collect under the skin and in the body cavity, and hemorrhages occur around the breast.

Calcium sulfate is an available source of both calcium and sulfur; however, the level added should not exceed 1% of the ration.

THE MICROMINERALS

Early investigators labeled the many elements which constantly occurred in living tissue in small amounts as "trace elements." Many of these trace elements were identified and quantitative amounts determined. Today, some investigators call trace elements "micronutrients." Many have been found to be essential dietary nutrients. Some possess metabolic activity, while others are vital to the enzyme processes of the living cell.

All forms of life require a number of inorganic elements for normal growth and reproduction. While virtually all elements of the periodic table have been found in poultry tissues, not all have been found essential. The elements found required in very small amounts are called microminerals or trace minerals.

The story of trace minerals in poultry nutrition is a recital of achievements of poultry nutrition research. Many diseases of poultry that caused heavy losses have been conquered. The need of the trace minerals has been established by rigidly controlled experiments.

Trace minerals play very important roles in the enzyme processes of the cells. A deficiency of some trace minerals results in metabolic disturbances that may exist undiscovered. Complicating these disturbances is the frequent interaction between two or more minerals, or between minerals and vitamins or amino acids (Table 13.4).

The trace minerals presently thought to be required by the domesticated fowl, and their functions, are shown in Table 13.4. During the period 1965 to 1975, a number of trace minerals became associated with

TABLE 13.4. FUNCTION OF TRACE ELEMENTS

Mineral	Essential Compounds or Enzymes
Iodine	Constituent of thyroxine
Copper	Amino oxidase, tyrosinase, ascorbic acid, oxidase
Manganese	Arginase, peptidases, kinases, decarboxylase, deoxyribonuclease
Iron	Cytochrome enzymes, catalase, peroxidase, hemoglobin
Zinc	Carbonic anhydrase, peptidases, lactic acid dehydrogenase, glutamic dehydrogenase
Cobalt	Constituent of cyanocobalamin (Vitamin B_{12})
Selenium	Free radicals control

Source: Patrick (1965).

the nutrition of the fowl, but additional data are needed to establish their essentiality definitely. Summaries of the functions of trace minerals in poultry have been prepared by many investigators. Some of the recently recognized minerals have been found to improve growth or retention of some organic substance. These reactions could be a toxic response because small amounts of toxic substances often cause an increase in growth or changes in tissue retention.

Manganese

Early investigators observed that chicks consuming rations containing oats, wheat, or rice by-products had fewer leg abnormalities than when corn was fed. Some natural calcium carbonate deposits gave better performance than the purified calcium carbonate or oyster shell. Each of these ingredients was found to be a good source of manganese.

The leg abnormality called hock disease, slipped tendon or perosis, was found to be related to manganese. An enlargement and malfunction of the tibial metatorsal joint results in the slipping of the gastro-nemous or Achilles tendon from the chondyles. The bones were also shorter and wider than normal bones but the ash content was normal. Most of the manganese is found in the bones.

Manganese was the first nutrient found required for prevention of perosis. Following this observation, zinc, choline and biotin were also found to be antiperosis factors.

Manganese is required for growth, egg production, hatchability, bone development and egg shell quality. Chicks hatched from eggs produced by hens which received a manganese-deficient ration developed shorter and thicker bones with a high incidence of perosis in the newly-hatched chicks.

Perosis is an abnormality of the protein structure and not one involving ash content. It cannot be cured, so it must be prevented. The protein abnormality resulting from a manganese deficiency influences the development of the bone in the embryo or during the first 21 days of life after hatching. After 21 days of life, the basic bone protein structure has formed. The ability of the chick's digestive system to absorb manganese after this period is not influenced by the high pH caused by feeding excess alkaline salts, such as calcium or magnesium carbonate.

Several minerals reduce the retention and absorption of manganese. When the pH of the duodenum or small intestine is increased above pH 6, the manganese becomes insoluble and is not absorbed. Excess calcium and magnesium or iron phosphate readily reduces the absorption of manganese.

Natural sources of manganese are oats, rice and wheat by-products. Manganese sulfate is the most common mineral used to provide this micromineral. Manganese is actively involved in several enzymes (Table 13.4). It is an activator of liver arginase, respiratory enzyme systems and cholinesterols. The alkaline phosphotase of bones increases in manganese-deficient chicks.

When formulating starting rations for any poultry, the calcium level must not be above the required optimum level for the first 21 days of life. If the calcium level increases the pH above 6 in the small intestine, the absorption of manganese is impaired. After the first 21 days of life, the absorptive nature of the starting chick or poult becomes similar to that of mature birds who can tolerate high levels of calcium for egg production.

Copper

Copper is one of the essential trace minerals and received early attention as a factor associated with blood formation, hemorrhages and bone deformities. Although anemia was initially considered to be a primary disturbance associated with a copper deficiency, today, connective tissue defects are recognized as the more important defect, and the anemia observed probably was caused by another nutritional problem that manifested itself when the copper-deficient ration was formulated.

Internal hemorrhages in copper-deficient chicks are often severe enough to cause death. When broiler-type chicks receive a purified diet deficient in copper, they grow slowly, and hemorrhages become highly apparent after 3 wk of age. Approximately 75% of the deficient chicks show massive blood clots in the pericardium or peritoneum. The walls of the aorta are discolored from hemorrhages within the tissue and a thickening of the aorta, accompanied by collagen disorientation, is apparent. Elastin, the protein which provides elasticity in blood vessels, and collagen, which imparts strength and rigidity, are not present in the same proportion in copper-deficient as in normal chicks (Table 13.5). The concentration of elastin in the deficient chick's aorta may be reduced by 50% and the lysine increased 3 to 4 times. The increase in lysine content of the elastin is due to the failure of the animal to form desmosine and isodesmosine. These structural crosslinkings involving 4 lysine molecules are not formed in appropriate amounts in copper-deficient tissue. Since the crosslinkages give elastin its elasticity and tensile strength, a copper deficiency results in weaker elastin which leads to rupture of the aortic walls.

The conversion of lysine during the production of desmosine or isodesmosine requires a copper enzyme (amino oxidase). The amino oxidase in the aorta from copper-deficient chicks may be only 25% of the normal. Copper is associated with several other enzymes (Table 13.4).

Copper is required for the utilization of iron in hemoglobin formation. Excess copper in the ration is toxic, although copper is readily excreted. Some poultrymen use copper sulfate to prevent or reduce algae or fungi growth in water. It is also fed to turkeys to help control microbiological growth in the crop.

In the presence of large amounts of ferrous sulfate, the copper will form copper sulfide, which is poorly absorbed. The assimilation and retention of copper are reduced when molybdenum and inorganic sulfate are fed in excess amounts.

TABLE 13.5. COMPARATIVE ANALYSIS OF AORTAS FROM NORMAL AND COPPER-DEFICIENT CHICKS

	Basal	Basal + 50 ppm Cu
Body weight (4 weeks), g	143.0	437.0
Aorta wt, mg	251.0	302.0
Aorta/body weight, %	0.133	0.069
Lipids, % dry	11.8	8.3
Aorta dry/fat-free wt, mg	40.1	56.4
Saline Extract, Based on Wet Weight		
N, %	0.61	0.35
Hydroxyproline, mg/g	1.27	0.35
Elastine Content[1]		
Formic acid residue % dry weight	19.5	47.5
Sodium hydroxide residue % dry weight	28.9	56.9
Amino Acid Compositin of Aortic Elastin		
Lysine	2.68	0.56
Desmosine	0.47	0.79
Isodesmosine	0.39	0.64

Source: O'Dell *et al.* (1966).

[1]Elastin is an insoluble protein characterized largely by its insolubility and resistance to chemical reactions. (Sodium hydroxide and formic acid are often used in its separation.)

Iodine

Iodine is one of the most highly-pooled minerals in the body. The only known role of iodine in the body of poultry is related solely to its function as a constituent of thyroxine and related compounds synthesized by the thyroid gland. It is absorbed from the small intestine and bound to the tyrosine molecules of thyroglobulin. Metabolically, it operates by a closed cycle. It is absorbed from the small intestine and pooled in the thyroid gland, excreted by the salivary glands and then reabsorbed from the small intestine.

When thyroglobulin is hydrolyzed, it yields thyroxine and iodinated amino acids.

An iodine deficiency has been reported in pigeons. The pigeon's crop is a "creative" organ which produces pigeon-milk for feeding the young squabs. There is a relation between the secretion of pigeon-milk and iodine requirements.

Supplementary iodine is not added to most poultry feeds.

Zinc

Zinc was recognized as an essential element in the metabolism of poultry, even when its specific functions were only vaguely understood. A deficiency of this element first appeared in poultry and swine that were fed rations containing mostly corn and soybean meal with above-normal high calcium levels. A deficiency also became observable on corn-sesame meal rations.

The deficiency symptoms in poultry include slow growth, shortening and thickening of the long bones, development of frizzled feathers, abnormal respiration and keratosis of the skin. The keratosis of the chick resembled the parakeratosis of swine. This skin condition is characterized by a breakdown of the outer cell walls before changes in the nucleus or inner membranes are observable.

The feeding of diets of the corn-soybean or corn-sesame meal type with above normal calcium levels resulted in a zinc deficiency. This occurred because the high calcium uptake caused an increase of the pH of the intestinal tract which reduced the availability of the zinc. An insoluble chelate with phytic acid was formed, and so the zinc became unavailable to the chick. The addition of ethylene-diamine tetreacetic acid (EDTA) to the ration resulted in the formation of a chelate with zinc that was more available than the phytic acid chelate. Soybean and sesame meals are good sources of phytates. When the calcium level is normal, a borderline zinc deficiency does not result in pronounced deficiency symptoms; however, at high levels, the deficiency occurs. The chick has a very sensitive digestive system to alkaline pH which becomes evident in manganese absorption. Both zinc and manganese become unavailable when excess calcium is fed to starting chicks or poults.

Zinc is widely distributed in the chick's body; however, most of it is accumulated in the bones and pancreas. In mammals, it is accumulated in the reproductive tract, as well as in the pancreas and bones.

Zinc is present in certain enzymes, carbonic anhydrase and phosphatases. It also is a component of insulin.

The requirement of zinc is approximately 60 ppm. Its requirement is increased when excess calcium is fed and also when the disease, coccidiosis, is present. Coccidiosis reduces the absorption of zinc.

Selenium

Selenium was found in 1931 to be the toxic element in forage crops responsible for "alkali disease" in cattle. In rather specific areas, the selenium content of the soil is such that the forage crops and many cultivated crops have a selenium content reflecting the presence of selenium in the soil. In general, forage crops concentrate soil selenium more than cultivated crops. The production of forage crops on seleniferous soils appears to increase the availability of selenium to cultivated crops. This increase in availability probably results from microbiological reactions.

Certain native plants accumulate high concentrations of selenium when they are grown on seleniferous soils. These are called indicator plants. The selenium content of a soil is hard to predict and also, when water leaches through seleniferous soils, if it will have a toxic level of selenium cannot always be foretold.

Selenium in Grains.—Selenium is found in the gluten of seleniferous wheat. The highest concentration is found in the bran and middlings, and lowest in the flour (Table 13.6). The selenium content of cereal proteins appears to follow sulfur; however, a wide variation in S:Se ratio exists. The proportion of organic and inorganic selenium compounds accumulated in plants depends upon the species; hence, it is genetically controlled.

Many studies have been made on the biosynthesis of Se^{75}-selenoamino acids. The greater amount of Se^{75} incorporated into the plant appears to be bound to the proteins by peptid links. Most of the Se^{75} in protein is freed by digestion. The incorporation of S^{35} (sulfate) and Se^{75} (selenate and selenite) into the glutenin and gliadin proteins by wheat demonstrated that S^{35} methionine and S^{35} cystine were formed but no seleno-substituted amino acids in leaf and wheat proteins were detected. The most common seleno-amino acids are formed in the indicator plants.

TABLE 13.6. SELENIUM CONTENT OF GRAINS GROWN IN SELENIFEROUS AREAS

	Selenium ppm	
Feed	Minimum	Maximum
Corn	1.0	20
Wheat	1.15	30
Oats	2.0	15
Barley	1.7	17
Rye	0.9	25
Wheat bran	—	88
Wheat midds	—	77
Wheat flour	—	53

Source: Rosenfeld and Beath (1964).

There are selenomethylcystine and selenocystathione. Both are uncommon in animal tissues, being found primarily in indicator plants. Selenium appears to be linked to glutathione peroxidose (GSH) to the extent that it is an integral and necessary part of the enzyme. Selenite or Se-amino acids appear to enhance the reducing ability of GSH. SH groups in GSH peroxidase appear to change their redox state during the catalytic process. Selenium may participate in the redox reactions. Some investigators feel that the major role of GSH peroxidase is the degrading of low levels of H_20_2 in cells. This peroxidase also acts on hydroperoxides of unsaturated fatty acids and thus protects cell membranes from oxidative degeneration.

Since Vitamin E prevents fatty acid hydroperoxide formation and the sulfur amino acids are precursors of GSH along with selenium's involvement in peroxide destruction, the intersorting of these substances is observed or postulated. The feeding of high levels of free fats, the low quality of some fats, the surface oxidation of fats and the catalytic effects of metals in the ration, may produce such a high level of peroxides that substances, such as selenium and antioxidants, are necessary to protect the cell membranes.

Vitamin E, antioxidants and selenium became of nutritional importance soon after the establishment of the soybean meal-corn-high-energy ration. Both Vitamin E and selenium investigators often use rations high in fish oil, unprotected fats and wood yeast (torula yeast). A typical Vitamin E-selenium chick ration contains:

Casein	6.0 g
Soybean protein	7.0 g
Torula yeast	10.0 g
Corn sugar	57.1 g
Cellulose	3.0 g
Stripped land	5.0 g

SELENIUM CYCLE POSSIBILITIES
(LOSS OF AVAILABILITY?)

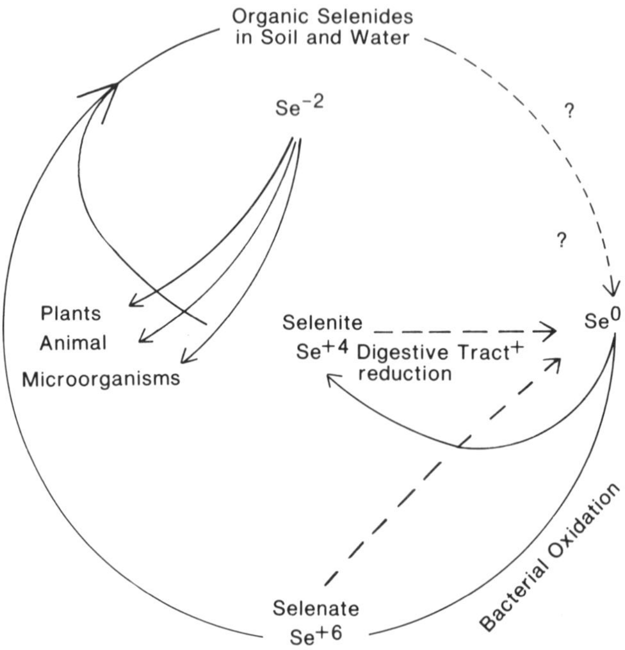

FIG. 13.2. ANIMAL AGRICULTURE MAY DEPLETE SELENIUM FROM SOILS IN 3 WAYS: BY VOLATILIZATION, IN THE ANIMAL ITSELF AND BY REDUCTION TO FORMS UNAVAILABLE TO PLANTS

Vitamin premix	5.0 g
Mineral mixture	6.04 g
Amino acid mixture	4.97 g

When seleno-amino acids are synthesized in the laboratory and fed to chicks, they seem to follow, in part, the metabolic path of the sulfur amino acids. The similarity between selenium and sulfur would suggest a similar metabolic path (Table 13.7 and Fig. 13.3).

Selenium, having a molecular weight of 79, would be expected to move more slowly than sulfur with a molecular weight of 32; however, some reactions with free radicals and volatilization from rations suggest that this element is very active, being more active than sulfur. The high toxicity of very low levels as compared to the low toxicity of high levels of sulfur, further suggests the reactionary nature of selenium.

TABLE 13.7. SIMILAR PROPERTIES OF SELENIUM AND SULFUR

Selenium	Valence	Sulfur	Valence
H_2SeO	−2	H_2S	−2
Se	0	S	0
Se_2O_2	4	SO_2	4
H_2SeO_4	6	H_2SO_4	6

Source: Data assembled from Handbook of Chem. and Physics (1966–1967).

2 8 P 5	2 8 S 6
2 8 18 As 5	2 8 18 Se 6

FIG. 13.3. POSITION OF SELENIUM AND CLOSELY-RELATED ELEMENTS IN THE PERIODIC TABLE

A careful study of its biochemistry shows that selenomethionine is a better methyl doner for choline biosynthesis in vitro than methionine. Selenium apparently has some affinity for methyl groups. When selenium satisfies the requirement to neuteralize a free radical, it possibly ties the methyl group to the structure with the free radical.

$$Ra + Se^- + CH_3 = Ra\text{-}Se\text{-}CH_3$$

A cycle for selenium has been proposed (Fig. 13.2). This cycle accounts for loss of availability.

FUNCTIONARY MINERALS

Functionary minerals help poultry to obtain better feed efficiency, growth and feather development. The bird can live a normal life without them but often, dietary structures or diseases influence the normal performance of animals in which functionary minerals provide some help.

Arsenic—A Functionary Mineral

Arsenic, one of the oldest recognized minerals, was known to the ancients and thought to be a kind of sulfur. It was first prepared in 1250, being considered a semi-metal.

Like phosphorus, arsenic exhibits allotropy, and 3 forms are ordinarily distinguished: (1) gray (2) black and (3) yellow. Both arsenic and phosphorus yield a hydrogen compound analogous to ammonia: phosphine $H\text{-}PH_2$ and arsinine $H\text{-}AsH_2$. Arsenic exists in the tri- and penta-valent forms. The primary, secondary and tertiary phosphines are weak bases and the tertiary alkye arsines are strong bases.

Arsenic-Phosphorus Exchanges.—Arsenic is often found in biochemical structures normally occupied by phosphorus. It apparently gains entrance into these positions by the phenomenon of molecular exchange because so far as determined on essentiality, arsenic is neither required, controlled nor associated with an enzyme system.

Some biochemical compounds containing arsenic:

DNA. Cock sperm was "marked" with radioactive arsenic and the DNA extracted and purified. The radioactive arsenic was found attached to the DNA.

Both phosphorus and arsenic have been reported to replace N in choline-forming arsenocholine. Arsenic can also replace phosphorus in carbohydrate metabolism forming 1 - arseno - 3 - phosphoglycenic acid.

Arsenic often stimulates growth in chicks receiving some fish meal or bone meal products. Analysis of these supplements revealed that those giving a response to arsenic contained high levels of vanadium. Arsenic is recognized as a mineral which reduces the toxic reaction of vanadium.

Fluorine

Fluorine is a member of the halogen group of the elements. It does not occur free in nature, and is widely distributed. Traces are generally found in rocks, thermal waters, bones, teeth, blood and plants. It is perhaps the most active element known.

Fluorine is a very powerful oxidizing agent and may be related to the assimilation of phosphorus. When fluorides are fed, they are rapidly absorbed from the intestinal tract and either deposited in the skeleton or excreted in the urine. Fluorine is always seeking an equilibrium in bone-body fluid content. The fluoride in bones is located in the apatite crystal.

It replaces the hydroxyl or carbonate on the crystal surface. The fluoroapatite is less soluble than hydroxyapatite. The only purpose served by the entrance of fluorine in the bone crystal of poultry is to remove it from the body fluids and thus reduce its toxic effects. It is very toxic and can block a stage in the glycogen cycle.

Bone hydroxyapatite becomes fluoroapatite by replacing an OH ion with fluorine.

$Ca_{10}(PO_4)_6F_2$	Fluoroapatite
$Ca_{10}(PO_4)_6(OH)_2$	Hydroxyapatite

Fluorine is a toxic contaminant in many raw rock phosphates. The fluorine is generally removed by heating the raw rock. It can become detrimental if released in the atmosphere. Phosphate rocks are generally heated to make the phosphate available to poultry.

Vanadium

Vanadium is a toxic mineral found in fish meal, bone and meat scraps. Some marine animals are capable of accumulating vanadium in high concentrations. The concentrations in some productions of fish meal and bone meal may be toxic and require another mineral, such as arsenic, to alleviate the toxic condition.

Vanadium is accumulated in some plants, the concentration being related to the species of plants and the chemical environment of the soil.

Molybdenum

Molybdenum is an active part of the enzyme, xanthine oxidase. This enzyme is found in all animals and is secreted in the milk of cows. An increase in the molybdenum content of the diet increases the production of xanthine oxidase.

Excess molybdenum is toxic and its toxicity can be reduced by feeding tungsten. Excess feeding of this mineral can reduce copper uptake. Supplemental copper, sulfate or sulfur amino acids also reduce the retention of molybdenum. Addition of molybdenum is not recommended.

Although the chick's body produces xanthine oxidase with molybdenum, data have not been obtained to demonstrate the essentiality of this mineral. There are several areas throughout the world, notably in the United States and Australia, where molybdenum is present at a toxic level. In these areas, a general recommendation of 20 ppm of copper prevents toxicity of 10 ppm of molybdenum.

Chromium

Chromium is a metal which does not occur free in nature. Its major industrial use is in the production of particular kinds of steel. It is considered toxic to poultry and has not been found active in any metabolic functions in poultry. It has been shown to act in glucose metabolism in rats and may be a cofactor with insulin. Chromium is rapidly excreted as trivalent chromium in which ascorbic acid in the blood rapidly changes chromium-6 to chromium-3.

REQUIREMENTS FOR NEW TRACE MINERALS

During the past 10 years, several trace minerals have been classified as essential. Growth, control of body content, response to special diets, and appearance are some of the factors considered. Enzyme response was also considered.

The growth, or metabolic response to a substance, does not prove its essentiality. Some substances fed at very low levels stimulate growth or enzymatic response, whereas others may show a toxic effect at low levels as well as high levels. When substances show adverse response at low levels, the law governing this is characterized as Von Euler's Law. This law was developed when it was observed that at a low level of irradiation, a small amount of a chemical or toxin or small amounts of physical stimuli will activate an enzyme system more than large amounts.

TOXIC MINERALS

Most functionary minerals, as well as the required microelements, are toxic at high levels. Some of the functionary minerals are fed at a near-toxic level. Arsenic is fed at such high levels that if the bird does not drink enough water, it will display a form of leg paralysis. The arsenic apparently prevents the kidneys from functioning.

The more liberal definition of an essential mineral places some of the functionary minerals in the required group. Increasing the growth by a trace mineral is not proof that the element is required. A toxic element can increase the activity of some enzymes if fed at a low level.

A mineral can be toxic by several reactions:

(1) Over-stimulation of an enzyme
(2) Prevention of an enzymatic response
(3) Prevention of the formation of a normal biological structure use, as bone crystal (Fluorine can influence the formation of the bone crystal.)
(4) Destruction of the cell by radioactivity
(5) Replacement of an essential element in an enzyme system

Contaminants as Toxic Minerals

Many of the macrominerals and functionary minerals are toxic if used in excess. Another class of toxics and contaminants are radioactive elements commonly called fission products or "fall-out" elements from nuclear energy chain reactions. These are:

Iodine as I^{121}
Cesium as Ce^{137}
Strontium as Sr^{89-90}

There are over 300 radionuclides formed during a nuclear chain reaction. These "fall-out" or fission products are not toxic because of metabolic interference in enzyme reactions as generally thought, but because of the high-energy particles or rays bringing about ionization of elements that cause molecular disintegration.

Iodine^{131} is absorbed and deposited in the iodine "pool" of the thyroid gland where it is used with stable iodine or decays into another element. Its destructiveness is due to the gamma rays emitted. This ray will destroy thyroid and other tissues. It has a radioactive half-life ($T\frac{1}{2}$) of 8.3 days, thus its destructive period is short.

Strontium$^{89-90}$ are bone seekers, and since they possess many characteristics similar to calcium, will be found in calcium deposits. Their radioactive half-life is 59 days for Sr^{89} and 28.8 years for Sr^{90}. They emit high-energy beta particles and can cause heavy damage to bone tissue when present in large amounts. Since the calcium biological half-life is approximately 28 days in chicks, one can expect a similar half-life for strontium.

Cesium^{137} is a soft tissue-seeker, and like potassium, is intracellular in location. It is readily complexed to the cellulose micelle and some silicates. Potassium is not pooled in the body and has a short biological half-life.

The rate of excretion of both radioactive strontium and cesium can be increased by feeding excess calcium and potassium. The formation of chelates with strontium and simple complexes with cellulose and silicates by cesium, aids in increasing the rate of excretion of these radioactive contaminants.

14

Nutrition of Water

Water is an indispensable nutrient, since no animal, plant or microorganism can live without it. All living things appear to require about the same concentration of water. The ready availability of water has caused many poultry nutritionists to overlook the real value of this nutrient. Its many properties account for its usefulness.

The nutritional biochemistry of water is complex and difficult to understand. The life process depends upon the peculiar properties of water and the probable way it is organized in the tissues and cells.

Water is not the simple structure that the formula, HOH, implies (Fig. 14.1). Many water molecules can be linked together by hydrogen bonds, thus forming a complex macromolecule. The bonding of the water molecule at the surface produces what is known as surface tension. Positioned in the cell is a form of water called "polywater." A part of the bound water "membrane water" and all of the surface water may become disorganized in such a way that the whole basic structure of the chemical organization in the cell is influenced by water absorption. The inner-water structure may be altered when a water deficiency occurs.

Although water is indispensable for chickens, we find that the most difficult problem in poultry management is controlling the water content of the poultry house environment. The high moisture content of the poultry house aids in developing a desirable environment for molds and bacterial growth which often cause diseases in young poultry. Also, the excess moisture can encourage decay of feed. Excess water, over 14%, encourages mold growth and oxidation of fat or fat-soluble substances in feed.

Watering equipment for poultry production is difficult to maintain and operate. Feed particles dissolved from the chickens' beaks and the microorganisms growing in the water continuously erode the metal troughs.

Water properties which allow it to serve in many body functions are:

(1) Living cells are lyophilic gels
(2) Water exists in three states:

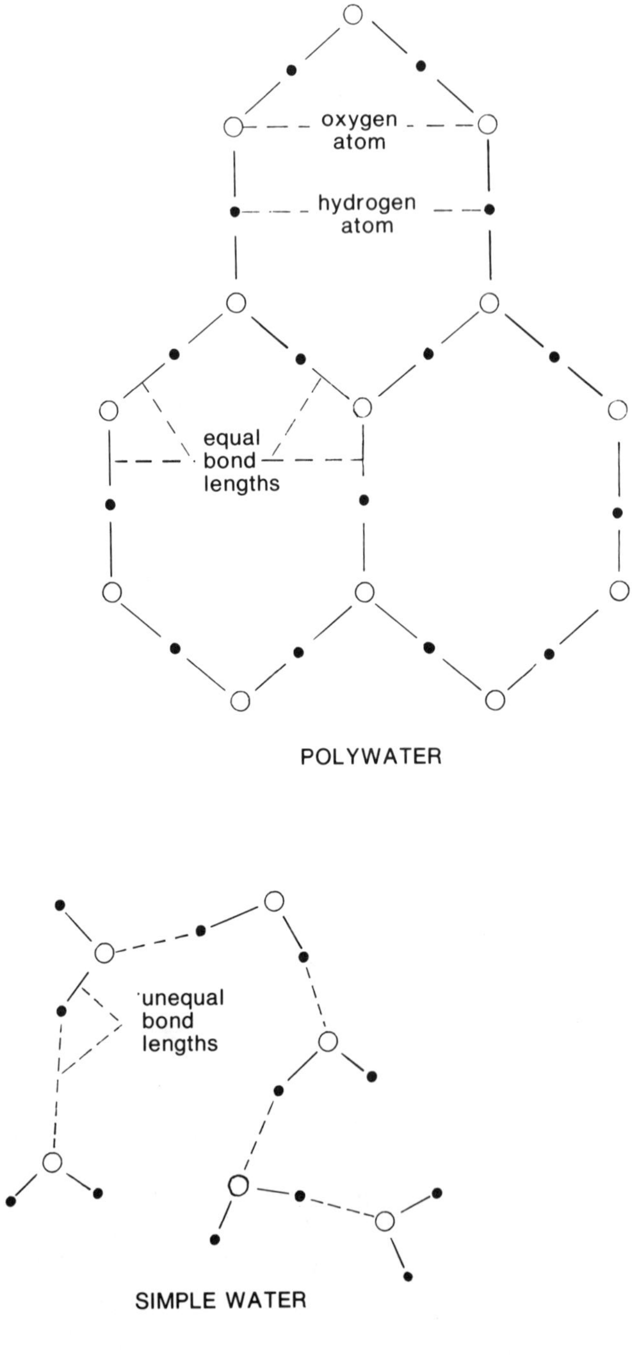

FIG. 14.1. SIMPLE WATER
Water is a complex macromolecule.

   (a) Liquid—definite volume—no shape
   (b) Gas (steam)—no fixed volume—no shape
   (c) Solid (ice)—definite volume—no fixed shape
(3) Water has no color or taste
(4) Water as a liquid has higher density than as a solid
(5) Water has unusually large specific heat—heats easily and cools slowly
(6) Water has unusually large heat of vaporization: 1 Cal. per degree of change up to 100°C. Then there are 540 Cal. from liquid to gas
(7) Water is a relatively stable substance
(8) It dissolves more substances than any other substance
(9) It is active in hydrogen bonds and hydration

The main functions of water are:

(1) Digestion—a hydrolytic process
(2) Translocation—all chemicals in the body
(3) Absorption of nutrients from digestive tract
(4) Excretion of metabolic wastes and unneeded substances
(5) Secretion of hormones, enzymes and other biochemical substances
(6) Heat exchange—aids in maintaining constant body temperature
(7) Cell shape—osmotic pressure
(8) Chemical solvent for biochemical reactions

Water is the universal solvent in the chicken's body, and performs many functions in the nutrition of the animal. These functions are performed either within the cell, the intercellular water or in the tissue fluids, extracellular water.

The intracellular water is considered to be distant chemically from the extracellular water. The fact that water seems to be always moving in and out of the cell while the water on the inside of the cell differs from the water on the outside, makes it hard to understand why chickens do not readily recover from a water deficiency. The answer may be that water near the cell's surface, or that which occupies or forms false membranes, is a complex structure that is not readily repairable.

WATER REQUIREMENTS OF POULTRY

Water requirements are influenced by many factors:

(1) Body Temperature.—The dissipation of body heat is through water evaporation. Nearly 40% of the water loss from the chicken's body is through respiration. The high body temperature of most poultry, 105° to 107°F, requires dietary energy sources to produce and hence, water to dissipate.

(2) Environmental Temperature.—Poultry requires more water in hot weather than (Table 14.1) in cold weather. Water evaporation from the watering equipment is greater in warm weather. One of the major factors increasing the water requirement in warm weather is lowered feed uptake. About 25% of the water required comes from metabolism of protein, fats and carbohydrates, and as the bird consumes less feed, additional water must be drunk to meet the daily requirements.

TABLE 14.1. INFLUENCE OF TEMPERATURE AND PROTEIN SOURCE ON FEED EFFICIENCY, WATER REQUIREMENT AND FECES OF BROILERS

	Type Protein 76°F			and	Temperature 60°F		
	Basal	Casein	S.O.M.		Basal	Casein	S.O.M.
Feed efficiency	3.1	2.5	2.9		3.5	2.8	3.3
Water/feed	2.5	2.0	2.4		2.0	1.5	1.8
Feed/feces dry	4.0	3.7	2.8		3.2	3.0	2.5

Source: Patrick (1955).
Casein (8% casein added to basal).
S.O.M. (15% soybean oil meal-44 added).

(3) Form of Nitrogen Excreted.—Most of the nitrogen excreted by birds is in the form of uric acid. The urine is excreted and passed into the lower gut where it is semidehydrated before being excreted with the feces. This allows the recovery of urinary water; however, additional water is required in synthesis of uric acid as compared to urea.

(4) Protein Source.—Protein sources, such as meat scraps and soybean meal, increase water requirements, whereas milk protein decreases the requirements. Some productions of fish meal increase the water requirements because of the high salt content.

(5) Genetic.—One of the main factors controlling the metabolism and requirements of water is the genetic makeup (Dunson and Buss 1968). An inbred strain of chickens was established which appeared to be normal in all respects except that excessive amounts of water, an average of 25.3% of their body weight per day, were consumed with similar excretion levels, while the normal controls consumed water equivalent to 7.2% of their body weight. Turkeys show similar water-metabolic characteristics.

Excessive drinking (polydespsia) and urinary output (polyuria) are present in many commercial flocks. These conditions have been recognized among birds placed in individual cages for biological studies and for commercial egg production.

(6) Fiber or Energy Level.—Poultry consuming high-energy feeds require less water than those receiving low-energy rations. The water formed per g nutrient metabolized is:

Fat	1.07 g
Starches	0.56 g
Proteins	0.34 g

For each 100 Cal. produced, the body forms 10–15 g water. Rations that are high in fiber increase the water requirements. Since more feces are produced on high-fiber rations and the moisture content of normal feces is similar, the bird will require more water in fecal production. (The moisture content of normal feces is approximately 75% and is apparently controlled by genetic factors.)

(7) Mineral.—Many minerals influence the water requirement in which salt (NaCl) has the greatest impact. Excess salt increases the water uptake and also the water content of the feces. The high-salt feces will not dry, but rather exchange their moisture with the drier litter, causing wet litter. The "salt" feces will also absorb moisture from the air and exchange this with the litter, causing increased wetness of the litter.

In general, any mineral fed in excess, or nutrient(s), such as vitamins, that cause excess mineral excretion, will increase water content of the feces and cause greater water uptake.

(8) Purpose of Animal.—Laying hens require more water than starting or growing birds. Hens in production require about 2 lb water for each pound of feed consumed and broilers (Table 14.2) require 1.5 lb water. Most of this increased requirement goes into the egg, which is about 66% water.

(9) Pathology.—Many diseases of the intestinal tract produce greater water uptake and higher moisture feces.

(10) Management.—Hens, or any poultry confined in cages, will consume more water than floor-managed birds. Many poultry producers limit the water uptake time to 15 to 30 min every 2 hr. The moisture content of caged birds' feces is higher than of floor-managed birds (Table 14.2).

WATER CONTENT OF POULTRY

The water content of poultry and livestock decreases with age. The White Leghorn's body water lowers from 68.7% at 2 wk of age to 55% at 32 wk of age (Table 14.3). The water content remains rather constant from 2 to 4 wk of age, then drops rapidly for the next 8 wk. In cattle, the water content remains relatively constant until the calves are eating solid feeds. The water is replaced with fat; therefore, the producer of meat should have his animals ready for market with the greatest amounts of body water and fat needed to produce a prime carcass.

TABLE 14.2. ANALYSIS OF FRESH MANURE FROM HENS UNDER 2 SYSTEMS OF MANAGEMENT

Management System	Number Birds	N %	P_2O_5 %	K_2O %	Average Moisture %
Cage	630	3.17	3.79	2.02	74.35
Loose housing	1013	3.18	3.29	1.84	70.16

Source: Patrick, Johnson and Hyre (1966).

TABLE 14.3. TOTAL BODY WATER OF WHITE LEGHORN PULLETS

Age of Pullet (Weeks)	% Plasma	% Body Water
1	95.5	85.2
2	96.3	68.7
3	96.1	67.1
4	95.8	68.9
6	95.8	59.2
8	95.5	65.9
16	95.1	48.7
32	94.6	55.0

NECESSITY OF AMPLE SUPPLY

Poultry apparently over-consume water. Their intake can be restricted 10 to 20%, but when restricted over 20%, laying hens may drop out of production at about the fifth day. Production will return to the former levels if the deprivation is for less than 36 hr. Week-old chickens deprived of water for less than 24 hr are not in serious trouble, but beyond that period, birds do not completely recover. Water and feed consumption figures for average conditions are summarized for broilers in Table 14.4. Broilers consume about one and one-half times as much water as feed. There is no surer way to stop egg production than to cut off the supply of water. Water must be kept from freezing in cold weather and an additional supply is needed on hot days. Birds do not like water that is over 80°F. Provision to keep water from freezing in the winter includes electrically-heated fountains, immersion-type heaters and light bulbs underneath the waterer.

Continuous-flow watering systems should be cleaned at least 3 times weekly with a stiff-bristled brush. Flush the troughs with a disinfectant, such as quaternary ammonia or cresylic acid, then carefully flush the disinfectant out of the troughs or it will poison the flock. Water treated with chlorine retards bacterial growth in water troughs. The amount and kind of various mineral salts in solution in water should be checked. The carbonates, bicarbonates, chlorides, and sulfates of sodium, potassium, magnesium and calcium occur most commonly. Certain sources of water

TABLE 14.4. WATER REQUIREMENTS OF BROILERS[1]

	Age in Weeks									
	1	2	3	4	5	6	7	8	9	Total
lb water per week	0.230	0.460	0.820	1.380	1.47	1.78	1.95	2.10	2.36	11.92
lb feed per chick	0.046	0.413	0.548	0.836	0.98	1.12	1.31	1.35	1.57	8.17
lb water per lb feed	1.960	1.120	1.480	1.640	1.50	1.59	1.49	1.57	1.48	1.46

Source: Patrick and Ferrise (1962).
[1]Live weight at 9 weeks 3.68 lb, feed efficiency 2.22.

contain such high percentages of minerals that they are unsuitable for poultry. Over 1000 ppm of various minerals cause problems. Very high levels of salt in the water cause prolapse.

HARDNESS OF WATER

Hardness of water is due to dissolved salts and is expressed in ppm of calcium carbonate or grains of hardness per gal. A grain of hardness is equivalent to 17.5 ppm. Soft water contains 15 to 50 ppm and is preferred by poultry; hard water contains from 150 to 200 ppm. High levels of iron in water can be a problem because of the staining characteristics of iron. Even 3 ppm can be objectionable. Excess iron in water will cause a rapid lowering of egg quality. Water treatment can remove calcium, magnesium and iron. A high nitrate content in water is serious, producing acute poisoning or interfering with the utilization of nutrients so that the animals are unable to convert carotene efficiently into Vitamin A. Thus the feed must be supplemented with a free form of Vitamin A. The presence of nitrites and ammonia in surface waters indicates contamination with drainage from barns and outhouses and the possibility of the presence of pathogenic microorganisms.

Some poultrymen control water consumption to keep the laying house as dry as possible and the droppings in better condition. Controlled watering systems are used to prevent overconsumption and reduce water wastage and also, to save as much as 4.5% of the feed over continuous-flow systems, since feed is washed away in the watering troughs. For trough watering systems, water is turned on for 30 min and off for 90 min. In other words, the birds have free access to water 30 min out of every 2 hr. This watering program will not affect egg production, providing the room temperature does not exceed 90°F and the humidity does not exceed 55%.

CONTAMINANTS FOUND IN WATER

Usually, drinking water contains very few contaminants; however, contamination occurs occasionally in certain parts of the country. Substances which prove toxic are fluorine, molybdenum, selenium, nitrate and other specific trace substances in relatively heavy concentrations. Water contamination may also occur from commonly-used household detergents. As little as 0.5 ppm detergent in the drinking water causes undesirable taste and foaming.

Infectious diseases can be transmitted through drinking water. Among these are coccidiosis, salmonellosis, leptospirosis, brucellosis and bovine tuberculosis. Ducks perform well with fairly high *E. coli* contamination in the water while chickens are depressed in weight under the same conditions. Turkeys have a higher tolerance for nitrates than cattle. Generally, acceptable drinking water for poultry should be free of hydrogen sulfide, seleniferous gases and ammonia, and be low in *E. coli* count, total nitrates and nitrites. Water temperature affects voluntary intake to a great extent. Poultry prefer water with the chill taken off when they are in a cold environment, and cool water when the surrounding air temperature rises above 75°F.

Water supplies are frequently tested to see if coliform bacteria are present. These bacteria are found in fecal material and are rather easy to identify. If coliform bacteria are found in a particular water supply, it is presumptive evidence that the water is being contaminated by fecal material.

AVAILABILITY OF WATER

There is an enormous quantity of water on the earth but relatively little of it is drinkable. Water in the ocean is not suitable for drinking since it is high in minerals. Minerals can be removed, but at great expense. Where water is impounded in a lake without an outlet, such as Salt Lake, salinity may develop up to 350,000 ppm. The palatability of water is greatly enhanced by the presence of carbon dioxide, as shown by the popularity of carbonated drinks. Rain water or water that is boiled has a flat taste, due to lack of carbon dioxide. There are 30 or more mineral salts present in surface or deep well waters, and of these, most are carbonates, bicarbonates, chlorides and sulfates.

Certain parts of the midwest have alkaline water which contains sodium and calcium carbonate at levels around 5000 ppm. Animals build up a tolerance for such waters, but feed efficiency is reduced in the process. Poultry is much more susceptible to minerals dissolved in their water than in their feed.

WATER ANALYSIS

Water that comes from open ponds or reservoirs must be tested for impurities and contamination. The only water that does not need testing is that from an approved municipal facility.

Well water should also be tested for contaminants. Some of the determinations made to evaluate water are as follows:

(1) Color.—Good water is colorless, odorless and tasteless. Soluble iron is visible and seems to influence the color of cooked egg yolks. Water that seeps through wood containing tannic acid will not be palatable. Tannic acid reduces digestibility of proteins.

(2) Hardness.—Salts of calcium and magnesium cause the water to be "hard" and may form scales or crystals which damage water supply valves. Magnesium tends to increase water uptake and also causes an increase in fecal water excretion.

(3) Trace Minerals.—Fluorine, iron, selenium, and molybdenum are often present in toxic amounts. Excess fluorine retards carbohydrate metabolism; iron causes discoloration in eggs; selenium is extremely toxic.

(4) Nitrogen.—The presence of organic nitrogen in water indicates decaying organic matter or fecal material. The presence of excess nitrates in water indicates surface water from highly-fertilized fields and the nitrates may be toxic to all animals. Excess nitrates influence carotene-Vitamin A utilization. Some shallow wells in highly-fertilized areas contain so many nitrates that they cannot be used for animals and plants which tolerate only low levels of soluble nitrates.

(5) pH.—The pH of a solution is a measure of its acidity or alkalinity. When above 7, it is alkaline; when below 7, it is acid. Most water is pH 7 to 7.2. If above 7.2, it should be tested for magnesium and calcium.

(6) Sulfur.—The presence of sulfur can be detected by its odor (rotten eggs). Water which is high in sulfur is not palatable and tends to cause hemorrhages under the skin, as well as the collection of fluids (edema). Many wells and ponds in coal-mining areas are high in both sulfur and iron. Water troughs become stained from the iron sulfates.

(7) Turbidity.—Water that contains small colloidal particles is generally turbid, causing damage to the valves and making the water unpalatable. Turbidity generally suggests surface water contamination. Turbidity and total solids should be determined if detectable by sight.

(8) Bacteria.—When bacteria are present, surface contamination or presence of organic matter, such as feces, is indicated. If present, the water should be treated with chlorine. Sometimes, wells become contaminated by bacteria following tree roots or gaining entry through broken walls. Treatment with a chlorinating solution often helps eliminate the bacteria.

Some common standards to use in evaluating water for poultry are:

Particulate	*Maximum Level*
Total dissolved solids	1000 ppm
Iron-soluble or colloidal	50 ppm
pH	7.6 to 8
Nitrates	40–50 ppm
Sulfates	250 ppm
Sodium chloride	500 ppm
Total alkalinity	400 ppm
Bacteria—*E. Coli*	not present

15

Nonnutrient Feed Additives

A feed additive is a substance or mixture of substances other than the basic feedstuff present in the feed. The term does not include chance contaminants. Feed additives make feeds better, safer and more efficient, and are an important "tool" of feed science. While not new, their increased use has focused public attention on them.

Feed additives are used in scientifically-controlled amounts. The amount necessary to perform the needed function is usually small. The safety of the substances is carefully established before they are used in feeds. Many are created by scientists in response to special problems. Some laboratory-created chemicals are also found naturally in foods. For instance, cheese naturally contains propionates, substances used to retard mold. Whether the additives come from foods or are synthesized, they are chemical in nature.

It is almost impossible to prepare high-quality poultry feeds without the use of chemical additives because they have become vital to modern feeds. Some, like vitamins, are essential for good health. Others are needed to protect crops from being consumed by insects. Still others are useful in improving color, flavor, texture and keeping-qualities of feeds.

Improvements in growth, egg production and hatchability result from enrichment of poultry rations with traces of various chemicals.

These additives include medications, mold inhibitors, hormones, and feed stabilizers, such as antioxidants. Such additives relate frequently to the environment: oldness or newness, sanitation or contamination of quarters; wire floors versus solid floors; litter versus no litter. These factors influence growth response and feeding results.

Chemical additives should be added to feeds by competent individuals, since these compounds may be harmful if fed in excess of recommendations.

MEDICATION OF FEEDS

Early poultry producers did not have specific drugs for common diseases. Many used "patented" remedies or local folklore remedies.

Coccidiosis was treated by feeding a flushing mash and elevating the temperature in the brooder. A flushing mash was made by combining 9 lb chick starter with 1 lb dry skim milk. This was fed for 2 days. Another flushing mash was prepared with 20 lb chick starter, 1 lb dry skim milk and 1 lb molasses.

During the early history of coccidiosis, only 3 forms were known: E. tenella; E. necatrix; E. maxima. Some chemicals were found effective when added to the diet. The addition of 5% flowers of sulfur to the feed was used. If not used for over 5 to 7 days, the chicks were given protection to withstand the damaging effects of the sulfur to the intestinal tract and destruction of the Vitamin D. The sulfur dioxide formed in the intestinal tract was very destructive to the intestinal walls and fat-soluble vitamins.

The feeding of specific drugs to control coccidiosis was initiated with the low-level feeding of sulfaquinoxaline. Several sulfa drugs were used until poultrymen observed that the drugs were becoming ineffective because "drug-resistant" strains were appearing. Also, new species of coccidia, formerly not pathological, were becoming destructive. Several drug manufacturers entered the field of coccidiostat production. The recommendation was soon made that a control program should change the coccidiostats rather than continually use the same drugs.

Sulfa drugs are also destructive to desirable intestinal flora which synthesize vitamins and decompose nondigested nutrients. Sulfanilamide reduces the synthesis of folic acid and causes anemia in chicks.

Many chemical structures have been found effective coccidiostats. A rather interesting drug that has been extremely effective is know as Amprol. This drug destroys coccidia by interfering with thiamin pyrophosphate (cocarboxylase). The coccidia organism requires this vitamin.

Table 15.1 presents some drugs used in the control of coccidiosis.

MOLD INHIBITORS

Mold inhibitors are chemicals added to poultry feeds to protect against mold development. Molding occurs when feed is too moist or is shipped or stored in adverse conditions of high humidity and high temperatures. Molding causes heating of feeds which results in destruction of vitamins and impairment of the availabilities of certain other nutrients. Some species of mold are definitely poisonous; others are harmless or beneficial, but it is safest to prevent the development of mold in feed for poultry.

Molds are found throughout the poultry environment. When humidity and temperature are high enough, molds will develop. One of the most common sources of molds is in peanut hulls. When used as litter, molds

TABLE 15.1. SOME DRUGS USED IN CONTROL OF COCCIDIOSIS

Drug	Manufacturer	Recommended Levels for (% Active Drugs, and lb per ton)	
Amprol	Merck	0.0125	1.0
Nicarbazin	Merck	0.0125	1.0
Sulfaquinoxaline	Merck	0.0125	1.0
Zoaline	Dow Chemical	0.0125	1.0
Unistat	Dr. Salsbury	0.06	2.0
Megasol	Lederle	0.0126	1.0
NFZ	Hess and Clark	0.0055	1.0
Buqinolate	Norwich	0.00825	1.6

often develop in the hulls and get into the lungs of young chickens, causing aspergillosis or brooder pneumonia. This disease is caused by the molds of the genus, aspergillus. The chicks are unthrifty, emaciated, and have pulmonary trouble resulting in rapid fatal pneumonia.

HORMONES

Hormones or hormone-like substances are used in poultry feeds to quiet birds and improve fattening. The most common hormone injected or implanted under the skin of chickens is diethylstilbesterol.

Estrogens are hormones which in nature, are produced by the ovaries of all females (hens) just before laying. Estrogens are responsible for the deposition of large amounts of fat in the muscle, under the skin, and in the visceral and abdominal sections. Synthetic estrogen-like compounds are available and are used in fattening both males and females. In older birds, this results in increased weight gains and a reduction in the number of pinfeathers on dressed carcasses.

Estrogen-like compounds added to some poultry feeds and diethylstilbestrol implanted in poultry affect fat deposition. Since stilbestrol is not effective when administered orally, dienestrol diacetate (Lipamone) is used in feeds to improve market finish. There is no improvement in feed conversion in broiler-age birds and little benefit to rate of growth. Males of broiler age fail to comb out and resemble pullets. The rather high cost of estrogen for feeding discourages widespread general use. This has been described as "chemical caponization" of males, although the effect is due to the presence of female hormones rather than the absence of male hormones.

Estrogens cannot be fed to poultry nor poultry used for food which have been implanted with pellets of this hormone. The Food and Drug Administration (FDA) does not approve its use.

THE GROWTH-PROMOTING ADDITIVES

The growth-promoting drugs were first introduced in feeds during the search for the identity of the Animal Protein Factor which was an Unidentified Growth Factor before it was recognized. Many substances of a nonrequired nature have been added to poultry feeds to promote growth or improve feed efficiency. This includes such products as iron soot from a blast furnace to strengthen the leg muscles of game cocks, and arsenic and strychnine to ward off miscellaneous evils. Today, the most commonly used growth-promoting drugs are antibiotics and arsenicals.

Arsenicals

Arsenicals, widely used as chemical additives in poultry rations, are 3-nitro-4-hydroxy phenylarsonic acid (3-Nitro) and P-amino phenylarsonic acid (arsanilic acid). These compounds improve growth and market grades of broilers and produce precocious combing with bright red combs and wattles. Capillaries are enlarged and engorged through the dilator effect of arsenic. Arsenicals probably function in a manner similar to antibiotics, although evidence is not as far developed on their mode of action. The amount of arsenic retained within the tissue is quite low—about the same as what is naturally found in some seafoods. Arsenicals must be removed from the feed 5 days before slaughter during which time practically all the chemical is excreted.

The ratio of arsenic-producing toxicity to that producing desirable results in poultry is fairly small, much smaller than in the case of antibiotics. Only simple precautions against swallowing and inhaling the arsenicals are required in the feed mill during operations of weighing and mixing of feed, but care should be exercised in the usage of arsenicals. It is claimed that arsenicals improve pigmentation, but this is controversial. Apparently, arsenicals do not promote an increase in pigmentation in conditions where disease is not present. Arsanilic acid is less toxic than 3-Nitro. Arsenicals are approximately equal to antibiotics in promoting growth.

Antibiotics and arsonic acids produce fairly comparable changes in the intestinal flora of chickens. Canadian researchers observed increases in lactobacilli counts and a fall in pH of fecal contents. Texas workers reported a marked drop in total clostridia per g chick feces. *Clostridium perfringens* is a possible inhibitor of animal growth, and arsenicals and antibiotics greatly decrease the numbers of this organism in the intestinal tract of turkey poults. More research, and particularly standardization of techniques, are needed to develop the promising leads that implicate specific bacterial groups.

Arsanilic acid has no unfavorable effects on egg production or hatchability when fed to pullets for 11 wk. Accumulation of arsenic in eggs and tissues is proportional to the amount in the ration and well below allowable levels. In general, toxicity appears to be related to the amount of arsenic deposited in tissue. Arsanilic acid is tolerated up to 0.1% in the diet of chicks and up to 0.02% in the diet of turkeys. Arsanilic acid and 3-Nitro offer partial protection against selenium at 10 ppm of the diet.

Arsenicals improve production on low-protein rations but not on normal protein rations. With turkeys, antibiotics and arsenicals generally pay for themselves when fed on a low-level basis. Like antibiotics, the addition of arsenic acids is more effective where conditions are suboptimal and egg production can be improved. Where production, egg size and quality, livability, and progeny performances are excellent, they give little or no response. Most experiments have been performed in conditions probably superior to those found in the field. Consequently, it is possible that their greatest value may be derived in more adverse and diverse conditions as with commercial production.

Poultry feeds may contain several chemical additives, including two or more antibiotics, an arsenical, a nitrofuran, an antioxidant and a coccidiostat. Most research work has shown benefits from adding each of these compounds, singly or in combination. The manufacturers of chemical additives have recommended either continuous or intermittent use of these drugs when disease strikes. Chemical additives make the biggest contribution during the very early growth of the birds.

Such chemical additives are also used in laying rations to stimulate egg production and the general health of the flock. Marked improvement in egg production from these drugs has usually been observed when production has dropped sharply. Less response is obtained when production is high.

This increased production is difficult to achieve if the birds are already laying at 70%. Before feeding these drugs, it is advisable to submit a few representative birds from the flock to a poultry pathology laboratory for diagnosis of the disease condition.

Some representative products are as follows:

Drug	*Manufacturer*	*Levels in Feed for Growth (G per Ton Feed)*
Aureomycin	American Cyanamid	10
Terramycin	Chas. Pfizer	10
NF-180	Hess and Clark	10
Zinc bacitracin	Commercial Solvents	4
Procaine penicillin	Merck, etc.	4

Gallimycin	Pfizer	10
Pro-Strep	Merck, CIBA, etc.	4
Tylan	Elanco Products	10
Arsanilic acid	Abbott Laboratories	90
3-Nitro phenylarsonic ac.	Dr. Salsbury	45
Sodium arsanilate	Dr. Mayfield Laboratories	45–90

FIG. 15.1. RHODE ISLAND RED COCKERELS WHICH RECEIVED DIET WITHOUT (left) AND WITH (right) 20 G AUREOMYCIN PER MT OF FEED; AT 6 WEEKS OF AGE CHICKENS WEIGHED 590 AND 765 G, RESPECTIVELY

Antibiotics

Years ago, Pasteur suggested that human diseases might be cured by marshaling microbe against microbe. Researchers looked for chemical agents from microbes friendly to man to destroy disease germs. Antiseptic chemicals of biological origin were the world's first antibiotics, but little was done with them until 1928, when Dr. Alexander Fleming, a bacteriologist at St. Mary's Hospital in London, England, had a profoundly important experience. One of his petri dishes containing bacteria had a clear ring in the center showing where germs in the vicinity of the invading substance had been destroyed. The interloper in the dish was a penicillin mold frequently found in the air. He extracted a chemical from the mold which had killed the *Staphylococcus* organisms and finally obtained a tiny quantity of a brownish secretion. During the course of World War II, Fleming showed that penicillin worked wonders on laboratory animals and humans.

Since Britain's production facilities were devoted to war work, the United States, not at war at the time, devoted its mass production techniques to further research. The penicillin was unstable and it was necessary to coax the mold into producing more penicillin. Finally, the mold was cultivated in large aerated tanks. A new strain of penicillin was then found on a moldy cantaloupe in a Peoria, Illinois, market. It thrived in tanks and eventually gave much more penicillin than the original strain.

With the success of penicillin, scientists set out to find other antibiotics capable of fighting diseases against which penicillin had proven ineffective. Molds, yeasts, actinomyces, bacteria or similar microorganisms are available in soil, lakes, streams, in the sea and in the air. No microbe is without enemies and none is invulnerable to attack.

Microbes are considered members of the plant kingdom, even though they are microscopic and without chlorophyll. Molds are more nearly akin to the layman's conception of plant life and look like trees, sending out thread-like branches of white, yellow, blue, green, gray, red, black or brown. Having achieved some growth, they are visible to the naked eye, as in the case of mold on bread. Molds reproduce themselves by means of spores, a process roughly equivalent to seeding in plants. Spores are 1/5000 in. in diameter and are dispersed by actual explosion. Survival of spores depends on the medium by which they are carried—wind, water, insect, bird or animal, and on the environment in which they land. The penicillin spores which were planted on Dr. Fleming's petri dish were able to flourish because the dish contained a nutrient on which it could feed.

Actinomyces are a good deal like molds in general appearance and are propagated in a somewhat similar manner. They form chains of spores like the beads in a necklace. Yeasts are colorless, egg-shaped plants produced by buds that develop into whole cells. Bacteria simply divide and produce new cells from the original ones. Under the microscope, they appear as dots, corkscrews or rods. In favorable conditions, some of the organisms multiply once every half hour. Bacteria may be dyed with a stain named after Professor Gram, a Danish bacteriologist. Those dyed are known as gram-positive, and include germs which cause boils, abscesses, pneumonia, venereal diseases, scarlet fever and rheumatic fever. Penicillin is effective against these diseases but is useless against the gram-negative bacteria which are not stained by the dye and cause diseases, such as typhoid, undulant fever and certain types of pneumonia.

Professor Waksman of Rutgers University and his students isolated from the throat of a chicken, *Streptomyces griseus*, which produced the antibiotic, streptomycin, effective not only against gram-negative bacteria resistant to penicillin, but against *Mycobacterium tuberculosis*, the organism causing tuberculosis, as well.

From all over the world have come hundreds of thousands of soil samples for evaluation. Molds which appear promising are allowed to grow in the presence of disease germs to determine whether or not they produce the "ring of inhibition," a signal that an antibiotic is at work. In 1949, a bit of earth from the Midwest was found to contain an actinomyces worthy of intensive investigation. The germ was named *Streptomyces rimosus* and the canary-yellow crystals of the antibiotic, terramycin, attack both gram-positive and gram-negative bacteria, as well as rickettsia and the large viruses which cause psittacosis.

Mode of Action of Antibiotics.—The method of discovery of antibiotics and their introduction into medicine and feeds have had a picturesque history. The theory that microorganisms, antagonizing each other on an agar plate, should yield drugs useful in the treatment of disease, recognizes the universal truth that there exist in the biological world many balancing forces which assure some sort of equilibrium between living things. This was recognized by Pasteur, who in 1873, proved that microorganisms commonly present in soil affected the anthrax bacillus in such a way as to render it unable to establish disease in animals. He clearly recognized that it is possible to inhibit selectively certain types of microbial growth by the use of proper chemical substances devoid of any inhibitory act on other microbial species.

Ehrlich and Wygant suggested that drugs act by competing in some of the essential metabolic processes of the parasite, hence interfering with its nutrition. Vigorous chemotherapy may bring about an almost complete elimination of the normal intestinal flora which is then replaced by other kinds of bacteria or fungi. There is ground for the view that the drugs enhance growth by inhibiting certain microbial activities which in some way or another exert a deleterious effect on their economy as a whole.

The manner in which feeding antimicrobial drugs increases the rate of growth of poultry and renders more efficient their utilization of food is not fully understood. Antibiotics are not nutrients in the ordinary sense. They are chemicals or drugs produced by living organisms which destroy other living organisms. Commonly used in the feed industry are procaine penicillin, oxytetracycline, chlortetracycline and bacitracin. They are used in 2 ways: (1) high level inclusion in the diet over a short period as a medicine to help cure a particular infection; and (2) low levels continuously in the diets of birds throughout the feeding period for improved poultry production. When antibiotics are used, it is imperative that the directions of the supplier be followed. The feed industry is particularly concerned with low-level usage.

Veterinarian use of antibiotics to treat diseases is outside the scope of this discussion; when antibiotics are put into feed at low levels, it is for growth promotion, not for the treatment of disease or for any medical reason.

When antibiotics were first used in feeding, researchers obtained differences of as much as 18 to 25% in the growth rate of birds. Now, it is unusual to get a 3% difference. Actually, the environment is improved due to the use of antibiotics. When first used, birds on the control, nonantibiotic rations started from a lower base; thus, percentage improvement was better. Evidently, antibiotic-fed birds are growing better than ever.

Poultry have achieved through the processes of revolutionary adaptation some sort of equilibrium with the microbial species in their environment, and in particular, with the microorganisms which they constantly harbor in their tissues. This equilibrium does not necessarily result in the best possible state of health; rather, it corresponds to a compromise between many conflicting necessities. The presence in the tissues of large numbers of relatively innocuous microorganisms helps in controlling the multiplication of other species with pathogenic potentialities. Many of the microorganisms normally present in chickens have mild deleterious effects which, although compatible with "normal" performance, interfere somewhat with the optimum performance of metabolic functions. Therapeutic measures which eliminate or attenuate these deleterious influences bring about, indirectly, greater metabolic efficiency.

Use of Zinc Bacitracin.—Low-level feeding with an antibiotic, such as zinc bacitracin, is usually all that is needed to control the disease level normally encountered in good poultry operations. This substance is mainly effective in the intestinal tract. Ulcerative enteritis may call for somewhat higher levels of this or other antibiotics. Once the invasion of the bloodstream by pathogenic organisms has occurred, high levels of highly-absorbable, usually broad-spectrum antibiotics are needed. These are administered in such a way as to promote high blood levels of the antibiotic, as by injection. The major growth-promoting effects of low-level feeding of antibiotics occur because of the beneficial effects of these antibiotics upon the microflorae of the intestinal tract. The mere fact that zinc bacitracin, a practically nonabsorbable antibiotic, is effective in growth promotion, should be ample evidence in support of this conclusion. Antibiotics vary in their stability and rate of absorption in the intestinal tract. No residues are present in meat or eggs with combinations commonly used as feed supplements.

Use of Chlortetracycline.—In 1950, researchers at American Cyanamid reported improved growth rate of young chickens when they were fed low levels of chlortetracycline (Aureomycin). In England, when birds were raised in quarters which had been in use for some time and exposed to infections which caused no visible pathology, growth of chicks was depressed. When an antibiotic was given, the birds grew normally. The antibiotic does not stimulate growth but lessens the conditions that retard growth. In this way, the birds approach their optimum efficiencies, for the more severe the growth depression, the greater the response to antibiotics.

TABLE 15.2. AVERAGE LEVELS OF AUREOMYCIN IN BLOOD SERUM OF POULTRY AS AFFECTED BY AUREOMYCIN LEVEL IN THE RATION

Aureomycin in Feed (g per ton)	Aureomycin in Poultry Blood Samples (mcg per ml)
0	0
25	0.002
50	0.0025
100	0.0045
200	0.009
400	0.170
500	0.200
1000	0.430

Aureomycin inhibits a wider variety of microorganisms and promotes growth in a wide variety of animals. Penicillin and bacitracin affect a more narrow group of bacterial types and have a smaller growth-promoting effect than that of Aureomycin.

Bacitracin and streptomycin are not absorbed, and thus do not reach the bloodstream if fed at the usual commercial levels. Penicillin, although absorbed when fed, is less effective than the tetracycline drugs in the prevention and control of diseases because of its narrow spectrum. It is restricted especially to gram-positive organisms.

The term, "narrow spectrum," is applied to penicillin and bacitracin and has been used to signify full activity against gram-positive bacteria and absence of activity outside this range. Neither assumption is completely true. The activity is uniformly strong but not complete throughout the gram-positive range, but is by no means restricted to that range. "Broad spectrum," as applied to the tetracycline group, implies strong activity throughout the range of the infective agents. This is not the case. The tetracyclines attack a considerable range of gram-positive bacteria, but this activity varies markedly from one organism to another.

When feed consumption starts to increase sharply or during outbreaks of specific diseases, such as paratyphoid, a decision should be made as to whether or not it may be more effective to use higher levels of effective drugs in the feed or drinking water. This could represent sizable savings if mortality is reduced or the time of infection is shortened.

The amount of antibiotics used in feeds is only a small fraction of the levels for treatment of disease. Growth stimulation is greatest during the first few weeks of life. Slower-growing chickens not receiving antibiotics gradually overtake the others with the result that at about 12 wk of age, the growth difference disappears. By the time birds are 20 wk old, there are no significant weight differences between those fed antibiotics and those that did not get antibiotics. Thus, the only advantage with respect to flock-replacement chickens is to decrease mortality and improve feed efficiency.

The efficiency of feed utilization is generally increased when chicks are given antibiotics, probably due to the improved rate of growth of the chicks, since the faster chicks grow, the better the feed utilization. The variability in chicks receiving antibiotics is somewhat less than in the control chicks not receiving antibiotics.

Dosage Levels of Antibiotics.—Two dosage ranges can be defined: low (20 to 50 ppm) and high (50 to 200 ppm). Most market production increases have been achieved at the higher levels, but profitability has not necessarily increased at the same time. Antibiotics are beneficial in reducing early chick mortality in disease situations and also in times of stress. Their use is indicated after any shock or upset, such as overheating, chilling, vaccination, worming or debeaking. They are fed 2 or 3 days prior to the period of stress and 2 to 5 days following stress, depending on the condition of the birds.

Reports are conflicting concerning the effects of antibiotics at low levels in laying rations. When adequate rations are fed, they exert no favorable effect, either on egg production or hatchability. Antibiotics enhance the utilization of calcium, phosphorus and manganese, possibly because the acidity of the small intestine and ceca is increased.

It is something of a mystery how antibiotics react on the microorganisms in the intestines of apparently normal animals and improve nutrition. Some scientists believe that it is mainly due to the weakening or destructive effect of antibiotics on the bacteria of the digestive tract. Reduction of harmful bacteria and encouragement of beneficial bacteria are involved, but not all scientists agree on this. The anatomy of the gut is changed with antibiotic feeding; the intestinal wall usually becomes thinner than if no antibiotics are fed. There has been much controversy as to the prudence of continuously using a drug which will destroy harmful microorganisms. If resistance is developed to it, the drug may not be effective in an epidemic.

Antibiotics do not significantly change the total number of microorganisms in the intestines, except for one kind (clostridia). Antibiotics spare the requirements for protein, certain vitamins and minerals, make the intestine more acid, improve absorption and stimulate bone growth. A certain level of contamination is necessary for them to be beneficial.

The modes of action of antibiotics may differ quite widely, depending upon the disease level. Antibiotics may act as follows: (1) favor the growth of nutrient-synthesizing, and inhibit the growth of nutrient-destroying microorganisms; (2) inhibit the growth of microorganisms that produce excessive amounts of ammonia and other toxic nitrogenous waste products in the intestine; (3) improve availability and/or absorption of certain nutrients; (4) improve feed or water consumption, or both; and (5) prevent actual pathological disease.

Antibiotics improve growth much more in the marginal diets than in more abundantly fortified rations. Feed efficiency is also improved. However, some baffling inconsistencies in growth response are obtained. A "new" environment (building or facility where no poultry has previously been raised) prevents manifestation of the antibiotic effect. Germ-free chicks also show no benefit. Where organisms from the contaminated environment were fed to the birds without antibiotics, growth is depressed. Thus, the antibiotic effect is related in some way to the microflora possibly through destruction of the harmful organisms or indirect stimulation of the beneficial ones which may synthesize some growth-stimulating factor.

Resistance to Antibiotics by Organisms.—Because the inside of the egg is sterile, experimentally-kept fowl can be reared with no demonstrable population of microorganisms in the gut or in the environment. Such birds permit investigation of digestion, nutrition and well-being in the presence or absence of microbial population. By such methods, it has been established that antibiotics exert their growth-promoting effects through their action on the gut population, since in the absence of microorganisms, they have no effect. Thus, the expression, "growth promoting," is a misnomer. Apprehension of possible future emergency from the development of resistant strains from long continued use of feeds containing antibiotics is widespread. If antibiotic-resistant strains of pathogenic microorganisms become established in farm livestock because they have been fed antibiotics, the possibility exists that an epidemic may occur with no therapeutic agents available for treatment. Hazard to the well-being of antibiotic-fed chickens arises from the potential therapeutic inefficiency of that type of antibiotic, should the chickens become diseased.

Some investigators have reported a lessened growth response to antibiotics following prolonged usage. They think this is due to the development of resistant strains of organisms. However, birds fed no antibiotics grow as well as those fed antibiotics, if raised in an environment where antibiotics have been used continuously for some time. Thus, it is possible that there may be a temporary eradication of depressive microflora rather than the development of resistance. Continuous use of antibiotics is necessary to prevent reinfestation of facilities with harmful microflora.

Possible antibiotic-induced bacterial resistance is of interest because of human health aspects. Tetracyclines increase resistance in bacteria belonging to the *Escherichia coli* and the *Salmonella* groups. The mechanism of resistance is a complex problem. A few inherently resistant cells in a bacterial population may take over when sensitive cells are killed off. The resistant strains are no more pathogenic than sensitive ones. A mixed population of organisms appears in the alimentary canal of newly-hatched chicks a few hours after they have been given food and water, and stable flora is established within the first 48 hr of life.

Growth-promoting effects of antibiotics are correlated, not simply with the elimination of *C. welchii*, but rather with the suppression of its toxin production. Growth of germ-free birds is depressed on contamination with *C. welchii*, an effect that is reversed by dietary penicillin. A mixed contamination of *E. coli, Streptococcus liquefaciens* and *Lactobacillus* does not affect the growth of otherwise germ-free chicks. These results suggest that the suppression of *C. welchii* is closely connected with the growth-promoting action of antibiotics. This is not likely, however, to be the full explanation.

Antibiotic supplementation has the same effects in rations consisting only of vegetable protein as in those containing animal protein. No significant difference in effectiveness between different antibiotics is observed during the growth of broilers to 8 wk of age. Antibiotic supplementation has no effect on brooding and hatching results but it leads to a slight increase in the body weight of the hens. Responses under average sanitary conditions and with balanced diets vary between 2 and 8% in growth and 2 to 4% in feed conversion. Antibiotics have also been shown to counteract marginal deficiencies and imbalances in the diet.

In summary, antibiotics inhibit or destroy organisms which produce subclinical infections. They suppress organisms which produce toxic reactions that cause slowing of growth of the host animal, and produce an increase in number or activity of organisms which synthesize certain growth factors that are eventually made available to the host.

The practice of feeding antibiotics to poultry is widespread, and so far, no deleterious effect on public health has been reported. Detectable amounts of aureomycin have been found in the bone marrow of broilers fed antibiotics at the higher levels (100 to 200 g per ton feed).

"Nutritional" levels of antibiotics are 2 to 10 g per ton complete feed. These have their greatest effectiveness in starting diets for chickens and poults. Whether or not antibiotics are used with other classes of poultry depends on the disease level and the amount of stress. Antibiotics are relatively low in cost and perform in a wide range of conditions.

Potentiation of Antibiotics.—The concentration of tetracyclines in the blood serum of poultry can be increased or "potentiated" by reducing the

FIG. 15.2. LARGE-SCALE SOPHISTICATED EQUIPMENT IS THE KEY TO ECONOMIC PRODUCTION OF ANTIBIOTICS

percentage of calcium in the ration and by substituting calcium sulfate and a soluble sodium phosphate for the commonly used calcium carbonate and calcium phosphate. Potentiation is limited by the necessity of supplying enough calcium to meet the animal's requirement. Within this limitation, the method can be used to potentiate the antibiotic level more than 2-fold. More effective potentiation up to 10-fold can be obtained through the use of certain sequesterants or chelating agents. However, none of these has thus far found practical application, due to cost. The inhibiting action of calcium on antibiotic absorption can be prevented by the use of terephthalic acid as a feed additive. However, its use in feed thus far has not been cleared by the United States Food and Drug Administration. Without potentiation, the effectiveness of tetracyclines against systemic diseases is sharply limited because of poor absorption into the bloodstream.

"Potentiation" increases the concentration of an antibiotic in the blood and tissues of animals by adjusting the composition of the ration. It is confined almost entirely to the tetracyclines: chloretetracycline (Aureomycin), oxytetracycline (Terramycin) and tetracycline, per se. Other commonly used antibiotics do not respond to potentiation by ration adjustments. The solubility of tetracyclines in the digestive tract is inhib-

TABLE 15.3. INFLUENCE OF CALCIUM, PHOSPHORUS AND TEREPHTHALIC ACID (TPA) LEVELS ON CONCENTRATION OF TETRACYCLINE ANTIBIOTICS IN CHICK BLOOD SERUM

% TPA in Feed	Antibiotic in Feed (g per ton)	% Calcium in Feed	% Phosphorus in Feed	Conc. of Antibiotics in Chick Blood Serum (mcg per ml) Aureomycin	Terramycin
0	100	1.18	0.8	0.036	0.15
0.375	100	1.18	0.8	0.092	0.201
0	200	1.18	0.18	0.065	0.193
0.375	200	1.18	0.8	0.232	0.373
0	100	0.18	0.8	0.130	0.237
0.5	100	0.18	0.8	0.209	0.400
0	100	0.18	0.18	0.260	0.433
0.5	200	0.18	0.8	0.484	0.668

ited by the presence of bivalent metals, including calcium, magnesium, iron, copper, manganese and zinc. It follows that the tetracyclines can be potentiated by decreasing the levels of bivalent metals in the ration or by adding to the ration certain sequestering (chelating) agents which inactivate the minerals after they reach the digestive tract. Of the inhibiting metals commonly present in the digestive tract, calcium is, by far, the most abundant, due to the use of calcium carbonate and calcium phosphate as feed additives.

Aureomycin produces significantly higher blood concentrations in low calcium rations than the same levels of Aureomycin incorporated into commercial broiler feeds containing commonly used levels of calcium. The rate of absorption of Aureomycin can be augmented by a calcium sequestering material, such as EDTA or its salt.

Lowering the calcium level or using a calcium salt, such as the sulfate, which does not dissolve in the acid portion of the gastrointestinal tract, yields a lower calcium ion concentration and thereby minimizes the interference of calcium with absorption of Aureomycin. Terephthalic acid enhances the blood concentration of antibiotics and permits the diseased animals to be more economically controlled.

Widespread use of antibiotics in animal feeds presents a potential public health hazard which has not, as yet, been fully evaluated. At the present time, there is no evidence linking use of antibiotics in animals to human disease. The public health hazard stems from the possibility of resistant organisms in animals being transferred to human beings. This raises the possibility of a microbial population occurring in human beings which would be partly, if not largely, resistant to normal therapeutic agents. Resistance to certain drugs which some bacteria possess can be transferred to other bacteria by conjugation, as well as genetically. In other words, one organism can, through conjugation, transfer the resistance it has to one or more drugs to other bacteria of the same species, or in some instances, to other species of bacteria. Regulations are being proposed to deal with this matter, particularly with the injectable preparations containing penicillin, streptomycin and dihydrostreptomycin, chlortetracycline, tetracycline, bacitracin and chloramphenicol. The basis for the proposed action is that the residues of these drugs remain in edible tissues of treated animals for an unknown length of time, sometimes for more than 75 days. Aqueous penicillin preparations must be labeled with a 5-day withdrawal time.

ANTIOXIDANTS

Antioxidants became a functionary part of poultry rations when fat was added to develop the high-energy ration. When free fat is added to a mixed ration, it readily becomes rancid and the peroxides formed are very destructive to Vitamins E and A and to any exposed fat. Antioxidants are also used to help stabilize alfalfa meal.

All free fat, Vitamin A and oils should have an antioxidant.

Antioxidants which help protect fats from rancidity and fat-soluble Vitamins A, D, E and K from destruction are:

Drug	*Manufacturer*	*Level in Feed (% Active Drug)*
BHT	Eastman Kodak	0.01
Santoquin	Monsanto	0.01

Antioxidants, such as butylated hydroxyanisole (BHA) and diphenylparaphenylelediamine (DPPD), added to fats and feeds, are used to help protect their quality. They are also added to dehydrated alfalfa meal to protect its carotene content. Antioxidants do not, per se, function as nutrients; they protect certain nutrients from destruction.

Antioxidants are chemical compounds which have the property of preventing the oxidation of another substance by preferentially accepting oxygen or electrons and thus breaking the chain reaction which otherwise would destroy the compound in question. They are commonly used to prevent the oxidation of fats or fatty acids. In 1924, it was demonstrated that Vitamin A was destroyed by oxidized fats, and in 1938, that the degree of destruction parallels the degree of unsaturation of the fat. The degree of oxidation is determined through what is know as the "peroxide number," a chemical term denoting a measure of the amount of oxygen absorbed by the fat.

Vitamin A is destroyed in the gut of the bird, as well as in the feed, in the presence of unsaturated fats. In the study of Vitamin E, research workers became aware that it acts as an inhibitor of fat autoxidation. It also protects carotene in the intestinal tract by virtue of its antioxidant properties.

Exudative diathesis in chicks is a nutritional disease, preferentially affecting the chick's adipose tissue. Exudative diathesis does not occur when the diet contains sufficint Vitamin E or when fat is not present in the diet. When Vitamin E is in sufficient quantities in the diet, no peroxide is demonstrated in the fatty tissues. Another nutritional disease of chicks occurring with the same type of diet which produces exudative diathesis, is encephalomalacia ("crazy chick" disease), first described in 1931. Vitamin E in the diet prevents this disorder, as do Vitamin C, nordihydroguariaretic and methylene blue, all of which have some antioxidant activity.

Antioxidants prevent the development of encephalomalacia, improve the utilization of Vitamin A from fish oils, aid in the development of good skin and shank color in broilers, and enhance the utilization of carotenoid pigments in pigmentation. Antioxidants, such a Vitamin E, are a natural product; synthetic antioxidants include BHT, BHA (butylated hydroxytoluenes) and ethoxyquin. The need for antioxidant protection is increased when highly-unsaturated fats or oils are included in the ration.

The interchangeable role of certain antioxidants and Vitamin E applies to turkeys, as well as to chickens. Some synthetic antioxidants possess Vitamin E-like properties, but not all the symptoms are considered to be indicative of Vitamin E-deficiency response to them. Among these is exudative diathesis in chicks which is chiefly a selenium deficiency, the factor 3 of Schwartz. The requirement for biological antioxidants, including Vitamin E, is dependent upon the content of polyunsaturated fatty acids or their oxidation products. Linoleic acid has been noted as a specific cause of encephalomalacia. The accumulation of auto-oxidizable lipids in muscle is a primary criterion in the production of dystrophy.

Early work indicated that DPPD was relatively nontoxic to chicks. However, it was found to induce prolonged gestation in rats and cause high mortality of both mother and pups. As a result of these effects on mammalian production, the United States Food and Drug Administration requested that feed use of DPPD be discontinued. BHA and BHT have a long history of safe use in fats and oils and are often added in combination to improve stability of fat-soluble vitamins.

Ethoxyquin is an antioxidant used for the preservation of carotene in dehydrated alfalfa meal. It is used at a level not to exceed 0.0125% in poultry feeds. The best known and most effective of the naturally-occurring antioxidants are the tocopherols (Vitamin E) found in cereal grains, grasses and other products, and also the phospholipids, including vegetable lecithin found in soybeans, raw oat flour, yeast, animal tissues, legumes, grasses and gums. Gum guaiac is an antioxidant used in lard, as well as liquid fats and dehydrated fats and foods.

Combinations of antioxidants sometimes prove more effective than the additive effect of the 2 individual antioxidants. This is known as the synergistic effect. Synergism is assumed to involve the chemical alteration of the antioxidant itself or the tying-up of certain metallic ions known to catalyze oxidative reactions.

Many antioxidants are used to help protect nutrients. Antioxidants become additional ingredients in the total feed mixtures and are subject to feed control regulations. They must be cleared for safety and usage because some of the most effective antioxidants are also very toxic to poultry. The use of antioxidants permits high levels of fat to be used in feeds without ill effects.

SURFACTANTS AND CHELATION

Surfactants.—Some surface-active agents possess the property of stimulating the growth of chicks. As in the case of other growth stimulants, their effect usually lasts only up to 12 wk of age. Some of the growth stimulants already described are surface-active. There are a great many surface-active compounds and much more research will be required before definite conclusions can be drawn concerning their value in practical poultry rations. Growth stimulation cannot be claimed generally for surfactants but is specific for particular compounds of this type. Some of these have been found to be effective; others, growth-depressing. In the case of effective products, growth increase has not been greater than that produced by antibiotics. No benefit is gained by feeding both an antibiotic and a surfactant.

Chelation.—Chelation plays a major role in increasing or decreasing the biological availability of metals. Chelation is the reversible reaction between polyvalent metal ions and polyamino-carboxylic acid to form within the molecules one or more ring structures incorporating the metal ions. The name is derived from the Greek, "chele," meaning claw, which describes the claw-like structure of the chelate molecule. Chelating agents are well-known industrially but their application in the field of nutrition is not well-developed.

As an illustration of chelation, copper as the cupric metal ion reacts with EDTA (ethylenediamine tetraacetic acid) in the following manner:

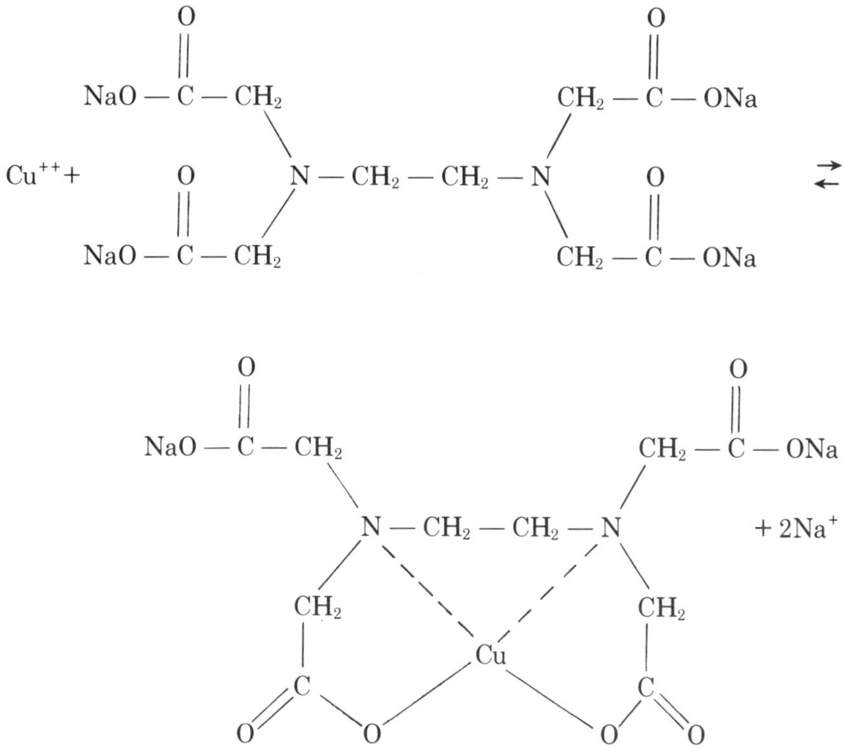

Chelates are of major importance in the regulation of many reactions taken for granted. The amount of calcium and phosphorus needed by osteoblasts for the deposition of bone is provided at the site of deposition. The ratio of exactly 2 parts calcium and 1 part phosphorus would precipitate at the pH of the blood, except for chelates.

Chelates from Antibiotics.—Tetracyclines and other antibiotics form chelates with calcium and other polyvalent cations which interfere with antimicrobial activity of the antibiotic. One of the primary functions of the polyvalent essential cations in metabolism is to form chelates between enzymes and substrates, and this is an important mechanism through which trace elements act as activators of enzyme reaction. The reaction compounds between the sugars and lysine which render lysine unavailable for incorporation into tissue proteins serve a useful purpose, since this type of sugar, amino acid compound is an excellent chelating agent. Thus, many chelating agents perform useful and even vital functions in the animal body. Others may cause drastic interference with

TABLE 15.4. RELATION OF pH TO ORDER OF CHELATION

	Strongly Complexed			Weakly Complexed				
pH 4.0	Cr		Cu		Co			
pH 6.5		Co	Cu	Zn	Ca			
pH 8.65		Co	Cu	Zn	Ca	Mg	Sc	Ba
pH 11.0		Co	Cu	Zn	Ca	Mg	Sc	Ba

Source: Prepared from VERSENESS, Bersworth Chem. Co. Tech. Bull. *2,1952*. Framingham, Mass.

metabolism. Until more research has been done, attempts to make practical use of chelates in animal nutrition may produce unexpected results.

Use of EDTA.– Natural materials also contain chelates which are as effective as EDTA in improving the utilization of trace minerals. EDTA not only improves the effectiveness of the zinc already present in the diet, but of the added zinc, as well. A casein-phytic acid complex in a chick diet causes severe growth depression which can be counteracted by the addition of zinc to the diet. One theory involves the absorption of the chelate itself. After absorption, the various enzymes and transporting systems within the animal may have stability constants sufficiently high to remove the metal from the chelate and make it effectively available to the animal. EDTA is absorbed by chickens and turkeys. Trace minerals are solubilized by the addition of chelating agents to the diet and absorbed in the chelate form.

Chelation is also recognized as important in the antibiotic field. The majority of antimicrobial compounds are metal-binding agents and part of their beneficial and detrimental properties may be related to this function.

While phytic acid has been implicated as a factor in the availability of zinc, it is not know whether phytic acid alone or in combination with protein is the detrimental moiety. The addition of EDTA to diets containing marginal levels of manganese and copper improves the utilization of these elements, too. On the other hand, the addition of EDTA to diets marginal in iron depresses growth of chicks and causes a lowered hemoglobin level, which can be overcome by the addition of higher levels of iron. It is certain that chelating agents can alter the availability of certain minerals.

Mode of Action of Chelates.– Much more work needs to be done to determine interrelationships of various mineral elements in the presence of chelating agents and to determine the stability constants of various components with which chelating agents must compete when they are added to biological systems. While organic chelates are relatively new to animal nutritionists, their effects on mineral availability have been known to plant and industrial scientists for many years. Chelation is used in medicine as a means of helping to free the body of toxic cations, such as lead, and it is possible that they may be of major importance in the regulation of many of the reactions which we now take for granted; for example, the calcium and phosphorus requirement for deposition of bone. It remains for the nutritionist to learn which chelates are beneficial and which are harmful or toxic.

Practically all metals which form stable di- or polyvalent ions can be chelated. The most common are the alkaline earths (calcium, magnesium, etc.) and the transition metals (copper, nickel, zinc, cobalt, etc.). For all practical purposes, the alkali metals (sodium, potassium, etc.) are unaffected. The pH of the system determines, to a great extent, which metals are chelated preferentially. Higher concentrations of the hydroxy ion will cause precipitation and metal hydroxides will react with the amphoteric metals to give various complex ions. In either instance, the metal chelate is destroyed or prevented from forming. Chelating products have low acute oral toxicity. In the concentrations used biologically, they are not irritating to the skin and eyes and, therefore, do not present handling problems.

Since the chelating agent is an organic chemical which surrounds metallic ions with multiple ring structures, it keeps the metal chemically inactive and holds it in solution. There are many present and promising applications to industry and nutrition. When used in solution, they inactivate practically any polyvalent metallic ions that they contact. They accomplish this by actually forming a new compound in which the metal ion becomes a member of a stable inner ring structure in the molecule so that no reversion or breakdown can occur. In nature, many compounds are attracted by a cation (an element with a deficiency of electrons) with which they can share their electrons, thereby forming a stable compound. Usually, the compound thus formed is highly insoluble in water, but nevertheless dissociates to a sufficient extent in the intestinal tract to allow absorption of the essential cations. This is usually brought about by hydrochloric acid from the proventriculus which tends to convert all the cations temporarily into the chloride salts, allowing a considerable degree of ionic dissociation. This results in good absorption from the intestinal tract.

The availability of zinc is an example of absorption as affected by chelation. Once the importance of zinc in the diet of the chick had been established, it was found that the zinc requirement was greater when a diet containing isolated soybean protein was fed than in one containing casein, gelatin or egg white. California researchers observed that EDTA, a chelating agent, markedly improved zinc absorption and reduced the amount necessary in the diet.

Organic chelates of certain mineral elements may be the most important factors controlling utilization of these elements. If an element is chelated by a compound which will enhance absorption or release it in ionic form at the intestinal wall, this may greatly improve absorption of the element. It may prevent conversion to an insoluble colloid in the intestine which would prevent absorption.

Chelates show exceptionally high stability in solution. Many chelates are highly absorbable and prevent the metal ions from forming an insoluble metal complex. This type of chelate is called a "sequestering" agent. Chelates in biological systems may serve to transport and store metal ions. They may be used in metabolism or they may interfere with utilization of the central cations.

As an example of the action of chelates in poultry nutrition, it was found that the availability of zinc for chicks fed a purified ration containing raw soybean protein is greatly enhanced by addition to the diet of EDTA. Phytic acid is a constituent of isolated soybean protein. The feeding of this isolated protein was in some way responsible for the unavailability of zinc for chickens and turkey poults. This growth depression could be corrected by the addition of zinc to the diet. This suggested that zinc phytate interaction decreases the biological availability of zinc. Phytic acid in the diet increases the fecal elimination of zinc about 2-fold. From in vitro studies, it was found that isolated soybean protein can bind zinc. The addition of sodium phytate to isolated protein increased the amount of zinc which was to be bound. EDTA or citric acid effectively removed the zinc from the soybean protein and brought it in solution. The reason for the improvement of growth on a practical deficiency of zinc by the addition of a synthetic chelation agent is not clear. It might function by transporting the mineral to a site in the intestinal tract more favorable for absorption.

INTERNAL PARASITES

Worms were a problem when the flocks were saved and hens were allowed to range in open yards or fields. Studies on range pastures showed that pasture was no better than bare lots for control of worm infection. Also, the effective destructiveness of 1 large dose of worms (eggs) was less than the same population given over a period of several weeks.

Built-up litter produced heavy worm infections in broilers, whereas birds on new litters remained free of parasites. Pullets reared on built-up litter should be wormed before being placed in the egg-production facilities.

Drugs for internal parasite control:

Drug	*Manufacturer*	*Level Recommended (lb per ton Feed)*
Piperazine	CIBA, Vineland	2.2
Wormal	Dr. Salsbury	20
Hygromycin	Elanco	1.5

Many treatments have been used for internal parasite control. One method used by early poultryman was lye-treated corn. This was replaced with tobacco dust in the feed. Later, nicotine sulfate was placed in pill form and each pullet or hen was given a "worm pill" during culling. The 2 common worms were capillary and round worms.

Blackhead (Histomoniasis).— Blackhead has been and remains one of the most severe diseases of turkeys. It nearly destroyed the turkey business in the states of Kentucky, Virginia and Minnesota, until Dr. A. B. Billings of the University of Minnesota developed a sanitation-disease-freeing field-rotation management plan that helped reestablish the production of turkeys.

Many substances were added to turkey feeds to control blackhead, directly or indirectly. Flushing mashes and ground tobacco were fed. The tobacco was fed to destroy the worms since they were considered to be active carriers of the organism. Also, some drugs were fed that increased the expulsion of cecal contents, since blackhead is a cecal organism.

Some drugs which have been used successfully in controlling or treating blackhead in turkeys are:

Drug	*Manufacturer*	*Level Used for Prevention (lb per ton)*
Carbosep	Whitmoyer	2
Histostat	Dr. Salsbury	2
Cyzine	Lederle	3
Hepzide	Merck	2

Before a successful treatment for control of blackhead was developed, the University of Missouri Department of Veterinary Medicine reported that removal of the cecum would prevent the blackhead organism from becoming established. The organism apparently cannot survive in other parts of the digestive tract. Although expensive, this method of control proved effective in preserving choice breeding stock.

The blackhead organism often gains entrance or lives in the round worms of chickens. The separation of turkeys and chickens is a "must" if the turkey operation is to be successful.

XANTHOPHYLL PIGMENTATION

Consumer Preference

Pigmentation of egg yolks and the skin of chickens has been emphasized in the United States, while it is considered an economic waste in England and France. English investigators find no reason to increase the cost of poultry by feeding a pigment to broilers which are to be broiled or fryed brown after being dipped in a batter.

Wholesale buyers of poultry in the United States offer a premium for deeply-colored broilers and roasters. Consumers prefer eggs that range in color from pale to deep yellow and object if yolks are orange-red. On the other hand, deep-orange yolks are desired for commerical use in egg noodles, yellow cake mixes and other products made for food processing.

Strong consumer preference for highly-pigmented broilers is associated with freshness and wholesomeness. The "golden glow" of the broiler sells more and brings better prices. Consumers in the United States, Mexico, Germany and Italy desire a deep-yellow skin in the dressed carcass, while in England and France, the skin must be white.

Xanthophylls (plural) refer to the entire group of hydroxycaroteniods, whereas xanthophyll (singular) refers to lutein. As a class of compounds, they have no nutritional value.

In 1922, Palmer proved that the xanthophylls in the diet produce yellow pigmentation of the skin and fat of chickens, while carotenes play no important part. All English breeds except the Cornish have white skin. These include the Dorking, Australorp, Sussex and Orpington. Thus, there are breed differences with respect to pigmentation. Growing pullets of yellow-skinned breeds store pigment steadily in the fatty tissues of the body. At the start of egg production, all pigment absorbed from the diet is diverted to the ovary, leaving the body as a component of the egg yolk. The laying pullet loses her tissue pigmentation in a definite pattern—skin first and shanks last. Regardless of the concentration of the xanthophylls in the feed, when egg production ceases, pigmentation returns to the various parts in the same order as it disappeared at the start of production.

Anything that interferes with the egg-feed ratio affects yolk color. The hen that lays an egg every other day puts more pigment in her yolk than the hen that lays an egg every day, even though both hens consume the

TABLE 15.5. RELATIONSHIP BETWEEN XANTHOPHYLL CONTENT OF DIET, β-CAROTENE-EQUIVALENTS, AND N.E.P.A. VALUES OF EGG YOLKS

Xanthophyll mg per kg Feed	β-Carotene Equivalents[1] (μ g β-CE/g Yolk)	Approximate NEPA Number
0	—	—
15	25	1.0
25	30	1.3
35	45	2.5
45	60	3.3
55	75	4.4
65	85	5.0

[1]A.O.A.C. method.

same diet. Hot weather reduces the efficiency of xanthophyll utilization. "Platinum yolk" and "lemon yolk" eggs occur occasionally. There is some question as to whether or not the condition can be transmitted from 1 group of birds to another. Since it can be corrected by feeding an antibiotic or furazolidone, it is attributed to a low-grade infection or to capillaria worms. *Eimeria maxima* is highly detrimental to skin pigmentation. Less difficulty with yolk pigmentation is experienced with layers in cages, which lends support to the idea that coccidiosis may sometimes be involved.

Most yellow and red pigments synthesized in vegetable materials are a closely-related group of chemical compounds known as carotenoids. Peppers and paprika contain capxanthin as well as xanthophyll, and are effective pigmenters. Some substances introduce abnormal colors which are not acceptable to consumers. For example, bixin and capxanthin produce red yolks; canthaxanthin produces orange-colored yolks. Annatto (achiote) contains bixin and xanthophyll. With normal feeding practices, 70% of the yellow color of egg yolk is due to xanthophyll, and most of the remainder is due to zeaxanthin.

Broiler pigmentation is measured by visual comparison or by extracting the yellow pigment from a sample taken from the toe web. A number of arbitrary scales are used for grading pigmentation. Results are expressed in terms of color equivalent to a specified carotene content in ppm of the sample. The eye detects colors which are not measured by a photoelectric colorimeter.

Most color in the shanks, skin and body fat of poultry is a single pigment, xanthophyll (lutein), although other hydroxycarotenoids are deposited. No nuritional function has been demonstrated for xanthophylls and their deposition in egg yolks is mainly a convenient means of disposal. Nonvitamin A carotenoids are known collectively as xanthophylls, and possess structures which birds are unable to convert into Vitamin A. They are absorbed from the intestinal tract and are deposited in fatty tissues and egg yolks as intact chemical compounds. Some are poorly absorbed or have different degrees of absorption. Absorption of carotenoids closely follows that of lipids.

In 1837, Berzelius extracted from autumn leaves a yellow pigment known as xanthophyll. In 1906, Tswett extracted pigments of nettle leaves and passed the solution through a tube of calcium carbonate. A series of yellow layers formed which he separated and showed to contain different xanthophylls.

Utilization of Xanthophylls

Xanthophylls are not stable compounds and can be lost from poultry feeds by oxidation. The amount of yellow pigment deposited in the skin of broilers is signficantly increased when an antioxidant is added to the feed and destruction in the feed is appreciably reduced. This is effective in preserving the xanthophyll in corn, as well as in dehydrated alfalfa meal. Minerals and other ingredients have prooxidant effects. Fat improves absorption of xanthophyll but has an adverse effect if it induces oxidative rancidity. Oxidation can occur upon storage in the intestine, or after absorption. To protect against oxidation, antioxidants are incorporated in poultry pigmenters.

Charcoal, mineral oil and sulfur in the diet interfere with deposition of xanthophylls in chicken tissue, as do cod-liver oil and sardine oil at high levels. Pigment suppression occurs with certain meat scraps, fishmeal and soybean meal samples. Some kinds of respiratory diseases and chronic intestinal parasite infestations have been implicated in pigment suppression.

Newly-recognized sources of xanthophylls are algae meal and flower-petal meal (Florafil). These are of interest because of the high concentration of xanthophyll. Florafil is a natural plant product. Petals from the blooms of a specially selected species of marigold grown in Mexico are a highly-concentrated source of biologically available xanthophyll containing 30 times as much as the same quantity of dehy. Florafil contains 5.4 g xanthophyll per lb. Corn endosperm oil is also a color additive. It is a reddish-brown liquid composed chiefly of glycerides, fatty acids, sitosterols, and cartenoid pigments obtained from the gluten fraction of yellow corn.

An equal quantity of xanthophyll from different sources does not produce identical results. The main factor involved is the digestibility of the ingredient which supplies the xanthophylls. The xanthophyll present in yellow corn, algae meal and Florafil is well-utilized by poultry. Lettuce with a high Vitamin E content shows good utilization of carotenoids for yolk pigmentation.

TABLE 15.6. AVERAGE XANTHOPHYLL CONTENT OF SOME NATURAL MATERIALS

Materials	Xanthophylls mg per kg
Marigold petal meal (*Tagetes erectal*)	6,000–10,000
Chlorella pyrenoidosa alga	4,000
Common algae meal (*Spongiococcum excentricum*)	2,200
Taiwan bee pollen	1,325
Seaweed (*Fucus serratus*)	920
Dried stinging nettles (*Urtica*)	800
Broccoli leaf meal	670
dried kao haole meal—Hawaii (called ipil-ipil in Philippines) (*Leucaena glauca* Linn)	660
Dehydrated clover meal	490
African red peppers, powdered	440
New York state pollen	440
Dehydrated alfalfa leaf meal (20% protein)	400–550
Seaweed (*Fucus vesiculosus*)	350
Mexico pollen	345
Seaweed (*Ascophyllus nodosum*)	340
Corn gluten meal (60% protein)	330
Spanish paprika	274
Ground annatto seed and hulls	265
Acidulated soybean soapstock	168–260
Dehydrated alfalfa meal (17% protein)	185–350
Coastal Bermuda grass	185–350
Dried chili peppers	185
Lakeweed meal	120
Corn gluten meal (41% protein)	90–180
Dried carrots	65
Dried sweet potatoes	55
Yellow corn	20–25

Pigmentation carries a price tag, so it is necessary to compromise between color and the cost of poultry meat and eggs. It is not necessary to have a continuous supply of xanthophylls in broiler feeds; the important time is during the final finishing period of a few weeks. Pigmentation of healthy birds of yellow-skin varieties can be obtained with 10 mg xanthophyll per lb feed. As the quantity of xanthophyll is increased in the feed, a plateau is reached at 25 mg.

Sources of Xanthophylls

Practical sources of xanthophylls are yellow corn and corn gluten meal. Yellow corn varies in xanthophyll content and cannot be depended upon to take care of xanthophyll requirements by itself. The supply of corn gluten meal is limited by the amount of corn milled and is frequently in short supply. Yellow corn contains chiefly lutein, zeoxanthin, zeinoxanthin and tryptoxanthin. Thus, the substances responsible for broiler pigmentation are part of a complex of related chemical compounds or their degradation products. The pigmentation of corn gluten meals varies according to the xanthophyll content of the corn used in its manufacture, and is principally a seasonal variation. Periodic xanthophyll analyses insure uniform pigmentation at minimum cost.

Green, leafy materials are excellent sources of xanthophyll. Alfalfa has 5 major xanthophylls which constitute 99% of the total xanthophylls present: lutein, cryptoxanthin, zeoxanthin, violaxanthin and neoxanthin. "Dehy" is the most desirable, with regard to xanthophyll content. 2.5% of 20% protein dehy is equal to 5% of the 17% protein grade. Pelleting and regrinding alfalfa leaves cause an increase in pigmentation. Alfalfa carotenoids and xanthophylls are absorbed, as well as those of corn or corn gluten meal.

A synthetic carotenoid beta-apo-8′-carotenal is well-utilized for pigmentation. Violaxanthin is ineffective as a yolk pigmenter, whereas zeoxanthin and beta-apo-8′-carotenol produce a pleasing golden glow. Canthaxanthin and isozeoxanthin are readily utilized by the hen, but yolk color has an undesirable orange tinge. Arsenicals and antibiotics have been noted by some scientists to enhance pigmentation. From the practical standpoint, it is the eye of the beholder that really counts.

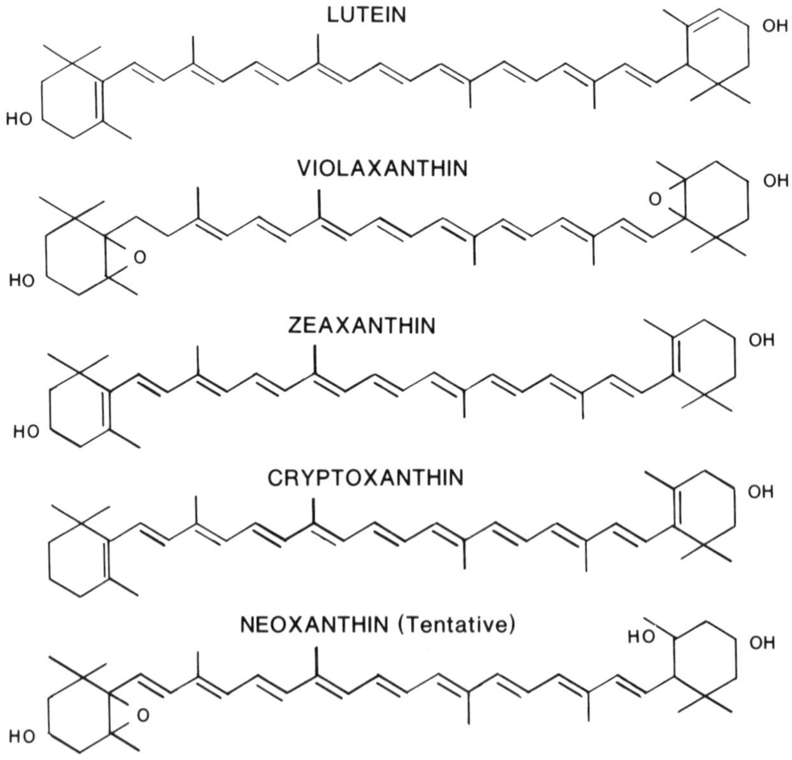

FIG. 15.3. STRUCTURES OF 5 MAIN XANTHOPHYLLS OF ALFALFA

16

Unrecognized Growth Factors

The vitamins known today were originally unidentified nutrients and recognized only by their effects on growth, performance and health of poultry. Once a new nutrient was identified, questions were raised as to whether or not all required nutrients were known. Factors for which scientists are still searching are those needed chiefly for economical performance and are recognized as U.G.F.

Throughout the study on U.G.F.'s, certain ingredients or concentrates were added as sources of U.G.F. The major ones were:

(1) Brewers' yeast
(2) Dried liver meal
(3) Dried milk
(4) Fish solubles
(5) Distillers' solubles
(6) Rice polishings
(7) Pasturage
(8) Animal protein

After many years of research, there is still a difference of opinion of how many growth factors there are. When breeding hens are placed on a corn-soybean ration without animal protein or some U.G.F. source, the hatchability may be slightly less, livability sometimes poor and the growth not uniform. Some strains appeared to be more responsive to U.G.F. supplements than others.

Today, the ingredient most often discussed as a U.G.F. source is distillers' solubles.

The identification of the need for a U.G.F. should involve a procedure which demonstrates that the diet is deficient in a U.G.F. This should involve a very rigid control program on biological value of nutrients in the different ingredients, the adequacy of the formula, disease conditions, management and possible presence of toxins. The following procedure should help establish the need for a U.G.F.

(1) The basic structure of the breeding ration should have characteristics similar to the starting-growing ration. If a corn-soybean structure is used for the breeding ration, then a corn-soybean structure should be used for the starting-growing ration.

(2) A well-planned disease-control program is desirable. During recent years, data have been presented showing that pathogens can influence absorption of specific nutrients.

(3) Many of the ingredients should have their biological value determined. This should include single ingredients and combinations of ingredients.

(4) A uniform management program should be used, as well as one administered by a qualified attendant.

(5) The breeding stock used to produce the experimental chicks should be maintained through 2 egg-laying periods. This is desirable because the response to U.G.F. is more clearly demonstrated during the latter part of the first, or during the second, laying period.

In general, if the breeding ration is adequate and hatchability does not decline, the chicks will show no need for U.G.F. under good management.

(6) A careful study should be made for mold toxins.

17

Nutrient Interrelationships

The nutritionist must avoid materials that might be harmful. He should become acquainted with possible sources of toxins and with the symptoms and lesions produced by these materials. Specific deficiencies can be created by the presence of unidentified substances occurring in natural feedstuffs. It is one thing to provide a ration containing all the required nutrients in pure form at required levels, and quite another to formulate a perfect ration from natural feedstuffs. In the latter case, not only must specific deficiencies be accounted for, but adjustment must be made for the presence of things known and unknown which alter the delicate balance of nutritional equilibrium. Growth-inhibiting factors are present in some ingredients.

ANTIMETABOLITES

In 1940, Woods discovered that the bacteriostatic action of sulfanilamide could be reversed in a competitive fashion by p-aminobenzoic acid. Thus, p-aminobenzoic acid, a close structural analogue of sulfanilamide, is an antimetabolite of the latter.

Characteristic signs of vitamin-deficiency diseases can be produced by the administration of a suitable structural analogue of a particular vitamin. The syndrome can be prevented by increasing dietary intake of the vitamin. The symptoms produced by the structural analogue occur while the animal is consuming an otherwise adequate diet. Though no deficiency is present in the ration, one is created in the animal by the antimetabolites. Since the signs resemble those which result from a simple deficiency, and since they are obliterated by increases in the intake of the vitamin, it is clear that the antimetabolite is creating a specific deficiency disease. Except for Vitamins A, D and B_{12}, antimetabolites of each of the vitamins have been produced and studied. For example, when pyrethiamin is fed to mice receiving an adequate, purified diet, typical manifestations of athiaminosis are called forth. These can be prevented by administration of extra thiamin.

Years ago, wheat flour was treated with nitrogen trichloride to improve quality. However, wheat treated in this fashion caused "running fits" in dogs. The toxic substance in the flour was identified as methionine sulfoximine. It arises from the reaction of nitrogen trichloride on methionine within the proteins of wheat.

The effects of methionine sulfoximine are overcome by methionine, and it is thus an antimetabolite of this amino acid. Although no ill effects occurred in humans as the result of consuming flour bleached with nitrogen trichloride, the flour industry promptly changed its bleaching agent.

Dicoumarol is the causative factor of the hemorrhagic sweet clover disease. Microbial action on coumarin normally present in sweet clover leads to the formation of a poisonous substance. Because the manifestations of the hemorrhagic disease are quite similar to the signs of Vitamin K deficiency, attempts were made to prevent the disease with this vitamin, and these proved successful.

Preliminary work has suggested that the time-honored association of a corn diet with the existence of pellagra may result from the presence of an antagonist of nicotinic acid in this food. Extract has been prepared from corn which depresses the rate of growth of mice maintained on a diet low in protein and nicotinic acid. The active component is a basic substance, possibly a pyridine compound. The deleterious effects of this extract are reversed by niacin or tryptophan. Although zein, the principal protein in corn, is lacking in tryptophan, corn itself has an appreciable quantity of this amino acid. Until the substance is isolated and its chemical structure established, it cannot truly be said whether or not it is an antimetabolite of nicotinic acid.

Poultry, cattle, sheep, horses, pigs and rabbits fed large amounts of rye have poor appetites leading to lower growth rates. The cause of the reduced feed intake is an agent soluble in oil, acetone and ether—a mixture of 5-N-alkyl resorcinols. Young animals are more susceptible to grain resorcinols than older animals. The decreased feed consumption of resorcinol-fed rats is not caused by an unappetizing taste of rye oil or rye resorcinols. There is no difference in harmfulness between fresh and 1-yr-old rye.

Raw wheat germ is said to contain a chick-growth inhibitor. The toxin is destroyed by heat or pressure-cooking of meal. The primary effect of the toxin is to block the absorption of fat, although protein retention is also involved.

Many plants contain antibacterial substances, many of which are highly volatile. Their value may lie more in their use as preventives of infectious diseases rather than their use in therapeutics. Further studies may assign an importance to their use in prevention of bacterial and parasitic infections which affect nutritional values.

Cattle, calves and horses are susceptible to laxicity after eating substantial amounts of bracken fern. After about 30 days, horses show incoordination, an increase in rectal temperature and hemorrhages under the skin as well as in some of the internal organs. Two toxic substances are present in bracken fern. One destroys thiamin and is responsible for symptoms seen in monogastric animals. Heating the dry bracken at 221°F for 18 hr does not completely destroy the thiaminase factor. The active substances are destroyed by steam at 266°F for 30 min; they are dialyzable and thermostable in acid solution and can be extracted from aqueous solution with butyl alcohol. From the standpoint of poultry, it would seem reasonable to eliminate fern from dehydrated green feed.

A hypothesis accounts for most of the facts concerning antimetabolites. Essential metabolites, such as the vitamins, hormones and amino acids, function by acting as substrates in enzymatic or semi-enzymatic reactions. For example, thiamin is known to act as substrate for the enzyme which synthesizes cocarboxylase (thiamin pyrophosphate). The antimetabolite exerts its effect because it, like the metabolite, can combine with specific enzymes. Being a structural analogue of the metabolite, it has most of those chemical groupings which are responsible for the combination. However, the antimetabolite is sufficiently different from the metabolite that the enzyme finds it impossible to form the usual product. The organism is thus deprived of a needed end product.

It cannot be assumed that a ration that is more than adequate in all nutrients will give maximum performance. Some natural feedstuffs contain phytate or oxalate which require higher levels of calcium and iron in the diet due to the fact that they form insoluble, indigestible precipitates with these elements. More of these minerals are necessary to supply the animal's requirements in addition to what is lost in the feces. Castor bean meal, though a good source of amino acids, is not suitable for use in feed because of toxic factors.

Certain barley samples support better growth when enzyme supplements are added. Antibiotics in feeds may affect the intestinal bacterial activity and improve performance. More than adequate amounts of nutrients may contribute to nutrient imbalance. A low-protein diet results in a reduced rate of growth; by supplementing this low-protein ration with threonine, one of the deficient amino acids, a reduced rate of gain results, rather than the increase which might have been expected. This effect is an amino acid imbalance.

INHIBITORS

By trial and error, growth inhibitors or toxic substances in some plants and animal products have been recognized some time. *De Materia Medica*, in the first century, compiled over 600 plants of therapeutic or poisonous character. During the period of rapid western expansion in the United States, poultry and livestock came in contact with many new plants and many reports dealt with illness or death of animals due to sorghums, lupines, castor beans and bitterweed, as well as to "loco" poisoning.

Kingsbury recently compiled extensive information on the poisonous plants of the United States and Canada. Information is summarized on the accompanying chart (Table 17.1).

Cottonseed Meal, Gossypol

One toxic component of cottonseed is gossypol, a yellow-colored substance which occurs in glands in all parts of the plant. The glands are oval-shaped and contain a semiliquid mass surrounded by a cellulose-like wall. Gossypol varies in color from yellow to orange to red to purple, and is synthesized in the roots of the cotton plant where it occurs at several times the concentration in seeds. In the glandless variety, something prevents the transport of gossypol from the roots. Pigment glands can be separated from the meal by a flotation procedure, using a mixture of organic liquids.

In commercial cottonseed meal, gossypol toxicity is reduced by the application of suitable heat and pressure. The released gossypol from the pigment glands combines with protein during heating to produce "bound" gossypol which is less toxic than the "free" compound. Binding of gossypol with protein involves the epsilon amino groups of lysine. Gossypol also reacts with glucose. Gossypol toxicity is also reduced by adding iron salts to rations containing cotton seed meal because this decreases the absorption of gossypol from the intestinal tract. Gossypol is 2,2′-bi-8-formul-1,6,7-trihydroxy-5-isopropyl-3-m-methylnaphthyl, and was synthesized in 1958 by Adams and his co-workers. Cottonseed meal also contains 3 pigments related to gossypol and derived from its heating. Very little is known concerning the biological significance of these pigments.

The animals most sensitive to gossypol toxicity are guinea pigs, rabbits, swine and dogs; rats and poultry are less sensitive. Mature ruminants are immune to gossypol, possibly because gossypol combines with soluble protein. Calves are susceptible to gossypol before the rumen flora is established.

Symptoms of gossypol toxicity are depressed growth rate and decreased utilization of feed. The younger the animal, the more susceptible it is. Animals continue to eat almost until they die. There is congestion and edema of the lung, liver, spleen and lymph nodes, and a flabby, dilated heart. Livers show intralobular necrosis. Seriously poisoned birds show

TABLE 17.1. PLANTS WHICH MAY CAUSE TOXIC REACTIONS WHEN INGESTED BY ANIMALS

Alkaloids	Cyanogenetic	Goitrogenic	Irritant Oils	Cardiac Glycosides	Saponins	Oxalates
belladonna	catclaw	chard	horseradish	pheasants-eye	corn cockle	beet
jimson weed	bahia	kohlrabi	wh. mustard	dogbane	tung tree	halogeton
henbane	mt. mahogany	wh. mustard	Indian mustard	Lily-of-the-valley	beech	sorrel
Crotalaria	florestine	rape seed	wormseed must.	foxglove	English ivy	purslane
viper's bugloss	mannagrass	bl. mustard	charlock	oleander	yellow pine	Russian thistle
heliotrope	velvet grass	kale	wild radish	be-still tree	alfalfa	greasewood
groundsel	hydrangea	broccoli	baneberry	squill	bounc. bet.	alocasia
hemlock	birdsfoot	cabbage	windflower		coffeeweed	J-in-the-pulpit
Indian tobacco	trefoil	Brussels sprouts	marigold			caladium
tobacco	cassava	rutabaga	buttercups			wild calla
prickly poppy	lima beans	Chinese cabbage				eleph. ear
celandine poppy	cherries	turnip root				dumbcane
fitweed	apple	soybean				philodendron
Dutch breeches	Sudan grass	flax				skunk cabbage
poppy	queen's delight					
bloodroot	poison suckleya					
ergot	white clover					
jessamine	arrowgrass					
manchineel	vetch seed					
African rue	corn					
false indigo						
Scotch broom						
goldenchain						
lupine						
mescalbean						
tomato						
potato						
staggergrass						
false hellebore						
death camas						
monkshood						
larkspur						

TABLE 17.1. *(Continued)*

Resinoid	Phytotoxins		High Nitrates		High Selenium	Photosensitization
milkweeds	prec. bean	pigweeds	cheeseweed	var. thistle	poison vetch	buckwheat
marijuana	tung tree	bishop's weed	sweet clover	nightshades	prin. plume	St. Johnswort
water hemlock	barbadosnut	tarweed	miner's lettuce	goldenrods	goldenweed	lechuguille
laurel	castor bean	bluegreen	witchgrass	sow thistles	woody aster	cult. rape
Labrador tea	black locust	algae	johnsongrass	horsebean	aster	lantana
chinaberry tree		beggar tick	horsebean	chickweed	saltbushes	sacahuiste
rustyleaf		rescue grass	popcorn flower	mustard	paint brushes	panic-grasses
Japanese pieris		plumeless	smartweeds	punct. vine	bat. roadflax	horsebrush
laurel, rhododendron		thistle	Calif. chicory	nettle	gumweeds	punct. vine
		pigweed	dock	crownbeard	hop-sage	water bloom
		lamb's quarters	Russian thistle	celery	snakeweed	oats
		Canadian thistle	annual sage	cucumber	tansy aster	milk purslane
		Rocky Mt. bee	elder	squash	beard tongue	molds
		poison hemlock	flax	carrot	ironweed	alfalfa
		bindweed	alfalfa	soybean		smartweeds
		jimson weed	radish	barley		summer cypress
		barnyard grass	rye	sw. pot. vines		Sudan grass
		goose grass		lettuce		clovers
		joe-pye weed		Sudan grass		vetches
		thoroughwort		wheat		anemone
		milk purslane		corn		poverty grass
		wh. ragweed				turkey mullein
		purple cudweed				squir. barley
		coast goldenbush				pea straw hay
		wild sunflower				blackberry vines
		fireball				foxtail grass
		prickly lettuce				needle grasses
		skeletonweed				crimson clover

dyspnea, with labored, audible respiration. Chicks develop leg weakness, edema and enlarged gall bladders, emaciation, anemia, and serous fluid in the pericardium and peritoneum. There is a reduction in feed intake and a cessation of egg production.

Hens fed cottonseed meal produce eggs that develop discolored yolks and whites during cold storage. Two groups of compounds are involved. Cyclopropene fatty acids impart a pink color to egg white and yolks; gossypol darkens the color of the yolk. The pigmentation of the egg content is not evident when the eggs are laid, but appears after about 4 wk of storage.

The cause of the discoloration of the egg content is an increased permeability of the vitelline membrane which surrounds the yolk. This is caused by malevolic and sterculic acids which are in cottonseed oil in the ratio of 2:1. The relaxed membrane causes a gradual increase in the pH of the yolk approaching that of the white. In the yolk, an increase in water content and a decrease in ammonia, iron and fat occur.

The pink color in the egg white results from iron diffusing from the yolk and reacting with conalbumin, a protein fraction of the white, to produce a complex with a characteristic pink color. It is possible to determine if eggs are from hens fed cottonseed meal by exposing the fresh eggs to ammonia. This immediately brings out the gossypol discoloration of the yolk.

Soybeans

Soybeans are of less nutritional value when raw than when properly heated. Mice are more sensitive to raw beans than are chicks and rats. The age of the animal influences sensitivity to the toxic factor. For the first 8 to 12 wk of life, chicks show a growth depression on a raw soybean ration but eventually, they gain in weight at a rate equal to those fed heated beans.

There is some controversy as to whether or not hens do equally well, fed heated or raw soybeans. Eggs from hens fed raw soybeans are as fertile and of the same yolk quality as those produced by hens fed heated soybeans; but they contain several times as many blood spots as the control eggs.

In processing soybeans, dry heat is not as effective as autoclaving for 15 to 20 min at temperatures ranging from 212° to 248°F. Maximum improvement in protein efficiency ratio (PER) occurs when beans are pressure-cooked. Raw soybeans in the ration cause the pancreas to enlarge severalfold and to reduce fat absorption.

The inhibitory substances in raw soybeans depress the utilization of methionine and cystine. Methionine or cystine, or both, added to a raw soybean ration, increase the growth rate of chicks. The decreased availability of methionine in raw beans is due to the slow rate at which it is released in the gut. Methionine does not completely compensate for the growth differential between raw and heated soybeans, nor does it prevent the increase in the size of the pancreas. The addition of other amino acids does not improve performance of raw soybeans.

Raw soybeans and navy beans contain a trypsin inhibitor which explains why heat treatment improves nutritional value. Animals fed raw soybeans excrete a higher percentage of ingested nitrogen and sulfur in their feces than do animals fed heated soybean rations. Some factor other than the trypsin inhibitor is responsible for the different values of raw and heated beans because, this being the case, the difficulty should be overcome when amino acids are used in the ration instead of the intact protein. However, this does not occur. Such studies rule out the possibility that raw soybeans owe their poor biological value to an interference with intestinal digestion of protein. As previously stated, chicks fed raw soybeans have enlarged pancreases. This suggests a compensatory mechanism for the reduced tryptic activity. Since there is no change in the concentration of proteolytic enzymes in the pancreases of birds fed raw soybeans because of their larger size, the total activity is increased.

Raw soybeans also have hemagglutinins which, it is claimed, may be partly responsible for the growth depression in animals fed raw soybeans. Since the activity of hemagglutinins is readily inactivated by pepsin, this fraction should be completely inactivated before the beans enter the intestine. The addition of antibiotics to a ration containing raw soybeans also enhances its biological value. Antibiotics may not act entirely by altering gastrointestinal flora but may inhibit excessive pancreatic secretion.

Raw soybeans and a number of plant products reduce the availability of such trace minerals as zinc. This effect has been ascribed to phytic acid, in the case of soybeans.

Linseed Meal

Chicks fed linseed meal grow at a slower rate than is normal. The growth-inhibiting substances in linseed meal are destroyed by autoclaving or by incubation with water or 50% ethyl alcohol. Drying (230°F) has no effect. The addition of pyridoxine hydrochloride to rations containing linseed meal increases weight gains, even though the ration initially contains an adequate amount of this vitamin. The highest concentration of the growth-inhibiting substance occurs in meal with the cotyledons containing most of the substance. There is practically none in the hulls isolated from flaxseed. A crystalline substance isolated from unheated linseed meal is soluble in water but insoluble in organic solvents, and contains 50% glutamic acid.

Raw Fish

Certain species of raw fish in the diet of certain animals interfere with the absorption of iron and destroy the thiamin in the ration. Raw coalfish, whiting, hake, carp and some other fish, from both fresh water and salt water, contain the substance. Mink become anemic and develop light-colored underfur; mortality is high. Boiling the fish prior to its incorporation in the ration overcomes the difficulty.

One factor is the enzyme, thiaminase, which is more concentrated in the viscera since eviscerated fish is not as toxic as the whole fish. Supplements of thiamin or cooking overcomes the effect of the thiaminase; another solution is feeding raw and cooked fish on alternate days.

Cycad

Cycad is a palm-like tropical tree that produces seeds in aggregates on the female plant. Natives on the island of Guam use the kernel for food and the outer hull of the seed as a confection. Humans and animals consuming cycad develop progressively incurable paralysis of the extremities. Etiology and effective treatment of the disease are not known. Natives recognize that the kernels are toxic and attempt to eliminate the toxic substances by soaking the kernels in water for 7 to 10 days. After drying, the powdered material compares closely with wheat flour in appearance, chemical composition and use. The toxic substance (methylazoxymethanol) is cycasin, a glucoside which makes up 2% of the unwashed dried kernel.

Chicks, rats, guinea pigs, horses, cattle and swine fed cycad develop severe centrolobular necrosis of the liver. In rats, hepatomas develop which metastasize to the kidneys, lungs, pancreas and intestinal tract. Both the hull and the kernel of the seed are toxic.

Alfalfa Meal

Levels of alfalfa meal in rations in excess of 5% inhibit the growth of chicks. Inhibitors can be removed by exhaustive extractions of the meal with hot water. The inhibitor(s) is soluble in 50 to 80% ethanol; acetone precipitates it from an aqueous alcohol solution. The agent is saponin which depresses growth, feed consumption and feed efficiency, as well as egg production. Saponins from alfalfa have hemolytic activity. The inhibitor is labile in water so that its activity disappears from alfalfa extracts while they still contain saponins. Cholesterol counteracts the effects of the inhibitor.

In addition to alfalfa, Ladino clover and black medic contain saponins, as well as an enzyme, pectin methylesterase (PME), which is responsible for cattle bloating. It reacts with pectin of forages, changing it to pectic acid and alcohol. The former reacts with calcium in the rumen and produces a sticky substance which traps gases formed in digestion. As the gas accumulates and the rumen swells, the animals become bloated. Alfalfa hay contains about 4 times as much PME as grass. The problem occurs most frequently in fields after a frost in the fall.

The estrogenic compound, coumestrol and the related coumestan, trifoliol; the estrogenic isoflavones—genistein, diadzein, biochanin A and formonetin; the dicoumaryl ether, daphnoretin; and the flavone, tricin, have been isolated and identified in alfalfa.

ANTAGONISTS

Vitamin B_6 Antagonists

When desoxypyridoxine is fed in the diet, skin lesions and other characteristic signs of Vitamin B_6 deficiency develop. When Vitamin B_6 is incorporated in the diet in normal amounts, the antagonistic action of desoxypyridoxine is prevented. One interesting difference, however, is noted. When Vitamin B_6 deficiency is induced by feeding desoxypyridoxine, there is no reduction in the Vitamin B_6 content of the liver and other organs as usually occurs when a deficiency is induced by a Vitamin B_6 deficient diet. In fact, the amount of pyridoxal phosphate remains high in the tissues, and the activity in the trans-aminases in the liver and heart and the glutamic-decarboxylase in the brain remains essentially unchanged unless the deficiency is extreme.

INTERRELATIONSHIPS AMONG MINERALS

One of the most confusing areas of nutrition is that of interactions and interrelationships. The complexity arises from the fact that interactions can occur among different mineral elements and between certain minerals and organic compounds. The site of these interactions ranges from the level of the intestinal tract to that of enzymatic action. If some nutrient is in excess, the animal often will perform abnormally.

Historically, one of the first relationships of nutritional significance was observed between calcium and phosphorus; magnesium was later shown to be related to the utilization of these 2 minerals. Calcium utilization is favorably influenced by an adequate level of dietary Vitamin D and phosphorus, and impaired by an excess of magnesium or oxalate in the diet. Diets high in protein significantly favor the absorption of both calcium and magnesium. Since a slightly acetic intestinal pH favorably influences the absorption of calcium, lactose (which promotes the establishment of an aciduric bacterial flora) is an important dietary factor in calcium utilization. In order to be utilized properly, calcium and phosphorus must be in the ration in a proper percentage. Excess calcium interferes with the utilization of zinc, manganese and other nutrients.

In the young, growing bird calcification requires intake of calcium, phosphorus and Vitamin D. Certain fatty acids exert a slight antirachitic value, thus reducing the requirements of Vitamin D. Proper balance of calcium and phosphorus minimizes the Vitamin D requirement. Similarly, an excess of Vitamin D permits calcification with border-line mineral levels or with an adverse ratio of calcium and phosphorus. Thus, it is impossible to state a specific requirement for one of these nutrients without defining the levels of the other two. Large amounts of iron, manganese and other minerals form highly-insoluble phosphates, and thus cause rickets. Phosphorus, present in most plant sources as a phytate, is only partially utilized for bone formation, but extra Vitamin D improves its availability.

ROLE OF TRACE ELEMENTS

Increased interest in the role of trace elements in poultry nutrition has been stimulated by reports on the possible nature of the active component in unidentified growth factor supplements.

A natural copper deficiency is encountered in peat and muck soils in Florida, England, the Netherlands, Australia, New Zealand, Czechoslovakia and Finland. Peat soil areas of the world produce native forages in quantities unmatched in other soil types and it is necessary to add minor minerals to the soil to maximize production. In addition to copper, such elements as manganese, zinc, phosphorus, potassium and boron are also needed, but copper provided the keystone that opened these areas to exploitation. Where fertilization is not used, "muck sick" animals developed symptoms, such as severe anemia (often blamed on parasites), rapid loss of weight, diarrhea, acra matricia, fragile bones, fecal abnormalities and quick death. There is good evidence that the interactions between molybdenum and copper are mediated by inorganic sulfate. Poultry's requirement for copper cannot be determined until the influence of other elements, such as molybdenum, is determined. Depending on the level of inorganic sulfate in the diet, the action of molybdenum upon copper may be quite different. The multiple interrelationships which exist between copper, sulfur and molybdenum, and phosphorus also include other mineral elements (Fig. 17.1).

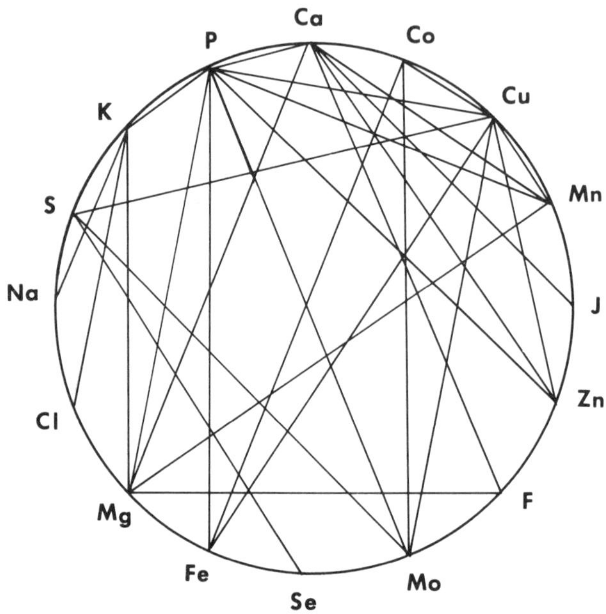

FIG. 17.1. MAJOR AND TRACE MINERAL DIETARY INTERRELATIONSHIPS
These influence specific requirements and tolerance to excesses.

Inverse Relationship Between Copper and Zinc

There is an apparent inverse relationship between copper and zinc. A high level of copper in the liver, such as occurs in copper toxicity, results in almost complete elimination of zinc from liver tissue. The mechanism involved is apparently within the animal body as opposed to the digestive tract.

In the case of trace minerals, copper and molybdenum counteract each other. Arsenicals counteract the toxicity of excess selenium. Chelating agents, including sugars and amino acids, have an important bearing on the availability of minerals, particularly zinc.

Copper is necessary for the utilization of iron. The need of copper for iron to be effective in the prevention of anemia is well-established, but little is known of the biochemistry through which this interaction works. It is apparent that no requirement can be stated for either element without specifying the level of the other. Copper is in some way primarily involved with erythrocyte numbers and survival time, and iron with hemoglobin concentration within the cell.

Molybdenum influences copper absorption, and sulfur or sulfate is involved in the copper-molybdenum interrelationships. Experimental data illustrate the complexity of the interrelationships existing between minerals. Minerals may interact not only amongst themselves, but also with the vitamins, proteins and other factors within the diet. Sites of mineral interrelationships may be in the diet or intestine and at the cellular level.

Importance of Calcium and Phosphorus Balance

Pullets fed growing rations with an improper balance of calcium and phosphorus that allows surplus calcium to be excreted, develop necrosis or degeneration of the kidneys at the time they go into the laying house. In addition, birds show lesions of dehydration, muscular hemorrhages, increased intestinal mucous, and such kidney changes as enlargement and loss of color. Urate deposits are present in the kidneys, joints and pleura, as well as in the protective membranes around the lungs. Microscopic examination of the tissues of birds with necrosis indicates that the parathyroid gland is wasting away. Mortality during both the growing and laying period is increased. A wide ratio of calcium and phosphorus starting at 12 wk of age causes poor egg production and mortality during the growing period. To reduce necrosis, starter and growing rations should not contain more than 1% calcium. Calcium can be increased somewhat in the late growing period without harm but pullets can build up calcium reserves very rapidly just before they begin to lay. It is desirable to change to high-calcium layer feed after production reaches 5 to 10%. Thus, feeding rations with wide calcium-phosphorus ratios during the growing period cause permanent damage to the kidneys. Therefore, a calcium-phosphorus imbalance sometimes shows up at a later time.

Relationship of Calcium to a Manganese Deficiency

Early investigators often observed a high incidence of perosis in diets high in calcium. Studies demonstrated that as the pH of the small intestines increased above 6.6, the manganese formed an insoluble complex and was not absorbed. When chicks or poults are first hatched, their digestive tract is very sensitive to excess calcium. This sensitivity remains until they are nearly 21 days of age. When rations are supplemented with bone meal, they seem to develop more perosis than when calcium carbonate or dicalcium phosphate is used.

Relationship of Calcium to a Zinc Deficiency

The feeding of vegetable protein diets containing a high level of calcium results in dermatitis and perosis. The active substance in the vegetable proteins responsible for causing perosis was found to be a chelating substance (phytic acid) which formed a chelate with zinc in the presence of excess calcium. Sesame seed meal, soybean meal and cottonseed meal are good sources of phytic acid.

Most of the phosphorus in plants is in the form of a phytate. The chick is unable to absorb phytin phosphorus, and when zinc is chelated by phytin, it too, becomes highly unavailable.

Availability of Phytin Phosphorus

The digestive system of poultry is not capable of efficiently utilizing phytin phosphorus because it does not possess an enzyme for freeing the phosphorus. The feeding of radioactive P^{32} phytin phosphorus to chicks demonstrated that the P^{32} was absorbed in the chicks' bones, yet the total phytin phosphorus in the feces showed that very little phytin phosphorus was absorbed. This proved that the P^{32} in the phytin phosphate molecule was exchanged with the P-ions in the digestive tract by the phenomenon of molecular exchange. A high percentage of -P- is not bound, to the extent that it cannot replace another -P-. The phosphorus in DNA, RNA and in milk casein is considered to be tightly bound.

When zinc becomes bound to phytin phosphorus at a pH above 6.5 -, the selective chelating power of EDTA in the presence of high calcium removes the zinc and becomes absorbed as soluble or sequesterated zinc EDTA.

Synergism Between Vitamins

As our knowledge of vitamins developed, it became increasingly evident that these substances are important components of enzymes which function in cyclic systems serving to metabolize carbohydrate, fat and protein. The amount of dietary fat affects vitamin requirements; high-fat diets reduce the thiamin requirement and raise the riboflavin requirement.

The choline requirement of the animal is increased as the fat content of the diet is increased. Synergism between vitamins occurs. Vitamin E promotes the efficient utilization of Vitamin A. Vitamins A and D are the only vitamins that produce distinctly deleterious effects from unbalanced overdosing, although the levels involved are beyond that which may be expected in routine usage. While it is true that malnutrition can be caused by excesses of certain nutrients, it is unlikely that reasonable uses of vitamins in the fortification of feeds would be detrimental.

Relationship of Niacin to Tryptophan

Niacin, through its 2 important enzymes, DPN and TPN, takes part in anaerobic and aerobic oxidation of glucose in fat metabolism, in the degradation and synthesis of amino acids and the oxidation of amino acid residues. It is easy to see, therefore, why the requirement of this vitamin is interrelated with those for a number of other essential nutrients.

On a niacin-low diet, excess threonine interferes with the formation of niacin from tryptophan in such a way as to limit growth markedly.

Some strains of hens are more capable of synthesizing niacin from tryptophan than others. The hens that cannot readily synthesize niacin have a dietary requirement of this vitamin.

Choline Synergism

Four-footed animals are able to synthesize choline from methionine. Early investigators reported that the essentiality of methionine for growth and nitrogen equilibrium was due to its methyl group; however, today, the sulfur-carbon skeleton, $-S-CH_2-CH_2-C-COOH$, is known to be the important part. This is because choline and betaine can supply methyl groups.

Choline can serve as a source of methyl groups and can be converted to betaine: Choline $\rightarrow$ Betaine Aldehyde $\rightarrow$ Betaine.

In poultry, the carbon chain in methionine is a very important part of taurine synthesis in young chicks. The response to synthetic methionine in young chicks is due, perhaps, to a need for more taurine than is normally synthesized by regular metabolic processes.

This ties methionine, choline, taurine, homocysteine and betaine together in a common synergistic pathway.

The requirement for choline is reduced when Vitamin B_{12} and folic acid are supplied in adequate amounts, since these are involved in the metabolic synthesis of methyl groups. Conversely, the dietary requirement for Vitamin B_{12} is influenced by the levels of choline and methionine. High-fat levels in feeds increase the need for choline, since choline is involved in fat transport in the body as a part of the phospholipid, lecithin. Very high levels of niacin can increase the need for methyl groups, also.

Action of Antibiotics

In general, the addition of the commonly-used antibiotics at the levels used for growth stimulation tends to spare the dietary requirement of the animal for certain B-complex Vitamins. This action may be due to the bacterial flora on the intestinal tract: (1) by increasing the numbers of microorganisms which synthesize vitamins; (2) by suppressing microorganisms which may actively compete with the host for certain nutrients; and (3) by suppressing microorganisms which may interfere with the absorption of vitamins from the intestinal tract into the bloodstream or possibly the reduction of microorganisms which produce toxic reactions that creating greater metabolic needs of the host for certain nutrients. Possibly, more than one of these modes of action of antibiotics is operative in most conditions. Continuous high-level feeding of certain drugs, including sulfaquinoxaline and arsanilic acid, increases dietary requirements of the chicks for Vitamin K.

18

Genetics and Nutrition

GENETICS AND NUTRITION

The geneticist has played a very important role in the development of our modern poultry. The first hens probably laid 12 eggs per year, whereas today, hens often produce a flock average of over 240 eggs. In 1930, the yearly flock average was 100 eggs compared to our present average. Many accomplishments have been made in egg size, feed efficiency, feathering, growth and maturity rate and in the meeting of nutrient requirements.

Breeding and Digestion

The digestive tract of chickens and other domestic birds is relatively short, as well as the time required for food to pass through it. The efficiency of digestion and the total capacity for intake and processing of foods is high and does not appear to be controlled by genetic factors (Table 18.1). Chickens, ducks, geese and pigeons have the ability to utilize food efficiently, even if fed by intubation. The digestion of reference nutrients by chickens is summarized in Table 18.2.

Metabolizable Energy

The major portion of the energy required by laying hens is for maintenance. Since there is a great weight difference among breeds of chickens, and maintenance requirement is a function of body weight, it becomes desirable to improve energy utilization. Published data do not show significant differences in basal metabolic rate (BMR) among the various breeds.

Selection for feed efficiency is a complex problem because so many interrelated factors are involved. Selection for growth has been accomplished and growth and feed efficiency are related. The rapidly-growing strains are the most efficient.

TABLE 18.1. TIME REQUIRED FOR FOOD PASSAGE IN CHICKENS AND TURKEYS

Chickens	Passage Time (min)
[1]Chicks (21 days of age)	142
[1]Chicks (42 days of age)	154
Pullets	190
Cocks	200
Hens—all mash	320
Turkeys	
[1]Poults	140
Hens (young)	140
Hens—laying	190
Old—toms and hens	240

Source: Patrick (1970).
[1]Determination made using chromium^{54}.

TABLE 18.2. THE DIGESTIBILITY OF REFERENCE SUBSTANCES BY THE CHICKEN

	%
Cornstarch	97
Sucrose	95
Glucose	97
Cellulose[1]	0
Casein	100
Fish protein	95
Isolated soybean protein	87
Corn oil	98
Soybean oil	98
Beef tallow	71

[1]Some nutritionists have reported that cellulose is partially digested. The reference cellulose was not digested.

The overall results of breeding for metabolic efficiency have been limited, and population size has not been large enough to produce acceptable evidence.

Water Metabolism

The water requirement of broilers is approximately 1.5 lb water per lb feed consumed and about 2 lb for laying hens. The moisture content of poultry feces when poultry are fed a medium-energy ration is about 72% for broilers and 75% for layers.

The water content of feces from hens in the Texas Random Egg Laying Contest was significantly different. The genetic basis for water metabolism of poultry was demonstrated by Dunson and Buss (1968). An inbred strain of chickens was established which seemed normal in all respects except that excessive amounts of water were consumed and excreted. The inbred strain's water consumption was 25.3% of body weight, whereas the normal controls consumed water equivalent to 7.2%

of body weight. The polydipsia appeared to be caused by a failure of the kidney to respond to the antidiuretic hormone. The finding that genetic defects occur in control of water intake and excretion in chickens sheds light upon the problem of wet droppings. With poultry-house moisture being the major problem in poultry management, investigations of polydipsia have proven of value in controlling this moisture.

Riboflavin

Riboflavin was one of the first vitamins for which geneticists attempted to find a strain of chickens having lower requirements. A strain was developed after 6 generations showing a genetic difference in riboflavin metabolism (Lamoreux and Hutt 1948). When 1 strain was fed a riboflavin-deficient ration, the chicks weighed 89.9 g at 5 wk of age, as compared to the chicks of another strain which weighed 49.4 g.

A study by Maw (1945) was made on the genetic influence on riboflavin utilization by chickens involving a riboflavin transport protein. Three pedigreed closely-related hens produced fertile eggs which would not hatch. A study of the riboflavin content of the eggs demonstrated a low riboflavin level. When riboflavin was injected into the eggs, the embryos developed normally. A recessive gene in the homozygous state was postulated. The riboflavin-binding protein was absent in the blood, egg yolk and egg albumen of the select strain.

Niacin (Nicotinic Acid)

Niacin requirements vary between species, and particularly among classes, of poultry. The requirement per kg feed by starting chickens is 27 mg, starting turkeys 70 mg, starting pheasants 60 mg, starting quail 30 mg and starting ducks 55 mg (NRC – 1977). The wide variation in niacin requirements of different species may be related to the efficiency with which tryptophan is converted to niacin. Selection of 2 strains of White Leghorn chickens resulted in 1 strain (Nesheim 1972) having a high requirement and the other a low requirement. This was based on their ability to convert tryptophan to niacin, because when both strains were fed niacin supplements, the growth rate was the same.

Thiamin

Several investigators have attempted, without success, to establish a genetic relationship regarding a thiamin requirement of White Leghorns and Rhode Island Reds. A difference appeared in the thiamin content of the eggs, those of the White Leghorns having the higher concentration of thiamin. A definite genetic difference was not established.

Vitamin D

Vitamin D requirements vary among species. The requirements of turkeys and pheasants are nearly twice those of chickens. A genetic difference was not established within a species.

Minerals

Breed differences in mineral requirements have not been sufficiently established to identify strains showing different requirements. Some investigators reported a manganese difference; however, the results were not confirmed.

Egg shell quality varies between individual strains, breeds and species. If a strain could be developed which possessed greater uptake and utilization of calcium, a strain producing better egg shells would be possible. When a hen lays over 15 to 20 eggs per clutch, the egg shells become thinner and many cracks occur. A high calcium-utilization strain would help correct this problem. Older hens do not absorb more calcium than younger hens; therefore, since the older hens lay larger eggs, the shells are thinner.

Proteins and Amino Acids

Protein utilization involves both digestion and metabolism of a complex structure. Each amino acid has a metabolic pathway that involves several enzymes. The protein requirements represent the cumulative effect of amino acids, temperature, vitamins, growth rate and energy on the chick. Fast-growing broiler chicks appear to convert a higher percentage of the dietary protein to tissue protein than slower-growing birds.

Genetic differences in protein requirements have not been fully established for growth or egg production. Some species seem to make better gains than others, but a genetic relationship seems to be more related to an amino acid than to crude protein.

Studies on the amino acids—methionine, lysine and arginine—have shown that genetic variations in amino acid requirements are possible. The lysine-arginine interrelationship studies show rather wide variations among chicks in ability to degrade lysine and to regulate body lysine pools. Whereas most amino acid levels in the blood remain relatively constant, it is possible to elevate the lysine level by feeding high dietary amounts of this amino acid.

An increase in arginine occurs when the dietary level of lysine decreases. The differentiation of the high-arginine and low-arginine strains of chicks appears to be due to the different response of chicks of the 2 strains to lysine-arginine antagonism.

There is a genetic difference between White Leghorn and Australorp chicks in growth response to methionine. White Leghorns have a higher requirement for methionine. Since the growth of White Leghorns was improved by both methionine and cystine supplements, and the Australorps responded only to cystine, it is probable that the Australorps convert less methionine to cystine than do the White Leghorns.

19

Adaptive Nutrition

Adaptive nutrition is defined as the biochemical-physiological changes in an animal which allow the usage of a ration that previously did not support normal body functions. Any adaptation must involve one or more enzyme systems and in some instances, a complete change in organ structure.

Adaptive nutrition is possible because all species of animals seem to have, within limitations, the same enzyme systems which have often become modified to function under a set of variable factors. Poultry genetics is sometimes considered a program designed to develop better enzymes, because the gene is a form of an enzyme.

THE RUMEN STOMACH

The transformation of the digestive tract of ruminants from an essentially simple non-ruminant to a ruminant state is actually adaptive nutrition. The non-ruminating stomach digests and absorbs similarly to a simple stomach, but the ruminanting organ actually is a microbiological cultural vat in which starch is converted to volatile fatty acids and cellulose is decomposed into volatile acids. Proteins are broken down into amino acids and the amino acids are, in turn, converted to volatile fatty acids, ammonia and minerals. The squamous, stratified epithelium lining the rumen is considered to be a poor absorptive tissue; however, upon changing to a ruminating stage, the volatile fatty acids are absorbed in large quantities.

The factors bringing about the change from a non-rumen to ruminant stage are highly dietary, although genetically, the organ is so structured that upon proper stimulation, the transformation can occur.

THE DIGESTIVE TRACT OF CHICKENS

When young chicks are fed raw ground soybeans, their digestive system does not free the amino acids as needed. As the chickens grow older, an adaptation is made, and the growing and laying chickens utilize the ground raw soybeans. This type of adaptive nutrition is very important because the farmer may be able to feed a small amount of raw soybeans throughout the growing period and thus bring about a better development of the digestive system toward the formation of the enzymes required to handle raw soybeans.

VITAMIN C

Studies on the Vitamin C consumption of peoples throughout North America indicated that the Vitamin C level in the blood of inhabitants of Greenland, and also the uptake, was very much lower than we recommend. Medical examination of the people of Greenland did not reveal Vitamin C-deficiency lesions, thus demonstrating that the body can adapt to lower levels of Vitamin C.

In 1976, many people ingested large doses of Vitamin C to protect them from the common cold, only to discover that when the massive dose was discontinued and the recommended daily dietary level resumed, they soon developed signs of a Vitamin C deficiency. Apparently, the body had developed a wastefulness of Vitamin C and was not able to readapt to the normal uptake. A similar condition has been observed in migrant citrus workers who drink excessive amounts of orange juice while harvesting the fruit, then return to normal uptake after the harvest, only to find Vitamin C-deficiency symptoms developing.

A somewhat similar condition develops in the redvented bulbul (*Pycon: onotus cafer*) which is a fruit-nectar-eating bird. Apparently, this species, and perhaps other fruit-nectar-eating birds, have developed a requirement for Vitamin C. When placed in captivity, they do not receive the nectar or high-fruit diets which are rich in this vitamin. Their adaptation to high levels of Vitamin C has changed their metabolic capabilities to become dependent upon a source of this essential nutrient.

After World War II, many military people, returned to the United States from overseas, noticed that they had difficulty digesting milk. A study revealed that when they had stopped drinking milk, enzyme-lactase production declined and the digestive system was no longer able to digest milk sugar (lactose).

IODINE

Birds, in general, do not have a high requirement for iodine, but those with a crop adapted to the production of "milk" during the feeding period, develop iodine-deficiency lesions. Pigeons, as well as doves, readily show signs of iodine deficiency. All are "seed-eaters" and produce "crop-milk." There seems to be a relationship between "crop-milk" producing and a requirement for iodine. However, it must be recognized that all seeds are considered low in iodine; hence, any seed-eater could develop an iodine deficiency.

UNRECOGNIZED GROWTH FACTORS (U.G.F.)

A U.G.F. stimulating growth or improving hatchability may indicate a requirement or need for controlling some over-active enzyme. It may be aiding in the readjustment of an adaptive condition which has placed the bird in an unbalanced state. Today, with a simple corn-soybean base ration that has been used for over 20 years in breeding programs, we may have developed an adaptive state which, when changed, results in the need for a U.G.F.

Within the framework of an adaptive state, a good nutrition status is one which develops a ration and uses it through the life of the chicken. When a corn-soybean base is used for over 20 years, an adaptive state probably exists.

20

Components of the Ration

FEED INGREDIENTS

Many feed ingredients are in use; however, ground yellow corn and soybean meal comprise from 60 to 80% of the rations that are fed to poultry. Individuals who are interested in poultry feeds and ration formulation should become thoroughly acquainted with both the official and tentative ingredients presented in the *Official Publication* of the Association of American Feed Control Officials (AAFCO). They should also take note of the identification numbers assigned each feed ingredient by the National Science Foundation (NSF), as these are permanent numbers which together with a systematic nomenclature and the analytical data for about 70 nutrients for the feeds or ingredients, are on IBM cards and magnetic tape. The AAFCC has adopted these permanent 6-digit reference numbers which appear after the definition of each feed or ingredient, to enable anyone to identify the information in the AAFCC handbook or the NRC feed composition report. This method allows one to convert the information into his own system for linear programming of diets, rations or feed mixtures. Nutrient composition data for approximately 6000 feeds or ingredients having permanent NRC reference numbers may be obtained from future NRC publications on feed composition. The NRC publications may be obtained from The National Research Council Feed Composition Academy of Science, National Science Foundation, 2101 Constitution Avenue, Washington, D.C. 20418.

When an ingredient in a mixture is not listed in the AAFCO or NRC publication, the appropriate investigator on the AAFCO Committee should be contacted, a term developed and the product defined. This procedure allows, without unnecessary delay, the usage of many new products or by-products having sound nutritional value.

The composition of 48.4 ground corn (NRC 4-02-861) and 84.7 soybean meal, dehulled and with the solvent extracted (NRC 5-04-612), should be well-visualized when formulating rations. A knowledge of suitable substitutes for these ingredients should all be acquired. The functionary analysis of these 2 ingredients is shown in Table 20.1. The term, functionary composition, refers to the nutrients given most consideration in the feed ingredient.

TABLE 20.1. FUNCTIONARY[1] COMPOSITION OF NO. 2 YELLOW CORN NRC 4-02-861 AND SOYBEAN MEAL NRC 5-04-612

	Yellow Corn	Soybean Meal
Dry matter	91.0	90.0
Crude protein	8.3	49.0
Ether extract	3.3	1.0
Crude fat	1.9	3.9
Calcium	0.01	0.27
Phosphorus	0.28	0.62
Choline (mg/lb)	240.0	2860.0
Xanthophyll (mg/lb)	10.0	1.0
Vitamin E (mg/kg)		40.0
Energy (kcal/kg)	3430.0	2440.0
Methionine (%)	2.24	1.39

Source: Patrick (1965).
[1]Functionary composition is a factor most often considered in ration formulation.

Poultry readily accept many ingredients because they do not appear to be influenced by either the taste or the smell of the feed. Chickens do not show measurable color preferences, whereas turkeys prefer some colors over others.

Today's rations are relatively simple compared to those used in 1940, and those which will probably be used in 1990. Ground corn and soybean meal make up approximately 80% of many poultry rations today. The rations in 1940 depended upon natural ingredients for nutrients, many of which were known as Unrecognized Growth Factors (U.G.F.). The only vitamin concentrate was cod liver oil for Vitamins A and D. Yeast, milk, liver meal and alfalfa meal were all used as U.G.F. sources. The following is a typical breeder ration for 1940.

	Breeder 17%
Ground yellow corn	48
Ground oats	5
Wheat midds	10
Wheat bran	5
Alfalfa meal	5
Molasses	1
Dried skim milk	2.5
Meat and bone scraps (55)	5
Fish meal (60)	5
Soybean meal (42% expeller)	5

Dried yeast	1
Dried liver residue	1
Vitamin A and D oil	0.1
Bone meal (steamed)	2.5
Ground oyster shell	2.5
Salt (NaCl)	0.4
Charcoal	0.5
Dried blood	0.5

A similar 17% breeder ration today would contain:

Ground corn	73.0
Soybean meal (50)	12.5
Meat and bone scraps (55)	2.5
Alfalfa meal (17)	2.5
Fat	0.5
Vitamin premix and antibiotics	0.5
Trace mineral premix	0.1
Calcium carbonate	5.0
Dicalcium phosphate	3.0
Salt	0.4

This ration in 1990 will perhaps contain:

Ground corn	20
Homing feed	10
Corn gluten meal	5
Wheat mill products	10
Ground milo	10
Soybean meal	7
Soybean screenings	10
Poultry by-products	2
Fermentation products	10
Alfalfa meal	2.5
Molasses	2
Fat	1
Dicalcium phosphate	3
Ground limestone	4
Ground dolomite limestone	1
Salt (NaCl)	0.4
Trace minerals premix	0.1
Vitamin premix and antioxidant	0.5
Amino acids and antibiotics	0.5
Pellet binder	1

The structure of the poultry ration in 1990 will depend on world exports. The export of grain is one of the major sources of the United States' foreign exchange payments. Imports of grain by China (P.R.C.) and Russia (U.S.S.R.) are increasing, although the U.S.S.R. has soil and weather requirements for expanded production of field crops if agronomical methods of growing and harvesting can be developed. South America should become a large exporter. The present purchase of farm land in the United States by foreign investors may change the United States' export policy.

The United States will continue to increase exports of grain and soybeans. In 1978, the United States produced 30% of the world's grain supply and used 20% of the world production. This nation's exports were approximately 60% of the total world export supply.

By 1983, the United States expects to export 367 million bu corn, 140 million bu soybeans and 444 million bu wheat over the 1977 export totals. The wheat and corn yields per acre here are expected to increase by 10 to 15% by 1983, with no increased yields per acre in soybeans. The large controlling factor in soybean yield improvement will be a shortage of irrigation water. The vast use of water by industrial operations and growing urban populations will greatly reduce the water available for agricultural purposes. Water and energy sources are fast becoming our industrial growth limitations. With these limitations, continued population growth and social and political readjustments, food will become increasingly important to world stability.

The production of by-products, after removing the portions to be used by the human population, will be given first-line attention. These will bring about government restrictions on usage of whole grains and seeds in poultry and livestock rations, and the development of new animal and human food sources from semiarid and tropical areas. Development of formulation restrictions will occupy most of the efforts of nutritionists.

Poultry nutritionists will consider the following when making decisions on formulation restrictions:

(1) Analytical Variations
(2) Ingredient Variations
(3) Research Conditions
(4) Research Interpretations
(5) Ingredient Nutrient Values in Formulation
(6) Commercial and Economic Application

The Association of American Feed Control Officials will make the decision in the development and usage of ingredient by-products. Today, this organization has research investigators covering the following classes of ingredients.

	Investigators
Alfalfa Products	Dale Lauback, Oklahoma City, Oklahoma
Amino Acids	Mary M. Sharpe, Tallahassee, Florida
Animal Products	D. J. Mitchell, Vermillion, South Dakota
Barley Products	Laszlo Torma, Bozeman, Montana
Brewers Products	William E. Simmons, Madison, WI
Chemical Preservatives	Donald W. McDaniel, Rockville, Maryland
Citrus Products	Robert F. Ridenour, Tallahassee, FL
Collective Terms	James Cassidy, Boston, MA
Cottonseed Products	Bily W. Southall, Richmond, Virginia
Distillers Products	Perkins S. Jones, Springfield, Illinois
Drugs & Feed Additives	William B. Bixler, Rockville, Maryland
Fats & Oils	Paul Caudill, Mississippi State, Mississippi
Feed Terms	Paul D. Crisp, Tallahassee, Florida
Fermentation Products	Darrell L. Sharpe, Jefferson City, MO
Flax Products	J. Gordon Hanna, New Haven, Connecticut
Grain Sorghums	Dr. Woodrow L. Martin, Jackson, MS
Lespedeza Products	Glen E. Horton, Las Cruces, New Mexico
Maize Products	Marvin Schreiber, Topeka, Kansas
Marine Products	J. R. Stevens, Augusta, Maine
Milk Products	Thomas A. McConnell, St. Paul, Minnesota
Molasses	E. A. Epps, Jr., Baton Rouge, Louisiana
Non-Protein Nitrogen Products	G. A. Pearson, Richmond, Virginia
Oat Products	Ralph Pay, Little Rock, AR
Peanut Products	M. D. Guthery, Montgomery, Alabama
Rice Products	I. J. Shenkir, College Station, Texas
Rye Products	Michael Purko, Laramie, Wyoming
Screenings	R. T. Wetherbee, Burlington, Vermont
Soybean Products	Duane H. Lieve, Des Moines, IA
Vitamins	William Y. Cobb, Raleigh, NC
Wheat Products	Jerry Sullivan, Oklahoma City, OK
Yeast	J. Allen Stine, Olympia, WA

21

Fats

FEED GRADE FATS AND OILS

Fats used in poultry feeds are obtained from:

(1) Meat slaughtering and processing
(2) Poultry dressing
(3) Rendering plant
(4) Vegetable oil meal plants
(5) Soap-producing plants
(6) Fat-fractionation plants

They are added primarily as a source of energy; however, they also make other contributions.

ADVANTAGES OF FATS IN FEEDS

Calories added in the form of rendered animal fat are well-utilized by poultry for meat or egg production. Fat contains 2.25 times as many calories per lb as protein or carbohydrate. On this basis, the increased caloric content per unit weight of feed obtainable through the addition of fat is particularly advantageous in the formulation of high-energy diets. No feed ration is better than its limiting factor. Proteins, minerals and vitamins must, therefore, be sufficiently high to permit dilution by the added fat. Fat is usually added at levels of 3 to 6% by weight of the complete feed formula. This provides a total of from 6 to 11% fat in the feed, and these levels have proven quite satisfactory. From the palatability standpoint, the addition of animal fats to feeds appears to have offered an effective answer to the common problem of dustiness of feeds which affects palatability. It may have more influence on growth results than is currently recognized. The stability of Vitamin A is greater in the presence of stabilized animal fats. The use of antioxidants in fats for feeds not only does not impair, but actually contributes to maintenance of vitamin stability. Many grades of tallow or grease can be utilized just as satisfactorily (Table 21.1) as choice white grease, providing safe-

TABLE 21.1. EFFECT OF FFA LEVEL AND FAT SOURCE ON FAT PERFORMANCE (8-WEEK RESULTS)

Fat Level	% FFA	Weight lb	Feed per Pound of Gain (lb)
No added fat	—	2.99	2.28
15% animal fat	7	3.10	1.94
15% animal and vegetable fat (primarily animal)	39	3.15	1.94
15% animal and vegetable fat (primarily animal)	79	3.15	2.08
15% vegetable fat	81	3.05	2.04

guards have been taken against rancidity and the fat has no objectionable odor.

While fat provides several desirable improvements in poultry rations, there are several critical disadvantages.

(1) Rancid fats destroy Vitamin A and carotene.
(2) Rancid fats destroy membrane permeability.
(3) Rancid fats increase the requirement for Vitamin E and the need for antioxidants.
(4) Rancid fats reduce the shelf life of dressed birds.
(5) Rancid fats influence the flavor and odor of dressed birds.
(6) Fat increases the need for Vitamin B_{12} and calcium.
(7) The feeding of excess fat can result in heavy accumulation of viscerol fat. This represents a loss to the consumer and often interferes with the rapid cooking of broilers.
(8) Many chlorinated hydrocarbons and aflatoxins accumulate in the fat; therefore, the reuse of fat can bring about greater toxicity because these chlorinated hydrocarbons and aflatoxins are poorly metabolized.

Although fats and oils are a desirable and economical source of energy, they are often destructive to the ration and produce a poultry product which is not always economical. The excessive fat accumulated in the body cavity represents a waste to the consumer; therefore, food preparers prefer dressed poultry without this excess flesh.

Poultry fat is relatively unstable and flesh removed from wings, necks, and other low muscle tissue parts which are high in fat has a very short shelf half-life.

A well-organized, high-calibre program should always be used. Table 21.2 presents some of the characteristics of inedible grades of animal fats. Fats are perhaps among the most dangerous ingredients added to poultry rations. When use of by-products from the milling industry increases, the need for fat to maintain desirable energy levels also increases because of the high-fiber content of many of the by-products. Specifications for animal fats appear in Table 21.3. Some of the desirable standards should include (Table 21.3):

(1) Amount of free fatty acids—Excess free fatty acids are not desirable. They indicate oxidation and poor stability.
(2) Antioxidants—All fats should contain an antioxidant.
(3) Color and odor—Many undesirable fats are dark and possess a rancid odor.
(4) Solids—Many bad fats contain black particles which suggest burning or filth. A clear, solid fat at room temperature without a rancid odor suggests a desirable ingredient.

TABLE 21.2. CHARACTERISTICS OF INEDIBLE GRADES OF ANIMAL FATS

Fats	Titer (Minimum)	F.F.A. (Maximum)	M.I.U. (Basis)	F.A.C. Color Maximum Untreated and Unbleached
Tallows				
Fancy	41.5	4	1	7
Choice	41	5	1	9
Prime	40.5	6	1	13 or 11B
Special	40.5	10	1	19 or 11C
No. 1	40.5	15	2	33
No. 2	40.5	20	2	37
No. 3	40	35	2	No color
Greases				
Choice white	37	4	1	11
A, white	37	8	1	15
B, white	36	10	2	19 or 11C
Yellow	36	15	2	37
House	37.5	20	2	39
Brown	38	50	2	No color

TABLE 21.3. SPECIFICATIONS FOR ANIMAL FATS ACCORDING TO GRADE

	Specifications			
Grade	Titer (min) (°C)	Free Fatty Acids (max) (%)	FAC Color (max) (no.)	Moisture, Impurities Unsaponifiable Matter (max) (%)
Tallows				
Fancy	41.5	4.0	7	1
Choice	41.0	5.0	9	1
Prime	40.5	6.0	13-11B	1
Special	40.5	10.0	19-11C	1
No. 1	40.5	15.0	33	2
No. 2	40.0	20.0	37	2
No. 3	40.5	35.0	none	2
Greases				
Choice white	36.0	4.0	13-11B	1
A white	36.0	8.0	15	1
B white	36.0	10.0	19-11C	1
Yellow	36.0	15.0	37	2
Brown	36.0	50.0	none	2

MIXING FATS WITH OTHER FEEDSTUFFS

Mills located close to the source of fat supply receive shipments in tank trucks, since the heat loss is usually low enough so that heating is unnecessary for unloading. The unloading of tank cars presents more of a problem, though, since the time involved usually causes solidifying of the fat during shipment, which necessitates heating the cars before unloading. The size and location of the heating coils vary, but normally, they consist of about 35-ft lengths of one-half to 2-in pipe. Most cars have a connection at the end or at the bottom. In heating the cars, 100 to 150 lb of steam are used, and the fat is heated to 175°F. Care is taken not to overheat, since expansion of the fat may cause full cars to overflow. Heating is continued until the bottom of the car is hot along its full length and there is no bridging of the fat in the upper portion of the car. This takes 4.5 hr, after which 75% of the car can be unloaded. The remainder is then heated until the ends of the car are too hot to touch; then the fat is pumped out. The overall time is usually 7 hr for cars having connections at the ends or bottom. Sometimes, unloading times can be shortened by recirculating the fat—pumping the hot fat from the bottom back into the top. A channel is formed in the unmelted fat, and increased circulation and agitation speed up the melting process.

Stored fat should be at as low a temperature as possible while keeping it fluid enough to handle. Where fat is used daily, a storage temperature of approximately 120°F is sufficient. Provisions are made in storage tanks to allow periodic bleeding off of water which may separate to the bottom of the tank. Storage tanks are cleaned out every 3 months. During shutdowns and over weekends, lines should be blown out and drained free of fat. To prevent fat from solidifying, lines should be steam-traced by being wrapped with copper tubing. The mixing of fat in feeds is accomplished in any type of feed mixer—horizontal, vertical or continuous. Where the batch-type mixer is used, measured amounts of fat can be run into the feed while the mixer is running. A simple method commonly used is a bucket with holes punched in the bottom so the fat is introduced slowly into the feed. With large mixers, the fat can be added directly in a small spray at a point where there is sufficient agitation of fat and feed before cooling takes place. In continuous mixers, fat is added along with the other ingredients but admitted at a point where most of the other ingredients have already been in contact with the mixing screw. In order to avoid fat balls in the feed, steam the fats before adding to the feed. A heat exchanger is provided with thermostatic controls to maintain a constant temperature. The required temperature varies from mill to mill and season to season, but will usually range between 140° and 200°F.

Where fat is received in drums, immersion-type heaters, wrap-around heating belts and metal case units can be used to melt the fats. All are portable and fairly simple to operate. The desirability of retaining pellet hardness and durability has led to the introduction of fats to pellets after they are formed. A simple, low-cost operation involves spraying hot fat onto the controlled and measured stream of cooled screened pellets which then pass into a retention drum until the fat cools and hardens on the pellets. The pellets are then either binned or sacked.

Fat should be stabilized when used for its energy content in poultry feeds. Fats aid in pelleting and are a source of essential fatty acids. Chickens can tolerate high levels of fat, but economics usually indicate levels of 2 to 4%. Animal tallow and grease, acidulated soap stocks and vegetable oil foods are all used in chick feeds and should be low in moisture and free from suspended solids. Fat products should be checked to be certain that they are not toxic.

Definition of Vegetable Fat or Oil

Vegetable fat (or oil) is the product of vegetable origin obtained by extracting the oil from seeds or fruits which are commonly processed for edible purposes. It consists predominantly of glyceride esters of fatty acids and contains no additions of free fatty acids or other materials obtained from fats. It shall contain not less than 90% total fatty acids, not more than 2% unsaponifiable matter and not more than 1% insoluble matter. If the product bears a name descriptive of its kind or origin, for example, "soybean oil" or "cottonseed oil," it must correspond thereto. If an antioxidant (or antioxidants) is used, the common name or names shall be indicated, followed by the word "preservative" or "preservatives."

Animal Fats and Oils

(American Feed Control Official Association.) Animal fat is the product obtained from the tissues of mammals and or poultry in the commercial processes of rendering or extracting. It consists predominately of glyceride esters of fatty acids and contains no additions of free fatty acids or other materials obtained from fats. It shall contain not less than 90% total fatty acids, not more than 2.5% unsaponifiable matter and not more than 1% insoluble matter. If the product bears a name descriptive of its type of origin, for example, "tallow," "lard" or "grease," it must correspond thereto (Table 21.3). If an antioxidant is used, the common name or names shall be indicated, followed by the word "preservative" or "preservatives."

Hydrolyzed Fat or Oil

Hydrolyzed fat or oil (feed grade) is a product obtained in the fat-processing procedures commonly used in edible-fat processing or soap-making. It consists predominately of fatty acids and shall contain not less than 85% total fatty acids, not more than 6% unsaponifiable matter and not more than 1% insoluble matter. Its source shall be stated in the product name, for example, "hydrolyzed animal fat," hydrolyzed vegetable oil," "hydrolyzed animal and vegetable fat." If an antioxidant (or antioxidants) is used, the common name or names shall be indicated, followed by the word "preservative" or "preservatives."

"Fat Product"

Fat Product is a fat product which does not meet the definitions for animal fat, vegetable fat or oil, or hydrolyzed fat or fat ester. It shall be sold on its individual specifications. If an antioxidant (or antioxidants) is used, the common name or names shall be indicated, followed by the word "preservative" or "preservatives."

The use of the term "feed grade" requires that the specific type of product will have been adequately tested to prove its safety for feeding purposes. In mixed feeds containing fats or fat derivatives, "feed grade" defined above, the term, "feed grade," may be omitted in the ingredient declaration.

Any mixture of two or more fats or fat derivatives, defined above, is to be identified by listing each component; for example, "animal fat and hydrolyzed vegetable oil."

Fatty acids are either saturated or unsaturated. Unsaturated fats have from 1 to 5 double bonds; safflower oil has many double bonds compared with the others. There are differences in the composition of these fats and oils.

Lecithin in Fats

Accompanying animal and vegetable fats is lecithin, a fatty substance containing nitrogen and phosphorus. Lecithin denotes a group of fatty substances, phosphatids, which occur throughout nature as an essential part of living organisms. Lecithin is easily absorbed and functions, in part, as an emulsifying agent to promote better absorption of fatty materials and other food elements. It is an excellent emulsifying agent which improves food utilization and enters into the structure of living cells.

Lecithin contains choline which prevents perosis in poultry. In the form of lecithin, choline is particularly effective. One of the substances contained in the commercial product is an inositol compound which functions in the growth of feathers. Inositol has a bearing on the transport of fats from the liver to other parts of the body. Commercial soybean lecithin contains ethanolamine and small amounts of biotin and tocopherol (Vitamin E). Lecithin has protective action against eczemas of nutritional origin and helps maintain the animal in a state of good health.

Crude Fat (Ether Extract)

The proximate analysis of ingredients and rations identifies all the ether-soluble substances as crude fat. This fraction contains:

(1) Fat (glycerides of fatty acids)
(2) Free fatty acids
(3) Cholesterol
(4) Lecithin
(5) Volatile acids
(6) Chlorophyll
(7) Alkali substances
(8) Resins
(9) Xanthophylls

The chlorophylls, xanthophylls, volatile acids, resins and alkalies are not classed as nutrients but may perform some functions of economic importance, such as are found with xanthophyll, the pigment responsible for coloring the egg yolk and the skin of yellow-skinned chickens.

The addition of fat to poultry rations began soon after the introduction of the corn-soybean ration. The corn-soybean ration was the foundation for the high-energy ration. These rations represent many nutritional achievements. Vitamin B_{12}, antioxidants and selinium became supplements for all poultry rations. This period of nutritional achievement is often called the "corn-soybean era."

Since fat can oxidize easily in feeds, a quality control program is necessary. Some common tests desirable for quality control are:

(1) Rancidity
(2) Free fatty acids
(3) Color and odor
(4) Melting point
    Tallows—melting points above 40° C
    Greases—melting points below 40° C

Fats and Oils (Official)

33.1 Animal Fat is obtained from the tissues of mammals and/or poultry in the commercial processes of rendering and extracting. It consists predominantly of glyceride esters of fatty acids and contains no additions of free fatty acids or other materials obtained from fats. It must contain not less than 90% total fatty acids, not more than 2.5% unsaponifiable matter and not more than 1% insoluble matter. If the product bears a name descriptive of its kind or origin, i.e., "tallow," "lard," "grease," it must correspond thereto. If an antioxidant is used, the common name or names must be indicated, followed by the word, "preservative(s)." (Proposed 1958, Adopted 1968.) NRC 4-00-409

33.2 Vegetable Fat, or Oil is the product of vegetable origin obtained by extracting the oil from seeds or fruits which are commonly processed for edible purposes. It consists predominantly of glyceride esters of fatty acids and contains no additions of free fatty acids, or other materials obtained from fats. It must contain not less than 90% total fatty acids, not more than 2% unsaponifiable matter and not more than 1% insoluble matter. If the product bears a name descriptive of its kind or origin, i.e., "soybean oil," "cottonseed oil," it must correspond thereto. If an antioxidant(s) is used, the common name or names must be indicated, followed by the word, "preservative(s)." (Proposed 1958, Adopted 1968.) NRC 4-05-077

33.3 Hydrolyzed Fat, Oil, Feed Grade is obtained in the fat-processing procedures commonly used in edible-fat processing or soapmaking. It consists predominantly of fatty acids and must contain not less than 85% total fatty acids, not more than 6% unsaponifiable matter, and not more than 1% insoluble matter. Its source must be stated in the product name; i.e., "hydrolyzed animal fat," hydrolyzed vegetable oil," "hydrolyzed animal and vegetable fat." If an antioxidant(s) is used, the common name or names must be indicated, followed by the word, "preservative(s)." (Proposed 1958, Adopted 1968.) NRC 4-00-376, 4-05-076

33.4 ________Ester________, Feed Grade is the product consisting of methyl, ethyl, or other non-glyceride ester of fatty acids derived from animal and/or vegetable fats. It consists predominantly of the ester and must contain not less than 85% total fatty acids, not more than 10% free fatty acids, not more than 6% unsaponifiable matter (2% for methyl esters) and not more than 1% insoluble matter. Its source must be stated in the product name; i.e., "methyl ester of animal fatty acids," "ethyl ester of vegetable oil fatty acids." Methyl esters must contain not more than 150 parts per million (0.015%) free methyl alcohol. If an antioxidant(s) is used, the common name or names must be indicated, followed by the word, "preservative(s)." (Proposed 1958, Amended 1962, Adopted 1968.) Reg. 573.640

33.5 Fat Product, Feed Grade is any fat product which does not meet the definitions for animal fat, vegetable fat or oil, hydrolyzed fat or fat ester. It must be sold on its individual specifications which will include the minimum percentage of total fatty acid, the maximum percentage of unsaponifiable matter, the maximum percentage of insoluble matter and the kind and amount of antioxidant(s) expressed as the common name or names, followed by the word, "preservative(s)," if any are added. (Proposed 1958, Amended 1966, Adopted 1968.) NRC 4-00-414

NOTE: The use of the term, "feed grade," requires that the specific type of product will have been adequately tested to prove its safety for feeding purposes.

In mixed feeds containing fats or fat derivatives, "feed grade" defined above, the term, "feed grade," may be omitted in the ingredient declaration.

NOTE: Any mixture of two or more fats or fat derivatives defined above is to be identified by listing each component; i.e., "animal fat and hydrolyzed vegetable oil."

33.6 Corn Endosperm Oil is obtained by the extraction of oil from corn gluten. It consists predominantly of free fatty acids and glycerides, and must contain not less than 85% total fatty acids, not more than 14% unsaponifiable matter, and not more than 1% insoluble matter. If an antioxidant(s) is used, the common name or names must be indicated, followed by the word, "preservative(s)." (Proposed 1963, Adopted 1968.) NRC 4-02-582 Reg. 8.322

33.7 Vegetable Oil Refinery Lipid, Feed Grade is obtained in the alkaline refining of a vegetable oil for edible use. It consists predominantly of the salts of fatty acids, glycerides, and phosphates. It may contain water and not more than 17% ash on a water-free basis. It is to be neutralized with acid before use in commercial feed. (Proposed 1964, Adopted 1968.) NRC 4-05-078

33.8 Corn Syrup Refinery Insolubles, Feed Grade is obtained in the refining of a corn syrup. It consists predominantly of the fatty fraction of corn starch together with protein and residual carbohydrate. It may contain water and not more than 7% ash nor less than 50% fat on a water-free basis. (Proposed 1964, Adopted 1968.) NRC 4-02-893

22

Carbohydrates

Starch is the most important source of energy for poultry rations. The most economical source of this carbohydrate is corn which often provides as much as 50% of broiler rations. Milo is used as a replacement for corn but cannot replace all of the corn because of the need for xanthophyll. Milo does not contain this pigment. Rice is equivalent to corn but is deficient in xanthophyll, and the cost is above other sources of carbohydrates. It is often used in specialty diets which are low in ash and fiber.

Molasses, oats and barley are excellent sources of digestible carbohydrates. Wheat is used very little in poultry rations.

MAIZE (CORN)

Maize is common corn and is sometimes called Indian corn. It was first discovered by Christopher Columbus when he explored Cuba in 1492. In the British Isles and most English-speaking countries, maize means all forms of corn. The word, corn, is used on the British Isles for such grains as wheat, oats, barley and rye.

Corn is the most valuable crop grown in the United States. Approximately 74 million acres (30 million hectares) are grown here annually.

Starches or Digestible Carbohydrates

The United States produced over 50% of the world's crop of corn with its 6.216 billion bu in 1976. No other crop equals corn in this respect. There was an increase of nearly 100% in production since 1962, when 3.6 billion bu were produced.

The introduction of hybrids in 1930 was greatly responsible for much of this growth. Theoretical yields of 400 bu or more per acre have been predicted, and actual yields of over 300 bu have been achieved. Corn leaves are very efficient in utilizing solar energy in photosynthesis. More total digestible nutrients are produced from corn than from any other crop. Productivity has been increased by improvements and changes in culture, such as earlier planting, higher plant populations of uniform stand, equidistant precision planting, adequate starter fertilizer, minimum tillage, irrigation, application of fertilizer during the growth of the crop through irrigation water or other means, and use of herbicides that result in a weed-free environment.

A major limitation to the attainment of higher yields is the length of time required from planting until the leaf canopy covers the ground area. Corn plants of the future will be single-cross, shortstalked, prolific hybrids, having a high grain to stalk ratio and resistance to lodging. Uppermost leaves will have a vertical orientation, lower leaves more horizontal. Tassels will be small and plants cone-shaped. There will be equidistant planting and high plant populations, capable of capturing a maximum of solar energy for photosynthesis.

High-lysine corn has been developed. Plant geneticists have known since the 1930s that the opaque-2 gene existed, but it was in 1964 that researchers discovered that this gene is responsible for the high-lysine content in corn kernels. Opaque-2 kernels are soft and floury, contain slightly more moisture at harvest and weigh less than normal dent kernels.

The yield for opaque-2 corn is considerably less than for straight hybrids. Accompanying the increase in yield of the straight hybrids was a decline in protein content and an increase in the carbohydrate fraction. Although the yield is lower in the opaque-2 corn, the biological value of the protein is much higher. The human populations which use corn as their chief staple grain experience a pronounced decrease in malnutrition when the opaque variety is used. Today, it is not economical to produce high-lysine corn for farm animals, but geneticists are working toward improving yield as well as protein content (Table 22.1).

The protein in conventional hybrids or open pollinated varieties is highly deficient in lysine and has poor biological value. Studies with chicks and rats have demonstrated that diets deficient in protein or in one or more amino acids are more fattening than adequate diets. When chickens are given diethyl stilbesterol implants and placed on a very high corn diet, they fatten readily, and the visceral fat is a very low melting-point fat. The yellow pigment in corn is also readily deposited in the body fat.

There are 29 official ingredients derived from corn which are approved by the American Association of Feed Control officials and given identification numbers by the NRC. Many of these may be used in poultry rations to replace part or all of ground yellow corn.

TABLE 22.1. PERCENT COMPOSITION OF NO. 2 YELLOW CORN AND HIGH-LYSINE CORN

	No. 2	High Lysine
Protein, crude	8.6	10.8
arginine	0.5	0.67
lysine	0.24	0.38
methionine	0.19	0.13
cystine	0.15	0.16
tryptophan	0.09	0.09
threonine	0.36	0.39
Xanthophyll (mg/lb)	9.0	9.0
Fat (ether extract)	3.9	4.1
Crude fiber	2.7	3.8
Metabolizable energy (ME/lb)	1560.0	1560.0

Source: Patrick (1965).

48. MAIZE (Corn)

Official

48.1 Corn Cob Fractions are obtained by the mechanical separation of one or more fractions of corn cobs. For identification purposes, the name of the fraction must be included parenthetically following the name of the product; i.e., Corn Cob Fractions (Hard Woody Ring and Beeswing). (Proposed 1958, Adopted 1964.) NRC 1-02-779

48.2 Corn Bran is the outer coating of the corn kernel, with little or none of the starchy part of germ. (Adopted 1931.) NRC 4-02-841

48.3 Corn Feed Meal is the fine siftings obtained from screened cracked corn, with or without its aspiration products added. (Adopted 1941.) NRC 4-02-880

48.4 Ground Corn is the entire corn kernel ground or chopped. It must contain not more than 4% foreign material. May also appear in the ingredient list of a mixed feed as Corn Meal or Corn Chop. (Adopted 1931.) NRC 4-02-861

48.5 Cracked Corn is the entire corn kernel ground or chopped. It must contain not more than 4% foreign material. (Adopted 1931.) NRC 4-02-854

48.6 Screened Cracked Corn is the coarse portion of cracked corn from which most of the fine particles have been removed. It may be fine, medium or coarse. It must contain not more than 4% foreign material. (Adopted 1941.) NRC 4-02-862

48.7 Corn Grits is the medium-sized, hard flinty portions of ground corn containing little or none of the bran or germ. May also appear in the ingredient list of a mixed feed as Hominy Grits. (Adopted 1941.) NRC 4-02-886

48.8 Corn Flour is the fine-sized, hard flinty portions of ground corn containing little or none of the bran or germ. (Adopted 1941, Amended 1960.) NRC 4-08-024

48.9 Ground Ear Corn is the entire ear of corn ground, without husks, with no greater proportion of cob than occurs in the ear corn in its natural state. May also appear in the ingredient list of a mixed feed as Corn and Cob Meal or Ear Corn Shop. (Adopted prior to 1928, Amended 1956.) NRC 4-02-849

48.10 Ground Ear Corn with Husks is the entire ear of corn with husks ground or chopped, with not greater proportion of cob than occurs in the ear corn in its natural state. May also appear in the ingredient list of a mixed feed as Corn and Cob Meal with Husks or Ear Corn Chop with Husks. (Adopted prior to 1928, Amended 1956.) NRC 4-02-850

48.11 Flaked Corn is obtained by running cracked corn which has been aspirated and properly tempered, over smooth flaking rolls and subsequently dried and cooled. (Adopted 1946.) NRC 4-02-859

48.12 Toasted Corn Flakes are obtained by running cracked corn which has been aspirated and properly tempered, over smooth flaking rolls, and subsequently dried, cooled, and toasted. (Adopted 1953.) NRC 4-02-860

48.13 Corn Gluten Feed is that part of the commercial shelled corn that remains after the extraction of the larger portion of the starch, gluten and germ by the process employed in the wet-milling manufacture of corn starch or syrup. It may or may not contain one or more of the following: fermented corn extractives, corn germ meal. (Adopted 1936, Amended 1960.) NRC 5-02-903

48.14 Corn Gluten Meal is the dried residue from corn after the removal of the larger part of the starch and germ and the separation of the bran by the process employed in the wet-milling manufacture of corn starch or syrup, or by enzymatic treatment of the endosperm. It may contain fermented corn extractives and/or corn germ meal. (Adopted 1936, Amended 1960.) NRC 5-02-900

48.15 Heat-Processed Corn is the product resulting from heating whole corn, under pressure, without removing any of the component parts. It may be ground, pelleted, or flaked. (Proposed 1961, Adopted 1964.) NRC 4-02-863, 4-02-864, 4-02-865

48.16 Hominy Feed is a mixture of corn bran, corn germ, and part of the starchy portion of either white or yellow corn kernels or mixture thereof, as produced in the manufacture of pearl hominy, hominy grits or table meal, and must contain not less than 4% crude fat. If prefixed with the words, "white" or "yellow," the product must correspond thereto. (Adopted 1935.) NRC 4-02-887

48.17 Dried Corn Syrup is a dried product from corn syrup, a purified concentrated aqueous solution of nutritive saccharides obtained from starch having a dextrose equivalent of 20 or more. (Proposed 1971, Adopted 1973.) NRC 4-02-892

48.18 Hydrolized Corn Protein is the product resulting from complete hydrolysis of isolated corn gluten, and after partial removal of the glutamic acid. (Adopted 1956.) NRC 5-02-901

48.19 Kibbled Corn is obtained by cooking cracked corn under steam pressure and extruding from an expeller or other mechanical pressure device. (Adopted 1958.) NRC 4-02-866

48.20 Dehydrated Corn Plant is the entire corn plant consisting of the ear, leaves, and stalk, which has been artificially dried and ground. (Adopted 1958.) NRC 1-02-768

48.21 Ground Corn Cob is the product resulting from grinding the entire cob. If it is designated as "Fine Ground," the entire grind must pass through a number 10 sieve and 33% of the total material must pass through a number 20 sieve. If it

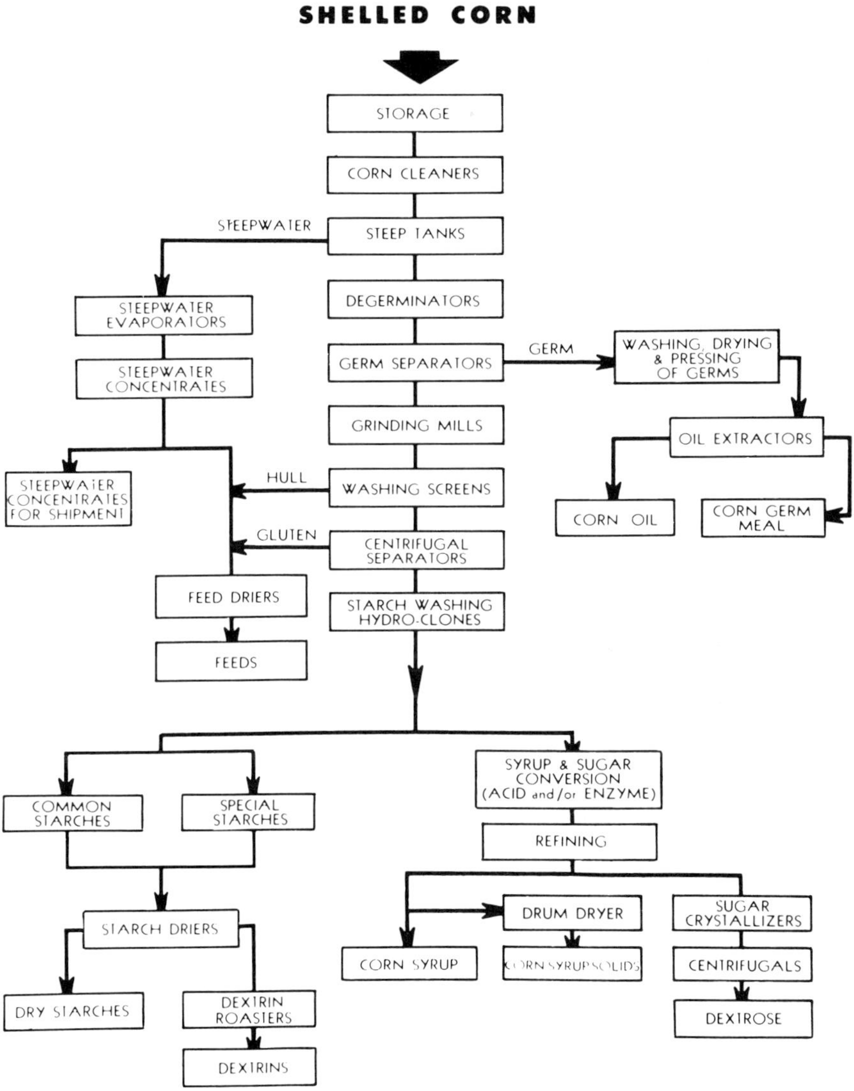

FIG. 22.1. SCHEMATIC OUTLINE OF THE CORN WET MILLING PROCESS

FIG. 22.2. DENT CORN IS A BIG-EARED, LARGE-KERNELED CORN EASILY IDENTIFIED BY THE DENT IN THE BROAD END OF THE KERNEL
It occurs in hybrid and open-pollinated varieties and most is yellow in color. It is the basis of poultry meat and eggs on the United States dining table.

is designated "Coarse Ground," the entire grind must pass through a number 4 sieve and 50% must pass through a number 10 sieve. If it is designated as "Dehydrated," it must contain not more than 10% moisture. (Adopted 1958.) NRC 1-02-780, 1-02-782

48.22 Corn Germ Meal (Dry Milled) is ground corn germ which consists of corn germ with other parts of the corn kernel from which part of the oil has been removed and is the product obtained in the dry milling process of manufacture of corn meal, corn grits, hominy feed, and other corn products. (Definitions combined 1960.) NRC 5-02-894

48.23 Corn Germ Meal (Wet Milled) is ground corn germ from which most of the solubles have been removed by steeping and most of the oil removed by hydraulic, expeller, or solvent extraction processes, and is obtained in the wet milling process of manufacture of corn starch, corn syrup, or other corn products. (Proposed 1960, Adopted 1961.) NRC 5-02-897, 5-02-898

48.24 Condensed Fermented Corn Extractives is obtained by the partial removal of water from the liquid resulting from steeping corn in a water and sulphur dioxide solution which is allowed to ferment by the action of naturally-occurring lactic acid producing microorganisms as practiced in the wet-milling of corn. (Proposed 1959, Amended 1960, Adopted 1961.) NRC 4-02-890

48.25 Malto Dextrins is a purified concentrated aqueous solution of nutritive saccharides, or a dried product derived from said solution, derived from starch having a dextrose equivalent of less than 20. (Proposed 1971, Adopted 1973.) NRC 4-08-023

48.26 Solvent Extracted Hominy Feed is hominy feed from which the fat has been extracted by the solvent process. (Proposed 1965, Adopted 1966.) NRC 4-08-025

NOTE: See 33.5 in Fats and Oils.

48.28 Condensed Fermented Corn Extractives with Germ Meal and Bran, Dehydrated. Condensed fermented corn extractives dried on the germ meal and bran that remains after the extaction of the larger portion of the starch, bran gluten, and oil by the process employed in the wet milling manufacture of corn starch or syrup. It may contain corn gluten meal to adjust protein content. (Proposed 1971, Adopted 1972.) NRC 5-09-333

48.29 Heat Processed Corn is the product resulting from heating whole corn under pressure (Steam Process) or with dry heat (Dry Roasted Process) without removing any of the component parts. It may be ground, pelleted, flaked, or crimped. (Proposed 1976, Adopted 1977.) NRC 4-20-874

GRAIN SORGHUMS (MILO, HEGARI, KAFFIR OR FETERITA)

Grain sorghums are produced throughout the world and are adapted to semiarid cultivation. The production in the United States was 510 million bu in 1962, 930 million bu in 1973 but only 724 million bu in 1976. The production of grain sorghums, oats and barley has been steadily decreasing in the United States.

The grain sorghums were used for many years as a scratch feed for poultry, usually by scattering in the poultry house litter or on the exercise ground.

Production in the United States grew slowly because early harvest methods were principally hand methods involving cutting the heads with a knife. The development of the combine and the introduction of new, shorter varieties which were conducive to combining, encouraged production. The development of varieties which would not be destroyed by birds, the establishment of the nutritional value and improved storage methods, further encouraged production.

Controlling or reducing the moisture content of sorghum grains is often difficult. To remove the moisture from within the grain requires that the vapor pressure of water in the grain be higher than that in the air. Therefore, the rate at which sorghum grain is dried depends upon the difference in vapor pressure between the water in the air and the grain. Sorghum grains require careful storage to prevent the return movement of moisture from the atmosphere to the grain. The grains may respire and release heat, causing a constant increase in temperature, damaging the grains. The higher the moisture content, the greater the respiration and heat release. When ground sorghum is incorporated into a formulated ration, it should be ground and immediately mixed in the ration, then fed as soon as possible, as by the use of bulk deliveries.

Sorghum grains do not contain the yellow pigment, xanthophyll; therefore, unless some yellow corn, corn gluten or alfalfa meal is used in the broiler- or egg-production rations, the pigment will not be present in the egg yolks or under the skin of broilers. Broilers lacking pigmentation, exported to various European countries, have been well-received, because these countries prefer unpigmented broilers. This has encouraged the usage of sorghum grains in the production of poultry for foreign export.

Grain sorghums are often more favorably priced than other grains, and therefore, become prime energy sources for poultry feeds in the American South and Southwest. Sorghum grain is equivalent to corn as a source of energy and protein but is deficient in the pigment, xanthophyll. The protein content varies more in sorghums than in most other grains. Therefore, to ensure the best use, each shipment should be analyzed for protein. The high-lysine varieties of sorghum are equal to the standard varieties, plus lysine.

The following sorghum ingredients are recognized by the American Feed Control Officials and have been given identification numbers by the NRC.

42. GRAIN SORGHUMS (MILO, HEGARI, KAFFIR, OR FETERITA)

Official

42.1 Ground Grain Sorghum is the entire product made by grinding the grains of grain sorghum. (Adopted 1947, Amended 1962, 1964.) NRC 4-04-379

NOTE: The word "cracked" must be substituted for the word "ground" in the above definition when the product is cracked instead of ground. NRC 4-04-378

42.2 Grain Sorghum Gluten Feed is that part of the grain of grain sorghums that remains after the extraction of the larger part of the starch and germ, by the processes employed in the wet milling manufacture of starch or syrup. (Adopted 1948, Amended 1950.) NRC 5-04-389

42.3 Grain Sorghum Gluten Meal is that part of the grain of grain sorghums that remains after the extraction of the larger part of the starch and germ, and the separation of the bran by the processes employed in the wet milling manufacture of starch or syrup. (Adopted 1948, Amended 1950.) NRC 5-04-388

42.4 Grain Sorghum Germ Cake or Grain Sorghum Germ Meal consists of the germ of grain sorghum grains from which part of the oil has been pressed and is the product obtained in the wet milling process of manufacture of starch, syrup, and other grain sorghum products. (Adopted 1948, Amended 1950, 1960, 1963.) NRC 5-04-377

42.6 Rolled Grain Sorghums is obtained by running whole grain sorghums over smooth flaking rolls after properly tempering, removing most of the fine particles, and subsequently dried and cooled. (Adopted 1953.) NRC 4-04-380

42.7 Grain Sorghum Grits consists of the hard flinty portions of sorghums containing little or no bran or germ. (Adopted 1959.) NRC 4-04-384

42.8 Gelatinized Sorghum Grain Flour is obtained from the endosperm of sorghum grain which has been gelatinized and reduced to a finely ground meal and must contain not more than 1% crude fiber. (Proposed 1965, Adopted 1966.) NRC 4-08-035

42.9 Partially Aspirated Gelatinized Sorghum Grain Flour is obtained from whole sorghum grain which has been partially aspirated and has been gelatinized and reduced to a finely ground meal and must contain not more than 2.5% crude fiber. (Proposed 1965, Adopted 1966.) NRC 4-08-036

NOTE: Any of the types shown parenthetically in the heading for this section may be substituted for the words "grain sorghums" in the above definitions. If the name of the type is given it must correspond thereto.

42.10 Grain Sorghum Mill Feed is a mixture of grain sorghum bran, grain sorghum germ, part of the starchy portion of grain sorghum kernels, or mixture thereof as produced in the manufacture of grain sorghum grits and refined meal and flour and must contain not less than 5% crude fat and not more than 6% crude fiber. (Proposed 1964, Adopted 1968.) NRC 4-04-385

WHEAT

Wheat is the most important food crop in the world. In 1976, world production was 412.4 million metric tons, and United States production was 58.432 million metric tons. Wheat grows over a wider area, and produces more tonnage than any other crop.

Wheat varieties of today are higher-yielding and are more resistant to many diseases and pests than the earlier varieties. They also respond to higher rates of fertilization and water. Improved varieties are being developed.

Wheat is the most universally grown cereal in the United States, as in many other parts of the world. Its composition varies with the variety and soil, and the cultural treatment received. The milling of wheat produces flour for human use and appreciable quantities of by-products for animal feeds. Wheat is primarily a source of carbohydrates or starch. By-products of wheat include: (1) bran, consisting almost entirely of the outer coatings of the wheat kernel; (2) shorts or brown shorts; (3) standard middlings or middlings consisting of the fine particles of bran and germ with very little Red Dog; (4) Red Dog, sometimes called "light shorts"; a product from the tail of the mill that consists, chiefly, of the aleuron layer with small particles of bran, germ and flour; (5) germ, consisting of the wheat germ separated in the milling process by grinding; and (6) flour, which normally goes into human foods.

Hard, red winter wheats are grown in Oklahoma and Kansas, and hard, red spring wheat comes from Montana and North Dakota. White wheat is obtained from Washington, and soft, red winter wheat from Indiana. Seventy two to 75% of the wheat is recovered as flour, 12 to 15% as bran, 7 to 8% as shorts, 1.5 to 4.5% as red dog and 1% as germ. Flour is lower in protein than the other by-products, and germ has the highest percentage of protein and the lowest yield. Shorts are the next highest in protein and are higher in protein than bran. The yield is second to bran. Protein content of wheat varies widely, influenced by environmental conditions, fertility and variety of wheat.

The metabolizable energy of different grains is of considerable significance in poultry feeding; wheat is lower than corn. Because of the large portion of grain contained in poultry feeds, the amino acids present in the protein are of importance. Most of the essential amino acids in wheat are readily available; however, wheat is deficient in lysine (Table 22.2).

In some Washington state tests, various strains were fed to turkeys along with a protein concentrate or balancer. Wheat was consumed in the greatest quantities, followed by corn, oats and barley. For many years, wheat has been used extensively in the Pacific Northwest; in fact, wheat constitutes essentially the only cereal grain used in feeding laying hens. Egg size, shell thickness and egg quality, as indicated by Haugh units, are essentially the same for all grains. Corn and wheat give identical values in lb of eggs per lb feed consumed, and have about the same effect on egg size. Oats, barley and milo give slower rates of increase in size than corn.

Wheat is a highly acceptable grain for use in poultry production, although slightly lower growth and feed efficiency are obtained in some instances when wheat is substituted for corn. Flake-milling by-products are effectively utilized in poultry production, and wheat makes an excellent replacement for either corn or milo in poultry feeds. The number of calories of metabolizable energy per lb is essentially the same for corn and wheat, but milo contains 5% less metabolizable energy than corn. The average percentage of protein in corn is 9, in milo 8 and in wheat 12.

A number of factors besides energy and protein content must be considered when wheat is used to replace corn or milo. These factors include amino acid balance, stickiness of ground wheat due to the gluten content, desirable color of egg yolks and the color of fat on broilers and market turkeys.

Protein in wheat is lower in the amino acids—leucine, lysine and alanine—and somewhat lower in threonine. Levels of glutamic acid and choline in wheat exceed those in corn and milo. Wheat contains gluten, which makes the ground wheat sticky when it becomes wet and gives the feed a tendency to adhere to the mandibles of chickens and turkeys. To overcome this tendency, pulverized oats, ground barley or wheat bran is used to coarsen the texture of the mash. Wheat is devoid of the xanthophyll pigments, but this can be overcome by the use of alfalfa meal or corn gluten meal, or their combination.

It has been reported that wheat germ has a deleterious effect on the growth of chickens and that this is associated with an increase in the fat content of droppings. It is claimed that these results are caused by a thermolabile toxin in the raw wheat germ which is destroyed by heating. Close observation during experiments where raw wheat germ was part of

TABLE 22.2. PERCENTAGE COMPOSITION OF HARD WHEAT AND WHEAT MIDDS

	Wheat	Midds
Protein, crude	12.5	16.1
arginine	0.6	0.81
lysine	0.4	0.57
methionine	0.17	0.15
cystine	0.22	0.16
tryptophan	0.16	0.16
threonine	0.28	0.43
Fat (ether extract)	1.8	4.5
Crude fiber	2.7	7.0
Metabolizable energy (ME/lb)	1360.0	890.0

Source: Patrick (1965).

the diet indicated that during and after feeding, the food formed a paste on the beaks of the birds in combination with salivary mucous and water and that these birds ate correspondingly less and grew more slowly. Birds clean their beaks on the wire of the floor, thus adding sufficient food fat to the droppings to account for the differences in fecal fat between the control and wheat germ groups. This simple explanation for the effects of raw wheat germ questions the idea of a thermolabile toxin.

Feeding "new" wheat to poultry is dangerous and results in a serious rate of mortality in flocks. It may involve feed dealers in lawsuits brought about by poultrymen to recover their losses. The higher the moisture content of the new wheat, the more dangerous it is to feed. Wheat which has been kept in a bin or elevator and undergone a "sweat," is safe to feed. New wheat causes the digestive system to become full and a decided enteritis and diarrhea are noticeable. Some believe that mortality is a direct result of toxins produced by fermentation of the new grain in the digestive tract. No specific organisms have been isolated, but symptoms of digestive troubles, such as profuse diarrhea and darkened comb, are observed.

Other grains, such as oats, fed in the milk stage before they are mature, cause complications similar to those caused by the feeding of new wheat.

Currently, wheat varieties are self-pollinated, comparable to an inbred line of crops, such as corn. Several problems related to the production of hybrid wheat are yet to be resolved. Lack of complete fertility restoration in the hybrid plants is the most troublesome problem; it results in poor set, and hence reduces yield. Another problem is the low percent set on the male sterile parent of the hybrid when planted in alternate strips with a fertility-restoring parent.This condition reduces the amount of hybrid seeds. One of the greatest uncertainties concerning hybrid wheat is that of economics. Can hybrid wheat be raised and sold at prices profitable to both the producer and farmer? Yield increases of 25 to 35%

occur as an advantage for hybrid wheat. Other advantages are its ability to stand unfavorable environmental conditions, such as low temperature, soil heaving and drouth. Hybrids are more vigorous in growth habits and better buffered genetically to withstand conditions than inbred varieties.

All the official wheat products can be used in poultry feeds.

93. WHEAT PRODUCTS

Official

93.1 Wheat Bran is the coarse outer covering of the wheat kernel as separated from cleaned and scoured wheat in the usual process of commercial milling. (Adopted prior to 1928.) NRC 4-05-190

93.2 Wheat Feed Flour or Wheat Flour consists principally of wheat flour together with fine particles of wheat bran, wheat germ, and the offal from the "tail of the mill." This product must be obtained in the usual process of commercial milling and must contain not more than 1.5% crude fiber. (Adopted 1949.) NRC 4-05-199

93.3 Wheat Germ Meal consists chiefly of wheat germ together with some bran and middlings or shorts. It must contain not less than 25% crude protein and 7% crude fat. (Adopted 1949, Amended 1953.) NRC 5-05-218

93.4 Wheat Mill Run consists of coarse wheat bran, fine particles of wheat bran, wheat shorts, wheat germ, wheat flour, and the offal from the "tail of the mill." This product must be obtained in the usual process of commercial milling and must contain not more than 9.5% crude fiber. (Proposed 1959, Adopted 1960.) NRC 4-05-206

93.5 Wheat Middlings consists of fine particles of wheat bran, wheat shorts, wheat germ, wheat flour, and some of the offal from the "tail of the mill." This product must be obtained in the usual process of commercial milling and must contain not more than 9.5% crude fiber. (Proposed 1959, Adopted 1960.) NRC 4-05-205

93.6 Wheat Shorts consists of fine particles of wheat bran, wheat germ, wheat flour, and the offal from the "tail of the mill." This product must be obtained in the usual process of commercial milling and must contain not more than 7% crude fiber. (Proposed 1959, Adopted 1960.) NRC 4-05-201

93.7 Wheat Red Dog consists of the offal from the "tail of the mill" together with some fine particles of wheat bran, wheat germ, and wheat flour. This product must be obtained in the usual process of commercial milling and must contain not more than 4% crude fiber. (Proposed 1959, Adopted 1960.) NRC 4-05-203

93.8 Defatted Wheat Germ Meal is obtained after the removal of part of the oil or fat from wheat germ meal and must contain not less than 30% crude protein. (Proposed 1960, Adopted 1962, Amended 1964.) NRC 5-05-217

NOTE: When "Ground Wheat Screenings" are added to any wheat product such screenings added must be limited to ground wheat screenings not exceeding the run of the mill; and screenings from outside sources must not be added. The declaration of "ground wheat screenings" must be made in the name and in the same size type as the product name itself; i.e., "Wheat Bran with Ground Wheat Screenings," "Wheat Shorts with Ground Wheat Screenings."

OATS

Oats are among the most important of the world's food crops. They belong to the same family of plants as barley, corn, rice, rye and wheat. The United States and Russia are the leading oat growers.

Oats are best suited to a cool climate; however, the southern states are producing autumn-sown oats which are of excellent quality and yield. The average yield of oats in the United States is approximately 60 bu per acre, with some select yields of over 150 bu per acre. The standard bushel of oats weighs 32 lb, but good heavy oats weigh up to 45 lb.

The production of oats has been on a rapid decline in the United States since 1962. The production in the United States was 562 million bu in 1976 as compared to 1012 million bu in 1962. This represents nearly a 50% reduction in 14 years. This reduction was, in part, a result of the usage of high-energy rations for poultry, the all-mash feeding of turkeys and attempts to reduce production costs.

Oats played a leading role in the feeding history of many farm animals. Heavy oats were used as the principal ingredient in mule feed (the heavy oats were called mule oats). The usage of oats in chick growing, pullet developers and laying feeds resulted in well-developed birds and a great reduction in cannabalism. Heavy oats, 36 to 44 lb per bu, became important in growing and finishing turkeys when the mash-grain feeding program was used. Turkeys readily eat light-colored, heavy oats in the warm weather, and as a result, turkeys can be "finished" for market any season of the year. Also, oats and barley-fed turkeys have more fat under the skin and less around the viscera than corn-fed birds. Since a "finished" bird must have fat under the skin and viscera fat is lost during processing, oats became of greater importance in making possible rapid finishing of turkeys for market.

Oats contain about 3 times as much cellulose and 7% less starch than corn. This makes the oat a low-energy ingredient, and hence corn not only contains more energy but is in general, cheaper, as well (Table 22.3).

Some feed manufacturers found that the fat was firmer in broilers receiving oats, barley or wheat than when corn was the only or primary grain used. The usage of some fats in some broiler feeds resulted in "greasy" chicken, highly objectionable in rapid-food-serving establishments. Today, many fresh broilers offered to the consumer contain

TABLE 22.3. PERCENTAGE COMPOSITION OF BARLEY AND OATS

	Barley	Oats
Protein, crude	11.5	12.0
arginine	0.55	0.8
lysine	0.4	0.5
methionine	0.12	0.15
cystine	0.22	0.22
tryptophan	0.1	0.16
threonine	0.3	0.44
Fat (ether extract)	2.0	4.0
Crude fiber	6.0	11.6
Metabolizable energy (ME/lb)	1200.0	1160.0

Source: Patrick (1965).

excess visceral and abdominal fat and make greasy-type dressed broilers. Although oats and barley have been eliminated from many poultry rations today, their use may be reestablished in the future. Poultry producers may find themselves in the same situation as pork producers in the early 1950s when consumers rebelled against overly-fat pork. Some of the solutions to the pork producers' problems were: (1) reduction in total digestive nutrients, (2) reduction in dietary fat, (3) increase in protein and cellulose content of the diet and (4) breeding.

The protein content of oats may vary from 11 to 25%, and yields of 150 bu per acre can be achieved.

Oats are an excellent source of pantothenic acid, but are free of xanthophyll, the pigment responsible for producing the yellow color in egg yolks and the skin of chickens.

The official AAFCO and NRC reference numbers for the ingredient list of oat products show 7 available ingredients. Oat hulls are often used in low energy growing pullet rations, whereas oats, either whole or pulverized, are used in egg production, breeding and broiler or turkey growing rations.

69-OAT PRODUCTS

Official

69.1 Oat groats are cleaned oats with the hulls removed. (Adopted 1931. Amended 1963.) NRC 4-03-331

69.2 Oat Hulls consists primarily of the outer covering of oats, obtained in the milling of table cereals or in the groating of oats from clean oats. (Adopted prior to 1928, Amended 1963.) NRC 1-03-281

69.3 Feeding Oat Meal is obtained in the manufacture of rolled oat groats or rolled oats and consists of broken oat groats, oat groat chips and floury portions of the oat groats, with only such quantity of finely ground oat hulls as is unavoidable in the usual process of commercial milling. It must not contain more than 4% crude fiber. (Adopted 1938.) NRC 4-03-303

69.4 Clipped Oat By-Product is obtained in the manufacture of clipped oats. It may contain the light chaffy material broken from the end of hulls, empty hulls, light immature oats, and dust. It must not contain an excessive amount of oat hulls. (Adopted prior to 1928.) NRC 1-03-269

69.5 Oats must contain a minimum of 80% sound cultivated oats and must not contain more than 10% wild oats; 3% heat damaged kernels of oats, other grains, and wild oats; and 5% foreign material. It must contain not more than 15% crude fiber. (Adopted 1958, Amended 1964.) NRC 4-03-309

NOTE: Foreign material must be all matter except cultivated oats, other grains, and wild oats, and must include oat clippings and detached oat hulls.

69.6 Mixed Feed Oats consists of a mixture of grain containing at least 30% of cultivated oats provided that the mixture consists of either (a) not less than 65% of cultivated and wild oats combined or (b) not less than 65% of wild oats. It must contain more than 25% of other grains, and not more than 10% foreign material which may include 4% fine seeds. (Adopted 1958, Amended 1964.) NRC 4-08-026

NOTE: Foreign material must be all matter except wild oats and grains for which standards have been established under the United States Grain Standards Act.

69.7 Oat Mill By-Product is the by-product obtained in the manufacture of oat groats, consisting of oat hulls, and particles of the groat, and containing not more than 22% crude fiber. (Proposed 1963, Adopted 1965.) NRC 1-03-332

BARLEY

Barley is one of the oldest cultivated crops and is the cereal grain most commonly used in malting. United States production of barley in 1975 was 8 million metric tons, and world production was 160 million metric tons.

In the Pacific Northwest, California and many European countries, barley is the principal grain produced for feeds. It is higher in protein and fiber and lower in energy than corn, but is considered good for turkey feeds. Its overall evaluation as a feedstuff is approximately 85 to 90% equivalent to corn. Its value in producing firm fat in animals is equivalent to heavy oats (Table 22.3).

The feeding value of barley can be greatly improved for turkeys or chickens by either water-treating the grain or breeding a fermentation-type enzyme preparation. Some varieties of barley produced in different areas of the United States respond to water treatment or enzyme supplements more than others. Grinding or crushing (rolling) of the grain improves its value.

Although in the United States most barley is fed to livestock, barley products are also desirable for poultry. The usage of barley hulls in poultry feeds is not recommended except in pullet developers, where a type of feed restriction is being used to delay maturity.

The following barley products are approved by the AAFCO and have reference numbers by the NSF.

12. BARLEY PRODUCTS

Official

12.1 Barley Hulls consist of the outer covering of the barley. (Adopted prior to 1928.) NRC 1-00-496

12.2 Barley must consist of at least 80% sound barley and must not contain more than 3% heat damaged kernels, 6% foreign material, 20% other grains or 10% wild oats. In this definition, foreign material must be all matter other than barley, other grains, and wild oats which are not separated from the barley in the proper determination of dockage. (Adopted 1959.) NRC 4-00-549

12.3 Pearl Barley By-Product is the entire by-product resulting from the manufacture of pearl barley from clean barley. (Proposed 1961, Adopted 1962.) NRC 4-00-548

12.4 Barley Mill By-Product is the entire residue from the milling of barley flour from clean barley and is composed of barley hulls and barley middlings. (Proposed 1961, Adopted 1962.) NRC 4-00-523

Barley is ground or rolled to obtain maximum feeding value. It is higher in niacin than corn. The value varies with its test weight per bu because high-weight barley contains less fiber and more starch. Whole barley is worth from 81 to 85% as much as whole corn. Barley is sometimes infected with scab.

About 80 to 85% of the total feed consumed by turkeys is eaten between 10 to 24 wk of age, and the greater portion of this is grain in some form. Whole and pelleted barley can be fed, and turkeys are produced of approximately the same weight, and requiring the same amount of feed to produce 1 lb of gain, as turkeys fed corn. The feeding value of barley can be greatly improved for young chicks or turkeys by either water-treating the grain or adding to the diet a crude enzyme supplement produced by fermentation. The magnitude of response is much greater for barley than for wheat.

Some varieties of barley produced in different areas of the United States respond to either enzyme supplements or water treatment, or both; others are unimproved. There is a correlation between areas of production in the United States and the type of response obtained. Washington state researchers found in 1957 that the feeding value of barley grown in the western states could be improved by certain enzyme supplements. Most of this work was involved with egg production. Chicks fed hull-less barley performed about the same as those fed regular barley, both types being inferior to corn, milo or wheat. When enzymes were added to a hull-less barley diet, growth rate and feed efficiency were increased and the droppings became less sticky. A protease enzyme may be one of the essential substances, but evidently, others are needed for maximum results. Apparently, chicks depend on enzymes that are either in or on the barley to help them digest the grains. The amounts naturally present in hull-less barley are not sufficient to allow the chicks to digest this grain as completely as can be done when more are added. When the grain is soaked at room temperature, these enzymes, and probably others produced by microorganisms during the soaking period, partially digest the grains; little more improvement is noted when more enzymes are added to the ration. When the barley is soaked at a high temperature, enzymes naturally present are inactivated and enzyme production by microorganisms during the soaking and drying period is minimized.

RICE

Rice is second to wheat as a world food commodity with a production of 343.781 million metric tons in 1976. A billion people depend on it as their major source of energy. Less rice than wheat is used in feed for poultry. At the International Rice Research Institute in the Philippines, selections have been made from crosses between the tall tropical indicas and the short-statured types from Taiwan. The yield potential of the new short-strawed type is several-fold that of established types because of its resistance to lodging and response to nitrogen fertilizer. New records in the productivity of rice are being achieved.

The new improved varieties of rice have a lower protein content than the native varieties and also a lysine deficiency similar to the native varieties.

Yields in wheat and rice are made up of 3 components: number of heads, kernel number and kernel weight. The new short-straw types respond to adequate moisture, nutrient reserves in the soil, and environment at the time when the growing point of the stem initiates the head or panicle and determines the potential kernel number. Optimum yield thereafter, is a result of limited lodging achieved by a short stiff straw. Yield capacity is conditioned by these variables.

The shape and form of plants have been altered genetically so they respond to fertilizer without lodging. Most old types of rice take 6 to 7 months to mature. New short-straw, high-yielding varieties mature in 4 to 4.5 months.

All official rice products except rice hulls and mill by-products are desirable ingredients for poultry feeds.

TABLE 22.4. PERCENTAGE COMPOSITION OF RICE AND RICE BRAN

	Rough Rice	Rice Bran
Protein, crude	8.0	13.0
arginine	0.7	0.99
lysine	0.3	0.54
methionine	0.17	0.22
cystine	0.1	0.09
tryptophan	0.12	0.22
threonine	0.31	0.5
Fat (ether extract)	1.7	13.3
Crude fiber	9.0	12.9
Metabolizable energy (ME/lb)	1215.0	1000.0

Source: Patrick (1965).

75. RICE PRODUCTS

Official

75.1 Rice Polishings is a by-product of rice obtained in the milling operation of brushing the grain to polish the kernel. (Adopted 1938.) NRC 4-03-943

75.2 Ground Rough Rice or Ground Paddy Rice is the entire product obtained in grinding the whole rice grain including the hulls. (Adopted prior to 1928, Amended 1959.) NRC 4-03-938

75.3 Rice Bran, Solvent Extracted is obtained by removing part of the oil from rice bran by the use of solvents and must contain not less than 14% crude protein and not more than 14% crude fiber. (Adopted 1951, Amended 1959.) NRC 4-03-930

75.4 Chipped Rice, Broken Rice, or Brewers' Rice is the small fragments of rice kernels that have been separated from the larger kernels of milled rice. (Proposed 1959, Adopted 1960.) NRC 4-03-932

75.5 Ground Brown Rice is the entire product obtained in grinding the rice kernels after the hulls have been removed. (Proposed 1959, Adopted 1960.) NRC 4-03-935

75.7 Rice Bran is the pericarp or bran layer and germ of the rice, with only such quantity of hull fragments, chipped, broken, or brewers' rice, and calcium carbonate as is unavoidable in the regular milling of edible rice. It must contain not more than 13% crude fiber. When the calcium carbonate exceeds 3% (Ca.—1.2%), the percentage must be declared in the brand name; i.e., Rice Bran with Calcium Carbonate not exceeding ________%. (Proposed 1963, Adopted 1964.) NRC 4-03-928

NOTE: See also 87.6 and 87.7.

MILLET

The feeding value of red and white proso millet is 95 to 100% that of corn in the starting, growing and laying rations of chickens. It is fed in starting rations to equal 44% of the all-mash ration. The supplementary values of grains when fed with red proso ranked in the order of oats, barley and corn when based on their rate of growth. The best growth or feed utilization occurs with equal amounts of proso millet, and either oats, barley, corn or wheat. Ground proso millet is found in growing rations to equal 52% of the mash ration, and up to 100% of the scratch-grain ration. Millet is fed in laying rations equal to 43% of the mash ration and 100% of the scratch-grain ration. Because of its small size, it is added to the laying mash unground, although, when this is done, 9% more feed is required to produce 1 dozen eggs. Neither egg production or hatchability is improved by increasing the amount of alfalfa leaf meal or meat and bone scraps.

Proso millet is a late-sown catch crop. It is grown on the Northern Great Plains and other areas having similar climate. Proso is distinguished from foxtail and barnyard millet, chiefly by the head or inflorescence (the seed-bearing part). It is known by such names as hog millet, broom corn millet and "hershey." Proso millet is classified as red or white, depending on the color of the hulls. The bran or seed coat is the same color in all varieties. The average yield of proso in pounds of grain per acre is less than that of either wheat, oats or barley, although, in rather dry seasons, proso often outyields green feeds.

Millet crops, on a small scale, are grown in several regions for bird seed, and are the major seeds fed to parakeets. The demand for millet for parakeet feeds is so great that it must be imported. The protein content of millet seed varies from 10 to 17%.

The typical millet plant is a tall, straight tuft of leafy spikes arranged around several stalks that are crowded with clusters of tiny seeds. The crop is ready for harvesting after 3 months. Varieties are named after their bushy stalks of seeds—broom corn, foxtail, finger and cattail. The common barnyard millet resembles its wild ancestor and might be taken for grass. The cattail millet can be grown in poor sandy soils where the climate is too hot and dry for any other cereal. Fields of millet are grown in certain parts of Kansas, Texas, Missouri and the Dakotas, to provide some hay for cattle and enough birdseed for our feathered pets. Millet is not used to any appreciable degree in poultry feeds.

RYE

Rye has a limited value in poultry feeds. When fed to young chickens, it causes sticky droppings which form balls on the chickens' toes. It can be used to replace 10% corn in the complete ration. The rye protein has a higher methionine and cystine content than either wheat or corn.

Some rye productions are infected with ergot fungus, and when present, should not be fed to the poultry.

Two official rye products are available for feed manufacturing. Most of the rye grain is milled for flour, and the rye mill run and rye middlings remaining after milling are generally cheap enough to use in poultry, hog and dairy rations.

78. RYE PRODUCTS

Official

78.1 Rye Mill Run is obtained in the usual process of the milling of rye flour from cleaned and scoured rye, consisting principally of the mill run of the outer covering of the rye kernel and the rye germ with small quantities of rye flour and aleurone, and must not contain more than 9.5% crude fiber. (Adopted 1946.) NRC 4-04-34

78.2 Rye Middlings consist of rye feed and rye red dog combined in the proportions obtained in the usual process of milling rye flour, and must not contain more than 8.5% crude fiber. (Adopted 1946.) NRC 4-04-031

SCREENINGS

Screenings may include weed seeds, cracked grains and foreign grains. Although most of these ingredients are used in dairy rations, grain screening are useful in poultry-growing rations. Grain screenings can replace whole grains. The screenings bearing the name of the predominating grain can replace that grain in the poultry ration. The chaff and/or dust should not be used in poultry rations. The following screening ingredients, 81.1, 81.2 and 81.4 as presented by the AAFCO, are desirable for poultry feeds; however, it is highly recommended that a careful standard be developed for usage of screenings.

81. SCREENINGS

Official

Screenings obtained in the cleaning of grains which are included in the United States Grain Standard Act and other agricultural seeds may include light and broken grains and agricultural seeds, weed seeds, hulls, chaff, joints, straw, elevator or mill dust, sand and dirt. It must be designated as Grain Screenings, Mixed Screenings and Chaff and/or Dust.

No grade of screenings must contain any seeds or other material in amount that is either injurious to animals or will impart an objectionable odor or flavor to their milk or flesh. The screenings must contain not more than 4 whole prohibited noxious weed seeds per lb and must contain not more than 100 whole restricted noxious weed seeds per lb. The prohibited and restricted noxious weed seeds must be those named as such by the seed control law of the state in which the screenings is sold or used.

All grades of screenings must bear minimum guarantees of crude protein and crude fat and maximum guarantees of crude fiber and ash. (Adopted 1953, Amended 1959, 1960.)

81.1 Grain Screenings must consist of 70% or more of grains, including light and broken grains, wild buckwheat and wild oats. It must contain not more than 6.5% ash. (Adopted 1953, Amended 1959, 1960.) NRC 4-02-156

81.2 Mixed Screenings is screenings excluded from the preceding definition. It must contain not more than 27% crude fiber and not more than 15% ash. (Adopted 1953, Amended 1954, 1960.) NRC 4-02-157

81.3 Chaff and/or Dust is material that is separated from grains or seeds in the usual commercial cleaning processes. It may include hulls, joints, straw, mill or elevator dust, sweepings, sand, dirt, grains, seeds. It must be labeled "chaff and/or dust." If it contains more than 15% ash the words "sand" and "dirt" must appear on the label. (Adopted 1953.) NRC 4-02-149

81.4 ________ Screenings must consist of 70% or more grains including light and broken grains, wild buckwheat and wild oats. It must contain not more than 6.5% ash. The predominating grain must be declared as the first word in the name. (Proposed 1974.)

MOLASSES

The production and usage of commercial molasses has been increasing. In 1976, the United States produced 406.3 million gal. and also imported 411.2 million gal. It competes with corn as a source of energy and is more advantageous to use at certain times. Where there have been surpluses, prices have been very low for long periods. Molasses is moved in large quantities from shore areas by ocean tanker, and in the United States, by tank car to feed mixers. Inedible molasses comprises cane blackstrap, beet molasses, hydrol (corn molasses), citrus molasses and wood molasses. They are by-products of raw sugar production, cane sugar refining, beet sugar and dextrose production, citrus canning and concentrate processing.

Molasses is substituted for carbohydrate ingredients in poultry feeds. Feeding value is 70% of 1 lb for corn. Molasses makes feed mixtures palatable, particularly feedstuffs, such as screenings. Molasses is also dried into molasses products, which contain from 40 to 75% molasses in combination with absorptive carriers, such as corn oil meal and bagasse pith. These products are handled in 50- and 100-lb bags as compared to 600-lb drums for liquid molasses. This permits the availability of small quantities of molasses to small users and eliminates the trouble and expense of heating and diluting liquid molasses. It also permits ease in mixing with other feed materials. Dried molasses costs substantially more than liquid molasses on a molasses basis.

The structure of the ration, age of the birds, temperature, humidity and production all play important roles in controlling the moisture of droppings produced by birds on diets containing molasses. Investigators, after considering these factors, must recognize that the moisture content of droppings is under genetic control. They must also determine if a dropping appears wet or firm due to the colloidal structure of the moisture in the droppings. Crude fiber and excess minerals play definite roles in shape, volume, moisture content and colloidal structure of the moisture in the droppings. In general, it requires an increase of over 2% moisture to change the water colloidal structure in droppings.

Tables 22.5 and 22.6 summarize studies on influence of age, and management and level of molasses on moisture content of droppings. A 6 wk test was conducted on laying hens with 2 levels of molasses in comparison with the control. Three age groups of hens were used on each treatment. Two and a half percent molasses resulted in an average solids analysis of 1.1% less than the control. Five percent molasses resulted in a reduction of solids analysis of 2.5% on the average. Differences between ages of hens and housing are indicated. Five percent molasses fed to the youngest birds in 6-hen cages resulted in droppings which appeared wet. The other levels did not produce typically "wet" droppings.

Results in the winter from pullets just coming into production indicated a 4.6% drop in solids content when the birds went from zero production (for the first 2 wk) to an average of 15% production (for the next 2 wk).

Feeding of 2.5% and 5% molasses to broilers resulted in decreases of 0.5% and 0.7%, respectively, in the solids analyses of rice hulls and shavings litter.

Types of Molasses

Industrial molasses are those used for purposes other than direct human consumption and include the following: (1) Cane blackstrap molasses is a by-product of the manufacture of sugar from sugar cane and usually contains slightly less than 55% total solids. (2) Refiners' blackstrap molasses is a by-product of the manufacture of white sugar from raw cane sugar. The latter molasses has approximately the same total sugars and chemical analysis as cane blackstrap molasses. Raw cane sugar and blackstrap molasses are manufactured in mills located in areas where sugar cane is grown. Refiners are separated geographically from raw mills, and thus blackstrap molasses production in the raw cane sugar stage and refiners blackstrap production in the refined sugar stage are separate products. (3) High-test (or invert) molasses is a product made from sugar cane without the usual granulation and extraction of sugar, and contains from 72 to 75% total sugars, less than 2% protein, and 82 to

TABLE 22.5. SOLIDS ANALYSES OF DROPPINGS FROM HENS IN CAGES RECEIVING MOLASSES IN THEIR RATIONS[1]

Diet	Age: Housing:	58 Weeks 6-hen Cages	75 Weeks Single Cages	94 Weeks Single Cages	Averages
Control		23.1%	28.6%	25.6%	25.8%
2.5% Molasses		21.9%	28.1%	24.2%	24.7%
5% Molasses		20.9%	25.2%	23.8%	23.3%
Averages		22.0%	27.3%	24.5%	24.6%

Source: Data obtained from Halloran (1965).

TABLE 22.6. SOLIDS ANALYSES OF LITTER FROM BROILER HOUSES WHERE MOLASSES WAS FED IN THE RATION

Diets	% Solids
Control	86.9
2.5% Molasses	86.4
5% Molasses	86.2

Source: Data obtained from Halloran (1965).

86% total solids. (4) Beet molasses is a by-product of the manufacture of beet sugar and contains from 48 to 53% total sugars, 6 to 10% protein and 80 to 85% total solids. (5) Hydrol is a by-product of the manufacture of refined corn sugar (dextrose) and contains 62 to 64% total sugars, less than 1% protein and about 75% total solids. (6) Citrus molasses is a by-product of citrus canning made by concentrating waste waters and contains 41 to 43% total sugars, slightly less than 14% protein and 70 to 73% total solids.

A newer product, "Wood molasses," technically hemicellulose extract, is said to be nutritionally equivalent to cane or blackstrap molasses. It is a by-product of the production of hard board and is the concentrated, soluble product that comes from the steam treatment of wood at elevated temperatures and pressures without the use of acids, alkalizers or salts. It contains both pentose and hexose sugars and their precursors and a total carbohydrate content of not less than 55%. Its 'Brix" (concentration and percent of sugar) is not less than 65%. So far, most of this molasses has been used in feeding beef cattle.

Louisiana and Florida supply all the United States cane blackstrap. Refiners' blackstrap is produced in Louisiana, Texas, Georgia, Maryland, Pennsylvania, New York, Massachusetts and California. Citrus molasses is produced primarily in Florida and hydrol is produced in Iowa, Illinois, Indiana and Texas. Puerto Rico and Hawaii are also important suppliers of molasses for United States use. Cane blackstrap is preferred by feed mixers because of fewer mechanical difficulties in the mixing process.

Feeders become accustomed to the sweet odor and the moist feel of feeds containing cane blackstrap. Some object to the taste of beet molasses and its laxative effect as compared to cane blackstrap. Hydrol does not have the sweet odor of cane molasses nor does it give feed a moist feeling. Because of its quality of more rapid penetration of feed materials, hydrol is sometimes blended with blackstrap to assist in marketing. The dextrose in hydrol crystallizes in transit and storage. At one time, it clogged mixing equipment; alkali is now added to arrest dextrose crystallization.

In the manufacture of beet sugar, the solution of sugar extracted from sliced raw beets is evaporated until a syrup is formed and the maximum amount of sugar has been crystallized. These crystals are then separated from the mother liquor by centrifuging. The condensed dark-colored liquid is molasses, containing 60% digestible nutrients of which 3.5% is protein and the remainder is carbohydrates. It contains 5% alkaline salts which have a laxative effect in poultry. Beet molasses has as high a feeding value as cane molasses. It is derived from Steffen's filtrate or waste waters where it is concentrated to approximately 60% solids. Glutamic acid and betaine may be isolated from Steffen's filtrate.

Hydrol came into use during World War II, when there was a shortage of blackstrap. Experiments have proven that there is little difference between hydrol and blackstrap, as far as nutrition is concerned. Citrus molasses differs in odor and taste from cane blackstrap and it is more difficult to mix. The citrus industry processes it primarily as a waste disposal measure. In this process, orange and grapefruit peels and pulp are run through a hammer mill. Water and juice are extracted from the material and the liquid is condensed in multiple-effect evaporators.

Blending Equipment

Molasses-blending equipment is common in many feed-mixing plants to benefit from the savings in molasses. Delivered by tank truck or car, it is used as a substitute for more expensive carbohydrate material. Storage capacity is needed to handle twice the size of the anticipated unit of purchase because of the difficulty in scheduling delivery. Some feed manufacturers produce highly-concentrated protein and mineral supplements with molasses for small mixers and farmers.

Molasses is a popular ingredient in feed mixes for improving palatability and it frequently is one of the cheapest sources of carbohydrates. There are some difficulties in mixing molasses in poultry feeds because it tends to ball into small masses. Individual birds may consume these masses in quantities large enough to disturb their digestive processes.

Although molasses is most valuable for its high content of readily available energy, it is also a source of highly available trace minerals. The molasses content usually is 2.5% or less in poultry rations to bind materials and provide a mild laxative effect. If more than 5 to 7% molasses is added to all-mash chick rations, there will be a slight decrease in growth, an increase in feed intake and a slight lowering of feed efficiency. Levels from 4 to 5% slightly lower feed efficiency. Levels from 2.5 to 3% are used in turkey-starter feeds and broiler rations without affecting significantly the average weights of the birds. Cane blackstrap decreases dustiness, improves appearance and odor and assists in pelleting operations.

In Hawaii, high concentrations of B-grade molasses are fed to chickens as a substitute for yellow corn in starting rations.

Five types of molasses are approved for usage in feeds and have been given NRC numbers.

63. MOLASSES

Official

63.1 Beet Molasses is a by-product of the manufacture of sucrose from sugar beets. It must contain not less than 48% total sugars expressed as invert and its density determined by double dilution must not be less than 79.5° Brix. (Adopted 1941, Amended 1960.) NRC 4-00-668

63.3 Citrus Molasses is the partially dehydrated juices obtained from the manufacture of dried citrus pulp. It must contain not less than 45% total sugars expressed as invert and its density determined by double dilution must not be less than 71.0° Brix. (Adopted 1952, Amended 1960.) NRC 4-01-241

63.5 Hemicellulose extract is a by-product of the manufacture of pressed wood. It is the concentrated soluble material obtained from the treatment of wood at elevated temperatures and pressure without use of acids, alkalis, or salts. It contains pentose and hextose sugars, and has a total carbohydrate content of not less than 55%. (Proposed 1965, Adopted 1966.) NRC 4-08-030 Reg. 573.520

63.6 Starch Molasses is a by-product of the manufacture of dextrose from starch derived from corn or grain sorghums in which the starch is hydrolized by use of enzymes and/or acid. It must contain not less than 43% reducing sugars expressed as dextrose and not less than 50% total sugars expressed as dextrose. It shall contain not less than 73% total solids. (Proposed 1967, Adopted 1968.) NRC 4-08-037

63.7 Cane Molasses is a by-product of the manufacture or refining of sucrose from sugar cane. It must not contain less than 46% total sugars expressed as invert. If its moisture content exceeds 27%, its density determined by double dilution must not be less than 79.5° Brix. (Proposed 1973, Adopted 1975.) NRC 4-04-696

23

Animal Proteins

During the early growth of the cattle, hog, and sheep packinghouse industry, meat scraps and tankage were the 2 principal protein products manufactured and available for animal feeds. Tankage was used in hog feeds and meat scraps in poultry feeds. The developing canned and dry dog-food industry found that pet foods were more palatable when meat products were used. This development soon consumed the better quality meat products, thus reducing their availability for poultry feeds. Many packing houses developed their own brands of pet food, which further consumed the better quality meat products that were formerly used in meat scraps. Also, new animal by-products were developed and today, the complete animal is used.

Packinghouse by-products are fat, protein and miscellaneous supplements. In processing, waste matter is "tanked" and heated with water under pressure, which serves to cook and sterilize the material, as well as to float off the grease. Some fatty raw materials are "dry rendered" or heated in the absence of water, and then pressed. Grease and tallow are classified as to color and free fatty acid content, and are marketed as white, yellow or brown grease and as various grades of tallow. Gelatin is manufactured from the white fibers of pork and calf skins, as well as from the ossein of bones.

The biological value of meat scraps varies from one packing plant to another. In general, meat scraps obtained from cattle-packing plants have a higher biological value than those from operations using hogs or "dead-animal" disposing services. Because of this variation in biological value, some feed manufacturers blend several different sources of meat scraps; the more sources, the more uniform the product. It is important to avoid long high-heat during processing to improve digestibility so that amino acids, such as lysine and tryptophan, will not be heat-denatured and destroyed. Processing techniques are being improved by the packinghouse industry and in the future, superior products will be marketed. There is much evidence that packinghouse by-products have a place in modern rations when the cost of nutrients is considered. The benefits seem to outweigh the disadvantages (Table 23.1).

TABLE 23.1. PERCENTAGE COMPOSITION OF MEAT MEAL AND MEAT AND BONE MEAL

	Meat Meal	Meat and Bone Meal
Protein, crude	53.0	50.0
arginine	3.5	3.7
lysine	3.0	2.75
methionine	0.75	0.65
cystine	0.7	0.35
tryptophan	0.34	0.3
threonine	1.75	1.7
Fat (ether extract)	7.1	9.9
Crude fiber	2.5	2.8
Metabolizable energy (Cal./lb)	910.0	900.0

Source: Patrick (1965).

ANIMAL PRODUCTS (OFFICIAL)

9.1 Blood Meal is ground dried blood. (Adopted 1926.) NRC 5-00-380

9.2 Meat is the clean, wholesome flesh derived from slaughtered animals and is limited to that part of the striate muscle which is skeletal or that which is found in the tongue, in the diaphragm, in the heart or in the esophagus, with or without the accompanying and overlying fat and the portions of skin, sinew, nerve, and blood vessels which normally accompany the flesh. If it bears a name descriptive of its kind, it must correspond thereto. (Adopted 1938, Amended 1939, 1963.) NRC 5-00-394

9.3 Meat by-products is the non-rendered, clean, wholesome parts, other than meat, derived from slaughtered mammals. It includes, but is not limited to, lungs, spleen, kidneys, brain, livers, blood, bone, partially defatted low-temperature fatty tissue, and stomachs and intestines freed of their contents. It does not include hair, horns, teeth and hoofs. If it bears name descriptive of its kind, it must correspond thereto. (Proposed 1973, Adopted 1974.) NRC 5-00-395

9.6 Whale Meal is the clean, dried undecomposed flesh of the whale, after part of the oil has been extracted. If it contains more than 3% of salt (NaCl), the amount of salt must constitute a part of the product name, provided that in no case must the salt content of this product exceed 7%. (Adopted 1933, Amended 1963.) NRC 5-05-160

9.7 Animal Liver bears a name descriptive of its kind, it must corespond thereto. Meal is obtained by drying and grinding liver from slaughtered mammals. (Adopted 1954.) NRC 5-00-389

9.8 Animal Liver and Glandular Meal is obtained by drying and grinding liver and other glandular tissues from slaughtered mammals. At least 50% of the dry weight of the product must be derived from liver. (Adopted 1954, Amended 1964.) NRC 5-00-390

9.9 Extracted Animal Liver Meal is obtained by drying and grinding the residue of mammal liver tissue from which a portion of the water-soluble material has been extracted. (Adopted 1944, Amended 1964, Amended 1975.) NRC 5-00-396

9.10 Poultry By-Product Meal consists of the ground, dry-rendered or wet-rendered clean parts of the carcass of slaughtered poultry, such as heads, feet, undeveloped eggs, and intestines, exclusive of feathers, except in such trace amounts as might occur unavoidably in good factory practice. It must contain not more than 16% ash and not more than 4% acid-insoluble ash. (Proposed 1958, Amended 1960, Adopted 1961.) NRC 5-03-798

9.11 Poultry Hatchery By-Product is a mixture of egg shells, infertile and unhatched eggs, and culled chicks which have been cooked, dried, and ground, with or without removal of part of the fat. (Adopted 1957.) NRC 5-03-796

9.12 Dried Meat Solubles is obtained by drying the defatted water extract of the clean, wholesome parts of slaughtered animals prepared by steaming or hot-water extraction. It must be designated according to its crude protein content which shall be no less than 70%. (Proposed 1961, Adopted 1962, Amended 1964, 1967.) NRC 5-00-393

9.13 Poultry Parts must consist of clean wholesome parts of dressed poultry characterized by large proportions of bone; such as neck, backs, and wings. If it bears a name descriptive of its kind, it must correspond thereto. (Proposed 1963, Adopted 1964, Amended 1967.) NRC 5-03-797

9.14 Poultry By-Products must consist of non-rendered clean parts of carcasses of slaughtered poultry, such as heads, feet, viscera, free from fecal content and foreign matter, except in such trace amounts as might occur unavoidably in good factory practice. (Proposed 1963, Adopted 1964.) NRC 5-03-800

9.15 Hydrolized Poultry Feathers is the product resulting from the treatment under pressure of clean, undecomposed feathers from slaughtered poultry, free of additives and/or accelerators. Not less than 75% of its crude protein content must be digestible by the pepsin digestibility method. (Proposed 1961, Adopted 1965.) NRC 5-03-795

9.40 Meat Meal is the dry-rendered product from mammal tissues, exclusive of hair, hoof, horn, hide trimmings, manure and stomach contents, except in such amounts as may occur unavoidably in good factory practice. It shall not contain added blood meal. It shall contain not more than 14% pepsin-indigestible residue*. Not more than 11% of the crude protein in the product shall be pepsin-indigestible*. It must be designated according to its protein content. If the product bears a name descriptive of its kind, composition or origin, it must correspond thereto. (Proposed 1971, Adopted 1972.) NRC 5-00-385

9.41 Meat and Bone Meal is the dry-rendered product from mammal tissues, exclusive of hair, hoof, horn, hide trimmings, manure and stomach contents, except in such amounts as may occur unavoidably in good factory practice. It shall not contain added blood meal. It shall contain more than 4.4% phosphorus (P) and not more than 14% pepsin-indigestible residue*. Not more than 11% of the crude protein shall be pepsin-indigestible*. It must be designated according to its protein content. If the product bears a name descriptive of its kind, composition or origin it must correspond thereto. (Proposed 1971.) NRC 5-00-388

9.42 Animal By-Product Meal is the dry product from animal tissues prepared for feeding purposes by tanking with live steam or dry rendering. It must be designated according to its protein content. (Proposed 1971.) NRC 5-08-786

9.50 Meat Meal Tankage is the dry product from mammal tissues, exclusive of hair, hoof, horn, hide trimmings, manure and stomach contents, except in such amounts as may occur unavoidably in good factory practice, especially prepared for feeding purposes by tankage under live steam or dry rendering. It shall contain not more than 14% pepsin-indigestible residue*. Not more than 11% of the crude protein in the product shall be pepsin-indigestible*. It must be designated according to its protein content. If the product bears a name descriptive of its kind, composition or origin it must correspond thereto. (Proposed 1971.) NRC 5-00-386

9.51 Meat and Bone Meal Tankage is the dry product from mammal tissues, exclusive of hair, hoof, horn, hide trimmings, manure and stomach contents, except in such amounts as may occur unavoidably in good factory practice, especially prepared for feeding purposes by tanking under live steam or dry rendering. It shall contain more than 4.4% phosphorus (P) and not more than 14% pepsin-indigestible residue*. Not more than 11% of the crude protein in the product shall be pepsin-indigestible*. It must be designated according to its protein content. If the product bears a name descriptive of its kind, composition or origin, it must correspond thereto. (Proposed 1972.) NRC 5-00-387

9.59 Hydrolized Poultry By-Products Aggregate is the product resulting from treatment under pressure of all by-products of slaughter poultry, clean and undecomposed, including such parts as heads, feet, undeveloped eggs, intestines, feathers and blood. (Proposed 1974) NRC 5-14-508

9.65 Extracted Animal Liver and Glandular Meal is obtained by drying and grinding the residue of mammal liver and other glandular tissue from slaughtered animals after a significant portion of water-soluble material has been removed. At least 50% of the dry weight of the product must be derived from liver. (Proposed 1975.) NRC 5-26-009

*AOAC Method 7.040 – 7.046.

Tentative

9.56 Spray-Dried Animal Blood is produced from clean, fresh animal blood, exclusive of all extraneous material such as hair, stomach belchings, urine, except in such traces as might occur unavoidably in good factory practice. Moisture is removed from the blood by a low-temperature evaporator under vacuum until it contains approximately 30% solids and a minimum crude protein of 85%. It is then dried by spraying into a draft of warm, dry air which reduces the blood to finely divided particles with a maximum moisture of 8%. It must be designated according to its minimum water solubility. (Proposed 1972. Amended 1976.) NRC 5-00-381

9.61 Egg Shell Meal is a mixture of egg shells, shell membranes and egg content obtained by drying the residue from an egg-breaking plant in a dehydrator to an end product temperature of 180°F. It must be designated according to its protein and calcium content. (Proposed 1975) NRC 6-26-004

POULTRY BY-PRODUCTS AND FEATHER MEAL

Poultry by-products meal and feather meal are desirable ingredients for poultry rations. They were developed when the modern processing plants appeared.

New York dressed poultry began to disappear from the markets around 1940, because of objections to off-flavors from the viscera. With the marketing of eviscerated, ready-to-cook poultry, the availability of a large and constant supply of offal attracted renderers. They were interested in it as a source of feed protein.

Poultry meat scraps (officially "poultry by-product meal") are made from offal (viscera, heads and feet) by conventional dry-rendering methods. They consist of the "clean, ground, dry-rendered parts of the carcass of slaughtered poultry such as heads, feet, undeveloped eggs and intestines, exclusive of feathers, except in such trace amounts as might occur unavoidably in good factory practice. It contains 16% ash and 4% acid-insoluble ash, 50–60% protein, the quality of which is higher than that of meat scraps." Poultry blood meal contains 65–70% protein and is a good source of histidine. The composition of feather meal is shown in Table 23.2.

Many firms are now engaged in producing feather meal. "Feather meal" (officially "hydrolyzed poultry feathers") results from the treatment under pressure of clean, undecomposed feathers from slaughtered poultry free of additives and/or accelerators. Not less than 70% of its crude protein content consists of digestible protein. Actually, hydrolysis makes no more than a minor contribution to the changes resulting from processing.

Feathers are processed in steam-jacketed equipment with horizontal-shaft agitation and top and bottom openings for loading and dumping. Jacket pressure is 60–90 psi; internal pressure is 30–40 psi; processing time is 30 min. The load is discharged from the cooker as soon as the pressure is released and is fed to a dryer. Agitation breaks up the material and little remains that is recognizable as feathers. After processing and drying, the material has an almost glassy brittleness which makes it easy to grind by impact or hammer mills. A light-colored meal is desired by the feed industry.

The quality of feather meal is evaluated by a 24-hr pepsin-hydrochloric-acid digestibility test, because it is most closely related to nutritional value. Color, microscopic appearance, friability and bulk density are used as rough estimates of quality at the time of processing. Cystine is the only amino acid in feathers that is significantly affected by processing with steam.

Feather meal contains 85% protein but is quite deficient in methionine, lysine, tryptophan and histidine. It can be used in poultry rations, pro-

TABLE 23.2. COMPOSITION OF FEATHER MEAL

Nutrient Name	Analysis	Units
Metabolizable energy	1078.00	Cal./lb
Productive energy	600.00	Cal./lb
Protein	86.40	%
Arginine	5.60	%
Glycine	6.80	%
Histidine	0.40	%
Isoleucine	3.80	%
Leucine	6.60	%
Lysine	1.75	%
Methionine	0.47	%
Meth. and cystine	4.80	%
Phenylalanine	3.80	%
Phenyl and tyrosine	6.20	%
Threonine	4.00	%
Trytophan	0.57	%
Valine	6.50	%
Total fat	4.40	%
Total fiber	1.50	%
Calcium	0.20	%
Avail. phosphorus	0.75	%
Sodium	0.70	%
Total Ash	2.80	%
Thiamin	0.05	mg/lb
Niacin	8.00	mg/lb
Riboflavin	0.90	mg/lb
Panthothenic acid	3.70	mg/lb
Vitamin B-12	39.00	mg/lb
Choline	400.00	mg/lb
Pyridoxine	2.00	mg/lb
Folacin	0.10	mg/lb
Biotin	0.02	mg/lb
Potassium	0.30	%
Magnesium	0.20	%
Pepsin digestibility	78.50	%

viding consideration is given to amino-acid balance in the feed. The high-protein content is of particular value in high-energy diets.

Serious amino acid deficiencies exist in feather protein, and its digestibility is poor because of its physical/chemical structure. However, feather meal of good quality is a satisfactory substitute for part of the soybean oil meal and cereal grains and gives good results in corn-soybean rations when they are adequately supplemented with sources of methionine. This is not surprising, since corn, soybean meal and feathers are all relatively low in this amino acid. The histidine in feather meal is not available to the chick; possibly, other amio acids are also partially unavailable. This failure to obtain maximum growth rates on diets dependent on feather meal protein is due to poor amino acid availability.

Poultry by-product meal contains protein having high biological value and is preferred to meat scraps. Most of the poultry by-product meal is used in poultry and turkey starting rations and broiler feeds. Dog food manufacturers are using large amounts of undried poultry by-products in pet foods.

Fish Products

Although the sea covers 75% of the globe, and living matter is found at all levels of the ocean, it provides only a small part of the food consumed in the world. The present annual harvest is 40 million metric tons of fish and fishery products. The most productive fishing areas are the continental shelves, especially in the Northern Hemisphere. Latin America produces 35% of the world's marine catch. Marine fish of 115,000,000 metric tons are potentially available annually, but a large portion is too thinly scattered to be economically harvested. Marine fish account for 80% of total production; fresh water species constitute the remainder. The clupeid (sardine, herring, menhaden and anchovy) contribute 11,000,000 tons and are taken with seines, traps and weirs. The next largest contribution is groundfish, totaling 6,000,000 tons; these inhabit the great offshore fishing banks, such as the Grand Bank of Newfoundland. The most famous of these are the Atlantic cod and the sea basses, fished with otter trawls. Tuna landing is 2.7 million tons and flounder, 1 million tons. To summarize, although the oceans have not yet been fully exploited, marine fish resources are not so vast as some believe. Probably less than 60 million metric tons could be available at present for harvest on an annual basis.

Fish are relatively good sources of many protective nutrients. Properly processed fishmeals are excellent protein supplements, supplying unusually high amounts of the amino acids—lysine and methionine— as well as the minerals—phosphorus, calcium, magnesium, iodine, Vitamin B_{12}, riboflavin, niacin and choline. Fish oils are rich in Vitamins A and D. A major role of fishmeal in poultry feeding is to extend supplies of vegetable protein by providing critical amounts of supplemental lysine and methionine.

Proteins of cereal grains and most other plant protein concentrates fail to supply the complete amino acid needs of poultry, due to lack of methionine, cystine and lysine. Soybean meal, widely used in animal feeding, is a good source of lysine and tryptophan, but is low in the sulfur-containing amino acids. Fish protein is an excellent source of all these amino acids. Fishmeal is decidedly superior to oil meals as a source of essential amino acids, as are meat scraps, which outrank oil meals as a source of lysine, the first limiting amino acid in grains. The limited supply of fish by-products is conserved by using in rations large amounts of the more abundant oil meals. The requirement for essential amino acids is progressively reduced as poultry grow older, and it is possible to meet the needs of mature birds with rations containing fewer animal by-products and higher levels of oil meals (Table 23.3).

Higher levels of fish products are used in Europe than are used in the United States, largely for economic reasons. The actual usage of fish

TABLE 23.3. PERCENTAGE COMPOSITION OF FISH MEAL AND CONDENSED FISH SOLUBLES

	Menhaden Fish Meal	Crab Meal	Condensed Fish Solubles
Protein, crude	60.0	36.0	30.0
arginine	3.6	1.7	1.1
lysine	4.6	1.4	1.4
methionine	1.7	0.5	0.5
cystine	0.5	0.4	0.3
tryptophan	0.6	0.3	0.1
threonine	2.7	1.0	0.6
Fat (ether extract)	9.4	1.8	7.8
Crude fiber	1.0	11.0	0.5
Metabolizable energy (Cal./lb)	1343.0	800.0	720.0

Source: Patrick (1965).

products in poultry feeds is determined by economics, but levels up to 10% fishmeal can safely be used in starting and laying feeds for poultry.

Differences in quality of various fishmeals are due to the percentage of protein and fat digestibility. Large variations occur in proximate analysis from species to species, in fish of the same species, and in different parts of the fish.The protein content of fish averages 19%, but varies from 6 to 28%; oil content varies from 0.2 to 64%; ash from 0.4 to 1.5%; and moisture from 28 to 90%. The cause of variation in the composition of fish is usually ascribed to such factors as geographical area and season, but actually, it relates primarily to the feed ingested by the fish, to metabolic rate and fish mobility.

Fish types are classified as to the amount of oil and protein they contain. Cod is considered a low-oil, high-protein fish; mackerel, medium-oil, high-protein fish; lake trout, a high-oil, low-protein fish; skipjack tuna, low-oil, very high-protein; and butterclams, low-oil, low-protein. Low-oil means under 5%, medium-oil 5 to 15%, and high is between 15 and 20%.

51. MARINE PRODUCTS

Official

51.14 **Fish Meal** is clean, dried, ground tissues of undecomposed whole fish or fish cuttings, either or both, with or without the extraction of part of the oil If it contains more than 3% salt (NaCl), the amount of salt must constitute a part of the brand name, provided that in no case must the salt content of this product exceed 7% (Adopted 1933.) NRC 5-01-977

51.24 **Fish Residue Meal** is the clean, dried, undecomposed residue from the manufacture of glue from non-oily fish. If it contains more than 3% salt (NaCl), the amount of salt must constitute a part of the brand name, provided that in no case must the salt content of this product exceed 7%. (Adopted 1933.) NRC 5-01-966

51.34 **Fish Liver and Glandular Meal** is obtained by drying the complete viscera of the fish. At least 50% of the dry weight of the product must be derived from fish liver and must contain at least 18 milligrams of riboflavin per pound (Adopted 1944, Amended 1945.) NRC 5-01-973

51.4 **Crab Meal** is the undecomposed ground dried waste of the crab and contains the shell, viscera, and part or all of the flesh. It must contain not less than 25% crude protein. If it contains more than 3% salt (NaCl), the amount of salt must constitute a part of the brand name, provided that in no case must the salt content of this product exceed 7%. (Adopted 1933.) NRC 5-01-663

51.5 **Shrimp Meal** is the undecomposed ground dried waste of shrimp and contains parts and/or whole shrimp. If it contains more than 3% salt (NaCl), the amount of salt must constitute a part of the brand name, provided that in no case must the salt content of this product exceed 7%. (Adopted 1933, Amended 1963.) NRC 5-04-226

51.6 **Condensed Fish Solubles** is obtained by condensing the stickwater. It must contain not less than 30% crude protein. (Proposed 1962, Amended 1963, Adopted 1964.) NRC 5-01-969

51.7 **Dried Fish Solubles** is obtained by dehydrating the stickwater. It must contain not less than 60% crude protein. (Proposed 1963, Amended 1964.) NRC 5-01-971

51.8 **Fish Oil** is the oil from rendering whole fish or cannery waste. (Proposed 1963, Adopted 1964.) NRC 7-01-965

51.9 **Fish Protein Concentrate—Feed Grade** is prepared from clean, undecomposed whole fish or fish cuttings using the solvent extraction process developed for the production of edible whole fish protein concentrate. It must contain not less than 70% protein and not more than 10% moisture. If the degree of fineness is stated, it must conform thereto. Solvent residues are not to exceed those established in Food Additive Regulations. (Proposed 1969, Adopted, 1970, Amended 1971.) NRC 5-09-334 Reg. 573.440

51.10 **Fish By-Products** must consist of non-rendered, clean undecomposed portions of fish (such as, but not limited to, heads, fins, tails, ends, skin, bone and viscera) which result from the fish processing industry. If it bears a name descriptive of its kind, it must correspond thereto. Any single constituent used as such may be labeled according to the common or usual name of the particular portion used (such as fish heads, fish tails etc.) (Proposed 1974, Adopted 1975.) NRC 5-14-509

MILK BY-PRODUCTS

Milk products have always been considered important feeds for poultry. They are regarded as a cornerstone on which to build nourishing diets. Liquid milk is a complex mixture consisting of an oil and water emulsion, stabilized by proteins absorbed on the surface of the fat globules. It contains proteins in colloidal dispersion, lactose in true solution, numerous minerals, particularly calcium and phosphorus, fat-soluble and water-soluble vitamins, enzymes and various organic compounds. Whole milk contains the following: 86 to 88% water; 3 to 5% lipids; 3.2 to 3.6% protein; 4.7 to 4.9% lactose; 0.7 to 0.8% ash; and 62 to 75% Cal. per 100 g. Most lipids of milk are in the fat globules and the surrounding membrane, but small amounts are also in the milk serum. The main proteins are casein and lactalbumin. Casein is in the largest amount and occurs as large, complex colloidal particles which contain considerable calcium and inorganic phosphorus, plus a little magnesium and citrate. Lactalbumin is insoluble in water and contains sulfur and very little phosphorus; beta lactoglobulins account for 50 to 60% of the total. Lactoglobulins consist of 2 proteins: euglobulin (true globulin) and pseudoglobulin (false globulin).

Lactose (milk sugar), the main carbohydrate, ranges in amounts from 4.7 to 4.9% and occurs in 2 forms—alpha and beta lactose. It is about 20% as sweet as sucrose, is relatively insoluble, and is slowly digested and absorbed from the intestine. Its presence in the intestine stimulates the growth of microorganisms that produce organic acids and synthesize B-vitamins. These acids check the growth of undesirable bacteria and favor absorption of calcium, phosphorus and magnesium.

Most milk is an outstanding source of calcium, a good source of phosphorus, a poor source of iron and magnesium and a variable source of iodine. The content of B-complex vitamins is fairly constant, since the microorganisms in the rumen of the cow synthesize these, regardless of vitamin concentration in the cow's diet. Milk supplements in complete feedstuffs make possible a well-balanced and nutritionally-adequate diet.

Most of the nutritional excellence of fresh whole fluid milk is also present in evaporated and powdered milk.

Whey, a by-product of the cheese industry, is produced at the rate of more than 5.4 million tons annually in the United States, of which 75% is disposed as waste, representing 204,000 tons sugar and 36,000 tons protein.

Dried whey varies in composition, depending on the cheese-making process from which it is derived. There are 2 types: (1) sweet rennet whey results from the making of cheddar, Swiss or other sweet-type cheese; and (2) cottage cheese whey is produced during the manufacture of farmer, pot and cottage-type cheeses. Lactose is made from wheys which have not been allowed to become acid. Casein is made from skim and buttermilk by coagulation methods; it contains little fat, ash, acid or moisture.

Milk is an excellent feedstuff, the use of which is chiefly for humans. Only off-quality milk is available for poultry feeds. When available, it is used in starting and breeding rations.

54. MILK PRODUCTS

Official

54.1 **Dried Buttermilk, Feed Grade,** is the residue obtained by drying buttermilk. It contains 8% maximum moisture, 13% maximum ash, and 5% minimum milk fat (Roese-Gottlieb method). The words "feed grade" are not required when listing as an ingredient in manufactured feed. (Adopted 1932, Amended 1964.) NRC 5-01-160

54.2 **Condensed Buttermilk** is the residue obtained by evaporating buttermilk. It contains 27% minimum total solids, 0.055% minimum milk fat for each % of total solids, and 0.14% maximum ash for each % of total solids. (Adopted prior to 1928, Amended 1944, 1964.) NRC 5-01-159

54.3 **Dried Skimmed Milk, Feed Grade,** is the residue obtained by drying defatted milk. It contains 8% maximum moisture. The words "feed grade" are not required when listing as an ingredient in manufactured feed. (Adopted 1930, Amended 1964.) NRC 5-01-175

54.4 **Condensed Skimmed Milk** is the residue obtained by evaporating defatted milk. It contains 27% minimum total solids. (Adopted 1930, Amended 1964.) NRC 5-01-172

54.5 **Dried Cultured Skimmed Milk** is the residue obtained by drying lactic acid bacteria cultured defatted milk. It contains 8% maximum moisture. (Adopted 1932, Amended 1964.) NRC 5-01-174

54.6 **Condensed Cultured Skimmed Milk** is the residue obtained by evaporating lactic acid bacteria cultured defatted milk. It contains 27% minimum total solids. (Adopted 1932, Amended 1964.) NRC 5-01-173

54.7 **Dried Whey** is the residue obtained by drying whey. It contains 65% minimum lactose. (Adopted 1934, Amended 1948, 1950, 1951, 1964.) NRC 4-01-182

54.8 **Condensed Whey** is the residue obtained by evaporating whey. The minimum % total whey solids must be prominently declared on the label. (Adopted 1944, Amended 1951, 1963, 1964.) NRC 4-01-180

54.9 **Dried Whey Solubles** is obtained by drying the whey residue from the manufacture of lactose after removal of milk albumin and partial removal of lactose. (Adopted 1944, Amended 1964.) NRC 4-01-189

54.10 **Condensed Whey Solubles** is obtained by evaporating the whey residue from the manufacture of lactose after removal of milk albumin and partial removal of lactose. (Adopted 1944, Amended 1964.) NRC 4-01-188

54.11 **Dried Hydrolized Whey** is the residue obtained by drying lactase enzyme hydrolyzed whey. It contains 30% minimum total glucose and galactose. (Adopted 1955, Amended 1964.) NRC 4-01-184

54.12 **Condensed Hydrolized Whey** is the residue obtained by evaporating lactase enzyme hydrolyzed whey. It contains 50% minimum total solids and 0.3% minimum total glucose and galactose for each % total solids. (Adopted 1955. Amended 1964.) NRC 4-01-183

54.13 **Condensed Whey-Product** is the residue obtained by evaporating whey from which a portion of the lactose has been removed. The minimum % total whey-product solids must be prominently declared on the label. (Adopted 1948, Amended 1964.) NRC 4-01-185

54.14 **Dried Whey-Product** is the residue obtained by drying whey from which a portion of the lactose has been removed. The minimum % lactose must be prominently declared on the label. (Adopted 1951, Amended 1952, 1964.) NRC 4-01-186

54.15 **Condensed Cultured Whey** is the residue obtained by evaporating cultured whey. The minimum % total cultured whey solids must be prominently declared on the label. (Adopted 1949, Amended 1964.) NRC 4-01-181

54.16 **Casein** is the solid residue obtained by acid or rennet coagulation of defatted milk. It contains 80% minimum crude protein. (Adopted 1946, Amended 1964.) NRC 5-01-162

54.17 **Cheese Rind** is obtained by cooking cheese trimming devoid of fat other than milk fat. (Adopted 1935, Amended 1964.) NRC 5-01-163

54.18 **Dried Milk Albumin** is obtained by drying the coagulated protein residue separated from whey. It contains 75% minimum crude protein on a moisture-free basis. (Adopted 1952, Amended 1964.) NRC 5-01-177

54.19 **Dried Whole Milk, Feed Grade** is the residue obtained by drying milk. It contains 8% maximum moisture and 26% minimum milk fat (Roese-Gottlieb method). The words "feed grade" are not required when listing as an ingredient in manufactured feed. (Adopted 1952, Amended 1964.) NRC 5-01-167

54.20 **Dried Milk Protein** is obtained by drying the coagulated protein residue resulting from the controlled co-precipitation of casein, lactalbumin, and minor milk proteins from defatted milk. (Proposed 1965, Adopted 1966.) NRC 5-08-044

54.21 **Dried Hydrolized Casein** is the residue obtained by drying the water-soluble product resulting from the enzymatic digestion of casein. It contains 74% minimum crude protein. (Proposed 1966, Adopted 1967.) NRC 5-08-055

AMINO ACIDS

Pure amino acids are often added to balance the requirements for methionine in soybean meal and lysine in grains. These are registered, and have official approval to be added to rations when needed.

Amino Acids (Official)

DL-Methionine Hydroxy Analogue Calcium is a product which contains a minimum of 93% racemic 2-hydroxy-4-methylthiobutyric acid calcium salt. The percentage alfalfa products means that the freshly cut alfalfa having a moisture of L-Methionine Hydroxy Analogue Calcium must be guaranteed. (Adopted 1959, Amended 1975.) NRC 5-03-087 Reg. 582.5477

DL-Methionine is a product which contains a minimum of 95% racemic 2-amino-4-methylthiobutyric acid. The percentages of L-Methionine must be guaranteed. (Adopted 1957, Amended 1975.) NRC 5-03-086 Reg. 582.5475

Glycine is a product which contains a minimum of 97% amino-acetic acid. (Adopted 1957.) NRC 5-02-127 Reg. 582.5191

L-Lysine is a product which contains a minimum of 95% α, ϵ-diaminocaproic acid. (Proposed 1966, Adopted 1969, Amended 1975.) NRC 5-08-092 Reg. 582.5411

L-Threonine is a product which contains a minimum of 95% α Amino-β-hydroxybutyric acid. (Proposed 1967, Adopted 1969, Amended 1975.) NRC 5-08-092 Reg. 582.5881

DL-Tryptophan is a product which contains a minimum of 95% 1-α-Amino-3-indolepropionic acid. The percentage of L-Tryptophan must be guaranteed. (Proposed 1967, Tentative 1969, Amended 1975.) NRC 5-08-093 Reg. 582.5915

DL-Lysine is a product which contains a minimum of 95% racemic α, ϵ-diaminocaproic acid. The percentage of L-Lysine must be guaranteed. (Proposed 1975.) NRC 5-14-501 Reg. 582.5411

24

Vegetable Proteins

Soybeans are legumes grown throughout the world for seed and forage. The corn belt in the United States is the most important area. The United States produces about 50% of the world's crop and the mainland of China produces nearly 30%, although production in South America is growing rapidly. World production of soybeans in 1976 was 60,118,000 metric tons, with a per-acre yield of 0.706 metric tons. The yield in the United States during the same year was 34,425,000 metric tons, with a per-acre yield of 0.696 metric tons.

Plant geneticists have developed suitable strains of soybeans for different soil types and growing seasons. Soybeans are very sensitive to length of day and season; therefore, many strains are needed to accommodate growing seasons from the southern to the northern extremities of the United States.

The soybean is the most interesting of the oil seeds. It has been rapidly expanding within the last few decades, principally in the corn belt of the United States and in other countries of the Western Hemisphere. There are many factors which have produced this rapid change. The soybean is an annual crop admirably suited to mechanized planting, cultivation and harvesting. Furthermore, soybeans can be handled and stored without much difficulty. They are milled and oil-extracted by hydraulic pressing, continuous-screw pressing or solvent extraction. The use of soybean oil in human foods helps "pay the freight" for the meal in poultry feeds. There are thousand of species of oil-bearing plants, but less than a score account for more than 90% of the fat produced.

SOYBEANS AND THE SOYBEAN ERA

The use of soybean meal with corn (the soybean-corn ration) ushered in a new era in poultry nutrition. The discovery that bone meal corrected the mineral deficiencies in soybean meal precipitated its use in poultry feeds. Its use grew rapidly.

FIG. 24.1. FROM A BOTANICAL CURIOSITY TO A MAIN INGREDIENT IN UNITED STATES POULTRY RATIONS, THE SOYBEAN IN 50 YEARS HAS HAD A SUCCESS STORY UNMATCHED BY OTHER FEED INGREDIENTS

Investigators soon found, however, that soybean meal was deficient in methionine, and that this deficiency was due to unavailability rather than to absence. During the development of processing methods, heating or toasting was found to improve the availability of methionine and also to improve the rate of digestion.

Investigations on corn gluten demonstrated that this protein was a good source of methionine in which soybean and corn gluten proteins were complementary. Corn gluten was found to be deficient in lysine and tryptophan, whereas soybean protein is rich in lysine.

Cooked cottonseed and sesame seed meal were found to be deficient in available lysine, and peanut meal was found to be deficient in both lysine and tryptophan. Following these disclosures, nutritionists recognized that, in general, all vegetables are deficient in one or more amino acids. Also, combinations of vegetable proteins with other vegetable proteins or animal proteins could be developed which would provide for the needs of poultry.

The rapid development of the poultry industry soon consumed the surplus animal proteins. The pet food industry also demanded animal proteins and was willing to pay a higher price than the poultry industry. This economic development led to the synthesis of methionine and lysine, and the usage of methionine in broiler feeds became a reality.

Further studies on deficiencies in soybean meal revealed that a factor in animal proteins (Animal Protein Factor) was active in supplementing vegetable protein diets. This factor was concentrated from fish solubles and crystallized. Today it is known as Vitamin B_{12}.

During the studies on the A.P.F., growth-promoting activity was found in residues from the antibiotic industry. These investigations showed that the growth-promoting substances were antibiotics.

A search for other factors possessing growth-promoting properties for vegetable protein diets disclosed that some arsenic compounds were active. Although not confirmed biochemically, arsenic was reported to increase skin pigmentation and feather development of poultry.

During this era, the following major contributions were made in poultry nutrition:

(1) Soybean meal is a good protein source for poultry.
(2) Methionine was found to be required in soybean meal-corn rations.
(3) Antibiotics were found to have growth properties in vegetable protein rations.
(4) Arsenic was found to have growth-promoting properties with broilers using vegetable protein rations.
(5) Animal proteins can be completely replaced in poultry rations.
(6) Vitamin B_{12} was discovered and synthesized.
(7) Methionine was synthesized and became available for feed usage.

The soybean is a native of Asia but has adapted well to American agriculture. Soja max, the soybean, is a time-honored crop, dating to the Materia Medica 4800 yr ago. A few soybeans were grown in the United States as botanical curiosities prior to 1907. Many different varieties were collected from the Orient and tried experimentally.

The soybean is a summer, leguminous annual with pods of 1 to 1.5 in. in length, containing 2 to 4 seeds. The seeds, when mature, are yellow, green, brown or black, and some are round, some oval and some flat. Large and abundant root tubercules permit this plant to build up the fertility of the soil. The plant grows from 2 to 6 ft in height; flowers are small, sweet-pea shaped, and white or purple in color.

Soybean plants are unique in their ability to produce a seed crop. If the first flowers are destroyed, the plant will reform, or attempt a second or third flower-budding to produce seed.

Soybean oil and soybean meal are the 2 major products obtained from soybeans. The oil is used primarily as a vegetable shortening for cooking and soup production, and in poultry feeds.

Soybean meal is a protein concentrate containing 44 to 50% protein. The protein in the ground seed meal has a poor biological value which can be improved by removing the oil and heating. Most soybeans are processed by the solvent process. The expellar or hydraulic process produces meals of variable protein content and often protein of inferior biological value.

TABLE 24.1. PERCENTAGE COMPOSITION OF SOYBEAN MEALS

	Solvent Soybean Meal	Dehulled Solvent Soybean Meal
Protein crude	45.0	49.0
arginine	3.4	4.0
lysine	2.4	2.35
methionine	0.65	0.7
cystine	0.64	0.70
tryptophan	0.66	0.7
threonine	1.8	2.2
Fat (ether extract)	1.0	1.3
Crude fiber	7.0	3.2
Metabolizable energy (Cal./lb)	1000.0	1100.0

Source: Patrick (1965).

Investigators often use urease activity as an index for protein quality in soybean meal. The higher the urease level, the poorer the biological value and the lower the availability of the methionine (Table 24.2). The availability of methionine in soybean meal is so often discussed that nutritionists give the impression that it is the only protein containing unavailable methionine. Table 24.2 shows data demonstrating that both peanut meal and meat scraps often contain unavailable methionine.

Soybean continuous-solvent extraction plants were brought to the United States in the 1930s from Europe. Preparation of soybeans for solvent extraction is as follows: soybeans are cracked, dried to 8–11% moisture content and heated to 140°–160°F. While the cracked soybeans are still hot and slightly plastic, they are passed through flaking rolls.

TABLE 24.2. THE RELATION OF UREASE ACTIVITY TO METHIONINE AVAILABILITY

	% Protein	Urease Activity	% Methionine	% Cystine	Bioassay Methionine	% Availability of Methionine
Soybean meal high urease	45.3	8.6	0.72	0.82	0.50	69
Soybean meal high urease	47.0	10.0	0.85	0.94	0.51	60
Soybean meal expeller 223°F Not toasted	45.0	1.1	0.85	0.76	0.78	92
Soybean meal expeller 192°F lightly toasted	44.5	1.9	0.80	0.76	0.71	89
Soybean meal extracted moderately toasted	43.7	1.1	0.79	0.83	0.63	80
Meat meal	85.2	—	1.53	0.94	0.74	49
Meat meal	59.7	—	0.60	0.90	0.22	37
Meat meal	56.0	—	1.12	0.67	0.51	46
Peanut meal	48.2	—	0.67	0.96	0.47	70
Peanut meal	48.7	—	0.54	0.73	0.42	78

Source: Almquist (1965).

Steam is added at the rolls so that the flakes hold together and thus produce less fines. Combined steps of drying, steaming and heating prior to flaking is termed tempering or conditioning.

Since soybean flakes are very fragile, they are usually transported to the extractor in drag-length conveyors, so they move without tumbling, which might happen in screw conveyors or bucket elevators. Extraction systems operate at atmospheric pressure so that seals are unnecessary. Three materials enter the extractor: flakes, solvent and air. The 3 products emerging from the extractor are mixtures of oil meal, solvent with oil, and solvent vapors. Each of these products requires treatment to separate the solvent. Meal dryers or desolventizers are of 3 general types: (1) horizontal steam-jacketed; (2) vertical direct steam; and (3) recyclization. The horizontal jacketed dryer has a central tubular shaft on which are fastened radially-projecting arms with progessing paddles fastened to the ends of the arms. Steam is introduced into the tubular shaft to increase the heating surface of the vessel. Direct steam is introduced near the bottom of the bank of dryers to displace the solvent vapor and allow complete effusion of solvent from the flakes.

In the case of solvent extraction, the maximum temperature at any stage of the process rarely exceeds 220°F. Although the meal is subjected to cooking action, very little browning occurs. Consequently, solvent-extracted meal is "toasted" at 250°–260°F at which temperature denaturation of the protein occurs and partial carmelization of the carbohydrates produces the golden-brown appearance. These flakes contain

25% of the material that will pass through a 16-mesh screen. Flakes and hulls which do not pass through the screen travel to the first grinder, where they are subjected to reduction.

Soybean meal finds a ready outlet for use in poultry rations, and thus finds its way to feeders via the feed manufacturer. Perhaps more than anyone else, the feed manufacturer has stressed the need for balanced rations and proper heat processing of soybean meal.

Soybean meal, or soybean oil meal, as it was originally called, refers to the ground residue that remains after processing soybeans to extract most of the oil that they contain. When raw soybeans are ground, the product is called ground raw soybeans, not soybean meal. In nonruminants, or animals with a single or simple stomach, raw soybeans are not satisfactory. Growth and metabolism trials at the University of Wisconsin showed that the application of heat to raw soybeans caused an increase in the availability of certain fractions of the protein molecule.

Effect of Heating Soybeans

Heat causes the methionine-cystine fraction to become available so it can be utilized by the animal. Autoclaving soybeans increase the nutritional value of the protein fraction in much the same way as cystine-methionine supplements raw soybeans. The rate, rather than the extent, of methionine liberation in soybean protein determines its biological value. Nebraska researchers have noted a trypsin-inhibiting substance in raw soybeans which can be extracted with water or dilute acid at pH 4.2. The substance has a proteolytic-inhibiting effect and reduces the growth of chicks. The material has been crystallized, characterized, and found to be a stable protein of the globulin type with an isoelectric point of 4.5. Improved nutritional values of soybean protein in raw flakes result after thoroughly washing with water and draining off the water extract.

No improvement in the nutritive value of protein in soybeans occurs from extracting them or defatting them with ether, hexane, carbon tetrachloride, benzene or ethanol. Autoclaving soybeans has been the most practical method of improving nutritive value, but autoclaving for too long or at too high a temperature impairs nutritive value. The material should not be overheated. The importance of proper cooking of commercial soybean meal is apparent in the case of meal that is to be used as a substantial source of protein in rations where rapid and efficient growth is a primary objective.

Tests for Nutritive Value

There are several tests to determine the nutritive value of the protein in soybean meal. If the material is soluble in water or has urease activity, it shows it has not been denatured and, therefore, not sufficiently heated.

A chick-growth test is often used to determine the biological value of the protein. A semipurified ration containing corn sugar, dextrin or starch, soybean oil, vitamins, minerals, and enough soybean test meal to provide 16% protein, is used. If a response to supplementary methionine is obtained with White Leghorn Cockerels, the meal is not properly processed.

Goiter is produced in chicks on rations containing 25% soybean meal. The goitrogenic factor of soybeans is partially inactivated by heat, and iodine counteracts the effect on the thyroid gland. The goitrogenic tendency of soybean meal is of no particular economic importance, since practical poultry rations containing soybean meal and iodine produce perfectly normal poultry. Soybean meal is decidedly superior to ground or whole soybeans, and is at least equal to cottonseed meal for growing birds. Soybean meal with sufficient added minerals is about equal to meat scraps and fishmeal, and somewhat better than gluten feed or cottonseed meal in rations for chicks. Protein, per se, is not needed as a supplement to soybean meal in chick-starter rations, provided they contain essential minerals and vitamins. A combination of properly heated soybean meal and grains supplemented with minerals and vitamins, supplies protein as high in quality as any combination that can be provided. Raw soybeans are definitely inferior to properly cooked soybean meal as protein supplements in chick-starter rations. As stated, soybean meal serves as a protein concentrate in laying rations, but it should not be in excess of 50% of the supplemental protein. The remainder should consist of fishmeal and meat scraps.

Soybean meal in rations for laying hens does not cause an objectionable discoloration of the egg yolks when stored, as in the case of cottonseed meal. Satisfactory hatchability of hens' eggs is obtained where rations contain soybean meal as a principal protein supplement. Soybean meal has been successful in starter and growing rations for turkeys when properly processed. It generally exceeds all other vegetable protein concentrates in starter mashes fed to chicks up to 8 wk of age when used as a principal protein supplement in rations otherwise adequate in essential minerals and vitamins. Best results are obtained when soybean meal is used in conjunction with fishmeal, poultry by-product meal or meat scraps.

The corn-soybean ration for poultry has been the basic structural foundation of feeding for over 20 years. The geneticist has fed the breeding birds a corn-soybean ration so that there is a genetic acceptance of this type of ration.

Although most of the soybeans produced in the United States are used in animal feeds, their usage in human foods is growing. Today, over 50%

of manufactured foods, from bread to meat extenders, contain some form of soybean products.

Soybean flour and soybean protein fibers are widely used in meat substitutes and meat extenders. Some of the more common uses of fibers are in vegetable protein sausage, sliced turkey, chicken loaf, sliced ham and ground beef. Soybean flour is used in many ways, for example, in wheat bread as a bleaching agent. Over 40% of prepared foods contain one or more fractions from soybeans, and their use is increasing. Besides the good biological value of the protein and its many physical properties, the fact that soybean flour or protein does not cause off-flavor or bitterness in foods will increase its usage in human foods. By the year 2000, man will be the largest user of soybeans, and poultry will use soybean by-products as a substitute.

The American Feed Control Officials have approved over 20 soybean products with official numbers by the NSF.

84. SOYBEAN PRODUCTS

Official

84.1 Ground Soybeans is obtained by grinding whole soybeans without cooking or removing any of the oil. (Adopted 1933.) NRC 5-04-596

84.2 Ground Soybean Hay is the ground soybean plant including the leaves and beans. It must be reasonably free of other crop plants and weeds and must contain not more than 33% crude fiber. (Adopted 1944, Amended 1964.) NRC 1-04-559

84.3 Soybean Hulls consist primarily of the outer covering of the soybean. (Adopted 1948.) NRC 1-04-560

84.4 Soybean Feed, Solvent Extracted, is the product remaining after the partial removal of protein and nitrogen free extract from dehulled solvent extracted soybean flakes. (Adopted 1948, Amended 1960, 1964.) NRC 5-04-613

84.5 Soy Grits or Soy Flour is the screened and graded product remaining after removal of most of the oil from selected, sound, cleaned and dehulled soybeans by a mechanical or solvent extraction process. It must contain not more than 3% crude fiber. (Adopted 1958, Amended 1964.) NRC 5-04-592, 5-04-593

84.60 Soybean Meal, Mechanical Extracted, is the product obtained by grinding the cake or chips which remain after removal of most of the oil from soybeans by a mechanical extraction process. Heat must be applied during the process. It must contain not more than 7% crude fiber. It may contain a non-nutritive inert, non-toxic conditioning agent to reduce caking and improve flowability in an amount not to exceed that necessary to accomplish its intended effect and in no case exceed 0.5%. The name of the conditioning agent must be shown as an added ingredient. The words "mechanical extracted" are not required when listing as an ingredient in a manufactured feed. (Proposed 1960, Adopted 1961, Amended 1964, 1967, 1968.) NRC 5-04-600

84.61 Soybean Meal, Solvent Extracted, is the product obtained by grinding the flakes which remain after removal of most of the oil from soybeans by a solvent extraction process. Heat must be applied during process. It must contain not more than 7% crude fiber. It may contain a non-nutritive inert, non-toxic conditioning agent to reduce caking and improve flowability in an amount not to exceed that necessary to accomplish its intended effect and in no case exceed 0.5%. The name of the conditioning agent must be shown as an added ingredient. The words "solvent extracted" are not required when listing as an ingredient in a manufactured feed. (Proposed 1960, Adopted 1961, Amended 1964, 1968.) NRC 5-04-604

84.7 Soybean Meal, Dehulled, Solvent Extracted, is obtained by grinding the flakes remaining after removal of most of the oil from dehulled soybeans by a solvent extraction process. Heat must be applied during the process. It must contain not more than 3.3% crude fiber. It may contain a non-nutritive inert, non-toxic conditioning agent to reduce caking and improve flowability in an amount not to exceed that necessary to accomplish its intended effect and in no case to exceed 0.5%. The name of the conditioning agent must be shown as an added ingredient. When listed as an ingredient in a manufactured feed, it may be identified as "Dehulled Soybean Meal." (Proposed 1960, Adopted 1961, Amended 1964, Proposed 1972, Adopted 1973.) NRC 5-04-612

84.8 Soybean Mill Feed is composed of soybean hulls and the offal from the tail of the mill which results from the manufacture of soy grits or flour. It must contain not less than 13% crude protein and not more than 32% crude fiber. (Proposed 1960, Adopted 1961, Amended 1964.) NRC 5-04-594

84.9 Soybean Mill Run is composed of soybean hulls and such bean meats that adhere to the hulls which results from normal milling operations in the production of dehulled soybean meal. It must contain not less than 11% crude protein and not more than 35% crude fiber. (Proposed 1960, Adopted 1961, Amended 1964.) NRC 5-04-595

84.10 Soy Phosphate or Soy Lecithin is the mixed phosphatide product obtained from soybean oil by a degumming process. It contains lecithin, cephalin, and inositol phosphatides, together with glycerides of soybean oil and traces of tocopherols, glucosides, and pigments. It must be designated and sold according to conventional descriptive grades with respect to consistency and bleaching. (Proposed 1958, Adopted 1961.) NRC 4-04-562

84.11 Heat Processed Soybeans (Dry Roasted Soybeans) is the product resulting from heating whole soybeans without removing any of the component parts. It may be ground, pelleted, flaked, or powdered. It must be sold according to its crude protein content. (Proposed 1960, Adopted 1964.) NRC 5-04-597

84.12 Soy Protein Concentrate is prepared from high quality, sound, clean dehulled soybean seeds by removing most of the oil and water soluble non-protein constituents and must contain not less than 70% protein on a moisture-free basis. (Proposed 1965, Adopted 1968.) NRC 5-08-038

84.13 Kibbled Soybean Meal is the product obtained by cooking ground solvent extracted soybean meal, under pressure and extruding from an expeller or other mechanical pressure device. It must be designated and sold according to its protein content and shall contain not more than 7% crude fiber. (Proposed 1969, Adopted 1971.) NRC 5-09-343

84.14 Condensed Soybean Solubles is the product obtained by washing Soy Flour or Soybean Flakes with water and acid at a pH of 4.2—4.6. The wash water is then concentrated to a solid content of not less than 60%. (Proposed 1969, Adopted 1971.) NRC 5-09-344

84.15 Ground extruded Whole Soybeans is the meal product resulting from extrusion by friction heat and/or steam, whole soybeans without removing any of the component parts. It must be sold according to its crude protein, fat, and fiber content. (Proposed 1974, Adopted 1975.) NRC 5-14-005

84.16 Soybean Protein Product, Chemically Modified, is the product resulting from the chemical modification of dehulled solvent extracted soybean materials (soy flour, soy flakes, soy grits, soybean meal, soy protein concentrate) utilizing acids, alkalies or other chemicals with or without heat processing, without removing any of the component parts. Chemicals used in this modification will be limited to those permitted for food processing as feed additives or those listed as GRAS. It must contain a minimum of 48% crude protein (as is basis) and may be in a ground, pelleted, flaked or powdered form. (Proposed 1976, Adopted 1977.) NRC 5-26-010

84.14 Soybean solubles, condensed, is the product obtained by washing soybean flakes with water and acid at a pH of 4.2—4.6 and/or water-alcohol mixture. The soluble portion is then concentrated to a solid content of not less than 60%. (Proposed 1976, Adopted 1977.) NRC 5-09-344

84.17 Soybean grits, full fat, is the granular material resulting from the screened and graded product resulting from the cracking or grinding of cleaned heat treated, dehulled, whole soybeans. It must be sold according to its protein count. (Proposed 1976, Adopted 1977.) NRC 5-20-875

84.18 Soybean flour, full fat, is finely powdered material resulting from the screened and graded product resulting from the cracking or grinding of cleaned, heat treated, dehulled, whole soybeans. It must be sold according to its protein content. (Proposed 1976, Adopted 1977.) NRC 5-20-876

Tentative

84.5 Soybean grits is the granular material resulting from the screened and graded product after removal of most of the oil from selected, sound, cleaned and dehulled soybeans by a mechanical or solvent extraction process. It must contain not more than 3.3% crude fiber. (Proposed 1976.) NRC 5-04-592

84.51 Soybean flour is the finely powdered material resulting from the screened and graded product after removal of most of the oil from selected, sound, cleaned and dehulled soybeans by a mechanical or solvent extraction process. It must contain not more than 3.3% crude fiber. (Proposed 1976, Amended 1977.) NRC 5-04-593

COTTONSEED MEAL (VEGETABLE PROTEIN)

Cottonseed meal is a by-product of the cotton plant. The United States, India, Brazil, the U.S.S.R., China and Egypt are the most important cotton-producing nations.

Cotton seeds contain a pigment, gossypol, which influences growth, reproduction and color of egg yolks. This pigment must be removed or destroyed if the cottonseed meal is to be used in poultry feeds. During the early development of the cotton industry, cotton seeds were processed by the hydraulic method, followed by the screw-pressed (expeller). In order to remove the fat and gossypol, the seeds are cooked, which results in a change in the protein structure. This results in a decrease in the availability of the lysine. Today, the solvent-processing method is widely used. Solvent extraction methods operate at lower temperatures than the expeller or hydraulic methods, and thus the lysine in solvent-extracted products is more available. Cottonseed meal is a by-product, rather than the primary product, of the cotton plant. It is geared to the production of cotton fiber, and cottonseed has only one-sixth of its value.

Cottonseed oil is an important edible oil, and methods for its processing are similar to those used for other oil seeds, except that the cotton seed, which has a fuzz, has to be cleaned (delinted) to remove the fuzz.

Four types of cottonseed meal are produced. Screw-pressed (expeller) meal accounts for 60% of the total. Direct solvent (hexane) and pre-pressed solvent meals each account for approximately 20% of the total. Hydraulically-processed cottonseed meal comprises less than 3%. Cottonseed meal contains: (1) oil, (2) free and bound gossypol (yellow polyphenolic pigment), (3) available amino acids, such as lysine, (4) protein and (5) fiber. Free gossypol values are lowest in screw-pressed and pre-pressed solvent meals with the highest values in direct-solvent meal. Most gossypol is removed, either lost or bound.

Gossypol in Cottonseed Meals

The protein and fiber contents of cottonseed meals are inversely related. There is considerable variation in oil, fiber, moisture, and protein quality of cottonseed meals, and thus significant differences in energy can be expected. There is a low availability of lysine, but tryptophan and methionine, plus histine, showed excellent availability. There is a need for lysine supplementation of many cottonseed meals when used in rations for growing poultry. Since the beginning of this century, it has been known that cottonseed meals contain gossypol, which has a poisonous effect when fed to livestock and poultry. The toxicity of cottonseed to nonruminants is related to the amount of free gossypol present in the cottonseed. Gossypol is present in the native cottonseed in many small pigmented glands. This gives cottonseed its characteristic color. Free gossypol is toxic. Bound gossypol reduced growth rate in chicks when fed at high levels, but there are other factors in cottonseed which inhibit growth.

Very low levels of free gossypol cause discoloration of egg yolks when the eggs are put into cold storage. Small quantities of gossypol are transferred to the egg and a slow chemical reaction occurs between the gossypol and the components of the egg yolk, resulting in the darkening of the yolk during prolonged cold storage. Cottonseed meal may constitute 50% of the protein source in chick and broiler rations if the dietary level of free gossypol is less than 0.01%. Some cottonseed meals furnish a small portion of the protein in laying-hen rations, if the resulting eggs are not stored. The effect of gossypol, the yellow polyphenolic pigment present in cottonseed pigment glands, upon the color, weight and production of eggs is dramatic. Free gossypol is believed to be the major biologically-active form, but bound gossypol may be released in the digestive tract.

High levels of gossypol in poultry rations depress weight, feed intake and feed efficiency; many times there is depressed hatchability of eggs and increased mortality. Additional deleterious effects are enlarged gall bladders and blood and bone changes. Substances are also present in cottonseed that produce pink-white discoloration, the cause of which has been shown to be a reaction of the conalbumin of egg white with iron diffused from the yolk. The primary chemical entity responsible for the discoloration is a cyclopropenoid ring. The intensity of the pink-white is governed by factors such as length of feeding period, storage temperature and amount of cyclopropenoid ingested. Accompanying effects are increased levels of stearic acid, and decreased levels of oleic acid in the egg yolks which make a stiffer yolk substance. Pink-white discoloration is usually accompanied by enlarged salmon-colored yolks. The author found this to be caused by the migration of conalbumin into the yolk due to relaxation of the vitelline membrane. Here, it is combined with the iron of the yolk to produce a salmon-pink color.

In future years, there is a likelihood of further development of glandless cottonseed which does not contain gossypol but does have fatty acids. Improvement in oilseed processing and increased knowledge of the biological effects of specific cottonseed components have led to increased usage of cottonseed meal, and new types of solvents for oil extraction are expected to lead to further improvement in cottonseed meal. At the end of World War II, 95% of the domestic cottonseed meal was produced by hydraulic processes and relatively little screw-pressed or solvent meal was available. All commercial cottonseed meals in rations are better for young growing animals when supplemented with L-lysine; older animals need less of this amino acid. After lysine needs are met in broiler rations, the most limiting amino acids in cottonseed meal are methionine, threonine, isoleucine and leucine. High-temperature cooking is the oldest and most widely used method for releasing free gossypol from the glands. After the gossypol is released, it is either removed or inactivated to make it nontoxic. Cottonseed meal with free gossypol levels below 0.04% is tolerated by chickens. The composition of 50% cottonseed meal is presented in Table 24.3.

TABLE 24.3. PERCENTAGE COMPOSITION 50% PROTEIN COTTONSEED MEAL

Crude protein	50.00	Methionine	0.77
Crude fat	1.00	Cystine	1.04
Crude fiber	8.00	Tryptophan	0.56
Ash	7.00	Glycine	2.50
Nitrogen solubility	75.00	Histidine	1.31
Calcium	0.20	Leucine	3.00
Phosphorus (total)	1.40	Isoleucine	1.88
Phosphorus (available)	0.40	Phenylalanine	3.05
Free gossypol	0.03	Threonine	1.77
Arginine	5.00	Valine	1.78
Lysine (total)	2.15	Metabolizable energy (Cal./lb)	1010.0
Lysine (epsilon amino)	1.90		

Source: Patrick (1965).

24. COTTONSEED PRODUCTS

Official

24.10 *Cottonseed Meal, Mechanical Extracted, is the product obtained by finely grinding the cake which remains after removal of most of the oil from cottonseed by a mechanical extraction process. It must contain not less than 36% crude protein. The words "mechanical extracted" are not required when listing as an ingredient in a manufactured feed. (Proposed 1964, Adopted 1966, Amended 1968.) NRC 5-01-625

24.12 *Cottonseed Meal, Solvent Extracted, is the product obtained by finely grinding the flakes which remain after removal of most of the oil from cottonseed by a solvent extraction process. It must contain not less than 36% crude protein. The words "solvent extracted" are not required when listing as an ingredient in a manufactured feed. (Proposed 1964, Adopted 1966, Amended 1968.) NRC 5-01-632

24.2 Cottonseed Cake, Mechanical Extracted, is the unground product composed of the kernel and such portions of the lint, hull, and oil as remain after removal of most of the oil from cottonseed by a mechanical process. It must contain not less than 36% crude protein. The words "mechanical extracted" are not required when listing as an ingredient in a manufactured feed. (Proposed 1964, Adopted 1966, Amended 1968.) NRC 5-01-623

24.30 Cottonseed Flakes, Mechanical Extracted, is the unground product, composed of the kernel and such portions of the lint, hull, and oil as remain after removal of the oil from cottonseed by a mechanical extraction process. It must contain not less than 36% crude protein. The words "mechanical extracted" are

*This product (when sold or distributed singly) may be additionally labeled with the following bold face terms when the requirements thereinafter are met;

________, Prime Quality must be free of mold, excess lint, and sour, musty, or burnt odors.

________, Off Quality shall be that which does not meet the prime quality requirements.

not required when listing as an ingredient in a manufactured feed. (Proposed 1964, Adopted 1966, Amended 1968.) NRC 5-08-820

24.31 Cottonseed Flakes, Solvent Extracted, is the unground product, composed of the kernel and such portions of the lint, hull, and oil as remain after removal of the oil from cottonseed by a solvent extraction process. It must contain not less than 36% crude protein. The words "solvent extracted" are not required when listing as an ingredient in a manufactured feed. (Proposed 1964, Adopted 1966, Amended 1968.) NRC 5-01-629

NOTE: The following levels of minimum crude fiber are adopted from cottonseed meals, cake, or flakes, of respective grade.

	Crude Fat		
Crude Protein	Mechanical	Solvent	Crude Fiber
36%	2.0%	0.5%	17%
41%	2.0%	0.5%	14%
43%	2.0%	0.5%	13%

24.4 Whole-Pressed Cottonseed, Mechanical Extracted, is composed of sound, mature, clean, delinted, and unhulled cottonseed, from which most of the oil has been removed by mechanical pressure. It must be designated and sold by its crude protein content. If ground, it must be so designated. The words "mechanical extracted" are not required when listing as an ingredient in a manufactured feed. (Proposed 1964, Adopted 1966, Amended 1968.) NRC 5-01-609

24.50 Low Gossypol Cottonseed Meal, Mechanical Extracted, is a meal in which the gossypol is not more than 0.04% free gossypol. The words "mechanical extracted" are not required when listing as an ingredient in a manufactured feed. (Proposed 1964, Adopted 1966, Amended 1968.) NRC 5-09-002

24.51 Low Gossypol Cottonseed Meal, Solvent Extracted, is a meal in which the gossypol is not more than 0.04% free gossypol. The words "solvent extracted" are not required when listing as an ingredient in a manufactured feed. (Proposed 1964, Adopted 1966, Amended 1968.) NRC 5-01-633

24.6 Cottonseed Hulls consists primarily of the outer covering of the cottonseed. (Proposed 1964, Adopted 1966.) NRC 1-01-599

PEANUT MEAL

Peanuts are produced throughout the world. World production in 1976 was 18,128,000 metric tons with the United States producing only 1,701,000 metric tons. Most of the peanuts in the United States are produced in the southern states, and both oil and peanuts are used chiefly for human foods.

Peanut meal is a vegetable protein having poor biological value. It is also often contaminated with molds which produce toxins, called aflatoxins.

Peanuts should be stored in a cool dry elevator and moisture should be maintained below 12 to 14%.

Poultry, in general, are very sensitive to aflatoxins. Ducklings are most sensitive, turkeys and poults moderately so, while broilers are only slightly sensitive. When broilers receive aflatoxins, they may develop hemorrhages, and soft bones (rickets) and mortality can be high. The toxic condition may appear unexpectedly and leave or disappear rapidly.

Aflatoxins are present in the fat fraction, and when absorbed into the chicken's body, concentrate in the liver. Tumors are often present in the livers of diseased birds.

Peanut meal is used in poultry rations to replace up to 5% of the soybean meal. Peanut oil is used in rapid-food serving places. The supply is limited, so only a very small amount is used in poultry feeds.

Peanut hulls are used as poultry-house litter and should not be used in poultry feeds.

The composition of peanut meal is shown in Table 24.4.

The AFCO has approved 7 peanut products.

TABLE 24.4. PERCENTAGE COMPOSITION OF SOLVENT PEANUT, SAFFLOWER AND LINSEED MEAL

	Peanut Meal	Safflower Seed Meal	Linseed Meal
Protein, crude	50.0	42.5	33.0
arginine	5.0	3.7	3.2
lysine	1.7	1.3	1.2
methionine	0.45	0.7	0.6
cystine	0.7	0.7	0.7
thryptophan	0.5	0.6	0.3
threonine	1.1	1.3	1.7
Fat (ether extract)	1.3	1.3	1.5
Crude fiber	3.9	15.0	9.3
Metabolizable energy (Cal./lb)	1150.0	920.0	640.0

Source: Patrick (1965).

72. PEANUT PRODUCTS

Official

72.1 Peanut Meal and Hulls is ground peanut meal with added hulls, or, is the ground residue after the extraction of part of the oil from whole peanuts by a mechanical or solvent extraction process. If solvent is extracted, it must be so designated. (Adopted 1954, Amended 1964.) NRC 5-03-655, 5-03-656

72.2 Peanut Skins is the outer covering of the peanut kernel, exclusive of hulls, as obtained in ordinary commercial processing. The product may contain broken peanut kernels. (Adopted 1946, Amended 1964.) NRC 1-03-631

72.5 Peanut Meal is obtained by grinding the cake, chips, or flakes obtained by the removal of most of the oil from peanut kernels by a mechanical or solvent extraction process. If solvent extracted, it must be so designated. It must contain not more than 7% crude fiber and only such amount of hulls as is unavoidable in good factory practice. (Proposed 1964, Adopted 1965.) NRC 5-03-649, 5-03-650

72.6 Peanut Hulls consists of the outer hull or the peanut shell. (Proposed 1965, Adopted 1966.) NRC 1-08-028

72.7 Peanut Meal and Hulls is ground peanut meal whether mechanically or solvent extracted, which contains a small amount of peanut hull residue or to which peanut hulls have been added. If the peanut meal is solvent extracted, it must be so designated. (Proposed 1976)

72.8 Ground Peanut Hay is composed of ground peanut leaves and stems from which the peanuts have been removed. (Proposed 1976)

72.9 Peanut Meal is a ground product of the shelled peanuts, composed principally of the kernels, with such portion of the hull, or fiber, and oil, as may be left in the ordinary course of manufacture by a mechanical or solvent extraction process. If solvent extracted, it must be so designated. It must contain not more then 7% crude fiber. (Proposed 1976)

CEREAL GRAIN PROTEIN CONCENTRATES MAIZE (CORN)

Corn gluten and corn-gluten feed are by-products obtained from the production of starch or related products in the corn industry. The protein is of low biological value but blends well with soybean protein. It is a good source of xanthophyll and methionine, both which are deficient in soybean meal. Most of the corn gluten products are used in broiler and egg rations. The composition of corn gluten is shown in Table 24.5.

TABLE 24.5. PERCENTAGE COMPOSITION OF CORN GLUTEN MEALS

	41% Protein	60% Protein
Protein, crude	41.0	60.0
arginine	1.4	1.9
lysine	0.8	1.0
methionine	1.2	1.9
cystine	0.8	1.1
tryptophan	0.2	0.25
threonine	1.4	2.0
Fat (ether extract)	2.5	2.5
Crude fiber	4.3	1.6
Metabolizable energy (Cal./lb)	1250.0	1690.0

Source: Patrick (1965).

The American Feed Control Officials have approved 2 gluten products. They are:

48.13 Corn Gluten Feed is that part of the commercial shelled corn that remains after the extraction of the larger portion of the starch, gluten, and germ by the processes employed in the wet milling manufacture of corn starch or syrup. It may or may not contain one or more of the following: fermented corn extractives, corn germ meal. (Adopted 1936, Amended 1960.) NRC 5-02-903

48.14 Corn Gluten Meal is the dried residue from corn after the removal of the larger part of the starch and germ, and the separation of the bran by the process employed in the wet milling manufacture of corn starch or syrup, or by enzymatic treatment of the endosperm. It may contain fermented corn extractives and/or corn germ meal. (Adopted 1936, Amended 1960.) NRC 5-02-900

GRAIN SORGHUMS (Milo, Hegari, Kaffir, or Feterita)

Grain sorghums are used in the production of starch and syrup. The part remaining after removing the larger part of the germ and starch is used to produce gluten feed or gluten meal. The protein content is similar to corn-gluten feed or meals with comparable biological value. The AFCO approved 2 gluten products.

42.2 Grain Sorghum Gluten Feed is that part of the grain of grain sorghums that remains after the extraction of the larger part of the starch and germ, by the processes employed in the wet milling manufacture of starch or syrup. (Adopted 1948, Amended 1950.) NRC 5-04-389

42.3 Grain Sorghum Gluten Meal is that part of the grain of grain sorghums that remains after the extraction of the larger part of the starch and germ, and the separation of the bran by the processes employed in the wet milling manufacture of starch or syrup. (Adopted 1948, Amended 1950.) NRC 5-04-388

OTHER OIL MEALS

Sesame-oil Meal

Only sesame-oil meal in the minor group of meals is produced in any appreciable amounts.

Sesame-oil meal is a by-product of the production of oil from sesame seeds which are raised in tropical countries. It contains as much protein as cottonseed meal and is high in calcium and phosphorus. Its protein is of lower quality, being very deficient in lysine, which is well-supplied by soybean meal. It is satisfactory for poultry feeding and supplements the protein of soybean-oil meal for chicks.

From 0.4 to 0.5% lysine is necessary to achieve optimum growth and conversion. Combined with soybean meal, sesame meal results in better chick growth than all-sesame rations.

Sesame meal contains enough phytic acid to form a chelate with zinc in the intestines, thus making the zinc unavailable. Excess calcium increases the formation of the zinc-phytate chelate; therefore, sesame meal should be included in rations with minimum calcium levels or those supplemented with zinc or ethylenediamine tetraacetic acid (EDTA), which forms an available zinc complex or chelate.

Coconut-oil Meal

Coconut-oil meal is a by-product of the production of oil from the dried meats of coconuts, the nuts of the coconut palm. It is made by the hydraulic or expeller process and contains less protein than corn-gluten feed. Its protein is of better quality than that in corn-gluten meal but not soybean meal. It should not be fed as the only protein supplement to poultry, but should be supplemented with some animal protein. Limited amounts of coconut meal are included in poultry rations when sufficient protein of better quality is provided. A level of 2 to 5% coconut meal is used in poultry rations on the Pacific Coast. A light-colored oil meal is preferred to a dark meal. The dark meal results from too high a temperature in expressing oil, and this reduces digestibility.

Babassu-oil Meal

Babassu-oil meal is produced from the hard-shelled seed of a Brazilian palm. It is similar in appearance and composition to coconut oil meal but a trifle higher in protein. It satisfactorily replaces 10% of the grain in poultry feeds.

Rapeseed Oil Meal

Rapeseed oil meal is the by-product of the production of oil from rapeseeds. It has varying amounts of glucosides from which mustard oils are formed in the digestive tracts of animals under certain conditions. These oils are irritating to the digestive system and can cause serious problems. This meal must be fed very carefully to avoid injurious results; not over 10% is used in poultry rations.

Navy Beans

Since 1920, it has been known that raw navy beans are injurious to animals and that autoclaving destroys the injurious effect. The beans also contain heat-labile trypsin inhibitors and hemagglutinin which are toxic to chickens. Sulfur-containing amino acids are the most limiting amino acids of navy beans, but raw beans do not support growth even when supplemented with all the amino acids in which the diet is deficient. Also, growth of animals not receiving raw bean diets is depressed by the addition of trypsin inhibitors from raw beans. The low nutritive value of raw navy beans is due to a methionine deficiency and the presence of a heat-labile toxic factor. Nevertheless, when beans are autoclaved, and supplemented with methionine, growth is as good as on the casein diet.

Velvet Beans

These are grown as a soil-improving crop in the United States and are often used as a livestock feed. They are toxic to pigeons; on the second day of feeding, birds show ruffled feathers and appear sleepy. Mortality commences on the fourth day. Feeding of diets containing velvet beans decreases performance of chicks as measured by reduced body gains and mortality; there is also decreased performance of laying hens. There is some factor in the beans that interferes with normal growth and egg production in chickens.

Mung Beans

These are sometimes used as a poultry feed. Each year, split and broken mung beans become available and are used in poultry feeds. They are ground and used in place of soybean meal in poultry rations and are also fed as a part of the scratch-grain mixture. Satisfactory egg production and hatchability are obtained when ground mung beans are fed as a source of vegetable protein. They have no detrimental effects on egg quality when fed as a source of protein in laying mashes for hens.

Safflower Meal

Decorticated safflower meal was first used in 1947. Research since that time shows that when adequately supplemented with lysine, safflower-meal rations give equivalent gains in growing chickens and turkeys compared to good control rations. Feed conversions are the same when rations are isocaloric. Decorticated safflower meal results in egg production similar to that obtained from soybean meal controls, when at levels of up to 15% of the rations. Feed conversions are also equivalent to the controls when feeds are isocaloric. Decorticated safflower meal has a metabolizable energy value of 774 Cal. per lb.

25

Fermentation Products

Fermentation products are those recovered from the brewers', distillers' and wood fermentation operations. The most valuable feed ingredient products of brewing are brewers' dried yeast and dried brewers' grains. Distillers' solubles and distillers' grains are the 2 most important feed ingredients produced by the distilleries. Torula yeast is produced by fermenting wood particles with a yeast found to use wood as a source of energy.

Brewers' and distillers' dried grains, brewers' dried yeast and distillers solubles are good sources of vitamins and proteins.

Torula yeast is deficient in Vitamin E and selenium. The biological value of its protein is being determined. It has been used for several years on the European continent.

Since the proteins of brewers' dried grains are derived chiefly from barley, and some rice and corn, they are deficient in some of the essential amino acids. Fortification of spent grains with yeast during the drying operation, if conducted in proper proportion, makes a superior feeding material. For poultry feeding, only small quantities should be incorporated in the ration because of their high-fiber content.

Brewers' dried yeast from malted grain fermentation constitutes a natural concentrated mixture of essential nutrients. Biologically-valuable proteins combined with high-potency Vitamin B-complex, important trace minerals and several unique "plus" factors, insure against deficiencies of known and still unrecognized nutrients. Brewers' dried yeast promotes vitality in the young and accelerates growth and feed conversion. It has natural antioxidants to protect against peroxide formation, and possesses Vitamin E-like activities, especially helpful in times of stress and during recovery from disease. Brewers' dried yeast is stable, free-flowing, available in various granulations and is a useful ingredient in poultry feeds, especially for turkeys and game birds.

Excessive body and liver fat develop in chickens maintained in cages for long periods of time. To prevent this, 10% brewers' dried grains are added at the expense of corn, milo and soybean meal. With this substitution, the gain in body weight and liver-fat accumulation during 50 wk is significantly less than with normal rations. Birds also grow faster, mature earlier and produce greater numbers of eggs. These results suggest that brewers' dried grains contain an unknown factor or that the biological availability of recognized nutrients is greater.

Distillers' solubles have wide usage in poultry feeds. Rye grains with solubles contain more arginine, lysine and tryptophan than the comparable corn product. The threonine content of both corn and rye grains with solubles is the same. Corn grains with solubles are richer in the remaining essential amino acids. Rye solubles are somewhat richer than corn solubles in all essential amino acids other than lysine. Both solubles contain the same amount of lysine.

On a protein basis, distillers' feeds are deficient in the same essential amino acids as the grains from which they are derived, chiefly lysine, tryptophan and methionine. However, the percentage of amino acids in the distillers' feeds is higher than that of the grains from which they are derived because the protein is concentrated by the manufacturing process in which starch is removed.

The percentage of various mineral constituents in distillers' feeds which make up the ash content is presented from a typical plant mash bill composed of 70% corn, 20% rye and 10% barley malt (Table 25.1). The solubles contain considerably more of the nutritionally-important elements—calcium phosphorus, manganese and magnesium—than do either light grains or the original grains from which they are derived.

Recent interest in distillers' feeds in poultry feeding has leaned largely toward supplying unidentified growth factors (UGF). The addition of

TABLE 25.1. PERCENTAGE MINERAL COMPOSITION OF THE ASH OF DISTILLERS' FEEDS (PLANT MASH BILL COMPOSED OF 70% CORN, 20% RYE AND 10% BARLEY MALT)

	Light Grains	Grains with Solubles	Solubles
Ash	2.0	6.0	8.0
Calcium	0.04	0.2	0.4
Phosphorus	0.21	0.8	1.6
Sodium	0.06	0.18	0.3
Potassium	0.09	0.8	1.8
Magnesium	0.06	0.3	0.8
Iron	0.01	0.03	0.05
Copper	0.007	0.007	0.008
Cobalt	0.00001	0.00001	0.00002
Manganese	0.002	0.006	0.01
Silicon	0.14	0.12	0.1
Sulfur	0.32	0.36	0.4
Chlorine	0.04	0.17	0.3

TABLE 25.2. CHEMICAL COMPOSITION OF DISTILLERS' FEED FROM CORN, RYE AND A TYPICAL COMMERCIAL MASH BILL

%	Light Grains			Dark Grains (Grains with Solubles)			Solubles		
	Corn, 90% B. Malt 10%	Rye, 85% R. Malt 15%	Typical M.B.[1]	Corn, 90% B. Malt 10%	Rye, 85% R. Malt 15%	Typical M.B.[1]	Corn, 90% B. Malt 10%	Rye, 85% R. Malt 15%	Typical M.B.[1]
Moisture	11	11	11	11	11	11	5	5	5
Protein	30	18–26	28	28	29	28	25	36–38[3]	27
Fat	11	6	10	9	3	8	6	1	5
Fiber	11	15	12	7	8	7	1–3[2]	1–3[2]	1–3[2]
Ash	2	2	2	6	6	6	8–9	8–9[3]	8–9
N.F.E.	35	44	37	29	43	40	52–55	44–49	51–54

[1]Typical mash bill of 70% corn, 20% rye, and 10% barley malt.
[2]Lower (1.0%) fiber content in solubles produced by use of centrifuges.
[3]Protein content of rye solubles decreases as mashing temperatures are increased.

TABLE 25.3. PERCENTAGE AMINO ACID CONTENT (DRY BASIS) OF DISTILLERS' FEEDS FROM VARIOUS MASH BILLS

	Grains with Solubles			Solubles		
	Mash Bill			Mash Bill		
	Corn, 90% B. Malt 10%	Rye, 85% R. Malt 15%	Typical Plant[1]	Corn, 90% B. Malt 10%	Rye, 85% R. Malt 15%	Typical Plant[1]
Essential						
arginine	0.9	1.2	1.0	0.6	1.4	0.9
histidine	0.8	0.7	0.8	0.7	0.8	0.7
isoleucine	2.0	1.6	1.9	1.5	2.0	0.7
leucine	2.4	2.2	2.3	1.4	1.9	1.6
methionine	0.5	0.4	0.5	0.4	0.5	0.4
phenylalanine	2.1	1.4	1.9	1.7	1.8	1.7
threonine	1.1	1.1	1.1	0.9	1.2	1.0
tryptophan	0.1	0.3	0.2	0.1	0.2	0.1
valine	1.8	1.7	1.8	1.4	2.1	1.6
Nonessential						
serine	1.9	1.3	1.7	1.2	1.7	1.4
tyrosine	0.7	0.5	0.6	0.6	0.6	0.6
glutamic acid	5.7	5.5	5.6	4.3	—	—

[1]70% corn, 20% rye, and 10% barley malt.

corn distillers' dried solubles to high-energy, corn-soy-fat-fish meal rations improves growth and feed efficiency even when the rations are maintained isocaloric and isonitrogenous. Distillers' solubles rank well in the list of high-energy ingredients. Their protein content is approximately 3 times that of cereal grains and 50% as much as soybean meal. (Their methionine content is equivalent to that of soybean meal.) Thus, addition of corn distillers' dried solubles improves energy and protein values of the feed and also provides added amounts of methionine, the first limiting amino acid in practical diets for both broiler and layers.

Distillers' dried solubles are a good source of vegetable oils and linoleic acid. This is important because of recent evidence that vegetable fats have special properties which appear to be essential for maximum growth rate and maximum egg size, especially in young pullets. Distillers' dried solubles are of value in helping nutritionists meet this linoleic acid requirement.

15. BREWERS PRODUCTS

Official

15.1 Brewers Dried Grains is the dried extracted residue of barley malt alone or in mixture with other cereal grain or grain products resulting from the manufacture of wort or beer and may contain pulverized dried spent hops in an amount not to exceed 3%, evenly distributed. (Adopted 1965, Amended 1972.) NRC 5-02-141

15.2 Malt Sprouts is obtained from malted barley by the removal of the sprouts which may include some of the malt hulls, other parts of malt and foreign material unavoidably present. It must contain not less than 24% crude protein. The term malt sprouts when applied to a corresponding portion of other malted cereals must be used in qualified form: i.e., "Rye Malt Sprouts," "Wheat Malt Sprouts," etc. (Adopted 1942, Amended 1964.) NRC 5-00-545

15.3 Malt Cleanings is obtained from the cleaning of malted barley or from the recleaning of malt which does not meet the minimum crude protein standard of malt sprouts. It must be designated and sold according to its crude protein content. (Adopted 1942.) NRC 5-00-544

15.4 Malt Hulls consists almost entirely of hulls as obtained in the cleaning of malted barley. (Adopted 1942.) NRC 1-00-497

15.5 Dried Spent Hops is obtained by drying the material filtered from hopped wort. (Adopted 1944.) NRC 1-02-396

15.6 Brewers Wet Grains is the extracted residue resulting from the manufacture of wort from barley malt alone or in mixture with other cereal grains or grain products. The guaranteed analysis shall include the maximum moisture. (Proposed 1971, Adopted 1974). NRC 5-02-142

Tentative

15.7 Brewers Condensed Solubles is obtained by condensing liquids resulting as by-products from manufacturing beer or wort. It must contain not less than 20% total solids, 70% carbohydrates on a dry matter basis and the guaranteed analysis shall include maximum moisture. (Proposed 1975.) NRC 5-26-011

27. DISTILLERS PRODUCTS

Official

27.1 Molasses Distillers Dried Solubles is obtained by drying the residue from the yeast fermentation of molasses after the removal of the alcohol by distillation. (Adopted 1943, Amended 1944.) NRC 4-04-698

27.2 Molasses Distillers Condensed Solubles is obtained by condensing to a syrupy consistency the residue from the yeast fermentation of molasses after the removal of the alcohol by distillation. (Adopted 1946.) NRC 4-04-697

27.3 Potato Distillers Dried Residue is the dried product obtained after the manufacture of alcohol and distilled liquors from potatoes or from a mixture in which potatoes predominate. (Adopted 1947.) NRC 5-03-773

27.4 ________ Distillers Dried Solubles is obtained after the removal of ethyl alcohol by distillation from the yeast fermentation of a grain or a grain mixture by condensing the thin stillage fraction and drying it by methods employed in the grain distilling industry. The predominating grain must be declared as the first word in the name. (Proposed 1963, Adopted 1964.)

27.5 ________ Distillers Dried Grains is obtained after the removal of ethyl alcohol by distillation from the yeast fermentation of a grain or a grain mixture by separating the resultant coarse grain fraction of the whole stillage and drying it by methods employed in the grain distilling industry. The predominating grain shall be declared as the first word in the name. (Proposed 1963, Adopted 1964.)

27.6 _______ Distillers Dried Grains with Solubles is the product obtained after the removal of ethyl alcohol by distillation from the yeast fermentation of a grain or a grain mixture by condensing and drying at least 3/4 of the solids of the resultant whole stillage by methods employed in the grain distilling industry. The predominating grain shall be declared as the first word in the name. (Proposed 1963, Adopted 1964.)

27.7 _______ Condensed Distillers Solubles is obtained after the removal of ethyl alcohol by distillation from the yeast fermentation of a grain or a grain mixture by condensing the thin stillage fraction to a semi-solid. The predominating grain must be declared as the first word in the name. (Proposed 1969, Adopted 1970.)

96. YEAST

Official

96.1 Primary Dried Yeast or Dried Yeast is the dried, non-fermentative yeast of the botanical classification Saccharomyces which has been separated from the medium in which propagated. It must contain not less than 40% crude protein. (Adopted 1955.) NRC 7-05-533

96.2 Active Dry Yeast is yeast which has been dried in such a manner as to preserve a large portion of its fermenting power. It must contain no added cereal or filler and must contain not less than 15 billion live yeast cells per gram. (Adopted 1951.) NRC 7-05-524

96.3 Irradiated Dried Yeast, Irradiated _______ Dried Yeast is the dried, non-fermentative yeast which has been subjected to ultraviolet rays in order to produce anti-rachitic potency. (Proposed 1958, Adopted 1959.) NRC 7-05-529

NOTE: When Irradiated Dried Yeast or Irradiated _______ Dried Yeast is used as an ingredient of proprietary feeds for four-footed animals, the name may be followed by a parenthetical phrase (Source of Vitamin D_2). (Adopted 1945, Amended 1959.)

96.5 Grain Distillers Dried Yeast is the dried, non-fermentative yeast of the botanical classification Saccharomyces resulting from the fermentation of grains and yeast, separated from the mash, either before or after distillation. It must contain not less than 40% crude protein. (Adopted 1955.) NRC 7-05-519

96.6 Molasses Distillers Dried Yeast is the dried, non-fermentative yeast of the botanical classification Saccharomyces resulting from the fermentation of molasses and yeast, separated from the medium, either before or after distillation. It must contain not less than 40% crude protein. (Adopted 1955.) NRC 7-05-532

96.7 Torula Dried Yeast is the dried, non-fermentative yeast of the botanical classification Torulopsis which has been separated from the medium in which propagated. It must contain not less than 40% crude protein. (Adopted 1955.) NRC 7-05-534

96.8 Yeast Culture is the dried product composed of yeast and the media on which it was grown, dried in such a manner as to preserve the fermenting activity of the yeast. The media must be stated on the label. (Adopted 1957.) NRC 7-05-520

NOTE: No reference to media in ingredient listing is required when yeast culture forms a component of a proprietary mixed feed.

96.9 Molasses Yeast Condensed Solubles is obtained by condensing to a syrup consistency the broth remaining after the removal of baker's yeast cells propagated on molasses. (Proposed 1973, Amended 1974.) NRC 5-14-009

Tentative

96.4 Brewers Dried Yeast is the dried, non-fermentative, non-extracted yeast of the botanical classification Saccharomyces resulting as a by-product from the brewing of beer and ale. It must contain not less than 35% crude protein. It must be labeled according to its crude protein content. (Adopted 1955, Amended 1975.) NRC 7-05-527

96.10 Brewers Liquid Yeast is the non-fermentative, non-extracted yeast of the botanical classification Saccharomyces resulting as a by-product from the brewing of beer and ale. It must contain not less than 35% crude protein on a dry weight basis. The guaranteed analysis shall include the maximum moisture. (Proposed 1976.) NRC 7-20-878

TABLE 25.4. COMPOSITION OF TORULA YEAST

Crude protein	48.0
arginine	2.6
isoleucine	—
lysine	3.8
methionine	0.8
cystine	0.6
theonine	2.6
tryptophan	0.1
Fat (ether extract)	5.1
linoleic acid	0.25
Crude fiber	2.0
Metabolizable energy (Cal./lb)	845.0
Productive energy (Cal./lb)	537.0
Calcium	0.55
Phosphorus	1.6
Ash	5.8
Riboflavin (mg/lb)	20.0
Pantothenic acid (mg/lb)	38.0
Niacin (mg/lb)	22.5
Choline (mg/lb)	1.3

Source: Patrick (1965).

26

Green Plant Producers

Feed manufacturers of today associate dried or dehydrated green plants with the term, alfalfa products. In the near future, a significant amount of dried or dehydrated green plants will include dry tropical plants and vegetable wastes. Already, some nations are making elaborate studies of leaves and stems from tropical plants as sources of protein for human food production. Multiple crop production and solar energy are the 2 major factors making this a strong possible food source.

Dehydrated green plants contain high-quality protein, rich in Provitamin A and Xanthophyll, and many water-soluble vitamins and minerals. Fifteen to eighteen percent of protein products are palatable to poultry.

ALFALFA MEALS

Alfalfa is known as a valuable feedstuff. It was grown in Iran in 500 B.C.; the Greeks and Romans recognized its value as a feed and soil improver, and were familiar with its perennial nature and with methods of its cultivation. "Alfalfa," in Arabic, means "best fodder." "Lucerne," its European name, originated from the fact that it was cultivated for many years around Lake Lucerne in Switzerland. Gold seekers, en route to California by way of Cape Horn and South America, brought alfalfa with them. A Chilean variety was introduced in California in 1850, and a German variety was introduced in Minnesota in 1857. In the United States, most alfalfa is grown in the West and Midwest. Originally, a crop was harvested primarily as a sun-cured hay. More recently, the artificial drying of alfalfa "dehy" has developed into an important commercial industry. It produces a uniform product with high nutritional quality. The first dehydrator west of the Mississippi was installed in 1931. Most "dehy" is used in chicken rations, but some is finding its way into turkey mashes.

Alfalfa breeding has been slanted toward greater yields, obtaining resistance to diseases, and selection of plants that have ability to produce more hay or seeds than others. Recently, there has been experimental work on propagating from individual clones. Buffalo alfalfa is purple-flowered, erect in its habits of growth, and quick to recover growth in the spring and after cutting. It is more resistant to leaf and stem diseases than other varieties. Most of the nutrients in alfalfa are in the leaves. Therefore, any variety containing a high percentage of leaves is higher in nutritive elements than varieties that drop their leaves before being harvested. Buffalo alfalfa is grown successfully across the United States between the thirty-first and forty-third parallels; in the Southwest, it competes with high-yielding, nonwilt-resistant strains.

Nutrients in Alfalfa

Alfalfa is valued as nature's outstanding source of vitamins, minerals and quality protein. Dehy (meal or pellets) is practically dust-free and available throughout the year at prices highly competitive with those of other ingredients. It offers values as an ingredient in breeder rations which require moderate- or low-energy feeds well-fortified with protein, minerals and vitamins. For turkey-breeder rations, it contributes important amounts of Vitamins E and K. As little as 2% dehy provides sufficient Vitamin K for the chick. The pigmentation of broilers and egg yolks is an important economic factor in poultry production and dehy is an important source. The use of grains with low xanthophyll content and the lowering of dehy in the diet coupled with the emphasis on high-energy diets, have caused alarm because of the resulting lack of pigment in broilers.

Carotene Stability Problem

The problem of carotene destruction during storage is being attacked in the United States by refrigeration and inert gas storage which are helpful but are not used, unfortunately, by small companies because of high cost. The use of antioxidants is most promising at the present time. An extremely important constituent of dehydrated alfalfa from the standpoint of the feed manufacturer is protein. Alfalfa meal has hitherto been limited to 15, 17 or 20% protein grades. As stated earlier, most of the nutrients of alfalfa are in the leaves, and the recent practice of separating leaf fractions from the stem may make commonplace an alfalfa meal at 25% protein or higher. Dehy, in fact, is sold on the basis of its protein content. Common meal is guaranteed to contain 17% protein. Protein content of unblended meal depends on the stage of maturity at which the alfalfa is dehydrated, as well as on weather and insect injuries. The amino acid and vitamin content of alfalfa are shown in Tables 26.1 and 26.2. It is seen that alfalfa meal contributes appreciable amounts of all the essential amino acids except methionine.

TABLE 26.1 AMINO ACID CONTENT OF ALFALFA MEAL

Protein Content	15%	17%	20%	22%
Amino Acids	%	%	%	%
Arginine	0.62	0.73	0.86	1.00
Lysine	0.63	0.76	0.89	1.04
Histidine	0.32	0.36	0.40	0.48
Methionine	0.21	0.24	0.31	0.35
Cystine	0.20	0.20	0.20	0.20
Methionine and cystine	0.41	0.44	0.51	0.55
Tryptophan	0.39	0.44	0.46	0.55
Glycine	0.71	0.86	1.00	1.13
Phenylalanine	0.78	0.83	1.06	1.16
Tyrosine	0.42	0.50	0.68	0.75
Phenylalanine and tyrosine	1.20	1.33	1.74	1.91
Leucine	1.07	1.27	1.50	1.70
Isoleucine	0.60	0.72	0.82	0.94
Threonine	0.63	0.76	0.87	0.97
Valine	0.74	0.92	1.01	1.16
Aspartic acid	1.72	1.87	2.09	2.30
Glutamic acid	1.48	1.72	2.06	2.28
Proline	0.80	0.89	1.02	1.12
Serine	0.66	0.74	0.85	0.90
Alanine	0.80	0.91	1.13	1.27

Source: American Dehydrators Assoc.

TABLE 26.2. VITAMIN CONTENT OF ALFALFA MEAL

Protein % Grade	15%	17%	20%	22%
Vitamins	mg per lb	mg per lb	mg per lb	mg per lb
Beta carotene	46.4	72.95	98.2	114.59
Xanthophylls	79.7	116.9	149.0	182.0
Thiamine	1.37	1.5	1.76	1.91
Riboflavin	4.81	5.59	7.04	7.9
Pantothenic acid	9.49	13.61	14.9	14.99
Niacin	19.02	20.8	24.83	26.68
Pyridoxine	2.93	2.86	3.6	3.52
Choline	703.7	689.2	734.6	841.3
Betaine	2110.0	2180.0	2540.0	2465.0
Folic acid	0.7	0.95	1.21	1.36
Vitamin K	4.48	3.95	6.66	3.86
Alpha tocopherol	44.5	58.1	66.73	68.55

Source: American Dehydrators Assoc.

The alpha-tocopherol content of dehydrated alfalfa meal varies widely. In 1 series, the content ranged from 2.8 to 14.1 mg per 100 g due to the stage of growth at cutting, elapsed time between cutting and dehydration, storage time and conditions, and many other factors. It is impractical to quantitate the contribution of biologically-available Vitamin E activity from the various tocopherol values in a feedstuff or ration. There might be incomplete assimilation of certain tocopherols from the ration. There may also be a 3- to 4-fold difference between the high and low values of standard ingredients.

The amount of carotene present at the time of dehydration varies considerably with the season and stage of maturity at which the crop is harvested. If harvested in early bloom, it contains 150,000 to 250,000 IU of Vitamin A potency per lb. This decreases rapidly if harvest is delayed until the crop is past early bloom. It is customary for the seller to guarantee carotene per lb. Other pigments in alfalfa are chlorophyll and xanthophyll. Chlorophyll is present in excess of 1362 mg, and xanthophyll, 192 mg per lb, and most of the vitamins are found in alfalfa meal.

Growing of Alfalfa

Alfalfa is a multiple crop. The important factors in maintaining a stand are to keep plants healthy and vigorous, and to allow the plant time between cuttings to accumulate food in the roots, permitting more vigorous growth. The critical period is in the fall, and it is necessary to plan the late cuttings in order to allow the fall top growth to supply the roots with plant foods for winter protection. The best management practice is to make the second cuttings in the early stages of growth and make the third and fourth after sufficient top growth is developed. In this way, the plant can build up food reserve in the roots for winter protection after the last cutting.

Carotene content in alfalfa is controlled by factors which rest on complex interactions between environment and heritable factors within the plant. Young alfalfa is higher in Vitamin A activity than alfalfa in the bloom stage. Potency is continually being built up until the bloom period of the plant. Commercial practices are to cut alfalfa in the early stages of growth to assure a high-quality product because of the high percentage of leaves. Recommended practice is to cut in the tenth bloom stage.

Production of Dehydrated Alfalfa

Dehydrated alfalfa is of much higher quality than sun-cured alfalfa. Alfalfa is dried rapidly by artificial means at a temperature above 212°F. Dehydrators are high-temperature units of rotating drum-type construction, usually 10 to 12 ft in diameter and 20 to 30 ft long. Gas is the usual fuel, but oil and coal have been used. In order to produce a good quality dehy, it is necessary to have a good quality raw material.

Alfalfa can be damaged by poor handling and dehydration. It is dried as soon after cutting as possible, since destruction of nutrients, particularly carotene, begins as soon as the alfalfa is cut. The length of time between cutting it and feeding it into the dehydrator is less than 1 hr. A pelleting machine, blending equipment, automatic weighing and sacking equipment, loading devices, coolers for cooling meal before sacking, spray equipment for adding oil to the dehydration and numerous other items are used in processing.

FIG. 26.1. A TYPICAL ALFALFA DEHYDRATING PLANT IN OPERATION

Pelleting makes it possible to handle dehy with bulk equipment and reduce labor costs. Dehy has been separated into fibrous stemmy parts and a finer leaf fraction richer in both protein and carotene. The smaller bulk of the leaf fraction permits it to be stored under refrigeration or in an inert atmosphere at less cost than the total meal; 80% of the carotene can be concentrated into a fraction containing 65% of the weight. A high-quality meal containing 200,000 IU per lb is separated into 2 fractions: a large fraction which has a carotene content of 246,000 IU per lb and contains 65% of the meal, and a smaller fraction with 114,000 IU carotene per lb containing 35% of the meal. The latter fraction is still a good meal and meets the maximum specifications for a guaranteed carotene content of 100,000 IU.

Although artificial drying has been practiced for many years, the commercial application of the principle of pneumatic drying dates back only to 1930. Fundamental factors that govern rate of moisture removal are: moisture distribution, temperature differential, agitation and particle size. Material having the shape of flakes is ideal to dry to low-moisture values in a pneumatic-type dryer because of the short distance for moisture migration. Dehydrated alfalfa is stored before being marketed, since production is limited to about 6 months of the year. High-quality meals are produced early in the season and used for blending operations later. Peak use occurs during the winter.

Some alfalfa meal is pelleted after being passed through the hammer mill. The most common size is $\frac{3}{16}$ in. in diameter. There is no significant dust problem, as pellets can be handled with grain weighing and loading equipment for bulk shipment. Pelleting requires careful control of moisture if the pellets are to be satisfactory. Pelleting has certain advantages over other methods: it increases xanthophyll availability and metabolizable energy, the dust problem is diminished, less volume is required for storage, and bulk loading is quite satisfactory. Sometimes a binder, such as bentonite, is used in quantities of less than 2%. Alfalfa is dried to a moisture content of 6 to 8% on a dry weight basis. Dehy picks up as much as 2 to 4% moisture from the atmosphere. The dust hazard in storage is eliminated, in part, by the use of a small amount of oil added to the meal. Oil has no stabilizing effect on the carotene in alfalfa.

Freshly-dehydrated alfalfa has a greater content of carotene than alfalfa which has been sun-dried. Properly dehydrated alfalfa contains 95% of the carotene present in the fresh alfalfa; sun-cured alfalfa contains 20 to 50% of the original carotene. Most meals lose Vitamin A activity during storage because of slow oxidation (Table 26.3). Dehydrating companies use an antioxidant, pellet the product and store it under gas. Since carotene destruction is an oxidative process, carotene loss is reduced if the oxygen surrounding the particles is replaced by a nonoxidizing atmosphere. The process consists essentially of placing the meal in large steel bins and flushing out the air with gas produced by the incomplete burning of natural gas or by nitrogen.

Destruction of carotene is very rapid after freshly-cut plants begin to wilt, so alfalfa is dehydrated as soon as possible. The loss of carotene during wilting and drying is caused by light and by enzymatic oxidation. Dehy loses 50 to 60% carotene in 3 to 6 months under ordinary storage conditions due to light, enzymes, heat and oxygen.

TABLE 26.3. LOSS OF CAROTENE FROM ALFALFA MEAL AT VARIOUS TEMPERATURES

Storage Temperature °F	Length of Time, Days	% Loss of Carotene
−9 to −15°	180	10
14 to −5°	180	14
34 to 43°	180	50
98°	16	38
140°	16	66
176°	16	98

Alfalfa meal also contains a considerable quantity (Table 26.4) of minerals essential to good nutrition. Alfalfa is usually fed at levels of 2 to 5% of the poultry rations. Higher levels of dehy in mixed feeds decrease weight gains slightly, which is not due entirely to fiber content. Alfalfa contains saponin, which has physiological effects on poultry and limits

TABLE 26.4. MINERAL CONTENT OF ALFALFA MEAL

Protein % Grade	15%	17%	20%	22%
Macrominerals (%)				
calcium	1.23	1.30	1.47	1.47
phosphorus	0.22	0.23	0.27	0.28
potassium	2.30	2.40	2.50	2.50
sodium	0.07	0.08	0.08	0.10
chloride	0.45	0.47	0.57	0.53
magnesium	0.26	0.26	0.32	0.32
sulfur	0.17	0.22	0.43	0.30
Microminerals (PPM)				
manganese	27.0	28.0	34.0	39.0
iodine	0.12	0.15	0.14	0.20
iron	230.0	309.0	281.0	306.0
copper	8.6	8.2	8.6	9.0
zinc	19.0	17.0	19.0	21.0
aluminum	226.0	314.0	267.0	295.0
barium	42.0	37.0	32.0	38.0
strontium	75.0	71.0	90.0	95.0
boron	38.0	44.0	47.0	49.0
chromium	2.0	2.3	3.0	3.0
selenium	0.50	0.60	0.50	0.54

Source: American Dehydrators Assoc.

the amount that can be placed in the ration. Toxic substances in alfalfa can be eliminated by exhaustive extraction of the meal with hot water. The extracted meal has little growth-depressing activity while the water extract produces marked inhibition of growth. The inhibitor has hemolytic properties; possibly, it is a saponin.

Dehy is also used in turkey rations. A coarsely-ground material is preferred to the finely-ground meal used for chickens. Up to 20% alfalfa is desired for finishing rations from 16 to 28 wk and for breeding operations.

Broiler raisers are vitally concerned with pigmentation of the birds which is due to the xanthophyll content of the feed. Twenty percent of protein dehy contains nearly 80% more productive energy than the 17% grade. Dehy is of higher quality now than ever before, and provides one of the most economical sources of nutrients known (Tables 26.5 and 26.6).

OTHER DEHYDRATED GREENS

Important cultivated grasses in the United States are timothy, Kentucky blue grass, Red Top, Bermuda grass, orchard grass, carpet grass, Canada blue grass, tall meadow grass, meadow fescue, foxtail millet, Rhodes grass, Napier grass, sheep fescue, red fescue, Para grass, Japanese millet, Italian rye grass, perennial rye grass, colonial bent, Dallas grass and Sudan grass. A good yield of high-protein, low-fiber, high-

TABLE 26.5. DEHYDRATED ALFALFA MEALS COMPARATIVE NUTRIENT VALUES

	17% Protein	20% Protein	25% Protein
Proximate analyses			
crude protein, %, minimum	17.0	20.0	25.0
fat (ether extract), %, minimum	3.0	4.0	5.0
crude fiber, %, maximum	25.0	22.0	17.0
ash, %, maximum	10.0	10.0	10.0
moisture, %	7.0	6.9	7.0
NFE, %	38.0	37.1	36.0
Energy (poultry) metabolizable, Cal. per lb	620.0	720.0	800.0

TABLE 26.6. THE AVERAGE XANTHOPHYLL CONTENT OF FEED INGREDIENTS

Ingredient	mg Xanthophyll per lb
Flower petal metal (Florafil)	4500
Dried algae meal	1000
Dried stinging nettles	370
Broccoli leaf meal	303
Dried koa haole meal	300
Dehydrated clover meal	223
African red peppers, powdered	200
Dehydrated alfalfa leaf meal (20% prot.)	180
Dehydrated alfalfa leaf meal (17% prot.)	85
Spanish paprika	125
Ground annatto seed and hulls	120
Dried chili peppers	85
Corn gluten meal	40
Dried carrots	30
Dried sweet potatoes	24
Yellow corn	10
Barley	2
Millet	2
Wheat	1.5
Oats	1.0
Soybean meal	1.0
Milo	0.7
Dried tomato pomace	0
Cottonseed meal	0

carotene feeds can also be obtained from a grass and clover mixture which is preserved by artificial drying followed by storage in bales. Other green feeds include: Kudzu, a perennial legume that produces large yields of hay in the South; Ramie meal, the tops from one of the truly ancient fiber sources grown in the Florida Everglades; dehydrated lettuce meal from California, the dried lettuce waste composed of the outside leaves and poorly-formed heads obtained as a by-product of the table market crop; and dehydrated celery tops, the waste from the vegetable growing operation (Table 26.7).

TABLE 26.7. PROTEIN, CRUDE FIBER, FAT, CAROTENE AND RIBOFLAVIN IN VEGETABLE WASTES

	% Protein			% Fiber				Carotene (mcg per g)			Riboflavin (mcg per g)		
	Leaves	Stems	Unseparated Leaves and Stems	Leaves	Stems	Unseparated Leaves and Stems	% Fat Leaves	Leaves	Stems	Unseparated Leaves and Stems	Leaves	Stems	Unseparated Leaves and Stems
Beet	31.9	17.0	—	5.8	12.9	—	6.2	460	36.7	—	16.6	5.9	—
Broccoli	32.8	—	—	10.5	—	—	6.1	670	—	—	22.9	—	12.4
Cabbage, Savoy	22.4	—	—	8.2	—	—	4.7	295	—	—	9.9	—	—
Carrot	18.0	9.8	14.4	9.1	16.9	12.4	4.0	274	122.0	240	9.1	5.8	8.3
Cauliflower	26.6	17.1	—	9.5	17.28	—	4.1	185	28.0	—	23.2	9.2	—
Collard	27.3	14.8	19.2	6.8	9.8	8.5	5.3	251	28.8	115	15.8	—	9.4
Corn, sweet	17.1	—	—	26.6	—	—	5.5	578	—	—	5.5	—	—
Kale	27.1	15.7	21.5	7.3	10.5	8.6	6.3	266	24.8	196	16.0	8.2	12.0
Lima bean	17.6	12.3	—	9.6	27.5	—	4.1	176	—	—	10.7	—	—
Parsnip	22.9	6.0	—	8.0	17.2	—	5.0	232	4.3	—	11.9	4.3	—
Rutabaga	31.5	18.5	—	6.3	14.9	—	6.5	257	13.4	—	20.9	8.5	—
Turnip	30.4	18.0	22.5	7.8	10.3	8.3	4.4	264	54.0	176	14.7	11.6	—
Spinach	32.0	22.5	30.4	6.8	9.3	6.3	4.1	314	120.0	272	14.6	8.5	13.3

VEGETABLE WASTES

Vegetable wastes occur at the farm, packing shed and processing plant, and consist essentially of discarded wastes including: vines, leaves, tops, roots, trimmings or peelings, as well as crops that are not harvested because of unsatisfactory quality, market conditions, and labor shortages. Because of their highly-perishable nature and rapid accumulation, they must be disposed of as quickly as possible. The leaves of the plants are, by far, the richest constituents, and these can be used in animal feeds as a source of carotene, protein and minerals. From the standpoint of abundance, occurrence in large concentration, and nutritional value, the leafy waste of the following vegetables have the greatest promise: beets, broccoli, cabbage, carrots, cauliflower, kale, rutabagas, turnips, spinach and tomatoes. The leaves are about twice as nutritious as the stems (Table 26.8). Broccoli leaves are particularly high in all nutrients. Other waste products used in poultry feeds are winery wastes, asparagus wastes, pineapple by-products and cull apples.

Inasmuch as vegetable crops are frequently treated with insecticides that contain toxic materials, such as lead arsenate or copper compounds, their residues must be within safe limits in the final product. Artificial drying of immature pasture herbage is done in a 2-tray batch dryer followed by baling, to give a product which retains its original green color.

The AFCO shows only 2 official alfalfa products; however, there are several grades which are separated by protein and fiber content. Poultry rations should use alfalfa meals containing 17% or more crude protein.

3. ALFALFA PRODUCTS

Official

3.1 Suncured Alfalfa Meal, or Pellets or Ground Alfalfa Hay is the aerial portion of the alfalfa plant, reasonably free of other crop plants, weeds and mold, which has been suncured and finely ground. If it is chopped instead of finely ground, it must be designated as "Suncured Chopped Alfalfa" or "Chopped Alfalfa Hay." (Adopted prior to 1928, Amended 1937, 1965.) NRC 1-00-104, 1-00-111.

3.2 Dehydrated Alfalfa Meal or Pellets is the aerial portion of the alfalfa plant, reasonably free of other crop plants, weeds, and mold, which has been finely ground and dried by thermal means. (Adopted 1928, Amended 1965.) NRC 1-00-025.

NOTE: 1: The following guarantees are recommended for the various grades of alfalfa meal and ground alfalfa hay:

For 13% Crude Protein, Crude Fiber not more than 33%
For 15% Crude Protein, Crude Fiber not more than 30%
For 17% Crude Protein, Crude Fiber not more than 27%
For 18% Crude Protein, Crude Fiber not more than 25%
For 20% Crude Protein, Crude Fiber not more than 22%
For 22% Crude Protein, Crude Fiber not more than 20%

TABLE 26.8. ESTIMATED YIELDS OF DRIED VEGETABLE WASTES

Raw Material	Yields, % of Raw Material	
	Leaves	Leaves and Stems
Beet tops	6–8	12–15
Broccoli-leaf waste	7–8	13–16
Carrot tops	11–12	20
Kale	10	18
Lima bean leaf fraction (from vine apron)	23	28
Pea vines (from vines)	—	15–19
Turnip tops	6–8	15–17
Spinach	8	12

Source: U.S. Dep. Agric. Eastern Reg. Res. Lab. Philadelphia.

27

Minerals

The American Feed Control Officials have approved more than 120 mineral ingredients for usage in feed manufacturing. Only the more common ingredients used in poultry feeds are presented in this section, although many other mineral products can be used when available.

Several functionary minerals and ores are presented in the general ingredient section. Recommendations on the usage of functionary minerals should be followed very closely. Some functionary minerals, such as arsenic and selenium, are toxic to both man and poultry.

The biological value of all phosphates should be established using the chick biological method. Phosphates should be tested for total phosphorus, fluorine and vanadium. The phosphorus should be over 90% available to chicks.

Calcium salts should be tested for magnesium and trace elements. The magnesium content should not exceed 2% and the calcium in calcium carbonate should be not less than 38%. Calcium sulfate is a desirable source of calcium, but not over 1% should be used.

More emphasis is being placed on mineral fortification of feeds today than ever before. Some of the major factors which have brought about the increased need for mineral fortification in poultry rations are:

(1) Changes in methods of processing feedstuffs.
(2) Increased egg production and the keeping of hens for 2 laying periods.
(3) Greater use of synthetic ingredients and less use of animal by-products which were good sources of the mineral elements.
(4) Greater crop yields and higher fertilizaton.
(5) Complex interaction of the various mineral elements.
(6) Confined rearing and feeding in cages and on slatted floors.

57. MINERAL PRODUCTS

Official

57.1 Bone Ash is the ash obtained by burning bones with free access to air, and containing a minimum of 15.3% phosphorus (P). The label must show a guarantee for calcium (Ca) and phosphorus (P). (Adopted 1935, Amended 1952.) NRC 6-00-401

57.2 Bone Charcoal is obtained by charring bones in closed retorts. It must contain a minimum of 14% phosphorus (P). It must be labeled with guarantees for calcium (Ca) and Phosphorus (P). (This product is sometimes referred to as "Bone Black," however, bone charcoal must be used in all labeling.) (Adopted 1938, Amended 1952, 1963.) NRC 6-00-402

57.3 Calcite is an acceptable source of calcium carbonate. It must be true to name and contain not less than 33% calcium (Ca). (Adopted 1952.) NRC 6-01-067

57.4 Oyster Shell Flour is an acceptable source of calcium carbonate. It must be true to name and contain not less than 33% calcium (Ca). (Adopted 1952.) NRC 6-03-481

57.5 Shell Flour is an acceptable source of calcium carbonate. It must be true to name and contain not less than 33% calcium (Ca). (Adopted 1952.) NRC 6-05-688

57.6 Chalk Rock is an acceptable source of calcium carbonate. It must be true to name and contain not less than 33% calcium (Ca). (Adopted 1952.) NRC 6-01-202

57.7 Calcium Carbonate, Precipitated, is an acceptable source of calcium carbonate. It must be true to name and contain not less than 33% calcium (Ca). Precipitated calcium carbonate must be used in all labeling. (Adopted 1952.) NRC 6-01-071

57.8 Chalk, Precipitated, is an acceptable source of calcium carbonate. It must be true to name and contain not less than 33% calcium (Ca). Precipitated chalk must be used in all labeling. (Adopted 1952.) NRC 6-01-201

57.9 Limestone, Ground, is an acceptable source of calcium carbonate. It must be true to name and contain not less than 33% calcium (Ca). Ground limestone must be used in all labeling. (Adopted 1952.) NRC 6-02-632

57.10 Calcium Carbonate is a product true to name which contains a minimum of 38% calcium (Ca). (Adopted 1946, Amended 1963.) NRC 6-01-069 Reg. 582.5191

57.11 Limestone, Magnesium, or Dolomitic, is an acceptable source of magnesium and calcium carbonate. The terms are synonymous and designate a native mineral composed of mixtures of magnesium carbonate ($MgCO_3$), and calcium carbonate ($CaCO_3$). It must contain not less than 10% magnesium (Mg) and must be declared as an ingredient as magnesium limestone or dolomitic limestone. (Adopted 1946, Amended 1952, 1965.) NRC 6-02-633

57.12 Phosphate, Defluorinated, includes either calcined, fused, precipitated or reacted calcium phosphate. It must contain not more than 1 part of fluorine (F) to 100 parts of phosphorus (P). The minimum percent of calcium (Ca) and phosphorus (P) and the maximum percent of fluorine (F) must be stated on the label. The term "defluorinated" must not be used as a part of the name of any product containing more than 1 part of fluorine (F) to 100 parts of phosphorus (P). The term "defluorinated phosphate" must be used, where appropriate, in labeling ingredient listings. (Adopted 1952, Amended 1965.) NRC 6-04-780

57.13 Iodized Salt is a common salt (NaCl) containing not less than 0.007% iodine, uniformly distributed. (Adopted 1942.) NRC 6-04-151

57.18 Bone Meal, Steamed is the dried and ground product sterilized by cooking undecomposed bones with steam under pressure. Grease, gelatin, and meat fiber may or may not be removed. It must be labeled with guarantees for phosphorus (P) and calcium (Ca). Steamed bone meal must be used in all labeling. (Proposed 1957, Amended 1964.) NRC 6-00-400

57.31 Salt is an acceptable source of sodium chloride. It must be true to name and contain not less than 95% sodium chloride. (Proposed 1973, Adopted 1975.) NRC 6-04-152

Loose salt shall contain not more than a total of 1.5% additives for non-nutritive purposes which may include free-flowing agents and/or anti-caking agents and/or humectants. Salt blocks shall contain not more than 0.5% of a non-nutritive additive which is used as a binder.

57.52 Calcium Gluconate—-The calcium salt of gluconic acid generally expressed as $C_{12}H_{22}CaO_{14}$ and its hydrated forms. Minimum calcium (Ca) must be specified. (Adopted 1975.) NRC 6-01-073

57.57 Calcium Sulfate— The calcium salt of sulfuric acid generally expressed as $CsSO_4$ and its hydrated forms. Minimum calcium (Ca) and sulfur (S) must be specified. (Adopted 1975) NRC 6-01-087

57.63 Copper Carbonate—The copper salt of carbonic acid generally expressed as $CuCO_3$. Minimum copper (Cu) must be specified. (Adopted 1975) NRC 6-01-703

57.71 Dicalcium Phosphate—A calcium salt of phosphoric acid generally expressed as $CaHPO_4$ and its hydrated forms. Minimum phosphorus (P), minimum calcium (Ca) and maximum fluorine (F) must be specified. It must not contain more than 1 part of fluorine (F) to 100 parts phosphorus (P). (Adopted 1975) NRC 6-01-080

57.77 Iron Carbonate—The iron salt of carbonic acid generally expressed as $FeCO_3$. Minimum iron (Fe) must be specified. (Adopted 1975) NRC 6-01-863

57.85 Magnesium Carbonate—A magnesium salt carbonate acid generally expressed as $MgCO_3Mg(OH)_2$ and its hydrated forms. Minimum magnesium (Mg) must be specified. (Adopted 1975) NRC 6-02-754

57.87 Magnesium Oxide—The oxide of magnesium generally expressed as MgO. Minimum magnesium (Mg) must be specified. (Adopted 1975) NRC 6-02-756

57.96 Manganese Sulfate—The manganese salt of sulfuric acid generally expressed as $MnSO_4$ and its hydrated forms. Minimum manganese (Mn) must be specified. (Adopted 1975) NRC 6-03-050

57.102 Potassium Chloride—The potassium salt of hydrochloric acid generally expressed as KCI. Minimum potassium (K) must be specified. (Adopted 1975) NRC 6-03-755

57.111 Sulfur—Elemental sulfur generally expressed as sulfur (S). Minimum sulfur (S) must be specified. (Adopted 1975) NRC 6-04-705

57.113 Tricalcium Phosphate—A calcium salt of phosphoric acid generally expressed as $Ca_3(PO_4)_2$. Minimum phosphorus (P), minimum calcium (Ca) and maximum fluorine (F) must be specified. It must contain not more than 1 part fluorine (F) to 100 parts phosphorus (P). (Adopted 1975) NRC 6-01-084

57.115 Zinc Carbonate—The zinc salt of carbonic acid generally expressed as $ZnCO_3$ and its hydrated forms. Minimum zinc (Zn) must be specified. (Adopted 1975) NRC 6-05-549

57.117 Zinc Oxide—The oxide form of zinc generally expressed as ZnO. Minimum zinc (Zn) must be specified. (Adopted 1975) NRC 6-05-553

57.119 Sodium Selenite—A sodium salt of selenious acid generally expressed as Na_2SeO_3 and its hydrated forms. Minimum selenium (Se) must be specified. (Adopted 1975) NRC 6-26-103

57.120 Sodium Selenate—A sodium salt of selenic acid generally expressed as Na_2SeO_4 and its hydrated forms. Minimum selenium (Se) must be specified. (Adopted 1975) NRC 6-26-014 Reg. 573.920

All premixes shall bear adequate directions and cautions for use including this statement: "Caution. Follow label directions. The addition to feed of higher levels of this premix containing selenium is not permitted."

28

Special-purpose Products

Phyllosilicates have been used for many years in poultry feeds. Bentonites have been found to have value as energy extenders (caloric efficiency factors), as control of moisture in droppings and as pellet-mill lubricants. The bentonites readily take up water approximately 11 times their weight. This quality of taking-up water makes poultry feces firm and allows the air to move freely through the droppings, resulting in more rapid drying. The bentonites also have the ability to adsorb heavy minerals, such as cesium and barium.

The 2 most common phyllosilicates used in poultry feeds are vermiculites and bentonites. The bentonites are volcanic ash that settled in depressions and valleys, altered through years of leaching. There are 2 types of bentonite: sodium, often called Wyoming or Western, which expands in water and carries sodium as its predominant exchangeable ion; and calcium bentonite, which has negligible swelling and carries calcium as its principal exchangeable ion.

The vermiculites do not absorb water, but when heated, expand, and are sometimes used as bulk in low-energy rations.

Bentonite can be used in poultry feeds at levels of 2 to 5%. The future cost of corn will encourage the use of bentonites in most poultry rations. Its value in controlling the moisture content of droppings, along with the formation of firm droppings, will help in poultry-house management (Table 28.1).

TABLE 28.1. PERCENTAGE WATER IN THE DROPPINGS OF CAGED LAYERS AS INFLUENCED BY DIETARY SODIUM BENTONITE AND SEASON OF THE YEAR

	Percentage Water in the Droppings		
Season of Year	Basal	2.5% Bentonite	5% Bentonite
Winter	79.6	77.3	75.5
Summer	81.5	78.1	76.3

Source: Quisenberry and Bradley (1964).

The vermiculates, like bentonites, are phyllosilicates but do not readily unite with water. Instead, they expand when heated. They improve energy, protein and feed efficiency.

The vermiculates are registered by the AFCO as a form of verxite.

87. Vermiculate

87.4 Verxite (exfoliated hydrobiotite), an additive, is a magnesium-aluminum-iron silicate containing a minimum of 98% hydrobiotite conforming to:

Verxite Granules is thermally expanded and has a bulk density of 5 to 9 pounds per cubic foot. NRC 8-08-993

It is used or intended for use in poultry feed at a level not to exceed 5% of the weight of the finished feed as a non-nutritive bulking agent for restricting calorie intake in pullet replacement feeds, or as anticaking or blending agent, pelleting aid, or non-nutritive carrier for the incorporation of nutrients in poultry, swine, dog, or ruminant feeds, in an amount not to exceed that necessary to accomplish its intended effect and in no case to exceed 1.5% of the dog feed or 5% of the final feed for other animals.

Verxite Flakes has a bulk density of 20 to 30 pounds per cubic foot. NRC 8-08-994

It is used or intended for use as an anticaking or blending agent in ruminant feeds in an amount not to exceed that necessary to accomplish its intended effect and in no case to exceed 1% by weight of the final feed for ruminants.

To assure safe use of the additive, the label of any feed additive supplement, feed additive concentrate, feed additive premix, or complete feed prepared therefrom shall bear, in addition to the other information required by the act, the name of the additive (verxite granules or verxite flakes) and when the additive is present in excess of 1%, a statement of the quantity of the additive contained therein and the term "non-nutritive" is juxtaposition therewith.

Verxite Grits contains a minimum of 80% hydrobotite. It has a bulk density of from 40 to 50 pounds per cubic foot. NRC 8-09-350

It is used or intended for use as a partial roughage replacement in ruminant feeds in an amount not to exceed that necessary to accomplish its intended effect and is in no case to exceed one percent by weight of the final feed. (Reg. 573.1000) (Proposed 1961, Adopted 1964, Amended 1968 and 1969).

SPECIAL PURPOSE INGREDIENTS

Some special-purpose ingredients can be added to a mash ration to facilitate pelleting. Two ingredients used for this manufacturing step are:

(1) Lignin Sulfonate

87.2 Lignin Sulfonate is either one, or a combination of, the ammonium, calcium, mangesium, or sodium salts of the extract of spent sulfite liquor derived from the sulfite digestion of wood or of abaca *(Musa textilis)* in either a liquid form (moisture not to exceed 50 percent by weight) or dry form (moisture not to exceed 6 percent by weight). It may be used in animal feed in amounts calculated on a dry weight basis, as: (1) A pelleting aid, in the liquid or dry form, in an amount not to exceed 4 percent of the finished pellets. (2) A binding aid, in the liquid form, in the flaking of feed grains in an amount not to exceed 4 percent of the flaked grain. (3) A surfactant in molasses used in feeds, as liquid lignin sulfonate, in an amount not to exceed 11 percent of the molasses. (4) A source of metabolizable energy, in the liquid or dry form, in an amount not to exceed 4 percent of the finished feed. (Reg. 573.600) (Proposed 1963, Adopted 1964, Amended 1970, 1971 and 1973.) NRC 8-02-627

(2) Bentonite

Bentonite is a phyllosilicate which is widely used in poultry feeds. It can absorb several times its weight in water and appears to aid in pelleting. Since it is hydroscopic and becomes swollen as well as absorptive, an improvement in energy and protein efficiency occurs. Its usage increases fecal production; however, the feces produced by poultry receiving bentonite are firmer and maintain their separate identity while drying.

87.12, 87.13 The value of bentonite as a caloric extender, and improver in growth and water content of droppings is summarized in Tables 28.1, 28.2 and 28.3.

TABLE 28.2. WESTERN BENTONITE (WB) AND MONTMORILLONITE CLAY (CAL.-MIN) AS CALORIC EXTENDER IN LAYING-HEN DIETS

Clay Additive Type %		Hen Day Production %	Egg Size (g)	Feed Efficiency	Mortality %
Control		74.1	60.4	2.6	12.5
WB	2.5	79.5	60.3	2.6	5.0
WB	5.0	80.0	60.9	2.5	2.5
Cal.-min	2.5	75.6	60.7	2.7	7.5
Cal.-min	5.0	78.7	61.1	2.6	7.5

Source: Quisenberry (1967).

TABLE 28.3. EFFECTS OF BENTONITE AND ENERGY LEVEL ON BROILER CHICK PERFORMANCE

Cal./lb Ration	Bentonite %	Av. Weight at 5 Weeks (g)	Feed Efficiency
770	none	531	2.38
770	2.5	548	2.23
870	none	622	1.98
870	2.5	623	2.06
970	none	640	1.93
970	2.5	6.50	1.96

Source: Combined data of Kurnick and Reid (1960), using New Hampshire and Delaware broiler chicks.

29

Miscellaneous Products

Each year, new products which have nutritional value are developed, but production is limited to certain areas.

These products may find usage as:

(1) Vitamin carriers
(2) Energy sources
(3) Proteins
(4) Minerals
(5) High fiber
(6) Unrecognized factors
(7) Digestive-efficiency factors

Although these products have official approval by the American Feed Control Officials, they should be used with caution. The following is a list of such products, many of which have been approved for usage in feed manufacturing:

Noncereal carbohydrate feedstuffs, such as yucca, manoic and cassava, are used in some countries in poultry rations to replace grains up to 20%. A disadvantage is the necessity for some kind of processing. Shrimp meal can replace equal quantities of fishmeal and is a very valuable feed if economics are in its favor. The lobster industry yields nearly 3 tons of offal for every ton of tails. It is processed into a valuable meal containing 28% protein and is high in minerals. Silkworm pupae are also sources of animal protein for poultry feeding, replacing equal amounts of fishmeal. Groundnut (peanut meal) gives satisfactory results, especially if its deficiency in methionine, lysine and minerals is balanced by other ingredients of the ration and the product is free of aflatoxins.

COCONUT, LINSEED AND HEVEA SEED MEAL

Coconut meal can be added to poultry rations up to 15 to 20% of the total if the lysine and energy contents of the total ration are well-balanced. Solvent-extracted meals are superior to those extracted by older processes. Linseed-oil meal has a growth-retarding and laxative effect when fed to poultry, so should be kept at levels of not more than 3 to 5%. Substantial amounts of hevea seed-cake meal are produced on rubber plantations. This meal has a composition similar to linseed-oil meal, but contains variable amounts of glucoside from which prussic acid (HCN) is split. This agent can be removed by washing and drying the finely-ground cake.

KELP

A machine has been developed to harvest and dry water milfoil and other dense growth from clogged bays or lakes. Aquatic weeds provide compost and mulch for commercial gardeners which can be used in feeds. The principle of the machine is a cutting bar which can be set at varying depths. The cut plants are carried over a series of rollers to press out excess water from the weeds which have been moved to a hydraulic lift on the rear of the machine. The entire unit is a self-propelled barge loaded with machinery.

TRITICALE

Triticale, a wheat-rye hybrid with promising agronomic characteristics, has a higher protein content than that of either parent. Preliminary tests indicate that its nutritional value for feeding chicks is below that of wheat and corn when used as the only grain.

GUAR

Guar is an annual, summer, row crop best suited for cultivation in the Southwest. It is grown on acreage diverted from feed grains. It is a nitrogen-fixing legume like alfalfa, but its seeds yield a vegetable gum when processed. The gum is used in foods and for other purposes. *Plantago obata* is best adapted to the Southwest.

CRAMBA

Cramba seed is a source of erucic acid, useful as a drying agent and as an ingredient of hydraulic fluids. It is adaptable as a spring crop in the Midwest and Pacific Northwest wheat areas and as a crop in Texas.

KELP MEAL

This is the dried, ground product of Pacific Ocean kelp. It is chopped, spray-dried or run through a vacuum dryer. Little is known of its real feeding value. Its principal use is as a mineral supplement.

LINSEED-OIL MEAL

Linseed-oil meal is made from flaxseed and is not as palatable as soybean meal or corn-gluten meal.

KENAF

Kenaf is a single-stock fibrous plant currently grown in Florida as a source of paper pulp. The top 36 in. of the plant is dehydrated, ground and used for its pigment. Kenaf contains less xanthophyll than alfalfa, but if it could be harvested from less mature plants, or if only the leaves could be used, this source of xanthophyll would be satisfactory for pigmenting broilers and egg yolks. Kenaf is as effective as alfalfa when both are used at equivalent xanthophyll levels.

HEVEA SEED CAKE MEAL

Dehulled hevea seed-cake meal can be included in poultry rations up to 5%. In the Far East, mung beans replace 20 to 40% of the animal protein in balanced poultry rations. Although other legume seeds could be incorporated into poultry rations, some are toxic even in small quantities and others, like bitter lupine, are unsuitable because they impart a bitter taste to the eggs. The toxic factor or growth inhibitor can be removed by washing and drying the ground seeds, as in the Philippines, with ipil-ipil seed meal.

KASHZ

A dry fermented sheep's milk product, "Kashz," is used in poultry feeding in the Middle East as a replacement for fishmeal. It is very high in protein and fat.

POTATOES

Cull potatoes may be cooked with a little salt and fed to poultry. One gallon of cooked potatoes is equivalent to 1 qt grain. They contain 23% as much dry matter as shelled corn. The dry matter is mainly carbohydrate.

BAKERY PRODUCTS

Stale bread and other bakery products are high in carbohydrate and have a composition similar to grains. Bakery products may be used to replace a part of the grains fed to poultry and be considered as energy substitutes. They have about 75% of the value of grain.

BUCKWHEAT

Buckwheat is unpalatable because of its dark, unattractive appearance and high-fiber content. Ground buckwheat or buckwheat middlings may be used to replace 10 to 20% of other grains or their by-products in poultry mash feeds.

CHARCOAL

Charcoal has very little more value than sawdust and there is no conclusive experimental evidence for justifying its use in poultry feeds. It can absorb some nutrients and make them unavailable to birds because charcoal is not absorbed.

TONICS

Tonics are supposed to stimulate appetite and increase feed consumption. Their value has not been proven by careful experimentation.

ALGAE

Algae have been used in livestock and poultry feeds. Grown on organic wastes at low cost, conversion efficiency is high. The ultimate cost of animal protein so produced is several-fold less than that of feeding conventional crops to animals. Waste material can be human sewage as well as organic industrial waste and livestock manure. The cost of algae culture can be minimized by growing the algae in open outdoor ponds. Nutrients for the algae are provided through the aerobic decomposition by bacteria of the organic matter in the sewage. Chlorella and Scendesmus are the 2 most abundant types encountered.

Harvesting of the algae crop involves 3 steps: initial concentration, dewatering and drying. The initial step yields a slurry having a solids content equal to 1 to 2% of the net weight. Dewatering results in a paste having a solids content ranging from 8 to 15% of the net weight. After dehydrating or drying, the moisture content of the algae product is reduced to 8 to 12%. Extensive studies are required before complete evaluation can be made of algae products as poultry feedstuffs.

ACTIVATED SLUDGE

Activated sludge has been dried as an adjuvant to poultry feeds. Activated sludge is a by-product of treated sewage and it sold frequently as an organic fertilizer. It has been found to contain extensive B-complex vitamins, especially Vitamin B_{12}, many important amino acids, and perhaps some unknown factors. Meat products of animals fed sludge is satisfactory.

60. MISCELLANEOUS PRODUCTS

Official

60.1 Dried Apple Pomace is the sound, dried residue obtained by the removal of cider from apples. (Adopted 1929.) NRC 4-00-423

60.2 Dried Apple Pectin Pulp is the sound, dried residue obtained by the removal of pectin from apple products. (Adopted 1929.) NRC 4-00-425

60.5 Buckwheat Middlings is that portion of the buckwheat grain immediately under the hull after separation of the flour. It must contain no more hulls than are obtained in the usual process of buckwheat milling, and must contain not more than 10% crude fiber. (Adopted 1944.) NRC 5-00-991

60.60 Coconut Meal, Mechanical Extracted, is the ground residue which remains after removal of most of the oil from dried meat of coconuts by a mechanical extraction process. The words "mechanical extracted" are not required when listing as an ingredient in a manufactured feed. May also be called "Copra Meal." (Adopted 1955, Amended 1963, 1968.) NRC 5-01-572

60.61 Coconut Meal, Solvent Extracted, is the ground residue which remains after removal of most of the oil from dried meat of coconuts by a solvent extraction process. The words "solvent extracted" are not required when listing as an ingredient in a manufactured feed. May also be called "Copra Meal." (Adopted 1955, Amended 1963, 1968.) NRC 5-01-573

60.7 Ground Almond Hulls is obtained by drying and grinding that portion of the almond fruit which surrounds the nut. It must be reasonably free of the nut shell and other foreign material. (Adopted 1953.) NRC 1-00-358

60.8 Dried Tomato Pomace is the dried mixture of tomato skins, pulp, and crushed seeds. If the pomace contains spices used in the production of the tomato product, this must be shown in the name as "Dried Spiced Tomato Pomace." (Adopted 1953, Amended 1964.) NRC 5-05-041

60.9 Yeast Dried Grains is the properly dried residue from the mixture of cereals, malt, and malt sprouts (sometimes cottonseed meal) obtained in the manufacture of yeast or vinegar, and consists of corn or corn and rye from which most of the starch has been extracted, together with malt added during the manufacturing process to change the starch to sugar, and malt sprouts (sometimes cottonseed meal) added during the manufacturing process to aid in filtering the residue from the wort and to serve as a source of food supply may also be listed as "Vinegar Dried Grains." (Adopted prior to 1928.) NRC 5-02-158, 7-02-159

60.10 Ground Straw is the ground product remaining after separation of the seed from mature forage plants. The source of the material shall constitute a part of the name of the product; i.e., "Ground Blue Grass Straw," "Ground Alfalfa Straw." (Adopted 1948, Amended 1964.) NRC 1-04-682

60.11 Ground Grass is obtained by drying and grinding grass which has been cut before formation of the seed. If a species name is used, the product must correspond thereto. (Adopted 1949, Amended 1964.) NRC 1-02-215

60.12 Dehydrated Garbage is composed of artificially dried animal and vegetable waste collected sufficiently often that harmful decomposition has not set in, and from which have been separated crockery, glass, metal, string, and similar materials. It must be processed at a temperature sufficient to destroy all organisms capable of producing animal diseases. If part of the grease and fat is removed, it must be designated as "Degreased Dehydrated Garbage." (Adopted 1954, Amended 1963.) NRC 4-02-092

60.130 Safflower Meal, Mechanical Extracted, is the ground residue obtained after extracting the oil from whole safflower seed by a mechanical extraction process. The words "mechanical extracted" are not required when listing as an ingredient in a manufactured feed. (Adopted 1954, Amended 1964, 1968.) NRC 5-04-109

60.131 Safflower Meal, Solvent Extracted, is the ground residue obtained after extracting the oil from whole safflower seed by a solvent extraction process. The words "solvent extracted" are not required when listing as an ingredient in a manufactured feed. (Adopted 1954, Amended 1964, 1968.) NRC 5-04-110

60.14 Cereal Food Fines consists of particles of breakfast cereals obtained as a by-product of their processing. (Adopted 1957.) NRC 4-01-199

60.15 Dried Bakery Product is a mixture of bread, cookies, cake, crackers, flours, and doughs which have been mechanically separated from non-edible material, artificially dried and ground. If the product contains more than 3.5% salt, the maximum percentage of salt must be part of the name; i.e., Dried Bakery Product with ______ % Salt. (Proposed 1962, Adopted 1967.) NRC 4-00-466

60.17 Buckwheat Hulls is the product consisting primarily of the outer covering of the buckwheat obtained in the milling of buckwheat flour. (Proposed 1963, Adopted 1968.) NRC 1-0-987

60.18 Guar Meal is obtained from whole guar beans after removal of most of the endosperm. If the product is heat treated, it may be designated as "heat treated" or "toasted." (Proposed 1966, Adopted 1968.) NRC 5-05-687

60.19 Dried Kelp is dried seaweed of the families Laminariacae and Fucaeae. The maximum percentage of salt (NaCl) and the minimum percentage of potassium (K) must be declared. If the kelp is sold as a source of iodine, (I), the minimum percentage of iodine must be declared.

If the product is prepared by artificial drying, it may be called "Dehydrated Kelp." (Proposed 1966, Adopted 1968.) NRC 1-08-073

60.20 Dehydrated Silage (ensilage) Pellets are pellets made from wholesome silage (ensilage) which has been dried by thermal means and formed into pellets by compacting and forcing through die openings by a mechanical process. The product should bear a name descriptive of the type of silage (ensilage) pelleted, such as "Dehydrated Alfalfa Silage (ensilage) Pellets," etc. (Proposed 1967, Adopted 1968.) NRC 3-08-812

60.210 Sunflower Meal, Dehulled, Mechanical Extracted, is obtained by grinding the residue remaining after the extraction of most of the oil from dehulled sunflower seed by a mechanical extraction process. (Proposed 1967, Adopted 1969.) NRC 5-04-738

60.211 Sunflower Meal, Dehulled, Solvent Extracted, is obtained by grinding the residue remaining after extraction of most of the oil from dehulled sunflower seed by a mechanical extraction process. (Proposed 1967, Adopted 1969.) NRC 5-04-739

60.220 Sunflower Meal, Mechanical Extracted, is obtained by grinding the residue remaining after extraction of the oil from whole sunflower seed by a mechanical extraction process. (Proposed 1967, Adopted 1969.) NRC 5-04-737

60.221 Sunflower Meal, Solvent Extracted, is obtained by grinding the residue remaining after extraction of most of the oil from whole sunflower seed by a solvent extraction process. (Proposed 1967, Adopted 1969.) NRC 5-09-340

60.23 Sunflower Hulls consists of the outer covering of sunflower seed. (Proposed 1967, Adopted 1968.) NRC 1-04-720

60.24 Paunch Product, Dehydrated, is a product composed of the contents of the rumen of slaughtered cattle, dehydrated at temperatures over 100° centigrade to a moisture content of 12% or less, such dehydration designed to destroy any pathogenic bacteria. It shall be dehydrated promptly after removal from the rumen to prevent decomposition. (Proposed 1969, Adopted 1970.) NRC 4-09-327

60.25 Rapeseed Meal, Mechanical Extracted. Rapeseed meal, mechanical extracted, obtained by grinding the cake which remains after removal of most of the oil by mechanical extraction of the seed from the rapeseed plant (Brassica). It must contain a minimum of 32% protein and a maximum of 21% crude fiber. The words mechanical extracted are not required when listing as an ingredient in a manufactured feed. (Proposed 1970, Adopted 1971.) NRC 5-03-870

60.26 Bagasse is that portion of the stalk of sugar cane, after removal of leaves and tops, remaining after extraction of the juice. (Proposed 1971.) NRC 1-04-686

60.27 Coastal Bermuda Grass Hay is the dried aerial portion of the perennial hybrid grass, Coastal Bermuda (Cynodon dactylon) (L.) (Pers.), reasonably free of other crop plants, weeds and mold, which has been cultivated as a crop and harvested during a period of active growth. If it is fully ground, it must be designated as "Dehydrated Coastal Bermudagrass Hay" or "Dehydrated Coastal Bermudagrass Meal." (Proposed 1971.) NRC 1-09-825

60.28 Dried Potato Products is the dried residue of potato pieces, peeling, culls, etc., obtained from the manufacture of processed potato products for human consumption. The residue may contain up to 3% hydrate of lime which may be added to aid in processing. (Proposed 1972.) NRC 4-03-775

60.29 Gelatin By-Products is the dried residue from the various process streams from the manufacture of edible gelatin. The total crude protein content will contain a minimum of 85% digestible protein as determined by the AOAC pepsin method 22.025-22.031. A 25% maximum of diatomaceous earth will not be exceeded. This product is for use in poultry feeds not to exceed 5% of the total rations. (Proposed 1972.) NRC 5-14-503

60.32 Taurine is a product which contains a minimum of 96% 2-Aminoethanosulfonic acid. It may be added to the food of starting and growing chickens so that the total taurine content of the feed does not exceed 0.054%. (Proposed 1971, Adopted 1972.) NRC 5-09-821 Reg. 573.980

60.35 Sugar Foods By-Product is the product resulting from the grinding and mixing of the inedible portions derived from the preparation and packaging of sugar based food products such as candy, dry packaged drinks, dried gelatin mixes, and similar food products which are largely sugar. It shall contain not less than 80% total sugar expressed as invert. It shall be free from foreign material harmful to animals. (Proposed 1976, Adopted 1977.) NRC 4-20-865

60.36 Beet pulp, dried, plain, is the dried residue from sugar beets which has been cleaned and freed from crowns, leaves, and sand, and which has been extracted in the process of manufacturing sugar. (Proposed 1976, Adopted 1977.) NRC 4-00-669

60.37 Beet pulp, dried, molasses, is the dried residue from sugar beets which has been cleaned and freed from crowns, leaves, and sand, and which has been extracted in the process of manufacturing sugar to which has been added (beet) molasses obtained in the extraction of sugar. (Proposed 1976, Adopted 1977.) NRC 4-00-672

60.38 Beet pulp, dried product, CSF, RNS, is the dried residue from sugar beets which has been cleaned and freed from crowns, leaves, and sand, and which has been extracted in the process of manufacturing sugar to which has been added the concentrated Steffen's filtrate obtained in the extraction of the sugar from the beets. (Proposed 1976, Adopted 1977.) NRC 4-00-675

60.39 Beet molasses, dried product, is the properly dried mixture of molasses and molasses dried beet pulp containing not less than 45% total sugar expressed as invert. (Proposed 1976, Adopted 1977.) NRC 4-20-866

30

Additives

Antibiotics are called feed additives when added in controlled amounts accompanied by a definite feeding program. Such a feeding program gives the amount, the name of the antibiotic and the withdrawal time.

The levels of antibiotics used are growth-stimulation level and medication level. The growth stimulation level is 8 to 20 g of the antibiotic per ton feed and the medication level is 100 to 400 g per ton.

For a complete understanding of antibiotic use, the American Feed Control Official publication should be consulted. The following antibiotics have official approval for use in poultry rations.

ANTIBIOTIC GUARANTEES ON THE LABEL

In terms of the amount of antibiotic present, equivalents, and standards are as follows:

Bacitracin: Bacitracin, zinc bacitracin, manganese bacitracin, bacitracin methylene disalicylate are expressed in terms of bacitracin activity which they contain. One gram of activity is equal to 1.0 gram of bacitracin master standard (42,000 units).

Chlortetracycline: The term chlortetracycline, may be used as an abbreviation for chlortetracycline hydrochloride, which is the standard.

Erythromycin: One gram of erythromycin base is equal to 1.08 gram of erythromycin thiocyanate.

Hygromycin B: On gram is equal to 1,000,000 units of hygromycin activity.

Neomycin sulfate: Label declarations should read as neomycin sulfate (providing X grams neomycin base).

Oleandomycin: One gram is equal to 1.0 gram of oleandomycin base.

Oxytetracycline: One gram is equal to 1.0 gram of oxytetracycline hydrochloride.

Penicillin: The term, penicillin, refers to sodium penicillin G (master standard). One gram of sodium penicillin G (master standard) is equivalent to 1.667 grams of procaine penicillin.

NOTE: In guarantees on a feed label, the form of the antibiotic may be shown in parenthesis. Example: Penicillin (from procaine penicillin) 6 grams per ton.

Penicillin (from procaine salt) 6 grams per ton. Bacitracin from zinc bacitracin) 10 grams per ton. The antibiotic shall be listed under the active drug ingredients on the label.
The following statements were current on August 3, 1977:

1. 2-acetylamino-5-nitrothiazole (Cyzine) for Turkeys. (a prior sanctioned drug)
Warning-Discontinue use 7 days before slaughter; use eggs from medicated birds for hatching purposes only.
2. Aklomide for Chickens.
Warning - Not to be fed to birds laying eggs for human consumption.
3. Aklomide plus Sulfanitran for Chickens.
Warning - Withdraw 5 days before slaughter. Not to be fed to laying chickens.
4. Alklomide plus Sulfanitran plus Roxarsone for Chickens.
Warning - Withdraw 5 days before slaughter; not to be fed to laying chickens; as sole source of organic arsenic; chickens should have access to drinking water at all times.
5. Aklomide plus Roxarsone for Chickens only.
Warning - Withdraw 5 days before slaughter; as sole source of organic arsenic; not to be fed to birds laying eggs for human consumption; chickens should have access to drinking water at all times.
6. Amino nitrothiazole for Turkeys.
Warning - Do not use in feed and drinking water simultaneously. Discontinue use at least 1 week before slaughtering birds for food to permit elimination of drug from the food.
7. Amprolium for Chickens and Turkeys.
For broiler chickens; for replacement chickens when immunity to coccidiosis is not desired.
Warning - Use as sole source of amprolium.
8. Amprolium and Ethopabate for Chickens and Turkeys.
Warning - Do not feed to laying chickens; Use as sole source of amprolium.
9. Arsenicals (Roxarsone, arsanilic acid, 4-nitrophenylarsonic acid, sodium arsanilate and carbarsone (Not U.S.P.)) for Poultry and Swine.
Warning - Withdraw 5 days before slaughter. Use as sole source of organic arsenic.
10. Buquinolate for Broiler, Fryer, Roaster or Replacement Chickens.
Warning - Do not administer over 75 grams per ton (0.00825%) to replacement chickens over 20 weeks of age.
11. Butynorate for Turkeys.
Warning - Withdraw 7 days before slaughter.
12. Chlortetracycline for Chickens.
500 grams per ton - Not to be fed continuously for more than 5 days.
Warning - Withdraw 24 hours before slaughter; not to be fed to laying chickens.
13. Nequinate for Roaster and Replacement Chickens.
Warning - Do not feed to chickens over 16 weeks of age.
14. Nicarbazin for Poultry. (A prior sanction drug.)
Warning - Do not feed to laying hens in production. Discontinue use at least 4 days before slaughtering birds for food to permit elimination of the drug from the food.

15. Nihydrazone for Chickens.
For broiler; for replacement chickens not over 14 weeks of age.
Warning - Poultry that have survived salmonella outbreaks should not be kept for laying house replacements or breeders unless tests show that they are not carriers. Not to be fed to laying chickens.
16. Nithiazide for Chickens and Turkeys. (A prior sanction drug.)
Warning - Do not feed to laying hens. Discontinue medication 24 hours before slaughtering birds for human consumption.
17. Nitromide plus Sulfanitran for Chickens.
Warning - Do not feed to laying hens. Withdraw 5 days before slaughter.
18. Novobiocin for Chickens and Turkeys.
Warning - Not for laying chickens, or turkeys; withdraw 4 days prior to slaughter.
19. Oleandomycin for Chickens and Turkeys.
Warning - Not to be used for laying hens.
20. Robenidine hydrochloride for broiler or frying chickens only.
Warning - Withdraw 5 days prior to slaughter. Do not feed to layers.
21. Roxarsone for Chickens and Turkeys.
Warning - Withdraw 5 days before slaughter. Use as sole source of organic arsenic.
22. Sulfadimethoxine plus Ormetoprim for Chickens and Turkeys.
A. For Broiler Chickens.
Warning - Withdraw 2 days before slaughter.
B. For Replacement Chickens.
Warning - Withdraw 2 days before slaughter; Do not feed to chickens over 16 weeks (112 days) of age.
C. For Turkeys.
Warning - Withdraw 2 days before slaughter; Do not feed to turkeys producing eggs for food.
23. Tylosin for Broiler Chickens.
Warning - Withdraw 5 days before slaughter.
24. Zoalene for Chickens and Turkeys grown for meat purposes only.
Warning - Not to be fed to laying hens.
25. Clopidol for Chickens.
For broiler chickens and/or replacement chickens up to 16 weeks of age, intended for use as caged layers.
Warning - For broiler chickens feed continuously as the sole ration; withdraw 5 days before slaughter if given at the level of 0.025 percent in feed or reduce level to 0.0125 percent 5 days before slaughter. For replacement chickens feed continuously as the sole ration; do not feed to chickens over 16 weeks of age; withdraw 5 days before slaughter.
26. Clopidol for Turkeys.
For turkeys grown for meat purposes only.
Warning - Withdraw 5 days before slaughter.
27. Coumaphos for Chickens.
Warning - Do not feed to chickens under 8 weeks of age nor within 10 days of vaccination or other conditions of stress; use as sole medication. Coumaphos is a cholinesterase inhibitor. Do not use this product in animals simultaneously or within a few days before or after treatment with exposure to cholinesterase

inhibiting drugs, pesticides, or chemicals; treatment of colored breeds of commercial layers should be avoided while in production since these breeds appear to be more sensitive to coumaphos than white breeds.

28. Decoquinate for Broiler Chickens.
Warning - Do not feed to laying chickens.

29. Dimetridazole for Turkeys.
Warning - Withdraw 5 days before slaughter. Do not feed to birds producing eggs for human consumption. Use as sole source of dimetridazole.

30. Erythromycin Thiocyanate for Chickens and Turkeys.
Warning - Chickens:
At 100 grams (g.) per ton - withdraw 24 hours before slaughter.
At 200 grams (g.) per ton - withdraw 48 hours before slaughter; do not feed to birds producing eggs for food purposes.
Turkeys:
At 10-20 grams (g.) per ton - do not feed to turkeys over 12 weeks of age.
At 200 grams (g.) per ton - not to be fed to turkeys producing eggs for food purposes.

31. Furazolidone for Poultry.
Warning - Withdraw 5 days before slaughter. Do not feed to laying birds. Do not feed to replacement birds over 14 weeks of age.

32. Ipronidazole for Turkeys.
Warning - Withdraw 4 days before slaughter. Do not feed to turkeys producing eggs for food.

33. Lasalocid Sodium - Withdraw 5 days before slaughter. For broiler and fryer chickens only.

34. Monesin (as monesin or the sodium salt) for Broiler Chickens.
Warning - Do not feed to laying chickens; withdraw 72 hours before slaughter. Finished chicken feed containing monesin as the mycelial cake shall bear an expiration date of 90 days after its date of manufacture.

31

Collective Terms

The American Feed Control Officials have developed 6 collective terms to provide a common simplified method for 1. Animal protein products, 2. Forage products, 3. Grain products, 4. Plant protein products, 5. Processed grain by-products and 6. Roughage products. For a complete identification of each ingredient referred to in each collective term, the official publication of the Association of American Feed Control Officials should be consulted.

COLLECTIVE TERMS

During the formulation of rations, substitutes are often used. Rather than a complete identity of the new ingredients, collective terms are used. An official term has been prepared for 5 collective terms recognizing the general classification of ingredient origin. Such terms give great flexibility in using by-products or some special buys.

22. COLLECTIVE TERMS

Official

Collective terms recognize a general classification of ingredient origin, which perform a similar function but do not imply equivalent nutritional values.

When a collective term is used, individual ingredients within that group cannot be listed on the label.

The control official shall be provided, upon request, the ingredients that are being used within each collective term by the manufacturer using collective terms on their labels.

22.1 Animal Protein Products may include one or more of the following:

- a. Animal products as defined in 9.1 to 9.17 incl., 9.40, 9.41, 9.42, 9.50, 9.51, 9.54, 9.55.
- b. Marine Products as defined in 51.14, 51.24, 51.34, 51.4, to 51.7 incl., 51.9.

c. Milk Products as defined in 54.1 to 54.21 incl.

22.2 Forage Products may include one or more of the following:

a. Alfalfa Meals defined in 3.1 to 3.3 incl., 3.5.
b. Entire Plant Meals defined in 39.3, 45.1, 48.20, 48.27, 60.11, 60.20, 72.3.
c. Hays defined in 3.1, 84.2.
d. Stem Meals defined in 3.4, 45.2, 72.4.

22.3 Grain Products may include one or more of the following:

a. Barley defined in 12.2.
b. Grain Sorghums defined in 42.1, 42.6.
c. Maize (corn) defined in 48.3 to 48.6 incl., 48.11, 48.12, 48.15, 48.19.
d. Oats defined in 69.5 and 69.6.
e. Rice defined in 75.2, 75.4 and 75.5.
f. Rye
g. Wheat

22.4 Plant Protein Products may include one or more of the following:

a. Algae Meals defined in 87.1.
b. Coconut Meals (Copra) defined in 60.60, 60.61.
c. Cottonseed Meals defined in 24.10, 24.12, 24.2, 24.30, 24.31, 24.4, 24.50, 24.51.
d. Guar Meal defined in 60.18.
e. Linseed Meals defined in 39.10, 39.11.
f. Peanut Meals defined in 72.5.
g. Safflower Meals defined in 60.130, 60.131.
h. Soybean Meals defined in 84.1, 84.4, 84.11, 84.12, 84.60, 84.61, 84.7, 84.13.
i. Sunflower Meals defined in 60.210, 60.211, 60.220, 60.221.
j. Yeasts defined in 96.1, 96.2, and 96.4 to 96.8 incl.
k. Rapeseed defined in 60.25.
l. Beans and peas

22.5 Processed Grain By-Products may include one or more of the following:

a. Brans defined in 48.2, 72.2, 75.3, 75.7, 93.1.
b. Brewers Dried Grains defined in 15.1.
c. Distillers Grains defined in 27.5, 27.6.
d. Distillers solubles defined in 27.1, 27.2, 27.4, 27.7.
e. Flours defined in 42.8, 42.9, 48.8, 84.5, 93.2.
f. Germ Meals defined in 42.4, 48.22, 48.23, 93.3, 93.8.
g. Gluten Feeds defined in 42.2, 48.13.
h. Gluten Meals defined in 42.3, 48.14.
i. Grits defined in 42.7, 48.7, 84.5.
j. Groats defined in 69.1, 69.3.
k. Hominy Feeds defined in 42.10, 48.16, 48.26.
l. Malt Sprouts defined in 15.2.
m. Middlings defined in 60.5, 78.2, 93.5.
n. Pearled defined in 12.3.
o. Polishings defined in 75.1.
p. Shorts defined in 93.6, 93.7.
q. Wheat Mill Run defined in 93.4.

22.6 Roughage Products may include one or more of the following:

a. Cobs defined in 48.1, 48.21.
b. Hulls defined in 12.1, 12.4, 15.4, 24.6, 60.7, 60.17, 60.23, 69.2, 69.4, 69.7, 72.6, 75.8, 78.1, 84.3, 84.8, 84.9.
c. Husks
d. Pulps defined in 21.1, to 21.3 incl., 60.1 to 60.3 incl., 60.8, 60.16.
e. Straws defined in 39.4, 60.10.

32

Tentative Miscellaneous

Dried poultry waste is a combination of dried feces and urine. Most poultry waste is used as fertilizer; however, the dried freshly-collected feces from commercial laying or broiler flocks have been used in poultry feeds. The wastes collected for several months in pits, and then dried, should not be used in poultry feeds.

All collections of poultry wastes should be dried in an oven at a temperature of not less than that required to kill all microorganisms and remove volatile odors. The composition of dried broiler manure is presented in Table 32.1 and its value in broiler rations is summarized in Table 32.2.

TENTATIVE

Dried Poultry Waste

A tentative ingredient, 60.31 Dried Poultry Waste (DPW), has been given careful study in many laboratories as a possible ingredient for poultry rations. Studies using 5, 10 and 20% dried poultry wastes in chick starting rations showed that (1) growth was irregular, (2) fecal production was increased, and (3) ash content of feces was increased. Dried poultry waste has proven of value in ruminant rations and as a fertilizer. This product may be contaminated with chlorinated hydrocarbons, pathogens and toxins. Unless properly dried, it will produce odors which may become noticeable. The composition of broiler (DPW) manure is shown in Table 32.1.

Tentative

60.31. Dried Poultry Waste (D.P.W.) is a product composed of freshly-collected feces from commercial laying or broiler flocks not receiving medicants. It shall be thermally dehydrated to a moisture content of not more than 15%. It shall not contain any substances at harmful levels. It shall be labeled to show the minimum percent protein, minimum percent fat and percent fiber. (Proposed 1973.) NRC 5-14-015.

TABLE 32.1. COMPOSITION OF DRIED BROILER MANURE[1]

Nitrogen	%	5.50
Phosphorus	%	3.00
Potassium	%	3.10
Calcium	%	2.88
Sodium	%	0.35
Magnesium	ppm	4638.00
Copper	ppm	19.00
Iron	ppm	0.598
Manganese	ppm	168.00
Zinc	ppm	49.10

Source: Williams (1973).
[1]Samples had average of 72.5% moisture.

TABLE 32.2. INFLUENCE OF FEEDING DRIED POULTRY WASTES (DPW) TO BROILERS ON PRODUCTION AND COMPOSITION OF FECES[1]

Composition	Basal Ration	10% DPW	20% DPW
Manure produced (%)	28.0	33.5	40.0
Calcium (%)	2.6	3.0	3.1
Phosphorus (%)	2.1	2.4	2.6
Sodium (%)	0.34	0.42	0.42
Potassium (%)	3.1	3.4	3.6
Magnesium (ppm)	4638.0	4412.0	4575.0
Zinc (ppm)	490.0	535.0	572.0
Manganese (ppm)	168.0	198.0	219.0
Iron (ppm)	600.0	700.0	744.0
Copper (ppm)	19.0	20.0	21.0
Molybdenum (ppm)	2.9	3.1	3.5

Source: Williams (1973).
[1]DPW used in this study was produced by broiler chicks fed the basal ration. The feces were collected every 48 hr or sooner and dried in a forced-air oven for 72 hr at 120°F.

33

Toxic Contaminants

Harmful agents may be present in a feedstuff or gain entry at any step of processing and handling from the source to the final destination. Seeds of poisonous plants can have deleterious effects, such as lowered feed value, even though present in less than toxic amounts. There are large numbers of such toxic plants and some may be present in small grains (mustard in oats, cockle in wheat, ergot in rye). Cocklebur and nightshade are often present in soybeans. Some lots of corn, shelled in the fields, are loaded with jimson weed, cocklebur, nightshade and milkweed seeds. Oats and barley are sometimes contaminated with wild flax and other noxious seeds. Sometimes good feeds are stored in conditions that promote rapid deterioration due to high moisture, high temperature, insect infestation and storage near active poisonous chemicals. Serious contamination is that of rodent contamination of stored grains which may contain such pathogenic organisms as *Salmonella* or *Leptospira*.

Disinfectants, fungicides, insecticides, and certain agents deliberately added to certain seeds, may cause disease or death. Surplus seeds treated with insecticides and fungicides have, on occasion, accidentally slipped into newly-harvested wheat and gained entrance into poultry rations with disastrous results. Chickens sickened, died or showed no effects at all, but stored the poison in their tissues or secreted it in their eggs, causing residues in food products destined for human use. How well poultry tolerate an excess intake of a given toxicant is complicated by many factors, i.e., the amount of the element consumed, the period of time over which it is consumed, the interrelationships with other substances competing in absorption and excretion, the rate of absorption and excretion, its accumulation in vital organs, the nutritional or disease condition of the birds, the chemical form of the element and the age and kind of fowl.

A poison is a substance which, upon ingestion, inhalation, absorption, application, injection or development within the body in relatively small amounts, produces injury to the body by its chemical action. Toxicity is the quality of being poisonous. Toxicity may be mild or severe, subacute or acute, or chronic. An acute toxicity may be lethal or fatal to the animal. The most prevalent kind of toxicity in poultry, however, goes unnoticed, and generally results in poor or mediocre performance in regard to growth, egg production, egg quality and efficiency of feed utilization. Many animals develop tolerance to toxins. Tolerance varies between species, form of toxin, age of animal and diet. Needless to say, diet deficient in protein often does not give the animal as much protection as do adequate rations.

Contaminants in feed ingredients and feeds can be divided into 6 groups:

(1) Minerals
(2) Mycotoxins (aflotoxins)
(3) Chlorinated hydro-carbons
(4) Filth
(5) Weed seeds
(6) Poisonous plants

MINERALS

(1) Toxic Trace Minerals

The toxic trace minerals most commonly observed are those located close to or adjoining the required elements in the periodic table or are trace elements which influence other required elements. These are:

vanadium	molybdenum
scandium	selenium
tin	fluorine
nickel	mercury
chromium	

When present in small amounts, these often stimulate certain enzyme systems. They may also help control the body burden of some minerals which readily become toxic. Another reaction is where excess excretion of a required element occurs. An example of this is excess excretion of copper as a result of feeding excess molybdenum. Minerals of this type are generally confined to limited areas. Fluorine is generally confined to aluminum-production plants, phosphate mining areas and defluorination of raw rock phosphates.

Although trace minerals are necessary for the proper nutrition of poultry, they can be toxic to the animal if an excess is consumed. The accompanying table shows the requirement, tolerance, and toxic level of trace minerals for poultry. It is obvious that there is a wide margin of safety between nutritional requirements and the toxic levels of the trace minerals, especially for zinc, manganese, iodine and copper. Since the addition of molybdenum or selenium to poultry rations is not permitted by the Food and Drug Administration, accidental poisoning by excessive levels of these minerals is unlikely. The addition of a small amount of trace minerals is sound; excessive levels may be disastrous.

Table 33.1 summarizes the toxic levels of many of the trace minerals. The nonrequired trace minerals are often fed at a near-toxic level to produce a biological response.

Selenium is found in the Dakotas and the eastern slope areas of the Rocky Mountains. Lead can become a contaminant in lead-mining and smeltering areas. Excess vanadium is often found in fish meals. Mercury has been reported as a contaminant of fish; however, analysis of dried fish obtained from old burial grounds and fresh fish suggests that no change in mercury content has developed.

(2) Fissionable Materials

Other kinds of mineral contaminants which could occur are those resulting from an atomic or nuclear explosion. The most common of these would be:

(a) Uranium
(b) Plutonium

These 2 elements are used in atomic reactions and could become contaminants as a result of spill, fallout or leakage. They are very toxic, as most heavy elements are found to be, and because of their molecular size, are poorly absorbed, and if absorbed, are very slowly excreted.

(3) Fission Products

Strontium 89-90
Cesium 137
Iodine 131

When uranium or plutonium fissions, many elements are formed which are radioactive, as well as toxic. The above radioactive elements of strontium, cesium and iodine represent a high percentage of the fission products. These elements could become contaminants of water, pastures, grains and air. Iodine 131 is the most difficult to remove since it is pooled and also functions in a closed cycle. However, due to its short half-life of approximately 8.2 days, the hazard soon disappears. The strontium

TABLE 33.1. REQUIREMENT, TOLERANCE LEVEL AND TOXICITY LEVEL OF TRACE MINERALS FOR POULTRY

	Requirement ppm		Tolerance Level ppm		Toxic Level ppm		Safety Index—Toxic Level-requirement Level	
Mineral Element	Growing Chicken	Hen	Growing Chicken	Hen	Growing Chicken	Hen	Growing Chicken	Hen
Cobalt	>5	?	5–10	?	>10	?	—	—
Copper	4	4	300	?	500	?	125	?
Iodine	0.35	0.30	180	50	180	50	514	167
Iron	75–80	40	?	?	?	?	?	?
Magnesium	500	900–1000	2000+	13,300	6400	?	12.8+	>11
Manganese	55	50	1000	1000	>1000	>1000	19–87	>20
Molybdenum	?	?	?	?	< 100	?	<50	?
Selenium	0.10	<2	<5	<8	< 10	8	50–100	?
Zinc	35	50–80	1000	1400	1500	?	43	>20
Fluorine	—	—	300–400	500–700	<700	?	—	—

<Less than.
>More than.

body load can be reduced by feeding a high level of calcium, and the cesium, by feeding a high level of potassium. Cesium can also be removed from the body by feeding diets high in crude fiber, or one of the phyllosilicates, such as bentonite or vermiculite. Feed ingredients stored in bins or closed bags probably would not become contaminated enough to pose a major threat to the health of the animal or to those consuming the animal products.

WEED SEEDS AND PLANT TOXINS

Toxicants frequently occur in weed seeds. Federal law prohibits interstate shipments of grain containing noxious weed seeds, but permits unrestricted shipment of feed grains, hay, straw, and feed mixtures containing seeds of such weeds. Noxious weed seeds come to the United States from all parts of the world. Many states have either no antiweed laws or ineffectual ones. Incorporation of screenings in formula feeds is considered particularly dangerous from the standpoint of spreading weeds. The present allowance of 2% dockage from grain values for weed content is virtually an invitation to grain handlers to see that 2% of screenings is present. Weed seed may be added to clean grain in order to ship the 2% screenings permitted by law. No feed man should be a party to any practice that is costing farmers several billion dollars per year. Methods should be established to make it impossible for weed seeds contained in feeds to germinate, and to render imports of screenings and whole grains harmless. It has long been known that seeds of many noxious weeds may be spread in the droppings of grazing animals. When seeds pass through cattle, conditions may cause an increase in promptness of germination.

Cockle

Cockle is usually found in wheat; the percentage runs from none to 2%. Corn cockle contains a toxic substance known as tithagin, a saponin-like glucoside listed as a hemolytic poison. It is destroyed by heating to 482°F. Young animals are more readily affected than older animals. One-quarter lb corn cockle per 100 lb live weight is harmful to the fowl; a toxic dose is 0.2% of the body weight. Corn cockle seed consumed to an amount of 0.005 to 0.01 g daily, poisons young chicks. Birds affected with the toxic principle present a generally listless and unkempt appearance, a pale comb, staggering gait, difficulty in breathing and marked diarrhea. Postmortem examination shows the presence of a yellow caseous lining of the crop, an amber gelatinous exudate next to the outer muscular coat of the crop, and similar material within the pericardium of the heart;

hemorrhages or congested areas in the fatty portions of the heart, a clear fluid surrounding the intestine and some congestion in the lungs and trachea. Body temperature and total blood count are not affected. When fed in the whole form, the cockle is not as toxic as when ground.

Ergot

Ergot is a fungus *(Clavicets purpurea)* which develops on rye plants and causes powerful uterine contractions and contractions of the muscles of the intestinal tract. It reduces and often interrupts blood circulation in the feet and legs. Symptoms include nervousness, sensitive skin, muscular trembling, incoordinated gait, and convulsions which may lead to death. Animals go off feed, become dull and depressed, and often their lower legs will be stiff and sore. There may be a dry gangrene of the comb, wattles and beak from poor circulation. Management measures consist primarily of using ergot-free feed.

Algal Poisoning

Algal poisoning has occurred in many parts of the world. The blue-green algae on a pond are without adverse effect, but when the algae bloom and turn the water red, they are highly toxic to animals. Animals show signs of convulsions, dyspnea, and paralysis of the extremities. Death occurs in 1 to 48 hr. The toxic principle is not well understood.

Limonin

Citrus seed meal, a by-product of citrus canning, is potentially a valuable ingredient for chicks since it has a 33% protein, 5% fat, 8% fiber and 6% mineral content. Experiments have demonstrated, however, that, as now prepared, this by-product is unsatisfactory for chicks. High mortality results during the first 3 wk, causing enlarged gall bladders and sometimes mottled liver, ascites, and congestion of the intestinal tract. The deleterious factor is a white crystalline compound, soluble in acetone and 95% ethyl alcohol, but insoluble in water and diethyl ether. Analysis showed it to be limonin, the bitter principle of the seed.

Lathyrus

Many varieties of peas and beans have been cultivated from earliest times as food for man and domestic animals. One exception has been *Lathyrus sativa*, which is harmful to farm animals. Large agricultural populations in some parts of India which used to subsist partially on

Lathyrus peas, were afflicted by paralysis of the lower limbs. In poultry, *Lathyrus* poisoning causes ascites, severe bone abnormalities and changes in the nervous system.

Gossypol

As discussed earlier, cottonseed products are a source of 2 toxic materials. Gossypol causes depressed growth rate in chicks, poor hatchability of eggs and discoloration of egg yolks. Cottonseeds also contain cyclic fatty acids which are capable of producing marked discoloration of the albumen of the eggs. Inasmuch as cottonseed oil is sometimes used in poultry diets as a source of energy, the possibility exists that this might become a field problem. Soybean meal, the very foundation of modern poultry feeds, cannot be successfully used without heat treatment to destroy an inhibitor. The toxic material contained in raw soybeans is of interest today because of attempts to use raw whole soybeans in feeds.

Crotalaria

A serious problem developed in certain flocks in South Carolina in 1959 due to *Crotalaria* toxicity and a change in the method of harvesting corn. *Crotalaria spectabilis* had been widely planted on the ill-advised recommendation that it be used as a hardy cover crop. As little as 0.2 lb *Crotalaria* seed per ton of complete feed is toxic to chicks, and growth retardation occurs at levels as low as 0.01%. The use of combines for harvesting corn brought about simultaneous harvesting of *Crotalaria*. When *Crotalaria* is consumed, excess fluid accumulates in the pericardial sac and peritoneal cavity. There is a blotchy hepatitis and occasional hemorrhages in the liver, heart and musculature.

Scurvy-like Condition

Some fishmeals stored for long periods induce a condition in chicks which has been described as similar to scurvy in appearance. It is believed to be an intoxication induced by nitrogenous bases. There is a profound increase in blood-clotting time.

Rapeseed Meal

Rapeseed meal contains a goitrogen, and attempts to inactivate this factor have not been successful. Alfalfa meal contains a saponin which causes growth inhibition. Aortic rupture in turkeys can be produced by feeding certain legumes, such as sweet peas, which contain beta aminopropionitrile (BAPN). Some weed seeds contain substantial urease activity. While not necessarily toxic in itself, it can cause a rapid loss of nitrogen from feed which contains urea. Feed manufacturers should guard against toxic weed seeds.

Oxidative Rancidity

Linoleic acid content is related to the tendency of low Vitamin E-diets to produce encephalomalacia. Unsaturated fatty acids cause Vitamin E-deficiency symptoms. This effect is unrelated to the oxidative rancidity of dietary fat. Another factor is the presence of harmful quantities of nitrates in feeds. High nitrate content interferes with the conversion of carotene to Vitamin A. All plants are likely to accumulate considerable nitrate if they are grown on high-nitrate soils. The stinging nettle and the elderberry accumulate nitrogen regardless of the type of soil on which they are grown.

A word of caution about heat treatment is also in order. It is generally recognized that excessive heat treatment impairs protein quality. Much of this may be due to the browning reaction between amino acids and reducing sugars.

Some feeds contain toxic materials which have been identified chemically and investigated pharmacologically. Among these are goitrogens, namely cabbage, cauliflower; cyanogens in various beans and nuts; solanins in potatoes, and neurotoxins in various species of seafood. Two organic seed protectants are Arasan and Orthocide.

PESTICIDES

It has been estimated that from 15 to 50% of all farm products in various regions of the world are destroyed by insects and other pests before being consumed. Chemical pesticides are used to prevent this waste. The use of chemical pesticides carries some degree of risk. Against this risk, however, the need to supply enough food for a rapidly-growing world population and to use every means available to conserve and protect for human use such crops as we are capable of producing must be weighed. Many alternatives to the use of chemical pesticides have been proposed by protectionists and ecologists. Many seem to value the needs of wildlife more than those of humanity.

Insecticides make up the lion's share of the pesticides produced. About 40% of the pesticides sold is for the farm; 40% is bought by other domestic outlets, and 20% is shipped abroad.

TABLE 33.2. RANGES AND AVERAGES OF INSECTICIDE RESIDUES FOUND IN FEEDS

Insecticide	% Contaminated	Range of Contamination (ppb)	Avg. Level of Contamination (ppb)
Gamma-BHC	100.0	5–79	16.0
Aldrin	86.1	1–16	2.6
DDT	98.0	10–448	80.5
Methoxychlor	45.5	10–580	74.0
Endrin	58.4	4–27	10.6
Chlordane	67.3	5–254	30.4
Dieldrin	88.1	1–16	4.9
Toxaphene	14.8	60–530	178.3
Heptachlor	8.9	2–13	4.8
Heptachlor epoxide	3.0	1–6	3.3

Source: Minyard and Jackson (1965).

FUNGICIDES AND HERBICIDES

Since the practice of treating seed grains with chemicals for the control of diseases has become quite general, there are often considerable quantities of treated grain which for one reason or another are not used for planting purposes. Experiments at Cornell University showed that of 3 chemicals used to treat corn seed, one caused a decided drop in egg production when fed to hens as one-third of the ration. Production dropped from 70 to 10%, 2 to 3 days after feeding. Two mercury compounds had no adverse effect. In research in Texas, 25% of a broiler flock exhibited hock and leg disorders (perosis and spraddles) when fed diets containing as low as 4 to 100 ppm TMTD. Growth rate was also markedly decreased. Laying hens dropped to zero production after being fed treated grain for 2 days.

A goodly part of the tremendous production of feed and food in this country is the result of the use of pesticides of one sort or another. Many of these chemicals are poisonous, whether eaten by grasshoppers, rats or humans. Pesticides are applied to feeds, and if people ingest enough of them, they have the potential of acting on them as they do on poultry, flies or cutworms.

Fungicides represent about 10% of the market and include soil fumigants, conditioners and animal dips. The introduction in the early 1940s of 2,4-D, although aimed at eliminating only broad-leafed plants, was the first large-scale effort to control weeds by other than mechanical tillage.

Herbicides have increased in recent years from 17 to 22% of the total.

FUMIGANTS

Ethylene Dibromide

Researchers in the South complained of a decreasing egg size in 1957 at a time of the year when it would normally be increasing. Investigation proved that the source of the problem was the fumigant used on oats. The fumigant contained ethylene dibromide, one of the components which has been shown by workers in Israel to cause reduction in egg size after feeding for a period of 7 wk duration.

Volatile Iodine

Volatile iodine can be used in water treatment and to kill worms. In the free form, it is very toxic.

Chlorine

Chlorine is used in many water purification plants. The directions of the manufacturer should be followed. It reduces microbiological growth in water troughs.

POISONOUS PLANTS

Poisonous plants may contain or produce toxic substances in amounts sufficient to harm animals. A plant may contain one or more of the following substances that are themselves toxic, such as alkaloids, resinoids, phytotoxins and oxalic acid.

A second category of substances is in itself harmless but decomposes to form toxic products, either before or soon after being eaten. For example, the nontoxic glucoside, amygdalin, produced in choke cherry, changes during digestion to form highly-toxic prussic acid.

A third group of substances are those formed by the action of microorganisms on plants or plant products. Fungi, in certain conditions, produce moldy hay or silage, forming decomposition products which may be toxic.

Fourthly, toxic substances may be absorbed from the soil and stored in harmful quantities in tissues of certain plants. Selenium and potassium nitrate are two of these substances.

Fifthly, there are substances that cause photosensitization. Certain plants contain a pigment which is absorbed through the intestinal wall into the circulatory system and is not eliminated. This substance causes the light or unpigmented areas of the animal body to be sensitive to light.

FUNGI, MOLDS AND AFLATOXINS

Fungi and molds are always present in natural feed ingredients. Some of them produce toxins called aflatoxins, a collective term referring to the toxic metabolites produced by several species of fungi and molds. The basic chemical structure of an aflatoxin is a coumarin-lactone-type structure. This structure is characteristic of many naturally-occurring physiologically active compounds, as well as aflatoxins.

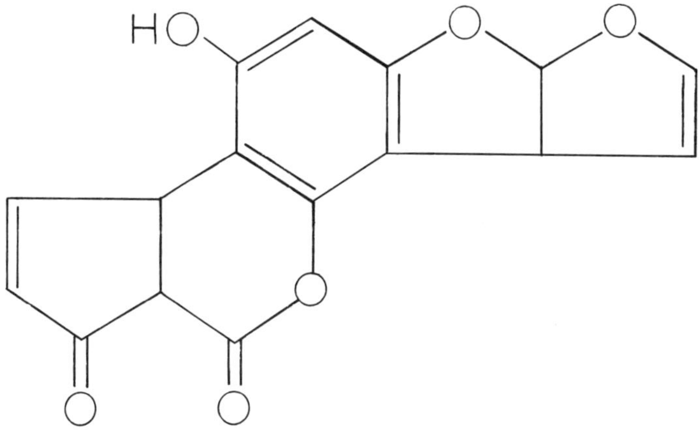

Molds are found everywhere, and have been a periodic problem for many years. In the early 1960s, this problem became prominent and received much attention. A total of over 900 research papers on aflatoxins were published from 1960 to 1967.

It should be pointed out that many molds and fungi are quite beneficial to the well-being of man. Without molds, we would not have Roquefort cheese, soya sauce, penicillin, or many of the other common antibiotics. It is interesting to note that penicillin, a mold toxin for several bacteria and other lower forms of life, is usually quite beneficial to man and most other higher forms of life. Therefore, molds cannot be universally considered as a broad group of undesirable organisms. Rather, they may be either or both beneficial and harmful to life, depending on the individual mold and its habitat. Nevertheless, the aflatoxins must be considered highly-undesirable contaminants of feed.

Distribution of Aflatoxins

Aflatoxins have been reported to occur in many agricultural commodities, including barley, Brazil nuts, cassava, cocoa beans, coconut oil cake, copra, corn, cottonseed, locust beans, palm kernels, peanuts, raisins, rice, sorghum, soybean meal and wheat. The wide occurrence of aflatoxins is not surprising since many toxic-producing molds are found universally. Environmental conditions of high interspatial relative humidity, and high, but not extreme, temperature, promote rapid growth of toxic-producing molds, such as A. flavus, and thus support the possibility of aflatoxin elaboration. Research shows that *A. flavus* is principally a post-harvest invader of cereals and peanuts but can be a pre-harvest invader of cotton seed. The prevention of aflatoxin contamination has been mainly directed at finding practical methods of preventing or minimizing mold development during growth, harvesting and subsequent handling of the commodity. To date, no practical commercial means of eliminating aflatoxins from contaminated commodities have been established. Several laboratory methods of extraction/inactivation have been extensively studied and may prove commercially feasible. Although no specific recommendations can be made regarding the prevention or control of aflatoxin production, the proper harvest, low moisture and low or stable fat seem to be factors worth consideration. Aflatoxins have been proven lethal to animals and to cell cultures when challenged in sufficient doses. They caused histological changes in cells when smaller doses were administered. They are active in many ways:

(1) They are active as a hepatocarcinogenic substance; they cause liver hepatomas in rats, ducks and trout; liver cirrhosis in humans.
(2) They suppress DNA synthesis and mitosis.
(3) They decrease plasma proteins.
(4) They lower hemoglobin (anemia).
(5) They cause chromosome breakage and translocation.
(6) They cause bile duct proliferation.
(7) They cause many pathological changes in the liver, kidney, stomach, pancreas, lungs and salivary glands.

Chronic toxicity has been observed in animals receiving meat and vegetable protein meals. Ducklings are the most sensitive, turkey poults are intermediate while chickens are relatively resistant although young broilers may develop hemorrhages. Pekin ducklings and New Hampshire chicks are the most sensitive poultry. Hogs are the most sensitive farm animals, with ruminants being highly insensitive.

Aflatoxins are generally present in high-fat-containing feedstuffs, and are rapidly absorbed from the gut and transported by the portal system to the liver, where they are partially metabolized. They may be excreted as a glucronide in the urine or a tauro-cholate conjugate in the bile. The bile is perhaps one of the principal routes of excretion, although sheep seem to excrete large amounts in urine.

Aflatoxins can be absorbed in and secreted in the milk of humans and cows to the extent that the consumption of the contaminated milk can cause cirrhosis of the liver in children and death in ducklings.

MYCOSES

Mycotic infections (mycosis) are invasions by pathogenic fungi of various tissues of a susceptible host. Mycotoxins are metabolites formed primarily by saprophytic fungi. In the first instance, the organism is the inciting agent; in the second, the preformed metabolite in the feed of an animal causes the trouble. The problem of toxic fungi appears to be associated more with management and environment than with feed, per se. The toxic compounds are polycyclic, unsaturated lactones and are stable to heat. Supplementation of diets containing moldy meals with certain amino acids tends to counteract the toxic effect. Whether or not the mycotoxins affect the availability of the amino acids remains for future researchers to determine.

Over a period of years, wet field conditions may exist during the latter part of the growing season and cause greater field infestation with fungal organisms in the developing corn plants. Field molds are of several types. One type, *Gibberella* or *Fusarium*, produces estrogenic substances. It is possible that these substances inhibit ovulation and egg production in the laying hen.

FEEDING MOLDY GRAIN

Should moldy corn be fed to poultry? As has been stated, most molds are not harmful in the quantities in which they sometimes occur in grains; hence, most moldy grains can be fed without bad side effects. However, some molds are harmful and will affect animals accordingly. The only sure way to determine whether or not moldy feed is toxic is to try it. This does not mean feeding it to the entire flock. "Trying it" means feeding it to a few birds away from the others for a 2-wk period. If the birds are normal after 2 wk, the grain can probably be fed safely. It may be wise to dilute the moldy grain with an equal amount of sound grain, when possible. The combination of a negative test on a few birds and the dilution of sound grains should make it safe to use. If the moldy grains produce an adverse effect, such as reduced growth, lower egg production or reduced feed consumption, feed the moldy feed to a non-sensitive animal. If the test birds die from eating the moldy grain, then the grain is not usable and should be destroyed because the toxin cannot be removed to make the grain safe. Sulfonamides and antibiotics have not proved effective against fungi.

Field molds which invade grain when the plant develops in the field require at least 22 to 25% moisture. Normally, these molds do not grow after harvest, except in corn stored on the cob. Poisons formed by the molds in the field are not destroyed by heating or cleaning the grain, etc.

They are gradually destroyed in stored seeds at moisture contents of 13 to 18%, and do not require free water to grow. They are mainly species of *Aspergillus* which do not invade seeds to any extent before harvest. Nearly all storage molds invade the germ of the seed. Germs of grains stored for a few years with a moisture content of 14 to 15% can be completely decayed. Such germ-damaged grain is referred to as "sick." Corn stored on the cob in piles and at high moisture content in cribs become decayed throughout the kernel. *Fusarium erium* and *Gibberella zia* complex are the principal cob-rotting molds. Corn that undergoes such spoilage is estrogenic and should not be fed to poultry.

RED STREAKING OF CORN

Red streaking of corn was observed in the Midwest in 1963, and since then, in several other states and in Canada. This condition is caused by a virus, but there is no apparent reduction in feeding value. When using a new supply of corn, it is desirable to replace the old supply gradually rather than to make an abrupt change. This applies particularly when molds have been observed prior to or after harvest and when using corn of unknown origin. The feeding value of questionable corn should be determined first on a few animals before mixing large quantities of feed. As a general recommendation, moldy corn should not be fed at more than 25% of the total feed. Corn unpalatable to younger stock, in some cases, can be used for older, growing, finishing stock.

CANDIDA ALBICANS MYCOSIS

Mycosis caused by *Candida albicans* invades and destroys the tissues of the upper digestive tract, particularly the crop and gizzard, frequently causing formation of whitish ulcers in the crop and less frequently in the mouth, upper and lower esophagus, and proventriculus. Pendulus crops are observed in mycotic infection of turkeys and chickens. Excess gas caused by fermentation contributes to distension and ultimate distortion of the crop muscles and tissues. Mycosis may follow on the heels of avitaminosis, due to decreased resistance. It could also complicate recovery from coccidiosis, and bacterial and viral disease. Outbreaks may be triggered by the prolonged use of antibacterial agents at high levels in the treatment of other diseases, such as Bluecomb in turkeys. This is due to alteration of the microbial balance in the digestive tract that allows fungi to become established in the lining of the tract. Overcrowding, poor ventilation, damp litter, hard-to-sanitize dirt floors, faulty water system, moldy feeds, and overall sloppy management and unsanitary conditions are factors which may contribute to mycosis, which strikes at any time of the year.

Mold-damaged grains, long regarded as acceptable for use in animal feeds, recently became suspect when attempts were made to explain certain unidentified field conditions. Fungi proliferate in the field, as well as in harvested grains and feedstuffs during storage. Damage from the fungi themselves is more important in the field; during storage of grains, mycotoxins are more so.

PREVIOUS OUTBREAKS OF MYCOSES

The most significant research on fungi has been conducted in the last decade, although older literature reveals many instances where moldy feedstuffs caused toxic conditions, altered nutrient value or reduced palatability.

In Georgia, during the 1940s, moldy corn caused the death of swine as a result of *Aspergillus flavus* and *Penicillium rubrum.* In the 1950s, a similar outbreak occurred in Alabama. In 1959, large numbers of turkey poults and ducklings died in England as a result of feeding contaminated peanut meal imported from Brazil. *Penicillium puberulum* was the cause of the problem. The toxin was carcinogenic when fed to rats.

The United States broiler industry has had sporadic occurrences of hemorrhagic syndrome over the years; it is suspected to be due to *Alternaris, Aspergillus clovatus, A. flavus* and *Penicillium purpurogenum.*

Aflatoxins or mold toxins produce a complicated deficiency picture resembling a deficiency of Vitamin K, Vitamin B_{12} and Vitamin D. Vitamin K is added to broiler ration as a safety factor for reducing or preventing hemorrages in broilers.

POLYCHLORINATED BIPHENYLS (PCB)

The PCBs—polychlorinated biphenyls—are chemically-resistant plasticizers which are used in plastics, resins, lubricants, varnishes, waxes and coolants. They also are, or were, used to improve *dispersal* of certain insecticides.

Levels of PCBs of 100 to 800 ppm were toxic to chickens. The first symptom is labored breathing, then excess abdominal fluids, poor growth or loss in weight, followed by mortality. As low as 50 ppm can reduce hatchability.

PCBs are very similar to many of the chlorinated pesticides. They will be found associated with fat structure in the body and their rate of turnover in the body will be influenced by the fat turnover. In general, they will have a slow turnover, and contaminated chickens should be destroyed and not used for food for any animal.

34

Quality Control

Quality control is the act of maintaining the nutritional value of a formulated ration. Since a ration can be no better than the quality of its individual ingredients, it is desirable to develop standards for maintaining and identifying the biological value of these feed ingredients.

FACTORS INFLUENCING QUALITY

Many factors influence the quality of an ingredient, and each must be recognized.

(1) Biochemical Value as Related to Genetics

Each plant or animal tissue has a biochemical structure which is controlled genetically. For example, our present corn varieties have a lysine deficiency, whereas the new opaque varieties have a lysine-adequate protein. Muscle tissue has a protein of high biological value and connective tissue protein is of low biological value.

(2) Soil or Field Conditions

Many plants tend to concentrate trace minerals in the seeds and leaves in proportion to the concentration in the soil. If the trace elements are present in large amounts in the soil, the plant will, in many instances, have a high concentration. This means that enough of the trace element can be present for plant growth, yet not be present in amounts for production of a concentrated feed. Some plants, too, will concentrate trace minerals whereas others will not, regardless of the soil composition.

Soil moisture plays a very important role in protein and mineral content. During drouths, the protein content is often lowered and the mineral content is increased. Table 34.1 shows that the composition of 2 grasses growing under different moisture conditions have different chemical composition.

TABLE 34.1. INFLUENCE OF CLIMATE ON HAY COMPOSITION

	High-moisture Soil Eastern Hay[1]	Low-moisture Soil Western Hay[1]
Crude protein %	11.70	5.87
Ether extract	1.46	1.89
Crude fiber	31.32	32.63
N.F.E.	48.76	46.55
Total ash	6.72	13.06
Sodium	0.06	0.02
Potassium	1.35	0.44
Calcium	0.92	0.32
Phosphorus	0.15	0.12
Magnesium	0.27	0.08
Silica	0.78	8.28

Source: Froman and Sauer (1962).
[1]Similar grass-seeding mixture and similar soil type and fertility.

(3) Planting Time

A crop must be planted in time to allow proper germination of seed, normal plant growth, necessary light units and maturity time. Generally, a plant with a reduced growing season yields a poor grain or seed, and the grain or seeds have high moisture content.

(4) Harvesting Methods

The moisture content of harvested grain is influenced by the stage of maturity and rainfall during the harvest period. In 1978, the moisture content of harvested soybeans near Columbia, Missouri, was as low as 9.2%, whereas in 1977, similar production was 12 to 14%. Immature corn having frost damage may have 16 to 18% moisture, whereas mature corn may have 14%. Corn stored as ear corn in open slatted cribs will have moisture content related to the local rainfall. All grains will seek a moisture equilibrium with the humidity in the atmosphere.

(5) Maintaining a Quality Control Program

A control of quality in feeds has many phases which must be maintained:

(a) On-farm Operations.—The desired variety must be selected, planted and harvested according to recognized procedures. The grain must be properly stored, or delivered directly to the feed mill.

(b) Mill-receiving Operations.—All ingredients received by the mill must be properly inspected. This responsibility should be assigned to the bulk-receiving operator. He should have standards developed which allow this operation to be routine.

(i) Moisture-testing All Grains.—Modern electronic moisture testers allow immediate moisture determination to be made.

(ii) Visual Inspection of Trucks and Hopper Cars.—The inspector should record color, odor, visual contaminants, mold masses or wet grain, and any other characteristic suggesting undesirable quality.

The following physical characteristics of bulk grain or other bulk ingredients should be recorded:

Color	Typical of ingredient, bright and uniform.
Odor	Clean, characteristic smell.
Moisture	No wet spots, caking, free-flowing. In general, if an ingredient, such as grain, contains 14% or more moisture, it will be placed in dryers.
Temperature	No evidence of heating. This can be detected by off-color, dark germs, or if wet, the hand can detect heating.
Texture	The size or shape of particles should be characteristic of the ingredient. Old, heated or moldy ingredients have a texture characteristic of the problem.
Uniformity	The particle size, color, texture and the overall appearance should be uniform.
Foreign materials	Dirt, stones, metal or sticks, mold clumps, or any material not characteristic of the standards for the ingredient must not be present.
Biological contaminants	Contaminants, such as manure, feathers, hair, urine and insects must not be present.

The inspection for maintaining the quality of bagged ingredients must be made:

Odor	A clean characteristic smell.

Bags	The bags must be clean, have no broken places, discoloration or hard surfaces. A hard and discolored surface generally indicates water damage, molds or rodent problems.
Temperature	If the sack feels hot it suggests high moisture, mold or water-damaged ingredients.
Set-up bags	When the bags or ingredients become wet, the content "sets-up" and forms a hard mass which molds and heats, and cannot be used in feeds.

Factors in Quality Control

The plant operator has several important housekeeping steps to perform so that the quality of the ingredients can be maintained.

(1) Feed Bin Housekeeping.—The feed bins should be cleaned at regular intervals. The bins storing ground ingredients should be cleaned every 30 days and those storing whole grain, every 60 days. A bin should be emptied before being cleaned.

(2) Clean, Dry Floors.—Wet or moist feeds allowed to accumulate in a mill are an excellent cultural medium for molds, bacteria and disease-producing organisms. A feed mill should have a continuous clean-up program.

(3) Trash, Scraps and Obsolete Broken Equipment.—Trash, such as damaged bags, wire, metal ties, string, boards or nails, must not be allowed to accumulate in the mill. Obsolete or broken equipment is a hazard to workers and can become a home for insects and rodents.

(4) Damaged Electrical Outlets or Lights.—All electrical outlets must have a mechanical ground, and dust-, insect- and moisture-proof covers. Broken recepticles and covers, switches and light bulbs should be immediately repaired.

(5) Windows and Fire Escapes.—All broken glass should be removed and replaced with safety glass. Also, the window frames or closures should be checked for ease of operation. When possible, entrance or accessibility should be provided. If the mill has two or more elevations (floors), a fire escape must be available.

(6) Sprinkler System.—A useable sprinkler system is necessary. It should be tested as recommended by the manufacturer.

BIOLOGICAL VALUE

Developing the Biological Value of Ingredients

Ingredient Composition—Proximate Analysis.—The Weende Experiment Station in Germany, over 100 years ago, developed a scheme for obtaining the approximate value of a feed or ingredient for feeding purposes without actually conducting a feeding trial with the feed. This approximation separated the feed or ingredient into 6 fractions. They are:

(1) Water or moisture
(2) Crude protein
(3) Crude fat or ether extract
(4) Crude fiber
(5) Nitrogen-free extract
(6) Ash or mineral matter

(1) Moisture

Determinations for moisture should be made using a well-selected sample. Several small samples (1 oz) should be taken, combined, thoroughly mixed, weighed and dried in an electric oven for an equivalent of 8 to 24 hours at 135°C. The percent moisture is obtaining by:

$$\frac{\text{Loss of weight during drying}}{\text{Weight of sample before drying}} \times 100 = \%\ \text{moisture}$$

(2) Crude Protein

The determination for crude protein is made from a small prepared sample (1 to 5 g). This determination is based on the possibility that most of the nitrogen in the sample comes from amino acids and that the average nitrogen content of the protein in the feed or ingredient is 16%.

The same is digested with concentrated sulfuric acid. The ammonical nitrogen is converted to ammonium sulfate $(NH_4)_2SO_4$. The excess acid is neutralized with sodium hydroxide and the ammonical nitrogen converted to ammonia hydroxide. The ammonia is removed by distillation and determined by titration.

The protein in the sample is obtained by multiplying the nitrogen by 6.25. The percent protein is obtained by:

$$\frac{\text{Amount of protein in sample}}{\text{Weight of sample}} \times 100 = \%\ \text{crude protein}$$

This method of nitrogen determination is commonly known as the Kjeldahl method.

(3) Crude Fat

Crude fat includes all that portion of a feed or ingredient soluble in ether. This fraction is sometimes called ether extract, or EE, and contains true fat, fat-soluble vitamins and pigments, phospholipids, waxes, etc.

This fraction is obtained by taking a dried, weighed sample and removing the fat in an ether extractor called a soxhlet extractor. The ether extract is evaporated and the residue weighed. This residue is crude fat.

$$\frac{\text{Weighed crude fat}}{\text{Weight of sample used}} \times 100 = \%\ \text{crude fat}$$

(4) Crude Fiber

Crude fiber includes cellulose, hemicellulose and lignin.

This fraction is obtained by taking a crude fat free sample and removing the protein, sugars, starches, minerals, and more soluble hemicellulosees. The residue remaining is ashed and the crude fiber obtained by subtracting the ash from the residue.

$$\frac{\text{Weight of crude fiber}}{\text{Weight of original sample}} \times 100 = \%\ \text{crude fiber}$$

(5) Mineral Matter or Ash

The ash includes the inorganic or mineral portion of the feed or ingredient.

Ash is obtained by taking a weighed dry sample and burning in a muffle furnace at 600°C for 4 to 6 hr. The ash is weighed

$$\frac{\text{Weight of ash}}{\text{Weight of original sample}} \times 100 = \%\ \text{ash or mineral matter}$$

(6) Nitrogen-free Extract (NFE)

The nitrogen-free extract or NFE includes starches, sugars and some of the more soluble hemicelluloses and lignin. The fraction is obtained by difference —that is, all the percentages in the above are added and taken from 100.

100 − (% moisture + % crude protein, % crude fat + % crude fiber + % ash) = % NFE

BIOLOGICAL TESTS

After obtaining the proximate analysis of the ingredients, some biological tests are necessary, protein tests and amino acid composition.

It is not necessary to determine the complete amino acid composition unless the protein has not been previously analyzed or the ingredient has been subjected to poor handling. When an ingredient has been inproperly handled, the complete amino acid composition should be obtained, as well as another test, such as (1) chick growth assay, or (2) plasma amino acid test or a nitrogen balance determination made.

The following are ways in which ingredients may be mishandled.

(1) Water damage—Ingredient can become overheated or moldy.
(2) Overheating—May be dryed at too high a temperature. High fat with moisture can result in overheating. Bacteria or fungi can cause overheating in the ground feeds.
(3) Immature grains may have inferior proteins.
(4) Refinement or manufacturing damage.

An example of this damage is wheat gluten prepared from wheat having a lower biological value than the crude protein in wheat, when based on chick growth studies; also, isolated soybean protein having a lower biological value than crude soybean protein. Apparently, during the manufacture of wheat gluten, the lysine becomes damaged or more unavailable, and in isolated soybean protein, the methionine becomes more unavailable for the same reason.

During the fractionation, the proteins of low biological value could become a part of the fraction directed to, or available for, poultry feeds.

MINERALS

Minerals may contain contaminants which are toxic, or the element desired may be unavailable. Mineral sources of calcium and phosphorus should have complete chemical analysis, and all phosphates should be assayed for available phosphorus, using the chick bone-ash method. Calcium sources are:

Calcium carbonates
High analysis limestone
Marble dust

Oyster shell
Coquina shell
Chalk

Dolomatic limestones are mixtures of calcium carbonate and magnesium carbonate.

If the magnesium is over 2%, this mineral source should not be used in egg-laying feeds at a level of over 1% in the ration. Excess magnesium produces wet droppings.

PHOSPHATES AND CALCIUM SOURCES

The phosphates used in poultry rations are generally calcium phosphates, and most are developed from rock phosphates. Bone meal is the only biological source used solely because of its calcium and phosphorus content.

A chick bone-ash test should be made on each source of phosphorus to determine the biological availability of the phosphorus. A good phosphate should be 90%-or-better available.

Each phosphate should be assayed for fluorine and vanadium because each of these minerals is toxic, and has been found present in some bone meals and rocks or colloidal phosphates at a toxic level.

The calcium sources should be assayed for manganese and zinc, because some sources contain enough of these minerals to provide a significant amount of these trace elements toward meeting poultrys' needs.

FATS

Some fats available to the feed trade have been collected from many sources and blended. Also, some fat sources, such as poultry fat, have been refed. To assure obtaining a desirable fat, several tests may be needed:

(1) Amount of peroxides present.
(2) Amount of chlorinated hydrocarbons present.
(3) If peanut oil is used, the amount of aflatoxins present.
(4) Free fatty acid level.
(5) Color or odor.

Table 34.2 shows the characteristics of some animal fats.

TABLE 34.2. CHARACTERISTICS OF INEDIBLE GRADES OF ANIMAL FATS

Fats	Titer (Minimum)	F.F.A. (Maximum)	M.I.U. (Basis)	F.A.C. Color Maximum Untreated and Unbleached
Tallows				
Fancy	41.5	4	1	7
Choice	41	5	1	9
Prime	40.5	6	1	13 or 11B
Special	40.5	10	1	19 or 11C
No. 1	40.5	15	2	33
No. 2	40.5	20	2	37
No. 3	40	35	2	No color
Greases				
Choice white	37	4	1	11
A, white	37	8	1	15
B, white	36	10	2	19 or 11C
Yellow	36	15	2	37
House	37.5	20	2	39
Brown	38	50	2	No color

VITAMINS

Quality control of vitamins centers around selection of carriers, vitamin sources and formulation.

Carriers.—Soybean meal was found by German investigators before World War II to be the best carrier of vitamins. They considered soybean meal to be active in the prevention of oxidation in which both fat- and water-soluble vitamins were stable for up to 6 months of storage.

Carriers using ground nut shells and distillers' grains have been found effective. These, and soybean meal, are free-flowing and do not readily take up moisture or develop high levels of surface energy.

Synthetic Vitamin A acetate and Vitamin D_3 are the preferred sources of these fat-soluble vitamins. The least desirable source is fish-liver oils. Many fish oils become rancid readily and the peroxides formed destroy Vitamin A.

The water-soluble vitamins and Vitamin K are stable, providing peroxides ar not present. Vitamin E acetate should be used, and an antioxidant added to both the Vitamin E and Vitamin A oils to protect them from peroxides.

Vitamin Formulation

Some vitamin premixes are formulated to be carriers of:

(1) Vitamins
(2) Antibiotics
(3) Antioxidants
(4) Pigments
(5) Trace minerals
(6) Amino acids

The trace minerals are very harmful to unstable vitamins, fats and fat-soluble pigments. Manganese and iron are two of the most destructive minerals in premixes. When minerals are added, they should be coated, or the vitamins and pigments coated. Gelatin and dextrin are sometimes used as coating agents.

QUALITY OF THE MIXED RATION

The quality of a mixed ration can be no better than the quality of the feed ingredients. The feed ingredients are not improved by storage and good storage and handling know no substitutes. Improper handling or poor storage can destroy the quality of an ingredient.

When only high biological-value ingredients and the desirable ration formula have been used, the quality of the mixed ration will then depend upon proper function and use of the mill machinery, along with proper weighing.

Grinding of whole grains, such as corn, milo, oats, barley or wheat, should be performed immediately or within 24 hr of mixing. Frequently, the moisture content of corn or milo is too high, over 14%, and when ground, the heat of grinding is great enough to activate spoilage of the ground material. This results in molding and caking.

The ideal method of handling a mixed ration is by the bulk method. When placed in closed bags, spoilage can readily occur if moisture is over 12 to 14%. A bulk bin should be emptied each time below the top of the funnel-shaped bottom, and completely emptied every month. Any cake or hard lumps should be removed and discarded.

If drugs or hormones are used for special feeds, the feed mixer should be cleaned after the drug-fortified feed has been mixed. Some drugs have electrostatic properties in which they are electrically attracted to metal surfaces carrying an electrical charge. This is best cleaned, or prevented by first adding the soybean meal and the drugs, and following with the ground corn. If fat is used in the ration, it should be added to the soybean meal before the drugs.

Separation of feed ingredients is often done in the feed trough or in the mixture as it comes from the feed mixer. So far, this separation is of minor importance to poultry, even when the mixture is fed as a mash.

There is basically no separation in pelleted feeds in the feed bin, and only an insignificant amount in the feed trough.

There is less separation in feeds containing added fat, corn and soybeans than in those containing wheat middlings, and fine alfalfa meal.

WATERING FEEDS

Some fed mills use water emulsions of fish-liver oils, and water solutions of drugs and hormones. Also, water is added to mixed rations containing less than 9% moisture to reduce dustiness. When water is added, it is generally absorbed on the outer surface. This results in rapid surface oxidation and discoloration. Feeds that become discolored are generally not palatable.

35

Nutritional Requirements

Determining the nutrients required by poultry, song and game birds was part of a very exciting era in poultry nutrition. Many "firsts" were found for new required nutrients using the chick and also many biochemical models were developed. Two of the most important firsts were Vitamin K and Vitamin B_{12}—both required by poultry and now playing significant roles in human medicine.

In early studies of nutrients, they were classed as unrecognized growth factors (U.G.F.), until their chemical identity was established by growth units with chicks. In the case of Vitamin K, blood-clotting time with chicks was used.

Some of the U.G.F. were first established with rats, but were included in poultry rations because of the reasonable certainty that they would be required.

Poultry perhaps have a higher requirement for each nutrient than do other domestic animals. The requirements are generally based on the highest requirement established by any 1 animal or strain of animals, because poultry are fed collectively.

Poultry require more nutrients than other animals and their requirements are often more specific. An example of this biological phenomenon is: 4-footed animals use all forms of Vitamin D, whereas poultry use only Vitamin D_3 (Activated 7-dehydrocholesterol). Poultry require Vitamin K because they do not have blood platelets. They require choline, whereas 4-footed animals synthesize choline from methionine. They also require glycine and more amino acids than 4-footed animals.

Eating regulations in poultry differ from those of other animals. Poultry eat to provide for energy needs, whereas protein seems to control food consumption in rats.

Poultry have a higher body temperature than other animals and accordingly, require a higher-energy ration.

They do not seem to have taste preferences and do not discriminate against colors. Texture or shape of food also exerts very little influence.

The digestive tract of poultry is one of the shortest among animals, yet processes comparatively larger amounts of feed than that of most animals. Some wild birds consume their weight in food daily, and when based on food production (eggs laid) are very industrious.

The eggs of birds are the most complete packaged food produced by animals.

BODY STORAGE OF NUTRIENTS

Many nutrition studies on nutrient requirements are short-term and use commercially-available birds. The breeding ration is generally fortified in those nutrients which the body will retain and which seem to give better livability. When making a biological assay, it is necessary that the bird become depleted during the early part of the study for a desirable response to be obtained. Some investigators deplete the birds before attempting to make a quantitative test. In this procedure, the birds are banded and individually weighed. They are then fed a depletion ration for 3 to 7 days and then reweighed. Those making the slowest gain are used in the requirement study. A depletion approach works for studies on nutrients that do not severely influence the absorptive nature of the digestive tract.

FACTORS INFLUENCING NUTRITIONAL REQUIREMENTS

There are many factors which influence the nutritional requirements of poultry. Each of these must be considered when developing dietary standards for ration formulation. Among these is the genetic factor.

The requirement for methionine of White Leghorns is more than for Austrolorps. The lysine-arginine requirement for some stains is also different. Heavy breeds require less protein than light breeds.

Hens which produce wet droppings on standard rations probably have a genetic form of polydipsia and water consumption must be restricted.

One of the causes of thin-shelled eggs is failure of the hen to absorb enough calcium; increasing the calcium uptake only partially corrects the condition because absorption is under genetic control.

In general, there is a specie difference for most of the nutrients.

Age of Animal

Poultry during the first 21 days of life have higher requirements for some nutrients than during the growth period, and also have requirements for other nutrients similar to those required for reproduction.

(1) The requirement for Vitamin A increases with age.
(2) The requirement for riboflavin is similar during starting and production of hatchery eggs and less during growth and production of table eggs.
(3) The trace-mineral requirement is similar throughout life but the calcium level is less during starting and growth than during egg production. The calcium requirement during egg production is 3 to 5 times more than during the starting and growing periods.
(4) The energy needs for broiler production is greater than for growing chickens.

Non-production hens must receive less energy than layers. The higher the egg production, the higher the energy needs.

Purpose of Bird

Breeding poultry must receive a high dietary level of vitamins than table-egg producers. The eggs must be fortified in vitamins to insure hatchability and livability of the newly-hatched birds. The 3 most critical vitamins needed in large amounts in breeding rations are riboflavin, Vitamin A and Vitamin B_{12}. The riboflavin requirements for desirable reproduction are at least 2 to 3 times that for egg production, and 20% more for Vitamin A and twice that for Vitamin B_{12}.

Poultry that are being fattened or finished for market, require less protein and more energy than do growing birds.

Temperature

When the environmental temperature increases, the requirement for protein increases and for energy decreases. Since there is a reduction in feed uptake in warm weather, all nutrients should be increased in proportion to the expected reduced feed intake.

BIOLOGICAL DEVELOPMENT

Parathyroid Gland

Developing pullets must not receive more than the average optimum calcium level if the parathyroid gland is to develop properly for control of calcium homeostasis during egg production. When the diet is high in calcium, the parathyroid gland does not need to control calcium metabolism; therefore, its functional possibilities decline. The egg-producing hen must have good parathyroid function for egg-shell development.

Reproductive Organs and Body Weight

Special feeds are used for sexual and physiological development. The rations are generally fed in a restricted feeding program. A common restricted uptake is often based on 80% of the ad libitum uptake of a complete all-mash feed. The ration fed in the restricted program would require 20% more vitamins and minerals than the ad libitum-fed ration. If the ration is not fortified by a 20% increase, a deficiency could occur in some birds.

The water requirement of broilers has been found to be a function of feed consumption, being 1.5 lb water per lb feed consumed. The laying hen consumes approximately 2 lb water per lb feed.

QUALITY OF NUTRIENTS

What is the difference between requirements established with:

(1) High biologically-valued ingredients?
(2) Optimum level of all required nutrients except that being investigated?
(3) Factors either fortified or deficient in sparing action?
(4) High or low plane of nutrition?
(5) Presence or absence of pathogens?
(6) Presence of toxins?

ESTABLISHING THE NUTRIENT REQUIREMENTS

Recommended nutrient requirements are established by evaluation of published results of nutritional investigations. Since poultry are fed collectively and published results involve many variables, the recommendations adopted are often averages of the higher-observed requirements. Also, these recommendations are nearly always based on a completely ad libitum-fed diet.

The National Research Council beginning in 1944 has put out several publications on the nutrient requirements of poultry. Since 1944, 7 revisions have been made, each developed from and built upon the fundamentals laid down in previous editions. These recommendations covered all nutrients except water. The recommendations arrived at by the NRC were intended for use by those concerned with formulation of the various poultry rations. Tables 35.1, 35.2, 35.3, 35.4, 35.5, 35.6, 35.7, and 35.8 cover the nutrient requirement of poultry as published in the Seventh Revised Edition 1977, Number 1, Nutrient Requirements of Poultry, The National Research Council. These are presented in this text with permission from the NRC.

Ration formulation should be preceded by studying the factors influencing the requirements and the development of a margin of safety.

NUTRIENT CALCULATION INVOLVEMENTS

The question often asked is, "How many nutrients should be involved in calculations to arrive at the formulation?" The simple answer would be, "Those that are likely to be deficient":

Vitamin A
Vitamin D_3
Vitamin K
Vitamin E
Riboflavin
Pantothenic acid
Choline
Niacin
Vitamin B_{12}

Protein
methionine
Energy level
Calcium
Phosphorus
Salt (NaCl)
Manganese
Zinc

As more by-products are used, sulfates, potassium, Vitamin B_6 (pyridoxine), lysine and tryptophan will be added. The by-products may be obtained from that part of the original ingredient which is low in an essential nutrient. For example: the polishings of rice are high in some vitamins where the remaining part of the rice is low.

Today, most rations contain an antibiotic and an antioxidant. These are not required biochemically but have desirable economic value.

TABLE 35.1. PROTEIN AND AMINO ACID REQUIREMENTS OF BROILERS

	Broilers					
	0–3 Weeks		3–6 Weeks		6–9 Weeks	
Nutrient	%[1]	g/Mcal	%[1]	g/Mcal	%[1]	g/Mcal
Protein	23.0	—	20.0	—	18	—
Arginine	1.44	4.50	1.20	3.75	1.00	3.13
Glycine + serine	1.50	4.69	1.00	3.13	0.70	2.19
Histidine	0.35	1.09	0.30	0.94	0.26	0.81
Isoleucine	0.80	2.50	0.70	2.19	0.60	1.88
Leucine	1.35	4.22	1.18	3.69	1.00	3.13
Lysine	1.20	3.75	1.00	3.13	0.85	2.66
Methionine + cystine	0.93	2.90	0.72	2.25	0.60	1.88
Methionine	0.50	1.56	0.38	1.19	0.32	1.00
Phenylalanine + tyrosine	1.34	4.19	1.17	3.66	1.00	3.13
Phenylalanine	0.72	2.25	0.63	1.97	0.54	1.69
Threonine	0.75	2.34	0.65	2.03	0.56	1.75
Tryptophan	0.23	0.72	0.20	0.63	0.17	0.53
Valine	0.82	2.56	0.72	2.25	0.62	1.94

Source: Natl. Res. Counc. (1977).
[1]Requirement for diets containing 3200 kcal metabolizable energy per kg.

TABLE 35.2. PROTEIN AND AMINO REQUIREMENTS OF EGG-TYPE AND MEAT-TYPE CHICKENS

Nutrient	Replacement Pullets[1] (Egg- or Meat-type)						Laying and Breeding Hens (Egg- or Meat-type)	
	0–6 Weeks		6–14 Weeks		14–20 Weeks			
	%	g/Mcal	%	g/Mcal	%	Mcal	% of Diet[3]	Daily Intake per Hen[4] (mg)
Protein	18.0	—	15.0	—	12.0	—	15.0	16,500
Arginine	1.00	3.45	0.83	2.86	0.67	2.31	0.8	880
Glycine + serine	0.70	2.42	0.58	2.00	0.47	1.62	0.5	550
Histidine	0.26	0.90	0.22	0.76	0.17	0.59	0.22	240
Isoleucine	0.60	2.07	0.50	1.73	0.40	1.38	0.5	550
Leucine	1.00	3.45	0.83	2.86	0.67	2.31	1.2	1,320
Lysine	0.85	2.93	0.60	2.07	0.45	1.55	0.60[5]	660[5]
Methionine + cystine	0.60	2.07	0.50	1.73	0.40	1.38	0.50	550
Methionine	0.32	1.10	0.27	0.93	0.21	0.72	0.27	300
Phenylalanine + tyrosine	1.00	3.45	0.83	2.86	0.67	2.31	0.8	880
Phenylalanine	0.54	1.86	0.45	1.55	0.36	1.24	0.4	440
Threonine	0.56	1.93	0.47	1.62	0.37	1.28	0.4	440
Tryptophan	0.17	0.59	0.14	0.48	0.11	0.38	0.11	120
Valine	0.62	2.14	0.52	1.79	0.41	1.41	0.5	550

Source: Natl. Res. Counc. (1977).

[1]The requirements shown are for reasonable growth and development at 20 weeks of age when rations are fed on an *ad libitum* basis. Most of the values are not determined but are based on values for young birds and extrapolated for older birds. Values under percentage are based on

[2]2900 kcal of metabolizable energy/kg.

[3]Based on 2850 kcal of metabolizable energy/kg diets.

[4]Based on a feed consumption of 110 g per day.

[5]Determined with corn-soybean meal diets. The requirement may be higher on wheat-based diets.

TABLE 35.3. VITAMIN, LINOLEIC ACID, AND MINERAL REQUIREMENTS OF CHICKENS[1] (IN PERCENTAGE OR AMOUNT PER KG FEED)

Nutrient	Starting Chickens (0–8 Weeks)	Growing Chickens (8–18 Weeks)	Laying Hens	Breeding Hens
Vitamin A activity (IU)	1500	1500	4000	4000
Vitamin D (ICU)	200	200	500	500
Vitamin E (IU)	20	5	5	10
Vitamin K_1 or equivalent activity (mg)	0.5	0.5	0.5	0.5
Thiamin (mg)	1.8	1.3	0.8	0.8
Riboflavin (mg)	3.6	1.8	2.2	3.8
Pantothenic acid (mg)	10	10	2.2	10
Niacin (mg)	27	11	10	10
Pyridoxine (mg)	3	3	3	4.5
Biotin (mg)	0.15	0.10	0.10	0.15
Choline (mg)	1300	500	500	500
Folacin (mg)[2]	0.55	0.25	0.25	0.35
Vitamin B_{12} (mg)	0.009	0.003	0.003	0.003
Linoleic acid (%)	1.0	0.8	1.0	1.0
Calcium (%)	0.9	0.6	3.25	2.75
Phosphorus (%)	0.7	0.4	0.5	0.5
Potassium (%)	0.2	0.16	0.1	0.1
Sodium (%)	0.15	0.15	0.15	0.15
Chlorine (mg)	800	800	800	800
Copper (mg)	4	3	3	4
Iodine (mg)	0.35	0.35	0.3	0.3
Iron (mg)	80	40	50	80
Magnesium (mg)	600	400	500	500
Manganese (mg)	55	25	25	33
Selenium (mg)	0.1	0.1	0.1	0.1
Zinc (mg)	40	35	50	65

Source: Natl. Res. Counc. (1977).
[1]The values in lightface type are estimates.
[2]With sucrose, diet requirement is 1.2 mg/kg.

TABLE 35.4. PROTEIN AND AMINO ACID REQUIREMENTS OF TURKEYS[1]

Nutrient	Age (Weeks)							
	♂ 0–4 ♀ 0–4	4–8 4–8	8–12 8–11	12–16 11–14	16–20 14–17	20–24 17–20	Holding	Breeding Hens
Metabolizable energy[2] (kcal/kg)	2800	2900	3000	3100	3200	3300	2900	2900
Protein (%)	28	26	22	19	16.5	14	12	14
Arginine (%)	1.6	1.5	1.25	1.1	0.95	0.8	0.6	0.6
Glycine + serine (%)	1.0	0.9	0.8	0.7	0.6	0.5	0.4	0.5
Histidine (%)	0.58	0.54	0.46	0.39	0.35	0.29	0.25	0.3
Isoleucine (%)	1.1	1.0	0.85	0.75	0.65	0.55	0.45	0.5
Leucine (%)	1.9	1.75	1.5	1.3	1.1	0.95	0.5	0.5
Lysine (%)	1.7	1.6	1.35	1.0	0.80	0.65	0.5	0.6
Methionine + cystine (%)	1.05	0.90	0.75	0.65	0.55	0.45	0.4	0.4
Methionine (%)	0.53	0.45	0.38	0.33	0.28	0.23	0.2	0.2
Phenylalanine + tyrosine (%)	1.8	1.65	1.4	1.2	1.05	0.9	0.8	1.0
Phenylalanine (%)	1.0	0.9	0.8	0.7	0.6	0.5	0.4	0.55
Threonine (%)	1.0	0.93	0.79	0.68	0.59	0.5	0.4	0.45
Tryptophan (%)	0.26	0.24	0.20	0.18	0.15	0.13	0.10	0.13
Valine (%)	1.2	1.1	0.94	0.8	0.7	0.6	0.5	0.58

Source: Natl. Res. Counc. (1977).
[1]Values in lightface type are estimates.
[2]Requirement for diets containing the stated levels of metabolizable energy.

TABLE 35.5. VITAMIN, LINOLEIC ACID, AND MINERAL REQUIREMENTS OF TURKEY[1]

Nutrient	% or Amount per kg of Feed		
	0–8 Weeks	8 Weeks[2]	Breeding
Vitamin A (IU)	4000	4000	4000
Vitamin D (ICU)	900	900	900
Vitamin E (IU)	12	10	25
Vitamin K_1 or equivalent activity (mg)	1	0.8	1
Thiamin (mg)	2	2	2
Riboflavin (mg)	3.6	3.0	4
Pantothenic acid (mg)	11	9	16
Niacin (mg)	70	50	30
Vitamin B_6 (mg)	4.5	3.5	4
Biotin (mg)[3]	0.2	0.1	0.15
Choline (mg)	1900	1100	1000
Folacin (mg)	1.0	0.8	1.0
Vitamin B_{12} (mg)	0.003	0.003	0.003
Linoleic acid (%)	1.0	0.8	1.0
Calcium (%)	1.2	0.8	2.25
Phosphorus (%)	0.8	0.7	0.7
Potassium (%)	0.4	0.4	0.4
Sodium (%)	0.15	0.15	0.15
Chlorine (mg)	800	800	800
Copper (mg)	6	4	6
Iodine (mg)	0.4	0.4	0.4
Iron (mg)	60	40	60
Magnesium (mg)	500	500	500
Manganese (mg)	55	25	35
Selenium (mg)	0.2	0.2	0.2
Zinc (mg)	75	40	65

Source: Natl. Res. Counc. (1977).
[1]The values in lightface type are estimates.
[2]From 8 weeks to market or prebreeding.
[3]Requirement may be 50% greater for wheat- or barley-based diets.

TABLE 35.6. NUTRIENT REQUIREMENTS OF PHEASANTS AND QUAIL[1] (IN PERCENTAGE OR AMOUNT PER KG FEED)

	Pheasant		Bobwhite Quail		Japanese Quail	
Nutrient	Starting	Growing (6–20 Weeks)	Starting and Growing	Breeding	Starting and Growing	Breeding
Metabolizable energy (kcal/kg)	2800	2700	2800	2800	3000	2800
Protein (%)	30	16	28[2]	24	24[3]	24
Lysine (%)	1.5	0.8	1.4	0.7	1.4	1.1
Methionine + cystine (%)	1.0	0.6	0.9	0.6	0.75	0.8
Glycine + serine (%)	1.8	1.0	1.6	0.9	1.7	0.9
Vitamin A (IU)	3000	3000	3000	3000	5000	5000
Vitamin D (ICU)	1200	900	900	900	480	1200
Riboflavin (mg)	3.5	2.6	3.8	4.0	4.0	4.0
Pantothenic acid (mg)	10	10	12.6	15	10	15
Niacin (mg)	60	40	31	20	40	20
Choline (mg)	1500	1000	1500	1000	2000	1500
Linoleic acid (%)	1.0	1.0	1.0	1.0	1.0	1.0
Calcium (%)	1.0	0.7	0.65	2.3	0.8	2.5
Chlorine (%)	0.11	0.11	0.11	0.15	0.15	0.15
Phosphorus (%)	0.8	0.6	0.65	1.0	0.65	0.8
Sodium (%)	0.1	0.1	0.085	0.15	0.15	0.15
Iodine (mg)	0.3	0.3	0.30	0.30	0.30	0.30
Magnesium (mg)	600	400	600	400	150	500
Manganese (mg)	90	70	90	70	90	70
Zinc (mg)	60	50	50	50	25	50

Source: Natl. Res. Counc. (1977).
[1]The values in lightface type are estimates. For nutrients not listed see requirements for turkeys as a guide.
[2]May be reduced to 20% at 6 weeks of age.
[3]May be reduced to 20% at 3 weeks of age.

TABLE 35.7. NUTRIENT REQUIREMENTS OF GEESE[1] (IN PERCENTAGE OR AMOUNT PER KG OF FEED)

Nutrient	Starting (0–6 weeks)	Growing (After 6 weeks)	Breeding
Metabolizable energy (kcal/kg)	2900	2900	2900
Protein (%)	22	15	15
Lysine (%)	0.9	0.6	0.6
Vitamin A (IU)	1500	1500	4000
Vitamin D (ICU)	200	200	200
Riboflavin (mg)	4	2.5	4
Niacin, available (mg)	55	35	20
Calcium (%)	0.8	0.6	2.25
Phosphorus (%)	0.6	0.4	0.6

Source: Natl. Res. Counc. (1977).
[1]The values in lightface type are estimates. For nutrients not listed, see requirements for chickens as a guide.

TABLE 35.8. NUTRIENT REQUIREMENTS OF DUCKS[1] (IN PERCENTAGE OR AMOUNT PER KILOGRAM OF FEED)

Nutrient	Starting and Growing Duck	Breeding Ducks
Metabolizable energy (kcal/kg)	2900	2900
Protein (%)	16[2]	15
Lysine (%)	0.9	0.7
Methionine + cystine (%)	0.8	0.55
Vitamin A (IU)	4000	4000
Vitamin D (ICU)	220	500
Riboflavin (mg)	4	4
Pantothenic acid (mg)	11	10
Niacin (mg)	55	40
Pyridoxine (mg)	2.6	3
Calcium (%)	0.6	2.75
Phosphorus (%)	0.6[3]	0.6
Sodium (%)	0.15	0.15
Manganese (mg)	40	25
Magnesium (mg)	500	500

Source: Natl. Res. Counc. (1977).
[1]The values in lightface type are estimates. For nutrients not listed, see requirements for chickens as a guide.
[2]Increasing protein level to 22% for the first 2 weeks will increase early growth.

FEED REQUIREMENTS

The feed requirements of poultry are influenced by many factors. Generally speaking, egg producing-hens consume at least 5 to 10% more feed than required. Also, a hen in heavy production, 70 to 80%, requires less feed per doz eggs than a poor producer and a large hen requires more feed per doz eggs than a small hen. Table 35.9 shows the relation of body weight to feed requirements.

The difference in growth and feed efficiency between male and female broilers is summarized in Tables 35.10 to 35.12. A 4-lb male broiler requires 2.49 lb feed to produce a 1-lb gain as compared to 3.14 lb feed for a female. This observation has led broiler producers to market the male and female separately.

The feed requirement of Leghorn hens is approximately 1 lb less per 100 hens daily than of medium-sized hens (Tables 35.13 and 35.14) at 26 wk of age. This saving of 1 lb of feed per day is significant over 1 season for a 10,000-hen flock.

TABLE 35.9. BODY WEIGHTS AND FEED REQUIREMENTS FOR MEAT-TYPE AND EGG-TYPE CHICKENS AND CAPONS[1]

Age (Weeks)	Body Weights (g) Leghorn[2] ♂	♀	Broiler[3] ♂	♀	±Cumulative Feed Consumption (g) Leghorn ♂	♀	Broiler ♂	♀
2	130	130	250	240	360	360	265	250
4	270	250	740	650	725	725	1110	975
6	450	430	1340	1190	1400	1350	2425	2175
8	690	540	2100	1760	2400	1900	4200	3600
10	940	725	2720	1990	3400	2750	6250	4750
12	1240	880	3240	2370	4900	3600	8125	6350
14	1350	1010	3640	2650	5500	4300	11,150	8100
16	1450	1155	3990	2880	6200	5100	13,650	10,225
18	1550	1315	4280	3060	7200	6300	16,650	12,350
20	1700	1450	4540	3230	8600	7600	19,575	14,890

Source: Natl. Res. Counc. (1977).
[1]Capons will follow the male closely except immediately following surgery.
[2]Egg strains (Leghorn types) are frequently restricted in the growing period.
[3]Broiler breeders are normally restricted to below these weights by reduced light and feed. Birds to be sold at broiler weights are fed for near-maximum rates of gain.

TABLE 35.10. MALE BROILER FEED CONSUMPTION BY GROWTH INCREMENTS[1]

Live Body Weight	Feed Consumption: To Add Last 0.1-lb Gain lb	Cumulative lb	Feed Conversion: To Add Last 0.1-lb Gain	Cumulative
2.8	0.22	5.19	2.19	1.85
2.9	0.22	5.40	2.22	1.86
3.0	0.22	5.62	2.24	1.88
3.1	0.23	5.85	2.26	1.89
3.2	0.23	6.08	2.29	1.90
3.3	0.23	6.31	2.31	1.91
3.4	0.23	6.54	2.34	1.93
3.5	0.24	6.78	2.36	1.94
3.6	0.24	7.02	2.39	1.95
3.7	0.24	7.26	2.41	1.96
3.8	0.24	7.50	2.44	1.98
3.9	0.25	7.75	2.47	1.99
4.0	0.25	8.00	2.49	2.00
4.1	0.25	8.25	2.52	2.01
4.2	0.25	8.50	2.54	2.03
4.3	0.26	8.76	2.57	2.04
4.4	0.26	9.02	2.59	2.05
4.5	0.26	9.28	2.62	2.06
4.6	0.26	9.54	2.64	2.08
4.7	0.27	9.81	2.67	2.09
4.8	0.27	10.08	2.69	2.10
4.9	0.27	10.35	2.72	2.11
5.0	0.27	10.63	2.74	2.13

Source: Natl. Res. Counc. (1977).
[1]North (1972).

TABLE 35.11. FEMALE BROILER FEED CONSUMPTION BY GROWTH INCREMENTS[1]

Live Body Weight lb	Feed Consumption: To Add Last 0.1-lb Gain lb	Feed Consumption: Cumulative lb	Feed Conversion: To Add Last 0.1-lb Gain lb	Feed Conversion: Cumulative lb
2.3	0.23	4.39	2.33	1.91
2.4	0.24	4.62	2.37	1.93
2.5	0.24	4.86	2.40	1.94
2.6	0.24	5.10	2.44	1.96
2.7	0.25	5.35	2.47	1.98
2.8	0.25	5.60	2.53	2.00
2.9	0.26	5.86	2.58	2.02
3.0	0.26	6.12	2.63	2.04
3.1	0.27	6.39	2.68	2.06
3.2	0.27	6.66	2.73	2.08
3.3	0.28	6.94	2.78	2.10
3.4	0.28	7.22	2.83	2.12
3.5	0.29	7.51	2.88	2.15
3.6	0.29	7.80	2.93	2.17
3.7	0.30	8.10	2.97	2.19
3.8	0.30	8.40	3.03	2.21
3.9	0.31	8.71	3.08	2.23
4.0	0.31	9.02	3.14	2.26
4.1	0.32	9.34	3.18	2.28
4.2	0.32	9.66	3.23	2.30
4.3	0.33	9.99	3.28	2.33

Source: Natl. Res. Counc. (1977).
[1]North (1972).

TABLE 35.12. STRAIGHT-RUN BROILER FEED CONSUMPTION BY GROWTH INCREMENTS

Live Body Weight lb	Feed Consumption: To Add Last 0.1-lb Gain lb	Feed Consumption: Cumulative lb	Feed Conversion: To Add Last 0.1-lb Gain	Feed Conversion: Cumulative
2.6	0.23	4.92	2.27	1.89
2.7	0.23	5.15	2.30	1.91
2.8	0.23	5.38	2.33	1.92
2.9	0.24	5.62	2.37	1.94
3.0	0.24	5.86	2.40	1.95
3.1	0.24	6.10	2.43	1.97
3.2	0.25	6.35	2.47	1.98
3.3	0.25	6.60	2.50	2.00
3.4	0.25	6.85	2.53	2.02
3.5	0.26	7.11	2.57	2.03
3.6	0.26	7.37	2.60	2.05
3.7	0.26	7.63	2.63	2.06
3.8	0.27	7.90	2.66	2.08
3.9	0.27	8.17	2.69	2.09
4.0	0.27	8.44	2.72	2.11
4.1	0.27	8.71	2.75	2.13
4.2	0.28	8.99	2.78	2.14
4.3	0.28	9.27	2.80	2.16
4.4	0.28	9.55	2.83	2.17
4.5	0.29	9.84	2.86	2.19

Source: North (1972).

TABLE 35.13. FEED CONSUMPTION PER 100 LAYING LEGHORN HENS PER DAY[1] (HEN-DAY BASIS)

Week of Egg Production	Feed Consumed Per 100 Hens per Day lb	kg	Cumulative per Hen lb	kg	Week of Egg Production	Feed Consumed Per 100 Hens per Day lb	kg	Cumulative per Hen lb	kg
1	17	7.7	1.2	0.5	27	23	10.4	43.7	19.8
2	21	9.5	2.7	1.2	28	23	10.4	45.3	20.5
3	23	10.4	4.3	2.0	29	23	10.4	46.9	21.3
4	24	10.9	6.0	2.7	30	23	10.4	48.5	22.0
5	24	10.9	7.6	3.5	31	23	10.4	50.1	22.7
6	24	10.9	9.3	4.2	32	23	10.4	51.7	23.5
7	24	10.9	11.0	5.0	33	23	10.4	53.3	24.2
8	24	10.9	12.7	5.8	34	23	10.4	55.0	25.0
9	24	10.9	14.4	6.5	35	23	10.4	56.6	25.7
10	24	10.9	16.0	7.3	36	23	10.4	58.2	26.4
11	24	10.9	17.7	8.0	37	23	10.4	59.8	27.1
12	24	10.9	19.4	8.8	38	23	10.4	61.4	27.9
13	24	10.9	21.1	9.6	39	22	10.0	62.9	28.5
14	24	10.9	22.8	10.3	40	22	10.0	64.5	29.3
15	23	10.4	24.4	11.1	41	22	10.0	66.1	30.0
16	23	10.4	26.0	11.8	42	22	10.0	67.6	30.7
17	23	10.4	27.6	12.5	43	22	10.0	69.2	31.4
18	23	10.4	29.2	13.2	44	22	10.0	70.7	32.1
19	23	10.4	30.8	14.0	45	22	10.0	72.3	32.8
20	23	10.4	32.4	14.7	46	22	10.0	73.8	33.5
21	23	10.4	34.0	15.4	47	22	10.0	75.3	34.2
22	23	10.4	35.6	16.1	48	22	10.0	76.9	35.9
23	23	10.4	37.2	16.9	49	22	10.0	78.4	35.6
24	23	10.4	38.9	17.6	50	21	9.5	79.9	36.2
25	23	10.4	40.5	18.4	51	21	9.5	81.4	36.9
26	23	10.4	42.1	19.0	52	21	9.5	82.8	37.6

Source: North (1972).
[1]Moderate weather.

TABLE 35.14. FEED CONSUMPTION PER 100 LAYING MEDIUM-SIZE HENS PER DAY[1] (HEN-DAY BASIS)

	Feed Consumed					Feed Consumed			
Week of Egg Production	Per 100 Hens per Day		Cumulative per Hen		Week of Egg Production	Per 100 Hens per Day		Cumulative per Hen	
	lb	kg	lb	kg		lb	kg	lb	kg
1	18	7.7	1.3	0.6	27	24	10.9	45.5	20.6
2	22	10.0	2.8	1.3	28	24	10.9	47.2	21.4
3	24	10.9	4.5	2.0	29	24	10.9	48.9	22.2
4	25	11.3	6.2	2.8	30	24	10.9	50.5	22.9
5	25	11.3	8.0	3.6	31	24	10.9	52.2	23.7
6	25	11.3	9.7	4.4	32	24	10.9	53.9	24.5
7	25	11.3	11.5	5.2	33	24	10.9	55.6	25.2
8	25	11.3	13.2	6.0	34	24	10.9	57.3	26.0
9	25	11.3	15.0	6.8	35	24	10.9	58.9	26.7
10	25	11.3	16.7	7.6	36	24	10.9	60.6	27.5
11	25	11.3	18.5	8.4	37	24	10.9	62.3	28.3
12	25	11.3	20.2	9.2	38	24	10.9	64.0	29.0
13	25	11.3	22.0	10.0	39	23	10.4	65.6	29.8
14	24	10.9	23.7	10.8	40	23	10.4	67.2	30.5
15	24	10.9	25.3	11.5	41	23	10.4	68.9	31.3
16	24	10.9	27.0	12.2	42	23	10.4	70.5	32.0
17	24	10.9	28.7	13.0	43	23	10.4	72.1	32.7
18	24	10.9	30.4	13.8	44	23	10.4	73.7	33.4
19	24	10.9	32.1	14.6	45	23	10.4	75.3	34.2
20	24	10.9	33.7	15.3	46	23	10.4	76.9	34.9
21	24	10.9	35.4	16.1	47	23	10.4	78.5	35.6
22	24	10.9	37.1	16.8	48	23	10.4	80.2	36.4
23	24	10.9	38.8	17.6	49	23	10.4	81.8	37.1
24	24	10.9	40.5	18.4	50	22	10.0	83.3	37.8
25	24	10.9	42.1	19.1	51	22	10.0	84.8	38.5
26	24	10.9	43.8	19.9	52	22	10.0	86.4	39.2

Source: North (1972).
[1]Moderate weather.

36

Feeding Programs

An understanding of feeding programs and feeding methods is important to have before formulating rations. Many programs and methods are used throughout the poultry industry; often, feeding methods are functions of the size, breed and purpose of the flock.

The first ground rations (mashes) were formulated for layers about 1908, by United States investigators who based their formulations on what layers were consumed when various feedstuffs were fed free choice (cafeteria style). This mash was fed as a supplement to grain, pasture and oyster-shell.

All-mash feeding came into being in 1920, when hatcheries and breeding flocks were being developed in small rural farm towns and fencing of poultry became necessary. Following this all-mash feed, a free-choice mash-grain program was developed. This saved about 50% of the grinding costs.

The development of mechanical feeders led to a complete change in feeding.

There are many ways to classify feeding programs; however, the one often used is based on formulated manufactured diets. These are:

(1) The all-mash complete program.
(2) The mash-grain program.
(3) The concentrate program.

These programs can be identified by 2 disclosures on the feed tag:

(1) Name of feed.
(2) Protein content.

The names and protein contents of the rations are summarized in Table 36.1 and a feed tag is shown in Fig. 36.1.

TABLE 36.1. NAME AND PERCENTAGE PROTEIN IN CHICKEN FEED MASHES

	Feeding Periods		
Feeding Programs	Starting Feed 1–6 Weeks	Growing Feed 6–18 Weeks	Reproduction Feed 18–Completion Weeks
Complete All-mash	20	16	16
Mash[1] (grain)	[2]	22	22
Concentrate	[2]	40	40

Source: Patrick (1965).
[1]Based on an equal parts of corn 9 and oats 11. (The numbers after a grain or protein concentrate designate the protein content.)
[2]The Complete All-mash feeding program is used for the starting period.

100 lb Net Weight

AVIAN

20%

BROILER GROWER

MEDICATED

Chlortetracycline for chickens. To be fed to broilers for nonspecific infection. Not to be fed for over 5 days.

WARNING: Not to be fed to laying hens. Withdraw 24 hours before slaughter.
Active Drug Ingredient:

Chlortetracycline—500 grams per ton.

GUARANTEED ANALYSIS

Crude protein, not less than	20%
Crude fat, not less than	4%
Crude fiber, not less than	5%
Calcium (Ca), not more than	1.1%
Calcum (Ca), not less than	0.9%
Phosphorus (P), not less than	0.5%
Salt (NaCl), not less than	0.3%
Salt (NaCl), not more than	0.5%

INGREDIENTS

Soybean meal, corn gluten meal, meat meal, dried fish solubles, yellow corn, oats, dehy. alfalfa meal, animal fat with BHT, DL-hydroxy analogue of methionine, choline chloride, riboflavin supplement, calcium pantothenate, niacin, vitamin B_{12} supplement, D-activated animal sterol, vitamin A acetate, menadione, defluorinated phosphate, calcium carbonate, salt (NaCl), manganese sultate, zinc oxide

DIRECTIONS

Feed as the sole ration to growing broilers.

Manufactured by

AVIAN FEED MILLS

Sunny, Colorado, 00001

FIG. 36.1.

FORMS OF FORMULATED RATION

The major factors determining the form of the formulated ration are the types of feeders and feed manufacturing equipment available. Hand feeders can use mash, as well as pellets or crumbles. Mechanical feeders and self-feeders cannot handle mash as well as pellets or crumbles do, because of bridging in the hoppers.

MASH FEEDING

Mash feeding is more economical for small flocks and for flocks on open ranges, where self-feeders are used. Mash feeding has another important advantage in that it makes for better contact between the poultryman and his flock. Mechanical feeders "remove" the poultryman from making frequent observations; therefore, early signs of disease or equipment breakdown are not observed, and major problems often result.

THE MASH-GRAIN SYSTEM

The mash-grain system is the most economical because only one-half the expenses of grinding and mixing are required. Management must provide more careful planning of feeders if this system is used. Leghorns tend to balance the consumption of mash and grain when equal numbers of mash or grain feeders are used. Heavy hens tend to overeat grain, so twice as many mash feeders as grain are needed.

The feeding of grain makes it desirable to feed granite grit. Studies have shown that granite grit pays for itself in better feed utilization and the development of a stronger, healthier digestive tract.

The feeding of a mash-grain feed declined following the development of automatic feeders. In the all-mash system, the bulk delivery trucks place the feed in the bins which feed directly into the mechanical feeders. With automatic feeders and waterers, 1 man can care for from 40,000 to 60,000 broilers from day one to marketing.

THE CONCENTRATE PROGRAM

The concentrate program for direct feeding is planned so that the first feed the birds receive in the morning is a high-protein concentrate, with grain being available throughout the day. This program generally involves a feed-selection ratio of 1 part concentrate to from 3 to 4 parts grain. The program calls for early feeding and full participation of each bird to obtain the daily needs of the protein concentrate.

The desirable program for the concentrate is to add the ground grains to the concentrate forming an all-mash or a mash-grain program. Many large poultry operations and some small feed mills purchase the concentrates from the larger mills and blend them with their home-grown grains.

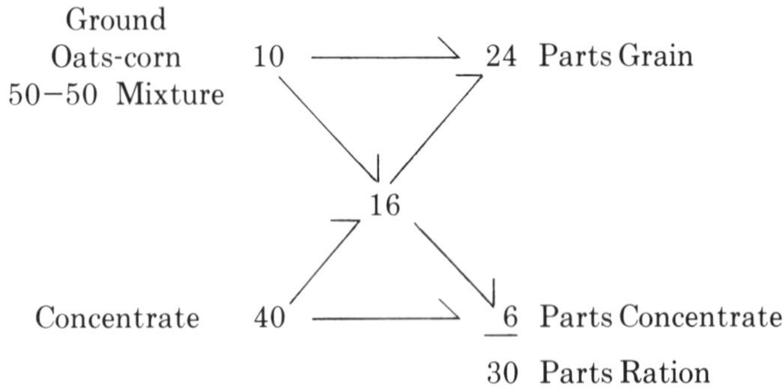

The All-mash Program (16) Using the Concentrate

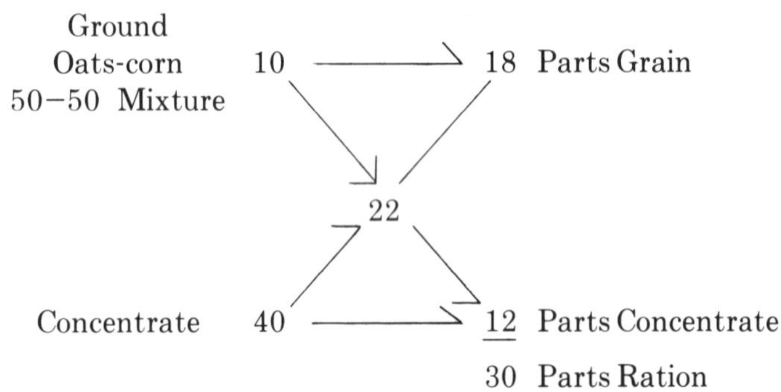

The Mash (22)-grain Program

THE WHOLE-GRAIN METHOD

This method gives free choice of grains, animal protein and sometimes, milk. Crushed oyster shell and granite grit are provided. This system was used in earlier times when the average egg production was approximately 100 eggs per laying year. Now, it is not recommended, although it gives fair results with Leghorns.

WET MASHES

Wet mashes were used for fattening poultry. Poultry today are grown so that they are fat when market time arrives. Wet feeds generally result in flies and unsanitary conditions.

FLUSHING MASHES

Before drugs were available for treatment of coccidiosis, many poultrymen fed young chickens or poults a flushing mash, adding 10% dried milk or molasses to a regular mash. This special feed was fed for 24 hr.

Some farmers had a flushing-mash feeding schedule every 60 days. This appeared to help the general condition of the poultry.

STARTING CONDITIONER

Chickens which become chilled soon after hatching or which are overheated in the incubator, develop digestive disturbances that result in pasted vents. Also, if the incubator does not perform properly during the last 4 to 5 days of hatching, the chicks will develop pasted vents.

Two methods of feeding help to reduce pasting. One method is to feed a coarsely-ground corn (fine corn chops) for the first 24 to 48 hours. This is often placed on top of the partially-filled feeders. Another method is to feed a small crumble made by crushing pelts formulated from coarse corn, soybean meal, minerals, vitamins and an antibiotic.

Coarse corn (corn chops)	76.0 lb
Soybean meal	20.0 lb
Salt	0.5 lb
Phosphate	2.0 lb
Trace minerals premix	0.5 lb
Vitamin premix	1.0 lb
Auneomycin	5.0 g

This mix is pelleted and the pellets are crumbled. It is fed 24 to 48 hr as a top dressing for the regular starting feed.

37

Ration Formulation

The chicken is a biological machine that operates at the highest temperature of the domesticated animals. The feed it consumes is used for growth, reproduction (eggs) and maintenance. Biochemically, it is highly developed and its nutritional requirements are very exacting. It lives a fast life and experiences physiological changes accompanied by biochemical changes which influence the nutritional requirements.

FACTORS GOVERNING RATION FORMULATION

The formulation of rations must recognize the chicken as a complicated biological machine with modifications for specific purposes. Before a ration is formulated, the following must be established:

(1) The Purpose of the Chicken.—Broilers, roasting ducks and turkey roasters would be considered meat birds and rations would be formulated for maximum growth and fleshing.

Female chicks grown for replacement of egg production would need formulated feeds to develop hens for egg production.

A breeding ration would stress production of hatching eggs with a higher fortification in some nutrients, such as Vitamin A, Vitamin B_{12} and riboflavin, than for production of table eggs.

(2) Recognizing the Strain (Genetic Influence).—Each strain has variations in the biological makeup which often express a difference nutritionally. Some of the recognized differences in amounts of required nutrients are in protein, arginine, lysine, methionine, Vitamin A, nicotinic acid, Vitamin D, pantothenic acid and water.

(3) The Feed-efficiency Graph.—Poultry have growth- and feed-efficiency responses characteristic of their breeding and purpose. A graph should be developed to allow the nutritionist to observe the highest feed-efficiency period in the age of the chicken, its period of beginning reduction of feed efficiency, and maturity or beginning of egg production. A

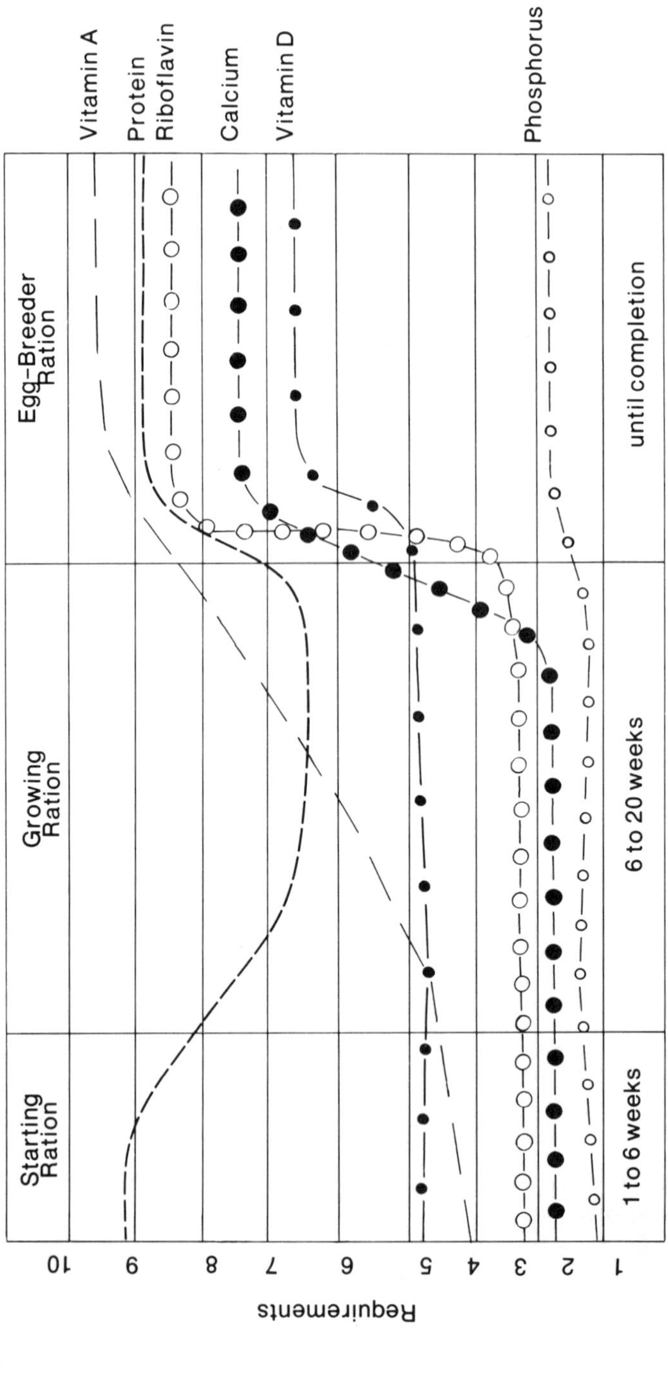

FIG. 37.1. GRAPHIC FIGURE SHOWING DIETARY LEVELS OF SOME NUTRIENTS IN THE STARTING, GROWING AND EGG-BREEDER RATIONS

feed-efficiency graph of Bronze turkey hens and toms is shown. A desirable time to sell turkeys or chickens is not later than when the feeds costs for the pound produced equals, or is less than, the value of the live weight selling price per lb (Fig. 37.1).

(4) Determining the Nutritional Requirements.—The recommended nutritional requirements are prepared from published data. These values are prepared after careful evaluation of the method of conducting the study, the factors probably influencing the results and the incorporation of the data in a pooled evaluation approach. Tables 1 through 8 in Chapter 35 show the nutritional requirements of broilers, egg- and meat-type chickens, turkeys, pheasants and quail, ducks and geese. The data in these requirement tables were prepared by the National Research Council and are presented with their approval.

Since the nutritional requirements summarized in Tables 35.1 to 35.8, inclusive, are from data published before 1977, it means that these requirements are made for breeds, strains and types of poultry available today. Also, the nutritionist should be aware of the everchanging strains of poultry and be prepared to make dietary changes when needed to provide more efficiently for the improved strains.

(5) Determining the Feedstuffs Available.—The basic ingredients used in most poultry rations are ground yellow corn and soybean meal. For over 25 years, these 2 ingredients have been used so extensively that the term, corn-soybean ration, is used. Also, some geneticists have stated that today's poultry are genetically adapted to the corn-soybean ration.

The American Feed Control Officials have developed an official list of collective terms for:

22.1 Animal Protein Products
22.2 Forage Products
22.3 Grain Feeds
22.4 Plant Protein Products
22.5 Processed Grain By-Products

Collective terms recognize a general classification of ingredient origin which performs a similar function, but they do not imply equivalent nutritional values. This allows the feed manufacturer to have more flexibility yet keep the registration of the ration established. Many of these products will increase in importance as the human population consumes more of the primary grains and seeds.

(6) Determine the Feedstuffs Needed.—The feed manufacturer first determines the feedstuffs available in his local area, then selects the desirable feedstuffs which will allow the manufacture of the ration.

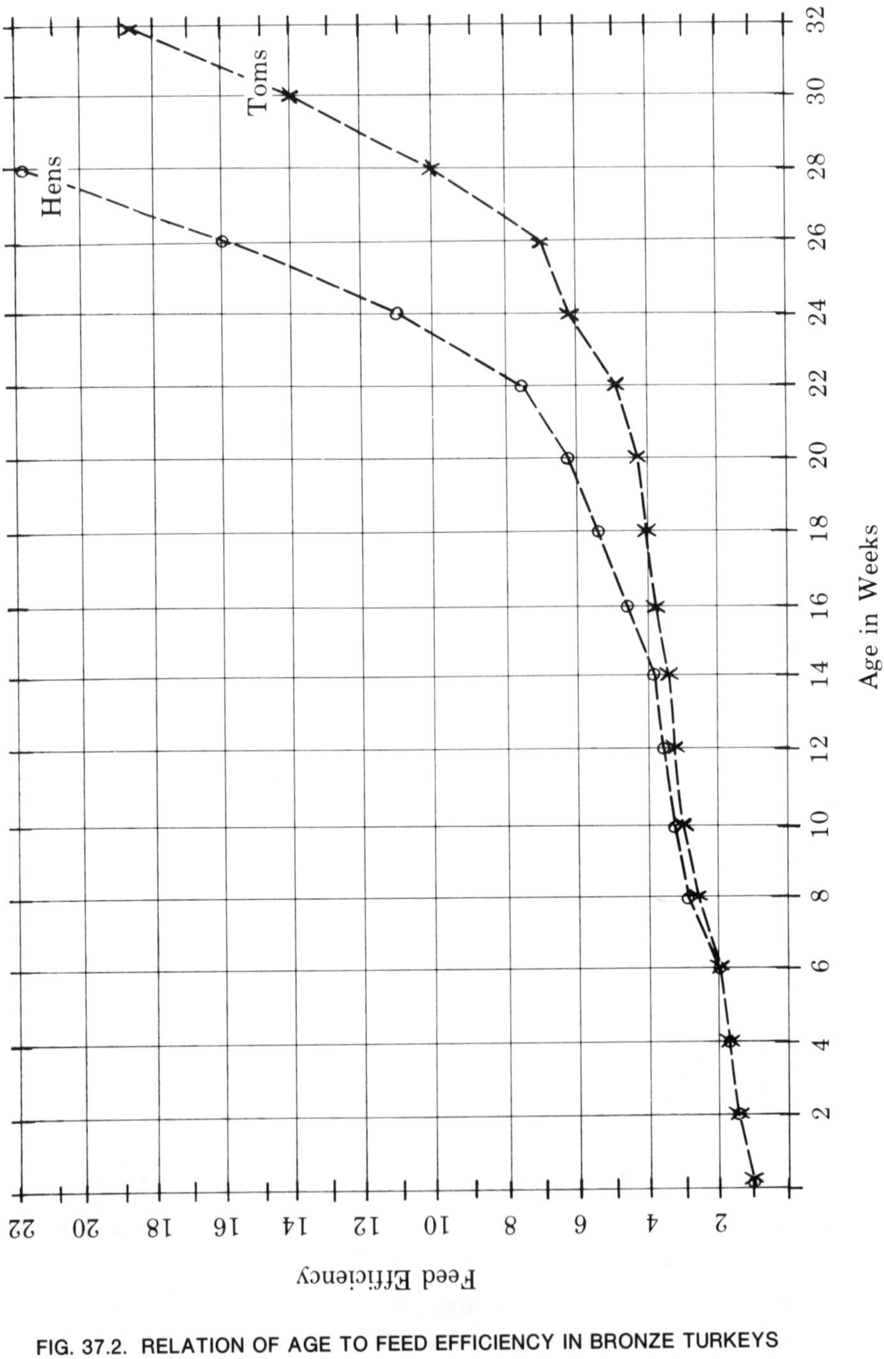

FIG. 37.2. RELATION OF AGE TO FEED EFFICIENCY IN BRONZE TURKEYS

Generally, vitamin premixes, antibiotics, antioxidants, minerals, fats and medication (drugs) are purchased. In selecting feedstuffs, it is desirable to check with the local feed dealer or manufacturer to determine the availability of the by-products or feedstuffs shown in the American Feed Control Officials under "collective terms." Many of the uncommon feedstuffs listed under "collective terms" may be equal to the primary ones, but be more available or more economical.

(7) Establish the Biological Value or Develop Quality Control.—The proper procedure is to purchase feedstuffs only from manufacturers who have a product produced under controlled conditions and whose description is similar to those shown in the Official Publication of the American Feed Control Officials.

The proximate analysis should be obtained on each purchase or shipment of each appropriate feedstuff. This analysis consists of 6 fractions:

(1) Water
(2) Ether extract
(3) Crude fiber
(4) Nitrogen-free extract
(5) Crude protein
(6) Ash

After determining the proximate analysis of the feedstuffs, tests should be conducted to determine the biological value of certain ingredients. The biological value may consist of digestibility, availability or growth response as well as chemical tests for undesirable substances.

(1) Calcium.—Quality control of ground limestone: 94 to 100% calcium carbonate with less than 2% magnesium carbonate.

(2) Phosphates.—Phosphorus 85 to 95% available, as determined by chick bone-ash method. Low in fluorine (defluorinated) and low in vanadium.

(3) Soybean Meal.—Chick growth responses using a protein assay ration: digestibility, feed efficiency and uniformity of growth are criteria used. Also, urease activity can be used (Table 24.2).

FATS

Rancidity and dressed appearance of broilers, free fatty acids, peroxides, color, odor, and general appearance values should be obtained.

FORMULATION OF RATIONS

The first ration one should formulate is one using a concentrate and one locally-grown or available grain (corn). Then, develop one using a concentrate plus two or more locally-grown or available grains.

Two methods of calculation should be used: (1) the square method and (2) the algebraic equation. After making these calculations, a ration should be developed using the requirements as set forth in Chapter 35 on "Nutrient Requirements."

The formulation of rations using a poultry concentrate and one or more carbohydrate sources is relatively simple, because the manufacturer of the concentrate has included all the nutrients except the carbohydrate source which is the locally-produced grain, such as corn. Formulating a ration from basic ingredients requires more ingredients, time, experience and many other specifications to ensure desirable results.

FORMULATION OF RATIONS

(1) Concentrate Program Using 2 Ingredients

Many small commercial feed mixers and large farms have the basic equipment needs to manufacture simple rations using one or more locally-produced grains with a commercial Poultry Concentrate.

(2) The Square Method Using 1 Ingredient and a Concentrate

Purpose: To calculate the amount of corn-9 and Poultry Concentrate-45 required to produce a 20% chick-starting ration (the number after the ingredients designates crude protein).

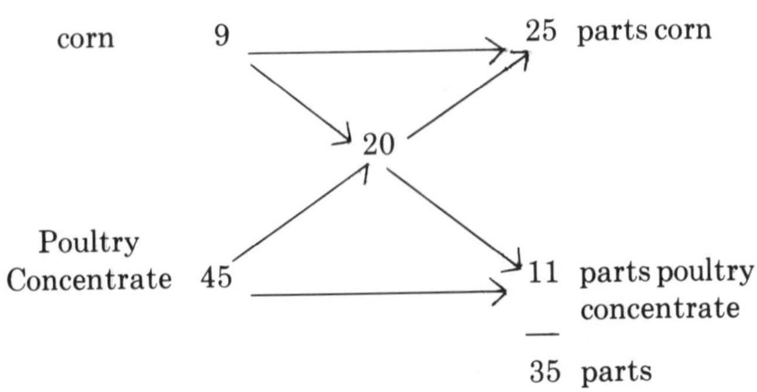

Percentage corn	= 25/36 × 100 =	69.4
Percentage poultry concentrate	= 11/36 × 100 =	30.6
Total %		100.0

The Algebraic Equations Method:

$$x = \text{lb corn } 9$$

$$y = \text{lb Poultry Concentrate } 45$$

$$x + y = 100\text{ lb}$$

$$0.09x + 0.45y = 20$$

$$\text{Subtract } 0.09x + 0.09y = 9$$

$$0.36y = 11$$

$$y = \frac{11}{0.36} = 30.6\text{ lb or \% concentrate}$$

$$x = 100 - 30.6 = 69.4\text{ lb or \% corn}$$

(3) Formulation of a Ration Using 3 Ingredients

Formulating a 20% starting ration containing 2 or more grains and a 45% Poultry Concentrate use a mixture of corn 9 (8 parts) and oats 11 (2 parts).

Oats 11 × 2 = 22 units protein

Corn 9 × 8 = 72 units protein

No. parts 10 = 94 units protein

1 part = 9.4 units protein in the corn-oats mixture

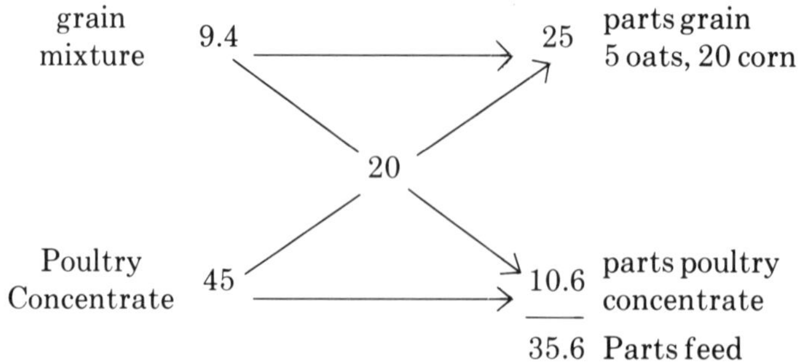

$$\text{Percentage oats} = \frac{5}{35.6} \times 100 = 14.0$$

$$\text{Percentage corn} = \frac{20}{35.6} \times 100 = 56.2$$

$$\text{Percentage poultry concentrate} = \frac{10.6}{35.6} \times 100 = 29.8$$

The Algebraic Equation:

y = lb concentrate

x = lb corn 9 + lb oats 11 per 100 lb mix

Using corn 8 parts + oats 2 parts = 9.4% grain mix

x + y = 100 lb of mix

$$
\begin{aligned}
& 0.094x + 0.450y = 20.0 \ \text{(lb protein/100 lb mix)} \\
\text{Subtract } & 0.094x + \underline{0.094y} = \underline{9.4} \\
& 0.356y = 10.6
\end{aligned}
$$

$$\text{Percentage poultry concentrate} \quad y = \frac{10.6}{356} = 29.8$$

$$\text{Percentage corn and oats} \quad x = 100 - 29.8 = 70.2$$

Percentage corn = 70.2 × 0.80 = 56.2

Percentage oats = 70.2 × 0.20 = 14.0

(4) Formulation of a Ration Using Available Ingredients with Premixes

Steps:

No. 1

Develop a ration – 18% grower for White Leghorn Pullets

No. 2

Obtain a vitamin premix
Obtain a salt (NaCl) trace mineral (T.M.) premix

No. 3 Available Ingredients[1]

Soybean meal (50)
Ground yellow corn (9)
Wheat middlings (14)
Ground oats (11)
Alfalfa meal (17)
Crushed limestone (40)
Dicalcium phosphate (17–30)
Salt (NaCl) trace mineral premix
Vitamin premix

After crushed limestone (40) is 40% calcium

After dicalcium phosphate (17–30) is 17% phosphorus and 30% calcium

No. 4 Specifications

Crude protein	18 %
Calcium	1 %
Phosphorus	0.6%

Ingredient and Premix Fixed Amounts

Animal Protein

Meat scraps 55 or	2.5%
Fish meal 60 or	2.5%
Poultry by-product meal 55	2.5%

[1]The numbers after the ingredients designate percentage protein in all except minerals.

Vegetable Protein

Soybean meal 50	open
Ground yellow corn 9	open
Alfalfa meal 17	2.5%
Wheat midds 14	5.0%
Ground oats 11	5.0%
Limestone 1% over 95% calcium carbonate	
Dicalcium phosphate 1% over 90% available phosphorus	
Salt (NaCl)	0.3%
Vitamin T.M. premix	10 lb per ton
Antibiotic	10 g per ton
Acereomycin or	
Terramycin or	
Bacitracin	
Antioxidants	
BHT	20 g per ton

Ground yellow corn 9	open
Soybean meal 50	open
Meat Scraps 55	2.5%
Ground oats 11	5.0%
Wheat middlings 14	5.0%
Alfalfa meal 17	2.5%
Limestone 40	1.0%
Dicalcium phosphate (17–30)	1.0%
Vitamin premix	0.5%
Salt	0.3%
Antibiotics	10 g
Antioxidants	20 g

No. 5

Determine the total protein and weight of fixed ingredients needed per ton

	Fixed Weight lb	Fixed Protein lb
Meat scraps 55	50	27.5
Ground oats 11	100	11.0
Wheat midds 14	100	14.0
Alfalfa meal 17	50	8.5
Limestone 40	20	
Dicalcium phosphate	20	
Salt and T.M.	6	

Antibiotics and anti-oxidants	Add	
Vitamin premix	10	
	356 lb	61.0 lb protein

No. 6. Gross Estimates

Protein needed 360 lb − 61 lb = 299 lb

Need 44 lb for final adjustment (assume it is corn)

Need a mixture of corn 9 + soybean meal 50 to supply 299 lb of protein. Use 1600 lb or corn + soybean − 1600 + 356 fixed ingredients + 44 corn = 2000 lb.

Calculate percentage protein in 1600 lb to yield 299 lb protein.

$$1600 \times x = 299$$
$$x = 299 \div 1600$$
$$x = 0.1869$$

Percentage protein = 0.1869 × 100 = 18.69

No. 7 Determine Amount Corn and Soybean Meal Needed

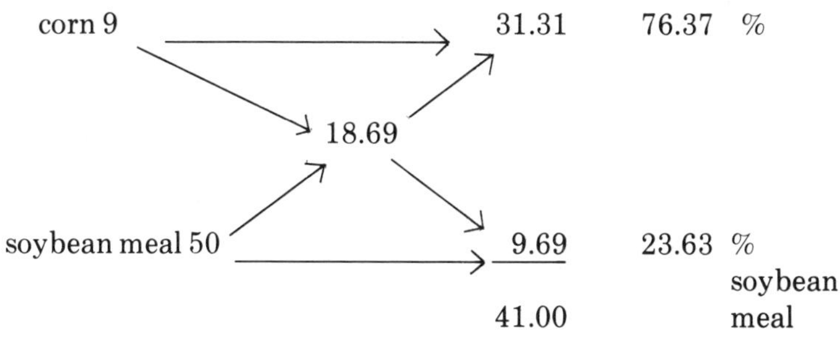

Amount corn = $\frac{31.31}{41}$ of 1600 lb = 1222 lb

Amount soybean meal = $\frac{9.69}{41}$ of 1600 lb = 378 lb

First Estimated Ration

	Weight lb	Protein lb	Calcium lb	Phosphorus lb
Ground corn 9	1222	110	0.45	3.7
Soybean meal 50	378	189	4.50	2.47
Fixed ingredients	356	61	15.41	7.2
Ground corn 9 (flexible)	44	4		
Total	2000	364	20.36	13.37

Ration is adequate in protein, calcium and phosphorus. The vitamin T.M. premix will provide adequate vitamins, trace minerals, antioxidants and antibiotics. Had more been needed they would have replaced ground corn (flexible).

No. 8 The Completed Ration

	lb
Ground yellow corn 9	1266
Soybean meal 50	378
Meat scraps 50	50
Alfalfa meal 17	50
Ground oats 11	100
Wheat midds 14	100
Limestone 40	20
Dicalcium phosphate	20
Salt	6
Vitamin and T.M. premix + antibiotics and antioxidants	10
	2000

The Vitamin Premix Contains

Aureomycin or Terramycin	10	g
Santoquin or BHT	20	g
Vitamin A (IU)	4,000,000	
Vitamin D_3 (ICU)	600,000	
Vitamin E (g)	10	
Vitamin K or equivalent (g)	0.5	
Riboflavin (g)	4	

Pantothenic acid (g)	10	
Niacin (g)	20	
Vitamin B_{12} (mg)	12	
Choline	240	g
Soybean meal to make	10	lb

The Vitamin and Trace Mineral Premix
(Add 10 lb per ton)

Vitamin A	4,000,000	IU
Vitamin D_3	600,000	ICU
Vitamin E	10	g
Vitamin K	0.5	g
Riboflavin	4	g
Pantothenic acid	10	g
Niacin	20	g
Choline	240	g
Vitamin B_{12}	12	mg
Manganese sulfate	4	oz
Zinc carbonate	4	oz
Selenium	90	mg
Soybean meal to make	10	lb
Aureomycin or Terramycin	10	g
Sanoquin	20	g

FORMULATION OF RESTRICTED FEEDS

Poultry require vitamins and minerals on a body-weight daily basis as well as a percentage of the ration. This means that if the requirements have been developed for a complete ad libitum-fed ration and the feed uptake is to be reduced 25%, then the vitamin-mineral content should be increased by approximately this percentage. If not increased, the poultryman may observe sporadic vitamin and mineral deficiencies. Also, if a disease results in lowered feed uptake, then the ration should be fortified to offset the reduced feed uptake. Moreover, some diseases influence a certain nutrient; in this instance, the nutrient should be added as a supplement. Vitamin A absorption is often influenced and zinc absorption is reduced in birds having coccidiosis.

Purified rations are generally used to establish the fundamental biochemistry of nutrients, the requirements of each nutrient, the biological value of some ingredients and the supplemental or compatability of ingredients. Once basic standards or functions have been established, a semi-purified ration is used. This allows a more applied value to be established before the information is used in developing the commercial-type ration.

REFERENCE RATIONS

Many rations have been developed by nutritionists following conferences. These rations have been well-formulated, and have served a useful function in improving the quality of formulated rations. The rations presented in Table 37.1 and composition of ingredients in Table 37.2, were developed by a New England Conference group of nutritionists, but the rations can be used as models for development of feeds for any area in the United States.

Table 37.3 summarizes the composition of rations used throughout the world.

COMPUTERS

Computers are electronic "brains" which are used in the feed industry to take ingredient costs, ingredient composition and nutritional requirements and formulate a desirable ration using the available ingredients. A ration is no better than the specifications fed to the computer. The computer must have the following information or support:

(1) Specific requirements for each nutrient known to vary in quality and amount in the ingredients.
(2) Ingredients whose nutrients have high biological value.
(3) Standards developed for substitutions such as are outlined in the tables on feed composition, and substitutions as recommended under ingredient limitations.
(4) Maximum and minimum amounts of ingredients.
(5) Exact amounts specified when possible.
(6) Premixes containing safety levels of the enclosed nutrients.
(7) Rations manufactured properly and handled, stored and fed according to the planned feeding program.

Guidelines for Computers

Guidelines are essential for ideal computer performance. The computer does not reason, advise or make decisions. The better the specifications, the better the ration. Some desirable specifications are:

(1) Use ingredients of high biological value.
(2) Formulate rations designed for starting chicks, pullet replacements,

broilers, turkeys, geese, ducks, wild birds, etc., or whatever is needed.

(3) Prepare and store vitamin, antibiotic and antioxidant premixes under ideal conditions.

(4) Test the ration with chicks before using it for farm production.

(5) Always, if possible, use *fresh ingredients* and *do not accumulate ground ingredients*. When possible, remove mixed feeds to the poultry operation directly from the mixer.

(6) Do not make erratic substitutions of ingredients. The fewer changes made in a formula after establishing its worth, the better the response by the bird and the easier on the mill operator. Some computer users emphasize how easily rations can be changed; however, several days are required to receive shipments, and erratic grabbing of "bargains" often requires changes not in the interest of the poultry operation.

Computers are used for formulating least-cost poultry feeds which should not be confused with lowest-priced feeds. The former is the lowest-cost formula that contains all the nutritional elements needed for maximum performance. The latter may not be the formula that will produce maximum performance which depends not only on the price of feeds but also on efficiency of utilization. Many factors influence daily feed consumption of laying hens. Among these are size and breed of hen, environmental temperatures, age of hen at start of egg production, type of housing, feeding space for hens, depth of feed in automatic feeders, whether or not the hens are properly debeaked, degree of crowding of hens in cages, availability of ample cool drinking water, disease level in the flock and energy content of the diet. Heavy breeds consume considerably more feed than light breeds, because heavier chickens require more energy for maintenance. A modern strain of small White Leghorns, however, receiving feed containing the same energy level under the same environmental conditions, would consume considerably less feed per day and, therefore, would require a higher protein level. Computer formulation of least-cost rations to contain the exact amounts of nutrients needed by hens, makes it absolutely essential that the poultrymen keep accurate records of daily feed consumption. This allows the feed manufacturer to adjust protein, mineral and vitamin levels to meet exactly the daily needs of the hens for maximum egg production and maximum efficiency of feed utilization.

It is beyond the scope of this discussion to outline detailed procedures for linear programming. In general, data are put on punch cards, then on magnetic tape. The latter is run through a computer which has been programmed to solve the problem within the restrictions placed upon it

TABLE 37.1. NEW ENGLAND CONFERENCE CHICKEN RATIONS(l)[1]

Ingredient		Starter	Grower[2]	Breeder 3.5–5 lb	Breeder 5–8 lb
Ground yellow corn (2,3)[1]		1,248	1,339	1,290	1,333
Wheat middlings		100	300	—	100
Yellow grease (16)		20	20	60	20
Soybean meal (dehulled)		394	203	283	182
Fishmeal, herring (65%) (4,5)		50	—	60	75
Meat and bone meal (50%) (5)		50	50	50	50
Alfalfa meal (17%) (100,000 A per lb)		50	50	50	50
Corn distillers' dr. gr. with sols.		50	—	50	50
Dicalcium phosphate (6)		10	12	9	5
Ground limestone, low magnesium (7)		21	19	141	128
Iodized salt		7	7	7	7
Antibiotic supplement		(8)	—	—	—
Antioxidant		(9)	(9)	(9)	(9)
Coccidiostat		(10)	(10)	—	—
Manganese (g) (11)		52	52	52	52
DL-Methionine		—	—	0.4	—
Zinc (g) (17)		—	—	16	16
Vitamin supplements (12)					
Vitamin A	(USP units)	1,000,000	1,000,000	4,000,000	4,000,000
Vitamin D_3	(ICU)	681,000	681,000	1,362,000	1,362,000
Vitamin B_{12}	(mg)	6	6	6	6
Choline chloride	(mg)	364,000	75,400	114,000	145,000
Riboflavin	(mg)	1,500	1,500	3,000	3,000
Calcium pantothenate	(mg)	3,000	3,000	6,000	6,000
Niacin	(mg)	10,000	10,000	20,000	20,000
	Totals (lb) (21)	2,000	2,000	2,000.4	2,000

Calculated Analysis (25)					
Metabolizable energy	Cal. per lb	1,350	1,340	1,350	1,315
Protein	%	20.00	15.00	17.00	16.00
Lysine	%	1.02	0.63	0.85	0.77
Methionine	%	0.35	0.25	0.33	0.30
Methionine + cystine	%	0.65	0.48	0.59	0.55
Fat	%	4.57	4.71	6.42	4.82
Fiber	%	3.02	3.43	2.55	2.84
Calcium	%	0.90	0.81	3.00	2.72
Total phosphorus	%	0.62	0.63	0.55	0.55
Available phosphorus (13)	%	0.38	0.35	0.48	0.47
Vitamins (units or mg per lb)					
Vitamin A activity	USP units per lb	4,444	4,520	5,992	6,040
Vitamin D	ICU per lb	340	340	681	681
Riboflavin	mg	1.90	1.75	2.57	2.61
Pantothenic acid	mg	5.11	5.08	5.96	6.06
Choline	mg	700.00	400.00	500.00	500.00
Niacin	mg	19.16	22.01	20.98	23.62

[1]See footnotes at end of table on pg. 447.

TABLE 37.1. CHICKEN RATIONS (1)[1] *(Continued)*

Ingredient		Layer			
		Caged Birds		Floor Birds	
		3.5–5lb	5–7 lb	3.5–5 lb	5–7 lb
Ground yellow corn (2,3)[1]		1,301	1,315	1,355	1,313
Wheat middlings		—	100	—	133
Yellow grease (16)		50	20	40	20
Soybean meal (dehulled)		303	240	322	312
Fishmeal, herring (65%) (4,5)		50	40	50	—
Meat and bone meal (50%) (5)		50	50	50	50
Alfalfa meal (17%) (100,000 A per lb)		25	25	25	25
DDGS or equivalent (1)[2]		50	50	—	—
Dicalcium phosphate (6)		10	8	4	4
Ground limestone, low magnesium (7)		154	145	147	136
Iodized salt		7	7	7	7
Antioxidant		(9)	(9)	(9)	(9)
Manganese (g) (11)		52	52	52	52
DL-Methionine		0.5	0.4	0.5	0.6
Zinc (g) (17)		16	16	—	—
Vitamin supplements (12)					
Vitamin A	(USP units)	6,500,000	6,500,000	5,000,000	5,000,000
Vitamin D_3	(ICU)	1,362,000	1,362,000	1,362,000	1,362,000
Vitamin B_{12}	(mg)	6	6	6	6
Choline chloride	(mg)	317,000	154,000	172,000	217,000
Riboflavin	(mg)	2,000	2,000	2,000	1,000
Calcium pantothenate	(mg)	4,000	4,000	2,500	2,000
Niacin	(mg)	20,000	20,000	20,000	20,000
	Totals (lb) (21)	2,000.5	2,000.4	2,000.5	2,000.6

Calculated Analysis (25)					
Metabolizable energy	Cal. per lb	1,340	1,302	1,350	1,300
Protein	%	17.00	16.00	17.00	16.00
Lysine	%	0.85	0.75	0.87	0.74
Methionine	%	0.33	0.31	0.33	0.29
Methionine + cystine	%	0.59	0.55	0.59	0.54
Fat	%	5.90	4.59	5.36	4.27
Fiber	%	2.25	2.56	2.16	2.59
Calcium	%	3.20	3.00	3.00	2.75
Total phosphorus	%	0.55	0.55	0.50	0.50
Available phosphorus (13)	%	0.48	0.47	0.43	0.41
Vitamins (units or mg per lb)					
Vitamin A activity	USP units per lb	6,004	6,020	5,288	5,240
Vitamin D	ICU per lb	681	681	681	681
Riboflavin	mg	1.98	1.98	1.92	1.38
Pantothenic acid	mg	4.81	4.96	4.05	4.10
Choline	mg	600.00	500.00	500.00	500.00
Niacin	mg	20.69	22.95	20.14	22.63

[1]See footnotes on pg. 447.
[2]Corn distillers' dried grains with solubles.

TABLE 37.1. BROILER RATIONS(1)[1] *(Continued)*

Ingredient		Starter (18)	Finisher (18)
Ground yellow corn (2,3)[1]		1,098	1,200
Yellow grease (16)		80	100
Soybean meal (dehulled)		540	385
Corn gluten meal (41%)		100	120
Fishmeal, herring (65%) (4,5)		110	100
Alfalfa meal (20%) (200,000 A per lb)		25	50
Dicalcium phosphate (6)		19	19
Ground limestone, low magnesium (7)		21	19
Iodized salt		7	7
Antibiotic supplement		(8)	(8)
Antioxidant		(9)	(9)
Coccidiostat		(10)	(10)
Organic arsenical supplement (19)		0.1	0.1
Manganese (g) (15)		75	75
Vitamin supplements (12)			
Vitamin A	(USP units)	1,816,000	1,816,000
Vitamin D_3	(ICU)	681,000	681,000
Vitamin K (20)	(mg)	1,000	1,000
Vitamin B_{12}	(mg)	12	12
Choline chloride	(mg)	454,000	390,000
Riboflavin	(mg)	3,000	3,000
Calcium pantothenate	(mg)	5,000	5,000
Niacin	(mg)	20,000	20,000
	Totals (lb) (21)	2,000.1	2,000.1

Calculated Analysis (25)			
Metabolizable energy	Cal. per lb	1,440	1,480
Protein	%	24.00	21.00
Lysine	%	1.33	1.08
Methionine	%	0.45	0.41
Methionine + cystine	%	0.81	0.73
Fat	%	7.07	8.18
Fiber	%	2.16	2.31
Calcium	%	0.85	0.80
Total phosphorus	%	0.63	0.59
Available phosphorus (13)	%	0.40	0.38
Vitamins (units or mg per lb)			
Vitamin A activity	USP units per lb	4,004	5,490
Vitamin D	ICU per lb	340	340
Riboflavin	mg	2.60	2.64
Pantothenic acid	mg	6.13	5.99
Choline	mg	800.00	700.00
Niacin	mg	21.81	21.46
Xanthophyll	mg	10.35	13.39

[1]See footnotes on pg. 447.
[2]Corn distillers' dried grains with solubles.

TABLE 37.1. TURKEY RATIONS (1)[1] *(Continued)*

Ingredient	Starter (0–8 Weeks)	Grower (8–16 Weeks)	Finisher (16–Mkt)	Breeder
			(22)	
Ground yellow corn (2,3)[1]	875	1,160	1,493	1,195
Wheat middlings	100	50	—	250
Yellow grease	40	50	50	20
Soybean meal (dehulled)	660	555	330	162
Fishmeal, herring (65%) (4,5)	115	—	—	100
Meat and bone meal (50%) (5)	100	50	50	50
Alfalfa meal (17%) (100,000 A per lb)	25	25	25	60
DDGS or equivalent (1)[2]	50	50	—	50
Dicalcium phosphate (6)	17	35	25	13
Ground limestone, low magnesium (7)	13	15	17	90
Iodized salt	5	10	10	10
Antibiotic supplement	(8)	—	—	—
Coccidiostat	(10)	(10)	—	—
DL-Methionine	0.1	—	—	—
Ethoxyquin	0.25	0.25	0.25	0.25
Manganese (g) (23)	30	30	30	30
Zinc (g) (24)	30	30	30	30
Vitamin supplements (12)[3]				
Vitamin A (USP units)	7,491,000	7,491,000	7,264,000	4,086,000
Vitamin D_3 (ICU)	1,700,000	1,700,000	1,700,000	1,700,000
Vitamin E (IC)	10,000	5,000	—	30,000
Vitamin B_{12} (mg)	6	6	6	6
Choline chloride (mg)	585,000	320,000	420,000	370,000
Riboflavin (mg	4,000	5,000	5,000	5,000
Calcium pantothenate (mg)	4,500	4,500	6,000	10,000
Niacin (mg)	42,000	46,000	48,000	50,000
Vitamin E (IU)	10,000	10,000	10,000	50,000
Totals (lb) (21)	2,000.35	2,000.25	2,000.25	2,000.25

Calculated Analysis (25)					
Metabolizable energy	Cal. per lb	1,310	1,360	1,440	1,285
Protein	%	28.00	21.32	16.18	17.00
Lysine	%	1.64	1.10	0.75	0.85
Methionine	%	0.48	0.34	0.27	0.33
Methionine + cystine	%	0.88	0.66	0.53	0.58
Fat	%	5.43	5.53	5.82	5.02
Fiber	%	2.73	2.60	2.29	3.38
Calcium	%	1.20	1.01	0.90	2.20
Total phosphorus	%	0.91	0.80	0.66	0.70
Available phosphorus (13)	%	0.65	0.55	0.45	0.61
Vitamins (units or mg per lb)					
Vitamin A activity	USP units per lb	6,016	6,340	6,576	6,427
Vitamin D	ICU per lb	850	850	850	850
Riboflavin	mg	3.28	3.51	3.38	3.72
Pantothenic acid	mg	6.31	5.82	5.87	8.52
Choline	mg	1,000	700	600	650
Niacin	mg	36.82	35.16	33.82	42.70

[1]See footnotes on pg. 447.
[2]Corn distillers' dried grains with solubles.
[3]Biotin and folacin may be required under certain conditions. If desired, they may be added at the rate of 100 mg of biotin and 1000 mg of folacin per ton of feed.

TABLE 37.2. VALUES USED TO DETERMINE THE ANALYSIS OF THE NEW ENGLAND COLLEGE CONFERENCE POULTRY RATIONS

Ingredient	M.E. Cal. per lb	Prot.	Fat	% Fiber	Ca	P	Vit. A IU per lb	Xanthophyll mg per lb
Alfalfa meal (20%)	640	20.0	2.6	17.0	1.90	0.22	150,000	150
Alfalfa meal (17%)	500	17.0	2.5	26.0	1.40	0.21	100,000	120
Barley, ground	1,190	11.0	2.1	6.0	0.07	0.36	—	—
Bone meal, steamed	450	6.5	0.6	2.6	31.30	14.50	—	—
Corn dry steepwater concentrate	1,170	31.0	3.0	—	—	—	—	—
Corn, yellow, ground	1,530	8.9	4.0	2.0	0.02	0.27	2,270	11
Corn DDG with solubles	1,050	28.0	8.0	7.0	0.19	0.35	1,100	—
Corn distillers' dried solubles	1,250	27.0	5.0	2.0	0.35	1.55	—	—
Corn ferm. solubles, dried extract	980	23.4	0.25	0.2	0.07	1.55	—	—
Corn gluten feed (yellow)	650	21.0	1.0	10.0	0.30	0.70	1,000	30
Corn gluten meal (41%)	1,480	41.0	2.0	3.0	0.07	0.37	12,000	60
Corn gluten meal (60%)	1,580	60.0	2.7	2.0	0.18	0.39	20,600	160
Dicalcium phosphate	—	—	—	—	23.00	18.50	—	—
Fat								
Animal tallow, feed grade	3,130	—	97.0	—	—	—	—	—
Fish oil	3,490	—	95.0	—	—	—	—	—
Hydrolyzed animal and vegetable fat	3,400	—	99.0	—	—	—	—	—
Poultry oil	3.720	—	98.0	—	—	—	—	—
Yellow grease	3,400	—	99.0	—	—	—	—	—
Fishmeal								
Canadian herring, 72%	1,450	72.0	9.0	0.5	2.40	2.20	—	—
Maine herring, 65% (10% fat)	1,420	65.0	10.0	0.5	3.00	2.30	—	—
Atlantic menhaden + 40% solubles	1,330	62.0	9.0	0.5	5.50	3.00	—	—
Atlantic menhaden (58–65% protein)	1,240	60.0	9.0	1.0	6.20	3.60	—	—
Peruvian	1,200	65.0	4.0	1.0	3.80	2.50	—	—
Fish solubles, condensed	575	31.5	4.6	0.5	0.15	0.63	—	—
Grain sorghums, Milo	1,500	10.5	3.0	2.5	0.04	0.32	—	—
Hominy feed (yellow)	1,370	10.6	5.8	5.1	0.08	0.52	1,600	—
Hydrolyzed poultry feathers	1,160	87.0	5.0	1.4	0.22	0.77	—	—
Limestone, ground, high calcium	—	—	—	—	38.00	—	—	—
Limestone, ground, low magnesium	—	—	—	—	35.00	—	—	—

Meat and bone meal	900	50.0	10.0	2.5	10.00	5.00	—	—
Molasses, cane	890	3.0	—	—	0.65	0.08	—	—
Oats, ground	1,170	11.0	4.5	11.0	0.10	0.36	—	—
Oyster shells, ground	—	—	—	—	38.00	—	—	—
Peanut meal	1,200	44.0	6.1	11.8	0.17	0.56	—	—
Poultry by-product meal	1,260	55.0	12.0	2.5	3.00	1.35	—	—
Sesame meal	820	42.0	7.0	6.5	2.00	1.30	—	—
Soybean meal (dehulled)	1,120	49.0	0.5	2.5	0.20	0.60	—	—
Soybean meal (44% protein)	1,020	44.0	0.5	5.2	0.30	0.60	—	—
Wheat, ground	1,410	12.0	1.9	2.4	0.04	0.39	—	—
Wheat, bran	510	15.6	4.2	9.0	0.11	1.21	—	—
Wheat middlings	890	16.0	4.5	7.5	0.08	0.93	—	—
Whey, dried	860	12.5	0.7	0.3	0.85	0.70	—	—
Yeast, dried	880	45.0	3.0	1.4	1.30	1.25	—	—

[1]Wherever substitutions are made in the rations, the total nutrient content should be adjusted to meet established requirements.

[2]200 to 400 lb of coarsely ground wheat or yellow hominy may be used to replace an equal amount of corn. If wheat is used, add 200,000 IU Vitamin A for each 100 lb of corn removed.

[3]There is usually some loss of Provitamin A activity in corn and alfalfa meal during storage. If stored ingredients are used, it may be advisable to increase the added Vitamin A level of the ration by 1000 or 2000 IU per lb. This can be accomplished by increasing the recommended supplement by 2,000,000 or 4,000,000 IU per ton of feed.

[4]If a high-salt fish meal is used, *omit* the added salt equal to that supplied by the fishmeal.

[5]Poultry by-product meal may be substituted for all of the meat and bone scrap and up to 50% of the fishmeal. Correct for calcium and phosphorus loss due to substitution of poultry by-product meal.

[6]Based on an 18.5% phosphorus product: Steamed bone meal or defluorinated rock phosphate may replace the dicalcium phosphate on a phosphorus basis.

[7]Based on 35% calcium limestone.

[8]An antibiotic may be used in these rations at the level recommended by the manufacturer.

[9]1.2-dihydro-6-ethoxy-2.2,4-trimethylquinoline (ethoxyquin) or butylated hydroxytoluene (BHT) are antioxidants used in the chick starter and broiler ration and ethoxyquin in the breeder ration at the 0.0125% level to help prevent the appearance of encephalomalacia (crazy chick disease) and to help prevent the destruction of the fat-soluble vitamins. If desired, either may also be added at the 0.0125% level (¼ lb per ton) to the other rations.

[10]A coccidiosis-controlling drug may be used in these rations at the level recommended by the manufacturer.

[11]This amount of manganese will be furnished by 0.5 lb of manganese sulfate or 0.21 lb of manganous oxide (70% feeding grades). An equivalent amount of manganese may be added from other acceptable sources.

[12]Caution should be used when high-potency vitamin mixes are involved. is recommended that 10 lb be the minimum amount of any item added to 1 ton of feed to ensure proper mixing. Thus, high potency vitamin, mineral, or drug mixes should be premixed with a carrier (such as corn meal) to such a dilution that 10 lb of the final mix will be added for each ton of feed mixed. Minerals and vitamins should not be premixed together.

[13]Available phosphorus has been taken as 30% of total phosphorus from plant sources for chicks, and 75% of total phosphorus from plant sources for adult birds. Phosphorus from other than plant sources are considered to be 100% utilized.

[14]For those persons wanting a restricted feeding program, feed the Grower at 80% or less of the amount normally consumed.

[15]This amount of manganese will be furnished by 0.7 lb manganese sulfate or 0.3 lb manganous oxide (70% feeding grades). An equivalent amount of manganese may be added from other acceptable sources.

[16]Stabilized fats may replace an equal amount of cereal grains to provide a higher energy level. Where maintaining body weight in layers is a problem, increase fat by 1% during the winter months.

[17]Approximately this amount of zinc will be furnished by 29 g of zinc carbonate or 20 g of zinc oxide. An equivalent amount of zinc may be used from other acceptable sources.

[18]To be fed as all-mash rations. Feed starting ration until birds are 35 days old.

[19]Based on 3-nitro-4-hydroxyphenylarsonic acid at a level of 45 g (0.1 lb) per ton. Other compounds that may be used at a level recommended by the manufacturer are sodium arsanilate or arsanilic acid.

[20]Based on menadione. Other compounds supplying equivalent levels of Vitamin K may be used.

[21]If an even 2000 lb is desired, adjust by removing or adding ground yellow corn.

[22]May be fed with grain after 20 weeks.

[23]This amount of manganese will be furnished by approximately 0.3 lb of manganese sulfate or 0.13 lb of manganous oxide (70% feeding grades). An equivalent amount of manganese may added from other acceptable sources.

[24]This amount of zinc will be furnished by approximately 53 g of zinc carbonate or 37 g of zinc oxide. An equivalent amount of zinc may be added from other acceptable sources.

[25]Any discrepancies in calculated analysis that occur in the decimal part of the figures are due to rounding errors built in by the computers used in the calculations.

TABLE 37.2. *(Continued)*

Ingredient	mg per lb				Arginine	Lysine	Methionine	Cystine	Tryptophan
	Ribo.	Panto.	Choline	Niacin					
Alfalfa meal (20%)	8.40	17.00	500	24.00	1.10	0.98	0.34	0.38	0.27
Alfalfa meal (17%)	6.70	16.00	400	22.00	0.76	0.86	0.30	0.33	0.25
Barley, ground	0.59	3.00	500	31.10	0.54	0.34	0.19	0.22	0.14
Bone meal, steamed	—	—	—	—	1.70	0.90	0.20	0.10	0.05
Corn dry steepwater concentrate	—	—	—	—	—	—	—	—	—
Corn, yellow, ground	0.73	2.09	227	9.08	0.44	0.21	0.19	0.16	0.08
Corn DDG with solubles	3.45	5.30	1,800	37.00	1.10	0.65	0.44	0.38	0.22
Corn distillers' dried solubles	7.00	11.00	2,500	68.00	0.94	0.84	0.50	0.37	0.18
Corn ferm. solubles, dried extracts	2.40	6.65	484	109.50	1.00	1.00	0.54	0.49	0.10
Corn gluten feed (yellow)	1.20	4.00	500	30.00	0.80	0.80	0.30	0.50	0.20
Corn gluten meal (41%)	1.00	6.00	750	13.60	1.34	0.74	0.98	0.64	0.22
Corn gluten meal (60%)	0.80	5.40	215	28.85	2.20	1.40	1.60	0.90	0.30
Dicalcium phosphate	—	—	—	—	—	—	—	—	—
Fat									
Animal tallow, feed grade	—	—	—	—	—	—	—	—	—
Fish oil	—	—	—	—	—	—	—	—	—
Hydrolyzed animal & vegetable fat	—	—	—	—	—	—	—	—	—
Poultry oil	—	—	—	—	—	—	—	—	—
Yellow grease	—	—	—	—	—	—	—	—	—
Fishmeal									
Canadian herring, 72%	4.50	8.00	2,200	35.00	5.30	6.50	2.10	1.00	0.90
Maine herring, 65% (10% fat)	4.00	4.00	1,800	40.00	4.80	5.70	1.90	0.80	0.80

Atlantic menhaden + 40% solubles	2.10	3.50	1.300	20.00	3.97	5.50	1.80	0.79	0.73
Atlantic menhaden (58–65% prot.)	3.10	4.00	1,230	38.59	3.97	5.50	1.80	0.79	0.73
Peruvian	4.00	4.50	1,600	40.00	3.60	5.30	1.70	0.75	0.70
Fish solubles, condensed	7.70	17.25	1,250	117.50	1.55	1.65	0.56	0.21	0.16
Grain sorghums, Milo	0.50	3.90	200	33.00	0.36	0.23	0.16	0.17	0.10
Hominy feed (yellow)	1.06	3.98	501	20.68	0.50	0.25	0.15	0.13	0.12
Hydrolyzed poultry feathers	0.90	4.25	400	11.00	5.75	1.75	0.56	2.65	0.57
Limestone, ground, high calcium	—	—	—	—	—	—	—	—	—
Limestone, ground, low magnesium	—	—	—	—	—	—	—	—	—
Meat and bone meal	2.10	1.50	990	21.40	3.15	2.60	0.58	0.57	0.31
Molasses, cane	1.23	15.58	366	14.40	—	—	—	—	—
Oats, ground	0.59	4.65	420	6.60	0.70	0.40	0.20	0.21	0.14
Oyster shells, ground	—	—	—	—	—	—	—	—	—
Peanut meal	2.40	25.05	795	76.20	4.40	1.33	0.47	0.71	0.44
Poultry by-product meal	4.50	4.00	2,720	18.00	3.75	3.70	0.99	0.98	0.45
Sesame meal	1.50	2.50	680	6.00	4.80	1.30	1.40	0.57	0.78
Soybean meal (dehulled)	1.20	6.50	1,150	13.50	3.53	3.14	0.69	0.72	0.63
Soybean meal (44% protein)	1.49	6.22	1,050	9.08	3.16	2.82	0.63	0.66	0.57
Wheat, ground	0.55	5.00	330	27.50	0.53	0.38	0.20	0.22	0.16
Wheat, bran	1.35	11.40	460	139.00	0.93	0.56	0.21	0.21	0.22
Wheat middlings	1.15	7.20	500	56.50	0.86	0.59	0.21	0.19	0.20
Whey, dried	10.90	21.00	700	8.50	0.38	0.97	0.28	0.34	0.21
Yeast, dried	16.00	50.00	1,500	200.00	2.10	3.23	0.78	0.48	0.56

TABLE 37.3. TYPICAL RATIONS AROUND THE WORLD

Ingredient	Israel	Singa-pore	Philip-pines	Thai-land	Bolivia	Colom-bia	Costa Rica	Mexico	Para-guay	Peru	Uru-guay	Queens-land	Hawaii
Corn	35.0	43.0	45.0	7.0	45.0	73.0	25.0	69.3	45.0	18.0	40.0	30.0	—
Oats, barley	—	—	—	—	—	—	—	—	—	—	20.0	—	—
Wheat, millet	—	—	—	—	—	—	—	—	—	—	—	—	—
Sorghum	25.7	—	—	—	20.0	—	30.0	—	—	—	3.0	24.0	—
Rice	—	—	—	8.0	—	—	—	—	—	—	—	—	—
Fat	1.0	20.0	—	—	—	—	—	—	—	—	—	—	—
Bran, midds	—	5.0	20.0	57.0	14.0	—	15.0	7.5	35.0	30.0	10.0	10.0	—
Gr. feeds, dr. alfalfa	5.0	—	5.0	—	5.0	3.0	—	3.0	3.4	5.0	—	3.5	30.0
Molasses	—	—	—	—	—	—	3.0	—	—	—	—	—	30.0
Meat scrap, blood meal and bone	—	6.0	—	—	3.0	3.0	—	2.0	7.0	—	7.0	30.0	22.5
Fishmeal	2.0	10.0	20.0	7.0	4.0	—	—	2.5	—	8.0	—	—	5.0
Skim milk, yeast	—	—	—	—	—	—	—	—	—	—	—	—	—
Veg. prot. meal	20.0	6.0	11.0	9.5	3.0	9.0	16.0	5.0	10.0	—	10.0	—	12.5
Dicalphos. or bone	1.5	1.5	—	—	3.5	0.5	—	2.5	—	5.0	1.5	—	—
Oyster shell	9.5	8.0	2.0	2.0	—	—	5.0	4.2	5.0	5.0	4.0	2.0	—
Salt	0.1	0.2	—	1.0	0.5	0.5	—	0.5	0.5	0.5	1.0	0.5	—
Vit., tr. min.	—	—	2.0	—	—	—	—	—	0.1	1.0	—	—	—
Grit	—	—	—	—	2.0	—	—	—	—	—	—	—	—

TABLE 37.3. *(Continued)*

Ingredient	Congo	Dahomey	Ghana	Kenya	Morocco	Ibadan	W. Nigeria	Sudan	Tanzania	Sahel	Burma	Cambodia	Ceylon	Allahabad
Corn	40.0	29.5	49.0	27.0	13.0	67.0	65.6	35.0	36.0	26.0	20.0	29.5	—	36.0
Oats, barley	—	—	—	13.0	46.0	—	—	—	—	30.0	—	—	—	—
Wheat, millet	20.0	—	—	—	—	—	—	—	—	10.0	—	—	—	12.0
Sorghum	—	35.0	—	—	—	—	—	34.0	6.0	—	—	—	42.0	—
Rice	10.0	—	—	—	—	—	—	—	3.0	—	19.0	20.0	—	—
Fat	—	—	—	—	—	—	—	—	—	—	—	—	—	—
Bran, midds	—	—	20.0	20.0	26.0	7.0	4.0	—	23.5	14.0	20.5	15.0	19.5	12.0
Gr. feeds, dr. alfalfa	7.0	3.0	3.0	8.0	4.0	—	3.5	—	4.0	5.0	—	3.5	2.0	—
Molasses	—	—	—	—	—	—	—	—	—	—	—	—	—	—
Meat scrap, blood meal and bone	4.0	—	11.0	20.0	4.5	5.0	1.0	10.0	19.5	—	12.5	—	—	—
Fishmeal	3.0	3.5	11.0	5.0	—	—	2.0	—	—	8.0	—	15.0	8.5	5.0
Skim milk, yeast	—	5.0	—	—	—	—	—	—	—	—	—	—	—	—
Veg. prot. meal	12.0	15.0	—	4.0	—	17.0	15.0	15.0	4.0	7.0	22.0	15.0	21.5	30.0
Dicalphos. or bone	0.5	1.4	3.0	—	3.5	2.0	2.5	1.0	2.0	—	1.5	—	—	—
Oyster shell	3.0	7.0	3.0	3.0	3.0	1.0	6.0	1.0	—	1.0	1.5	2.0	—	2.0
Salt	0.5	0.3	—	—	—	1.0	0.4	0.5	0.5	1.0	0.5	—	0.5	—
Vit., tr. min.	—	0.3	—	—	—	—	—	0.5	0.5	2.0	2.5	—	—	3.0
Grit	—	—	—	—	—	0.5	—	—	—	—	—	—	6.5	—

FIG. 37.3. THE COMPUTER IS BEING USED BY THE FEED INDUSTRY TO SOLVE COMPLICATED PROBLEMS IN FEED MANUFACTURE

by the programmer. The programmer sets up a schedule which relates the various bits of information to the problem at hand.

Linear Programming

From what has been stated heretofore, it is evident that there is no one ideal formulation for a poultry feed. Values are based on the nutrients that natural feeds contain. On the basis that there is complete interchange of nutrients among sources, the computer is used to determine least-cost formulas. Linear programming (LP) is technically a mathematical procedure for obtaining a value-weighting solution to a set of simultaneous equations. Many solutions to each series of equations are possible, but when the factor of cost is applied, there is one least-cost combination. The computer takes information provided by the nutritionist and, being capable of making thousands of calculations in a very short time, determines solutions. Computer input reflects the knowledge and ability of nutritionists. The computer is incapable of correcting human errors, so LP results will be no better than the values entered into the machine.

Application to Feed Manufacturing

Application to feed manufacturing is as follows: In choosing feed ingredients to combine in the mix, there are basically 2 aspects with which to be concerned. (1) The mix must meet quality standards, and (2) the cost must be as low as possible. This can be looked at in mathematical terms. A simple mix problem of 6 quality specifications and 10 ingredients could have millions of feasible solutions. Linear programming has changed the dimension of the problem so drastically that the mathematical model of the mix problem may be meaningful to you. Effort is exerted upon the solutions that are of economic interest for a given set of ingredient prices. This is what is called "value weighting." Essentially, the procedure is that of beginning with a solution to the mix problem and considering only a relatively few alternate mixes at one time. If any one of these alternatives is better than the first, the computer selects the one that is best for that set and proceeds to compare it with another set of alternate solutions. This is repeated until no alternate solution lower than the present solution can be found. It is possible by linear programming to determine: (1) a least-cost combination of feedstuffs to meet specific nutrient requirements; (2) acceptance or rejection of ingredients based on their cost and nutritive value; (3) effect of variations in nutrient content of feedstuffs on their economic value; (4) final cost of a diet and the cost contributed by each ingredient; (5) relationship between requirements for any specific nutrient and the cost of the feed formulation; (6) optimum nutrient density of a diet for maximum economic return to the feed formulator; and (7) cost of replacing any ingredient not in the solution and assistance in ordering feed ingredients.

Necessary Information for Computer

The following information must be supplied to the computer: (1) nutrient requirements or restrictions; (2) available nutrient content of the feedstuffs; and (3) relative costs of the feedstuffs. Except for trial-and-error methods, a hand desk-calculator solution of values for ingredients to be used in solving a set of requirement equations is impractical. Eventually, values could be worked out which would satisfy the equations. However, the chance of finding a least-cost solution by hand solution within several working days is remote. A mathematical technique developed for solving a system of formulas has called for the development of more extensive and more exact specifications for feeds.

Computer Limitations

Electronic computers are extremely efficient but lack initiative. The computer does what it is told. It is no miracle gadget. Persons knowledgeable in programming are available and problems are posed. Given the right information, the computers calculate the least-cost formula in the time for a coffee break, but do not care if the formulas are palatable or even completely usable by the chicken. Consequently, the person formulating the feed and the computer must work together. The nutritionist specifies what is necessary and arranges in a set of specifications, information that will provide the computer with the exact limits within which it is free to operate. It is necessary to specify maximum levels of some feed ingredients that are unpalatable, high in fiber, not readily available in large quantity or are subject to other disadvantages. They may be related to toxicity of the material, as in the case of cottonseed meal or alfalfa meal. It is necessary to specify minimum levels of some ingredients that provide unidentified nutrients, enhance palatability or have other desirable characteristics. These are established on the basis of the nutritionist's experience in prior use.

Nutrient Composition Requirement

Knowledge of the nutrient composition of feedstuffs is necessary—average values from tables can be used if the supplies to be used are average. In the case of specific batches of feedstuffs, it is more accurate to have values for the given lot of material. The other information used in linear programming is the price at the point where the feed is to be mixed, and includes transportation costs to that area. Certain ingredients which are in constant use, for example, salt and antibiotics, need not be programmed. From the standpoint of amino acids, methionine and lysine are the two most apt to be deficient. Tryptophan and glycine may be deficient under certain conditions. The remainder of the essential amino acids are ordinarily adequate when most combinations of feeds are used. Since excesses of some amino acids are detrimental, a maximum must be set in these instances.

"LEAST COST FEEDS"

W.V. Waugh of the United States Department of Agriculture was among the first to see the potential of the mathematical procedure of linear programming developed by George B. Danzig in 1947. By 1957, several firms were experimenting with this technique. Among the leaders were G.L. F., McMillen Feed Mills and Nutrena Mills. In 1958, Dr. Hutton published a series of articles on "The Use of Linear Programming in Formulating 'Least cost' Feeds." This development expanded rapidly among the major feed manufacturers who had sufficient volume to justify the added expense, and in many cases, had electronic accounting systems, computers and trained personnel available within their own companies.

Formula costs can sometimes be reduced by $1.00-$3.00 per ton when calculated on the computers rather than by hand, depending on the number of ingredients used and the fluctuation in their prices. Savings realized through linear programming have usually been passed on to the customer through lower prices.

Ultimately, there will be formula changes due to price fluctuations. Formulas may be run as often as desired due to changes of ingredient costs. Feed formulas can be changed and in the hands of the mixer within 30 min. As stated, one problem with computer rations is that there is no easy way to assess palatability. If it is impossible to obtain full consumption of formulated rations, the feeds will not work. Maximum feed consumption is necessary for the feed to be valuable. If feed consumption is limited due to poor palatability, the results will be no better than if nutritional deficiency limits growth. Nutritionists must minimize or be able to assess the sources and magnitude of any variables. In order to make a smooth operation for the computer, certain problems must be mastered. With respect to personnel training, employees must be instructed in the data needed and how they are presented. All data must be double-checked at every step by personnel before placement into the computer. Solutions of feed formulation problems must be prepared so that when they reach management levels, the answers are simplified and concise, requiring minimum paper work. Unless an emergency arises, feed formulas should be changed only when conditions are favorable.

Computers have already made large contributions to the poultry and feed industry, not only in the formulation of feeds but also in the field of genetics. This is only a beginning. The cost of computer services has dropped drastically in relation to the work that they can do. Modern computers can calculate faster, store more information in their memories, and print results at a higher rate of speed than ever before.

OTHER COMPUTER APPLICATIONS

In addition to the feed formulas, computers are used for breeder egg operations and broiler operations. Flock history, as well as management studies, disease, housing, and equipment are sorted and stored automatically on a random access disc by the computer for future use. The computer can provide hatching-egg producers with flock statistics weekly. Future uses for computers are unlimited. Management will have at hand information which would be too cumbersome to produce by traditional methods. Information on egg supply, production, forecasts, cumulative averages of disease loss, comparative performance of breeds, when layers can be expanded sooner than otherwise in relation to finished broiler output, anticipated market price, the best time to hedge corn and at what price; all this provides invaluable aid in helping management make major decisions on future capital investments. The computer can help in providing sophisticated forecasting techniques on prices, sales mix and cash flow needs.

Computors are priced so that businesses may own or rent small units. Costs vary from a few thousand to several hundred thousand dollars. A computer offers to all levels of management unusual speed in collecting, assimilating and reporting data to facilitate decision-making. It can be used to assist in matters of budget, finance, personnel, inventory, purchasing, analysis, and long-range projections. Computers operate by prodigious numbers of high-speed additions and subtractions, as well as storage and quick retrieval of immmense amounts of data. However, these functions do not constitute thinking (the use of judgement and imagination). That is where nutritionists fit into the picture. As the prices of ingredients change, the new costs are fed into the computer and it automatically alters the formula to maintain the nutritional value at the lowest possible price.

In one instance, on the older computer equipment, computer calculations were made on 252 formulas containing from 8 to 10 requirements and using from 19 to 32 ingredients. Machine time for this job was 26 hr, an average of just over 6 min per formula. Linear programming techniques of formulating feed mixes should be at least given consideration by a majority of the feed manufactures, since in most cases, it would permit them to sell a product of equal or superior quality at a lower price.

COMMERCIAL USE OF COMPUTERS

The computer is a tool for accelerating calculations. Some companies use computers extensively. One such company is engaged in feed manufacturing and soybean processing with a chemical division concerned with the production of isolated proteins and protein concentrates, a grain division and a purchasing division. The research department is involved in areas including nutrition research which is conventional, work on drugs and diseases by veterinarians, and effects of environment on animal performance.

This company uses a nutritional standard for every feed, including: (1) a complete breakdown of all the nutritional factors or levels; and (2) the composition of the ingredients. This is probably one of the weakest areas in formulation. The feed industry uses by-products, and these are not uniform in quality and composition, so it is necessary to be conservative in the values that are applied. Costs of ingredients are obtained from the purchasing departments.

This is how their nutritional standard might look. Protein, fat, fiber, minerals, etc., amino acids, and xanthophyll to control pigmentation in the carcass of a broiler or to control the pigmentation of an egg. Much is not calculated. Vitamins E and K are practically ignored. Vitamin K is in alfalfa. Forced ingredients (those which are to be added irrespective of cost) are used as follows: 3 mg per lb Vitamin B_{12}, irrespective of quantity present in the natural ingredients. The values and availability of some of these vitamins, as well as composition, leave much to be desired. Therefore, they are ignored and put in at a forced level.

What would it mean to calculate Vitamin D in most ingredients? It would only make the problem more complex and no useful data would be gained; so a certain level of Vitamin D is forced. Vitamin A is not considered to any great degree because a stabilized product is used at a certain level. Consideration is given to such things as the effect of an ingredient on palatability. If too much alfalfa is put in a ration, birds will not eat as much. At times, it is necessary to restrict minimums because of feed registration. In other instances, one must eliminate certain ingredients on the basis of cost but a minimum amount is needed to satisfy feed tag purposes.

If too much fishmeal is added to a turkey ration, there may be flavor problems in the carcass. This is controlled by ingredient restriction. Certain restrictions are made because of production problems. A pellet cannot be made with too much of certain ingredients; so restrictions are made to limit their use. Cost data comes from the purchasing department and the analytical composition of ingredients is submitted to the computer area from the research department. Programmers put all this information together.

FEED FORMULA ECONOMICS

An ingredient replacement cost sheet is produced for every plant every week. It gives the ingredient code and the costs of the particular ingredient. Persons in the computer area will run a whole series of formulas every week, primarily for high-usage formulas. If there is much change, the formula will be changed. Formulas are changed for even a very few cents per ton.

Broiler performance is affected by energy and protein levels in the diet. Interrelationships of protein and energy, amino acid level and availability, and other nutritional facts must be taken into consideration.

Rations are designed to place an economic value on protein and energy and their interrelationships. Protein quality and other nutritional factors should be held constant as a function of energy. As a result of deciding this experiment using the least-cost approach, it is possible to draw some rather valid conclusions of where one should operate and get the least cost per lb of broiler meat. This gives the general ideas of how some of the problems are approached. Where economics is a rather major consideration, this is important. The main thing is to gather the maximum number of facts and correlate them.

If the right ingredients are used, the computer will indicate feeds with nutritional variations less than those obtained with hand-calculating. Formulas used to change at least twice per year. With the new alfalfa in the fall, the feed was greener in color. With the computer, there was too much variation in color, giving the impression that the feed was changed.

38

Feed Manufacturing

Feed accounts for over 60% of the cost of producing poultry. It involves many facets of the poultry industry. Feed mills are of many sizes and located throughout the country. Most of the feed manufactured is poultry feed.

LOCATION OF THE FEED MILL

During the period from 1940 to 1960, many companies consolidated their operations and constructed "super" mills near major shipping or supply points. Today, many of these large mills are idle and are considered the "tombstones" of a fast-changing industry.

The feed mill of today is small, automated, and located close enough to the poultry farm operations for "bulk-delivery." Small prefabricated storage facilities have encouraged the use of locally-produced grains and minor feed ingredients. Most feed ingredients are delivered to the mills by large trucks. Many trucks making grain deliveries are operating as a return-haul operation. This reduces the cost to the mill. The trucks operating under Federal Control generally haul ingredients, such as soybean meal, which have "set" freight rates.

BUYING OR MARKETING INGREDIENTS

Most feed mills effectively use the services of feed brokers and jobbers. The reliable broker has the best knowledge of availability of feed ingredients, costs, regulations, and delivery to anyone in the feed business. He is paid per carrier load and each shipment is a cash transaction.

MARKETING GRAIN THROUGH A GRAIN EXCHANGE

The grain exchange is a public marketplace where grain supply and demand meet—where processors, storage elevator operators and expor-

TABLE 38.1. FACTORS AFFECTING PRICES OF GRAIN IN THE UNITED STATES

Supply	Demand
(1) Carryover of stocks	(1) Domestic
(a) Government	(a) Processors
(b) Nongovernment	(b) Livestock producers
(2) Production	(c) Storage operators
(a) Amounts placed under loan	(d) Speculative demand
(b) "Free" production	(e) Other
	(2) Export
Other (related to supply and demand)	(a) Production in major exporting and importing countries
(1) Government loan rate	(b) Processors
(2) CCC sales policy	(c) Livestock producers
(3) Government and private estimates and forecasts	(d) Storage
(4) Hedging pressure and trading systems of speculators (futures markets are directly involved but they affect cash, also).	(e) Speculation
	(f) Food for peace authorization
	(g) Trade regulations
	(h) International trade agreement
(5) War scares	(i) Other
(6) Other	

ters go to purchase the farmers' grain. Its trading activities are governed by rules and regulations of the exchange itself and it is licensed by the United States Department of Agriculture. It neither buys nor sells grain nor sets prices. There are exchanges in Chicago, in the corn belt; Minneapolis, in the spring-wheat country; Kansas City in the winter wheat-producing area; exchanges are also located in St. Louis, Omaha, Toledo, Peoria, Indianapolis and other cities.

Commercial marketing of grains begins at the country elevators which accumulate grain supplies from farmers and assemble carlot shipments. Grain is shipped to the terminal markets in care of an exchange member. His job is to see that the car of grain is sold to the highest bidder and for this service he receives a commission set by the exchange. In some exchanges, there are 2 markets: the cash market and the futures market. The cash-grain receiver sells to processors, terminal elevator operators and exporters. Processors buy for their own needs, which may be flour, cereal or feed. Terminal elevators are reservoirs of the grain trade which buy the surplus for storage and distribution at a later date. Exporters buy for the foreign trade.

Exchange Operations

Thus, the grain exchanges maintain a market where buyers and sellers of grain meet and carry on their business. Members are from all phases of the grain business who follow the same rules and code of ethics to main-

FIG. 38.1. GRAIN EXCHANGE IN CHICAGO WHICH PROVIDES A CASH MARKET AND FUTURES MARKET
The grain exchange is a marketplace where supply and demand for grains meet. It is licensed by the U.S. Department of Agriculture and governed by definite rules and regulations.

tain membership. It is governed by 15 directors, a president, and 2 vice presidents who serve without pay.

Trading in commodity markets is done by outsiders whose callings range from professional speculators to consumers who hope to make a killing on a small investment. Buying a commodity in the futures market is obligating yourself to accept delivery of a certain quantity at an agreed-upon price at some particular time in the future. You deposit "earnest" money with your broker and promise to accept delivery. You do not pay for it until the commodity is actually delivered. Your broker sends your order to a broker on the floor of the exchange who buys the lot from another broker who has a client who wants to sell. The price is hammered out in the open in back-and-forth bidding. A clearing house of member firms associated with each exchange guarantees that buyers and sellers will perform their obligations. A peculiar aspect of the futures contract made by the buyer and seller is that it is rarely fulfilled. Only a very small percentage of commodity futures deals are ever consummated by a transfer of the product itself. Only a few traders in the market actually want to accept or make delivery of the grain at the agreed-upon price. They can do so by not offsetting their original buy or sell order. They will pay or be paid for the grain when it is delivered.

To the outsider, commodity futures seem to be nothing more than an elaborate guessing game. This is a valid conclusion, yet they do perform certain marketing functions, chiefly by providing firm supplies of grains and protecting against unfavorable price changes. Hedging operations have the aim of protecting the businessman against adverse price changes. There are standard contracts as to size for each commodity. For example, corn is traded in units of 5000 bu No. 2 yellow corn. Other prescribed grades can be delivered against the contract at fixed premiums and discounts on the standard grade.

Commodity futures trading is a thrilling and sometimes profitable game, but there are drawbacks. An outsider who has no inventories to hedge would participate in the market as a speculator. Commodity futures react to innumerable forces. Unlike a security, a futures contract cannot be salted away in a safety deposit box. There are markets in about 30 commodities, from butter to zinc. From the standpoint of feeds and feeding, there are markets in corn, oats, barley, grain sorghums, wheat, rye, soybeans, fishmeal and soybean meal.

Futures Market Trading

Futures market trading is a method of buying and selling grain in specified quantities for delivery at a later date. It is conducted in trading pits at the exchange, one for each kind of grain handled. Buyers and sellers shout out their offers and bids and at the same time hand-signal the number of bushels offered or bid, and the price. Delivery months are commonly July, September, December, March and May. In July, winter-wheat crops are moved to market. In May, there is a cleanup of the old crop.

Prices are reached through open, competitive bidding; daily price fluctuations are a reflection of changing supply and demand. Supply is influenced by weather, plant and insect disease, drouth, floods and many other forces. Demand is influenced by government support and loan programs, the level of export movements, the flour demand by bakers, seasonal activity and war.

Hedging

Hedging is a grain term meaning price insurance. Millers, country elevator operators, processors, merchandisers and exporters of grain can protect themselves, through hedging, against price risk, and thus operate on narrower margins of profit. To hedge, a person does 2 things at once: He operates in both the cash and futures market at the same time, but in opposite directions. If he buys X bushels of cash wheat, he sells X bushels of futures wheat. When he sells the feed or flour he makes from X bushels of wheat, he then buys X bushels of futures wheat. Selling the feed or flour cancels the cash deal, and buying the futures wheat cancels the futures contract.

In the meantime, any rise or decline is automatically taken care of, and the miller knows when he buys the grains what he will have to charge for his product to break even. Without hedging, he could not foretell the price until he is ready to sell. The above example is what may be called a "perfect" hedge; it cannot always be accomplished; no insurance affords 100% protection. Actually, the hedging transactions are rather complex business operations, and the example given here is only intended to illustrate the fundamentals.

The futures markets serve both producers and consumers of grain. From the farmer's point of view, this system means a year-round market, available price information and low middlemen costs. Hedging is only possible in a speculative futures market. A speculator performs a very important function in grain marketing and makes hedging possible. The speculator is a professional risk-taker who takes price risks in the hopes of making a profit. Thus, many different business operations are involved in bringing grain from the farm to the table in the form of cereal or bacon and eggs.

Role of Large Feed Manufacturers

Larger feed firms also have greater opportunity to substitute feedstuffs according to changing market prices. Since they are engaged in applying new research findings to practical feeding operations, much effort is directed to finding the best combinations of ingredients for each feeding operation. Feed mills with few or no facilities for testing mix to customer specifications put out untested products. These mills may overcome the research deficiency simply by overformulating, that is, putting in more of certain critical ingredients than are required. Overformulating is expensive. Both formulation and manufacture of feeds have become increasingly complex. Formulae of some feeds now call for 40 or more ingredients in carefully controlled proportions. This complexity has benefited those with technical knowledge, allowing them to substitute ingredients in response to price and quality changes without changing nutritional results. For example, corn begins to lose Vitamin A after about 6 months in storage. Proper inspection and testing can detect this loss and substitutes can be used to rectify the loss in Vitamin A content.

FEED TECHNOLOGY

The feed industry, uses many terms with definitions far removed from some of those found in the dictionary. Obvious examples include pellets, concentrates, mash, flaking, dressing or scalping. Feed ingredients are defined adequately in the Official Publication, Association of American Feed Control Officials, and in this book.

Feed-mill design requires the solving of hundreds of related individual problems. The end result is a coherent unit that meets all needs. Factors that contribute to satisfaction are: flexibility, profitability, efficiency, quality, low first cost, appearance, location, expandability, storage capacity and low maintenance. Flexibility means being able to adjust to highs and lows of sales and remain efficient. Mill designs are not standardized. Probably no industry exercises so little control over raw material, yet is required to produce products within such close tolerances.

For many years, feed production consisted of the manual blending of ingredients to make a crude ration. Machinery and methods from the milling of flour, namely grinding, sizing and separating, were incorporated into feed production. The production of feed became as complicated as flour milling, but so dissimilar that the two were completely separated in production facilities. Roller mills, sifters, screw conveyors and elevator legs were augmented by batch- and continuous-mixing lines, hammer mills, pelleting equipment, and all other facilities included in the modern feed plant.

Only in recent years have the tools of scientific management been applied to feed production. Once it is decided what to put into the feed, it is more than just taking a couple of scoops and shovels and putting it in. What about incorporation of microingredients that are put into feeds for purpose? They are expensive, and some return for the money invested if required.

A very concentrated material is put in feed at a level of 1 part in 400,000 and this is difficult to do. Chemical manufacturers supply a diluted Vitamin B_{12}, such as 100 or 200 mg per kilo, a considerable reduction. In the feed plant, another dilution has to be made to 2 to 5 lb per ton, so that very active trace ingredients will be distributed uniformly in the completed feed.

Feed Elevating and Conveying Equipment

Feed elevating or conveying equipment, such as the portable auger, vertical auger and cup elevator, are used. Steel bins, round top or flat, and storage and small bulk feed bins with hopper bottoms, are commonly used. Some examples of farm feed equipment are hammer mills, roller mills, metering or weighing boxes, timer equipment and scales for trucks and feed wagons. Mechanization of farm feeding improves working conditions and contributes to a higher standard of living for the producer. A few basic types of equipment do nearly all the bulk movement jobs. Increase in size of farm units to remain competitive requires considerable outside capital. Such financial needs have been met by traditional sources: commercial banks, government sources and feed manufacturers. Credit expansions, particularly in poultry, reached their peak in 1958. Again, the larger feed manufacturers, through their dealers, have been in a better position to offer credit to feeders. Credit has been extended for farm-mixing and bulk-handling equipment in return for guaranteed purchase of the manufacturer's concentrate. Ironically, feed manufacturers have been pushed by competitive pressure to finance the very things which compete with their business.

Role of Federal Government

The Federal Government has played a major role in agriculture since the early 1930s. Its activity has centered primarily on controls for wheat and other food crops, and less stringent control programs for corn and other feed crops. Production control and acreage restriction have been extended to corn, and legislation has been passed that would permit extension of production controls to turkeys. Production controls are based on voluntary "participation" elected directly by the farmers involved. The Government, in 1968, dumped corn in the midst of the corn harvest. This was interpreted by trade observers as a means of disciplining farmers who did not sign up for restricted corn acreage. The commercial broiler industry has had recurring problems of oversupply and prices below cost. Integration has been an effort by feed manufacturers to preserve a share of the poultry feed market. The problem of overproduction results from the ease of entry which would make monopoly restrictions impossible. The trend toward increasing controls is vigorously opposed by feed manufacturers as a hard, cold business proposition. There is reason to believe that controls would actually benefit the feed manufacturer. The ultimate objective of agricultural controls is a redistribution of income in favor of the farmer. One of the by-products is to widen the margin of agricultural products. The prices would produce wider profits; the profitability of manufactured feeds would improve proportionately.

The traditional market of large manufacturers has been small or medium-sized farmers who buy feeds on service and programs, as well as on the price. Activities to benefit independent feed dealers, who have suffered from the trend toward integration and direct selling, would benefit large feed manufacturers. Of course, the feed industry is only a part of total animal agriculture. Problems of oversupply and talk of government control are not new in agriculture. Possibly, the government will attempt to stabilize these animal products, and if it succeeds, there is a strong chance that controls may spread to the remainder of agriculture.

Feed Ingredient Storage

Feed ingredients can be stored in bulk bins or bags. A small feed mill may operate with as few as 2 large bins, one for soybean meal and one for shelled yellow corn and several small bins. One small bin for ground yellow corn, one for whole oats, one for ground oats and one for each feed manufactured. The large bins should hold 40 to 60 tons and be equipped with screw elevators, and the small bins should hold 4 to 10 tons and have clam openings.

Each feed mill should have 1 bulk corn bin and 1 bulk oats bin connected to the hammer mill. The hammer mill should be connected with the ground corn and ground oats bins.

The weighing of ground ingredients or other ingredients can best be accomplished by the useage of a weigh-buggy, Fig. 38.2, or a platform scale.

Many fixed ingredients, such as meat scraps, alfalfa meal, vitamin premix, minerals and wheat products can be purchased in paper bags containing 10, 25, 50 or 100 lb. Often, the ration can be formulated to contain these amounts.

Many types of bins are used. Plywood is used for inside square bins. Metal square bins are also used inside mills. Square bins are more desirable for inside use than round bins. The round bins are generally more desirable for outside use because they do not develop water leaks as readily as square bins (Fig. 38.3).

Broiler and egg producers use both square and round bins made with metal or fiberglass for bulk storage for the growing operation. Fiberglass seems to be desirable for not over 3000 lb.

Feed Preparation Operations

Particle Size.—Square-hole, wire-mesh sieves are common devices for measuring particle size. If very small particles are being measured, microscopes equipped with cross-hairs and graduated mechanical stages are used. Very fine particles can be measured by sedimentation techniques. Particles are suspended in liquid or air and allowed to settle. These techniques work best if the particles are essentially spherical or cubical. Express results of a particle-size analysis by stating the size screen through which all material passes. For example, a specification for grinding might call for all materials to pass an eight-to-the-inch mesh sieve.

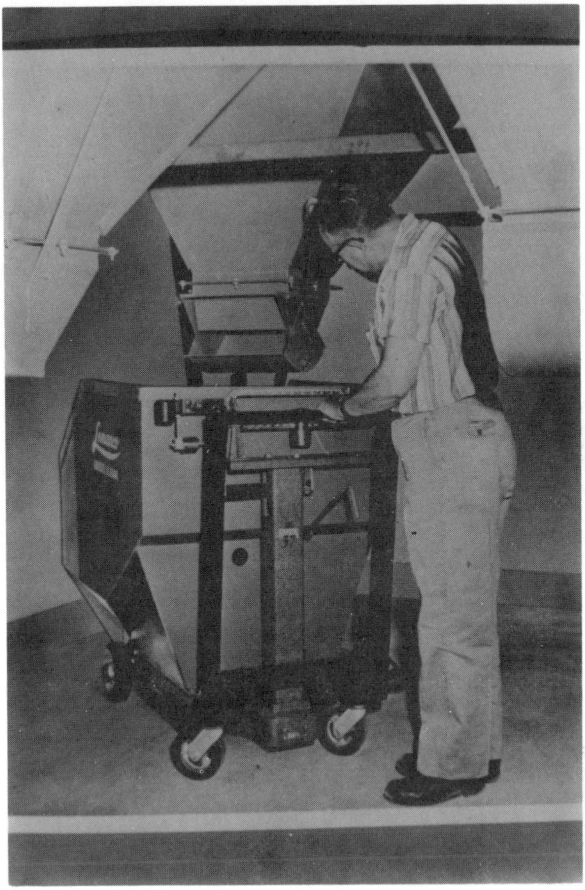

Courtesy of Leach Mfg. Co.

FIG. 38.2. WEIGH-BUGGY

Size Reduction.—Crushing is accomplished by applying a compressive force to particles of feed. The roller mill is most frequently used and produces a product relatively uniform in size, containing only a small amount of finely-pulverized material. If a product is plastic, as steamed grain, the crushed product has the appearance of a flake. Such steamed and rolled material is called "flaked" or "crimped," depending upon whether smooth or corrugated rolls are used. The double roller mill used for this purpose consists of 2 serrated rolls rotating in opposite directions at the same speed. If the rolls have a speed differential, cutting and shearing will take place. If the rolls are operated at the same speed, the

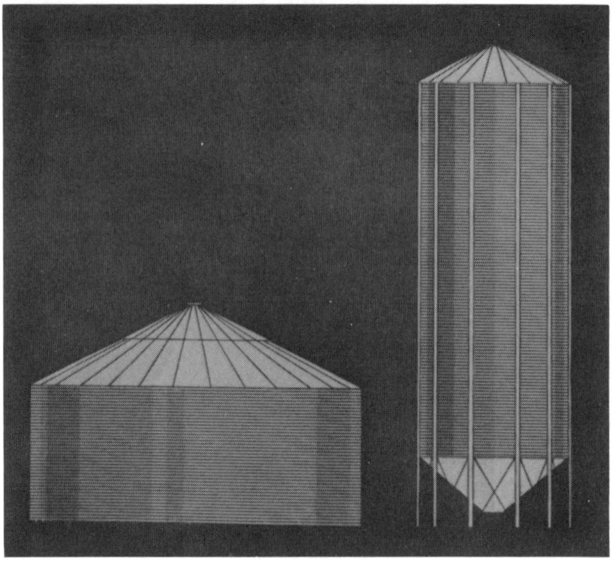

FIG. 38.3. ROUND BULK BINS

TABLE 38.2. GENERAL GRINDING RECOMMENDATIONS FOR CHICKS AND HENS

Ingredient	Chicks	Hens
Corn	Medium to coarse	Medium to whole
Sorghum grain	Medium to coarse	Medium to whole
Oats	Medium	Medium to whole
Barley	Fine	Fine
Wheat	Medium	Medium to whole
Hay	Fine	Fine

reduction is mostly by crushing. Ball mills, percussion mills and jaw crushers are mills that also use the crushing principle. Shearing is a combination of cutting and crushing; the action is similar. A rotary-type knife and stationary bar cutter use the shear principle.

Grinding

Several types of grinders are used in securing the proper feed ingredient size. The most popular is the hammer mill, and every feed mill has 1 or more units (Fig. 38.4). The capacity can be from 1 to many tons per hr. Particle size can be varied and almost any ingredient can be pulverized.

FIG. 38.4. HAMMER MILL CUTAWAY TO SHOW HAMMERS

Mixing Requirements

A perfect feed mixture is one in which the particles are uniformly distributed throughout. The feed mixer must cause the particles to be randomly distributed. The most important particle characteristic is size, but shape, electrostatic charge, friction and resilience are also involved. From a practical standpoint, particle sizes should be similar. The recommendations of the mixer manufacturer as to operating speed and capacity should be followed. Mixing time requirements should be determined by analysis. Efficiency can be greatly improved by proper loading of the mixer. Ingredients having the greatest volume should be added first, then liquid material, such as fat, fish solubles, molasses, soybean oil or water. For mixing critical drugs and additives, the stage blending process should be used. This means a batch containing all small items is mixed prior to the final mix. All ingredients used in small quantities thus receive the benefits of premixing.

Feed Mixing.—Feed mixing disperses ingredients so that each small unit of the whole has the same proportion of each ingredient as the original formula. Many combinations of ingredients with differences in physical properties are blended. Differences include particle size, shape, density, coefficient of friction, resilience, electrostatic charge and viscosity.

"Mixed" means blended, which implies uniformity. The objective of mixing is to scatter dissimilar parts until they are blended. During the operation, there is the possibility that particles which have been mixed will become separated.

Solids have no inherent mobility of their own; however, mixing of solids is often described in terms used chiefly for liquids. The following 3 possible mechanisms cause mixing: (1) convection—the transfer of groups of particles from 1 location in the mass to another, (2) diffusion—distribution of particles over a freshly developed surface, and (3) shear—slipping planes of solids within the mass. These actions are accomplished by gravity and mechanical forces. The position of particles within a container is determined by chance, and the effects of chance accumulate until the material is mixed. Particle segregation is due to differences in physical properties of materials and the design of the mixer.

Ingredients should be ground as uniformly as possible. Crimped or rolled grains will not mix well with a finely-ground material. The most desirable material for pelleting passes at least a 7-mesh screen.

Particle size is more important than other factors in causing segregation. Improvement in mixing made by decreasing particle size can be measured quantitatively by statistical methods. Where very small amounts of solid microingredients are added, the required particle size is very small. Certain ingredients are unstable in finely-divided form and may acquire electrostatic charges.

Tests of Mixing.—Small particles of dyed salt are sometimes used as tracers to measure mixing action. Mixing by V-type mixers permits more rapid distribution radially than longitudinally; materials having a lower angle of repose go to the ends first. Mixing formula feeds in conventional equipment involves so many variables that the exact nature of the distribution of component particles is quite obscure. Mixer performance is evaluated by sampling at the point of discharge. A small quantity of additive of the same size and approximately the same bulk density as the diluent is easy to mix properly. Additives considerably smaller than the diluent will segregate during the process of mixing and discharge, appearing in high concentration during the first part of the sampling sequence. Partial replacement of the diluent with material intermediate in size between additive and diluent is effective in preventing segregation of small particles of the same bulk density as the diluent. Addition of vegetable oil retards segregation of high-density additives due to the fact that particle friction is increased.

The senses are used to evaluate feed quality. Molasses and dehydrated alfalfa meal are detected at 5, 10 and 15% levels by color, odor and texture of the mixture. The percentage of crimped oats can be approximated at low levels. The quantitative recovery of an addition by counting or sieving is impractical. Chemical analysis is the most reliable method.

Feed Mixers

Needless to say, modern-day feed mixing did not always exist. Birds searched for insects, seeds, leaves, stones, water, and ate fecal material to provide the nutrients required for the life processes. Since they could function normally on scavenger-collected feeds, poultry operators placed their own ingredients in separate troughs, giving the birds free choice. This was a step toward the development of adequate feeds and feeding programs.

The free choice or "cafeteria" had containers for grains (milo or sorghums, corn, oats or wheat), meat and bone scraps, grit and oyster shell, skim milk and water. Hens produced less than 100 eggs per year and fryers required 12 to 14 wk for maturity. Standards for growth, egg production, feed efficiency and livability were not used.

The Scoop Shovel.—The first mechanical feed mixer was the scoop shovel. The ingredients were ground with a burr or hammer mill. They were proportioned to equal observed amounts generally consumed in the free-choice system. These quantities were mixed on the granary floor with a scoop shovel. This method used the following mixture:

Wheat bran or shorts
Ground corn
Meat and bone scraps
Dried skim milk
Salt
Bone meal

Oyster shell and grit were given in open feeders.

The Letz Feed Mill.—This feed mill ground the "scoop-shovel" mixture together. The grains were added whole, not ground, before making the "scoop-shovel" blend. This gave a well-ground and well-mixed feed that encouraged the addition of more ingredients.

The Vertical-Horizontal Batch Mixer.—Nutrition studies at the Agricultural Experiment Stations developed more adequate formulated rations and the size of the poultry flock became larger. Poultry breeders were breeding for egg production as well as for show purposes. Some breeders were advertising sale stock that was producing over 180 eggs per year. Layer flocks of less than 50 hens increased to as many as 500. Better housing, feeding, processing, management and merchandising were changing poultry production from a "pin-money" business to a family business.

The development of the vertical feed mixer followed the development of the screw elevator. Screw elevators are easy to construct, cheaper than other elevators and highly adaptable to many operations involving the movement of semisolid to whole grains. The basic structure of the vertical feed mixer is shown in Fig. 38.5. This type of mixer is the most common, and generally has a capacity of 0.5 to 5 tons.

Courtesy of Jay Bee Mfg. Co.

FIG. 38.5. A VERTICAL FEED MIXER

The horizontal mixer uses paddles or open spirals with 1 set moving the contents forward and the other moving backwards. These strong, rigid paddles or open spirals can agitate or blend heavy materials, such as fertilizers, mineral mixtures, ground feed ingredients or whole seeds, including incorporated semisolids, like whey, fish solubles and molasses.

The horizontal mixer is more expensive than the vertical, but will outperform it. This type of mixer is often used in continuous-operating automated feed mills (Fig. 38.6).

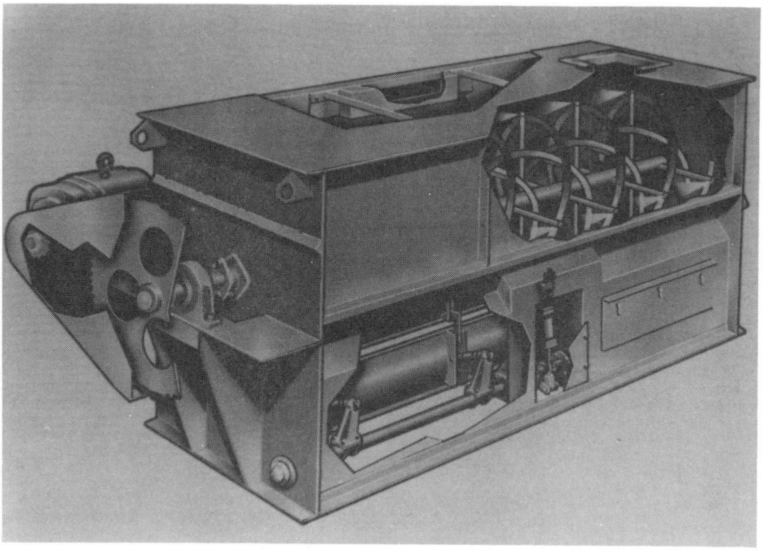

Courtesy of Sprout-Waldron Div., Loppers Co., Inc.

FIG. 38.6. A HORIZONTAL FEED MIXER

The vertical feed mixer is the preference of farm and small feed operations or custom blenders. The horizontal mixer shown in Fig. 38.6 is a drop-bottom ribbon mixer (batch-type) designed to give quick discharge. A panel at the center and the rear allows rapid escape of displaced air in the surge bin below when dumped. Optional air jets scour the interior of the mixer vessel to prevent cross-batch contamination—especially important in poultry formulation with its many medicants.

Batch Mixing of the Ration

Although there are some general rules used in mixing a ration, certain principles should be developed after the ration is formulated. The following procedure can be used with success:

(1) After formulating the ration, the ingredients must be checked and the needed ingredients obtained.
(2) The corn should be ground just before the ration is to be mixed. A coarse grind of corn should be used—it must never be finely pulverized. The coarse grind encourages development of a strong digestive tract, costs less to grind and is less likely to heat after grinding.
(3) The soybean meal is weighed and dumped into the mixer. The mixer is set in motion.
(4) Vitamins, antibiotics, antioxidants, drugs if needed, and trace minerals are added to the mixer (2 min mixing time).
(5) If fat is used, it should be added at this time (3 min mixing time).
(6) Ground corn is added (1 min mixing time).
(7) Other carbohydrate sources are added (1 min mixing time).
(8) Other protein sources are added (1 min mixing time).
(9) Remaining minerals and ingredients are added.
(10) The mill should run for 8 to 10 min after the last ingredient is added.
(11) The mixed ration can be bagged, "dumped" in bulk bins or trucks, or run into the pelleting machine, then prepared for storage or delivery.

Continuous Feed Mixers

Large feed mills and automated mills have mechanical weighers or proportioners that deliver the ingredients to a continuous-collector belt or screw-open trough. The collectors deliver the ingredients to a continuous-horizontal mixer that continuously mixes the ingredients and places the mixed feed in bulk trucks, bins or bagging facilities.

This type of system is set to deliver an exactly proportional amount of each ingredient, and the total amount of ingredients to be delivered is also preset. Many automated mills making early morning bulk deliveries have the feed mill preset; the truck driver selects the tonnage and turns on the mill. After the preset tonnage of feed is delivered to the bulk trucks, the mill automatically stops.

Feed Delivery

The use of paper, cotton cloth and synthetic bags was the most common method to store and deliver feeds and feed ingredients. Today, most manufactured feed is delivered in bulk-feed trucks (Fig. 38.7). These trucks are equipped with either screw- or air-elevators which place the feed directly into the farm-feed bins. This method reduces labor costs and wastage.

Courtesy of Henderson Mfg. Co.

FIG. 38.7. BULK FEED TRUCK

Pelleting

Most poultry feeds are pelleted. Approximately 60% are fed as pellets or crushed pellets (crumbles).

Pellet mills can be purchased in several sizes and types. Practically all pellet mills are of the ring-die and extrusion type. The ring die, mounted in a vertical position, rotates. The feed is forced and pushed through the holes in the die by rollers. This action is known as extrusion (Fig. 38.8).

Steps in Pelleting Feeds.—There are 6 important steps involved in feed processing to form pellets or crumbles: (1) formulation, (2) conditioning, (3) formation in the die, (4) drying and cooling, (5) granulating or crumbling, and (6) screening and cleaning. Of the various requirements for high-quality pellets, the most important is binding quality, since without it, pelleting and processing into crumbles is difficult. Starches are highly desirable since they break down into sugars in the presence of heat and moisture and form a very effective binding agent. The moisture content also affects pelleting, and should be from 15 to 18% for the best results. Steam is a practical way of adding moisture and heat in pelleting operations; the pressure of the steam has little effect upon the pelleting.

Pellet Binders.—Ground corn and soybean meal serve as binders, while high-fiber feeds tend to reduce the hardness of pellets, as well as to cause

Courtesy of Calif. Pellet Mill Co.

FIG. 38.8. A PELLET MILL

the steel die to become hot and wear excessively.

Special binders are often used. These are a part of the ration formula. Many types of binders are used to improve pellet durability and pelleting efficiency. Thick dies and high-conditioning temperatures increase pellet durability and efficiency.

There are many binders which contribute to better pellet quality—lignosulfonate, bentonite, soft phosphate, gelatinized starches and molasses. There is a growing demand by feed manufacturers for molasses to add as a pellet conditioner in certain formulations. Molasses is delivered to the conditioner in a state of low viscosity, so it mixes uniformly with the mash. Molasses at 100°F is pumped, controlled, and metered into the conditioner. Steam is introduced in the molasses line 3 ft from the discharge at the conditioner. The steam traveling at a high velocity relative to the molasses, mixes with it and heats it, adding moisture by condensation. When it hits the dry mash, it penetrates almost completely, instead of merely coating the particles.

Pellet Size and Crumbles.—Many different sizes and shapes of pellets are manufactured, and each requires a different die. The thickness of a die is 10 times the diameter of the pellet. A good binding agent makes a harder pellet than any amount of pressure applied in pelleting. After pellets have been formed, the next operation is to dry and cool them. Heat and moisture are removed before the pellets can be stored; this takes 7 min for pellets of one-fourth in. in diameter. Pelleting produces homogenous feeds, reduces dustiness and dust losses, gelatinizes starches and brings out the natural oils in grains.

Crumbles are often made by crushing large three-fourths in. pellets. Since the costs for steam and electricity are less for large pellets than for the small one-fourth in. size, it is often possible to produce crumbles as cheaply as, or more cheaply than, pellets.

Advantages and Disadvantages of Pelleting Feeds.—Advantages of pellets include the following: (1) feed waste is minimized (feed spilled from hoppers can be picked up again in floor-type operations but not in cage operations); (2) each pellet is a balanced ration so selection of ingredients is prevented; (3) less total feed is required; (4) palatability is improved; (5) less feed storage and feeder space is required as a result of increased density; (6) any growth inhibitor is destroyed; (7) pellets are readily adaptable to automatic feeders; and (8) Samonella contamination is somewhat reduced.

Disadvantages of pelleting are: (1) added cost of manufacture; (2) possible destruction of some micronutrients; (3) increased water consumption with associated problems of litter condition; and (4) increased cannibalism or feather picking.

On the scientific level, the reasons for the efficiency of pellets are far from being fully understood. One reason why pellets do well compared to conventional mash is that they are 25% heavier than mash. Because of this greater density, the birds consume less feed to get the same amount of nutrients; therefore, there is less waste. With turkeys, there is a small difference in intake between complete feeds and free-choice protein concentrates and milo. There is some seasonal trend, since more grain is consumed in cold weather. For turkeys from 8 wk of age to market, there is no significant difference between pelleted and mash forms. More compact feed increase consumption, especially in the case of bulky materials. It is difficult to handle bulky ingredients by automation. Pellets also require much less space in storage than mashes. Obvious advantages are that pellets can be handled easily in bulk-like grains, dry and dusty sacking in the mill are eliminated and dust losses around the mill are kept to a minimum.

Making Crumbles.—Crumbles are made from pellets. The amount of fines in crumbling is determined by the quality of the pellet and the way it is bound together. Sharp, corrugated rolls are used. The final product, made from regrinding pellets to crumbles, is more compact, free-flowing and uniform than conventional meal, as well as less dusty.

Important factors for a good crumbling operation include: (1) cool, dry, high-quality pellets; (2) uniform feeding to the rolls; (3) proper adjustment of the rolls; (4) maintenance of a sharp corrugation to permit a cutting rather than a crushing action and (5) 10% of oversize particles for return to the crumble roll. The addition of fat after pelleting rather than before, improves the quality and durability of high-fat pellets, and usually these are the only pellets with over 4% fat that can be made.

The expansion methods of cooking grains are of increasing interest. The coarse structure and bulk density resulting from expansion are just as important as the cooking. The expander consists of a tube containing breaker screws and worm shaft with interrupted flights. At the discharge end of the tube, there is a flat plate containing a number of small openings through which the material is extruded. The number and size of these openings can be varied for different materials. The only pretreatment is grinding. Water and sparge steam are injected into the tube through valves. The material is cooked by the steam and heat from the friction of the rotating shaft. The pressure exerted within the tubes keeps moisture in a liquid state, but immediately upon discharge, it vaporizes, which causes the materials to become puffed or expanded. In the process, water is injected to bring the moisture level up from 10 to 20%. Enough steam is injected to raise the moisture content from 15 to 35% and the temperature between 120° and 190°F. The pressure exerted upon the material is 100 psi. At the die, the material is heated mainly by shaft friction from 200° to 420°F and placed under pressure from 200 to 1500 psi. Retention time in the expander is about 30 sec. Cooking, therefore, is relatively short. Starch is the ingredient which allows the material to puff, and puffing is influenced by the amount of starch present. A typical cereal grain can be expanded 4 diameters. Enzymes are deactivated by the expansion.

The Kansas State University has reported on developing and feeding pellets. Steam-pelleting a corn-soybean meal did not result in daily gain or feed efficiency much different from a diet pelleted without steam-conditioning.

Steam-pelleting at 176°F reduced the mechanical friction between the meal and the die-hole surface, and thus reduced the electrical energy required to pellet 1 ton feed by two-thirds. Pellet hardness was increased by steam-conditioning.

Pelleting has proven of great value in reducing volume and dust in alfalfa. This method of feed-handling has assured greater uniformity.

Feeding heavy pellets to ducks has increased their growth and reduced feed required to produce a marketable product.

Pellets, mash, reground pellets and steamed-pelleting give similar total metabolizable energy (T.M.E.) (Tables 38.3 and 38.4).

TABLE 38.3. THE MEAN T.M.E. VALUES OF 7 DIETS FED AS MASH, STEAM PELLETS AND REGROUND PELLETS

Diet	Mash[1]	Pellets[1]	Reground Pellets[2]
1	3.64 ± 0.05	3.65 ± 0.02	3.65 ± 0.04 (3)
2	3.95 ± 0.01	3.92 ± 0.04	3.80 ± 0.09 (3)
3	3.66 ± 0.06	3.63 ± 0.07	3.58 ± 0.12 (2)
4	3.65 ± 0.02	3.73 ± 0.05	3.58 ± 0.02 (2)
5	3.74 ± 0.04	3.72 ± 0.03	3.96 ± 0.03 (2)
6	3.73 ± 0.02	3.73 ± 0.06	3.68 ± 0.03 (3)
7	3.67 ± 0.05	3.66 ± 0.02	3.70 ± 0.08 (3)

Source: Sibbald (1977).
[1]Data are means, with standard errors, of 4 determinations expressed as kcal/g of dry matter.
[2]Data are means of the number of determinations in brackets.

TABLE 38.4. THE MEAN T.M.E. VALUES OF 3 DIETS FED AS MASH, PELLETS AND REGROUND PELLETS

Diet	Mash[1]	Pellets[1]	Reground Pellets[2]
1	3.38 ± 0.03	3.40 ± 0.03	3.47 ± 0.01 (3)
2	3.88 ± 0.03	3.81 ± 0.02	3.78 ± 0.02 (3)
3	3.76 ± 0.05	3.73 ± 0.02	3.73 ± 0.06 (4)

Source: Sibbald (1977).
[1]Data are means, with standard errors, of 7 determinations expressed as kcal/g of dry matter.
[2]Data are means of the number of determinations shown in brackets.

Effects of Grinding and Pelleting.—Grinding and pelleting had no consistent effects on the apparent metabolizable energy (A.M.E.) values of cereal grains fed to growing chicks.

Mean T.M.E. Values of Whole and Cold Pelleted Grains[1]

	Whole Grain	Pellets
Barley	3.45	3.48
Wheat	3.55	3.66
Oats	3.92	3.79
All grains	3.64	3.65

THE EXPANSION OR STABILIZATION OF THE FEED INDUSTRY

The day of the huge feed mill is over; today's small automated mill is a vital part of an integrated poultry operation. The total tonnage of feed

[1]Data expressed as kcal/g of dry matter.

manufactured for poultry increases yearly. The broiler and turkey operations account for most of the increased tonnage.

Name brands are becoming more established in marketing poultry meats and eggs and less important in feeds.

Bulk deliveries are cheaper and more efficient than feed-bag delivery.

The small feed mill can take advantage of local grains, many minor by-products and local off-season labor; and since the small mill uses mostly truck delivery, proximity to a large railroad facility is unnecessary.

The availability of vitamin and mineral premixes at less cost per unit and quality control, gives the small mill good support. Nutritional information is available from Agricultural Experiment Stations and Consultants.

39

Feed Mill Design

The feed industry has grown from a hand-operating system to the mammoth mill of yesterday and is now returning to a small or medium sized automated mill. The automation of mills, along with advancements in nutrition and a quality control program have made many changes in the poultry and feed industry.

A well-designed and operated feed mill is essential in the production of desirable feeds (Fig. 39.1). A poorly-designed mill can result in excess labor costs, poor feed mixing and unsanitary conditions.

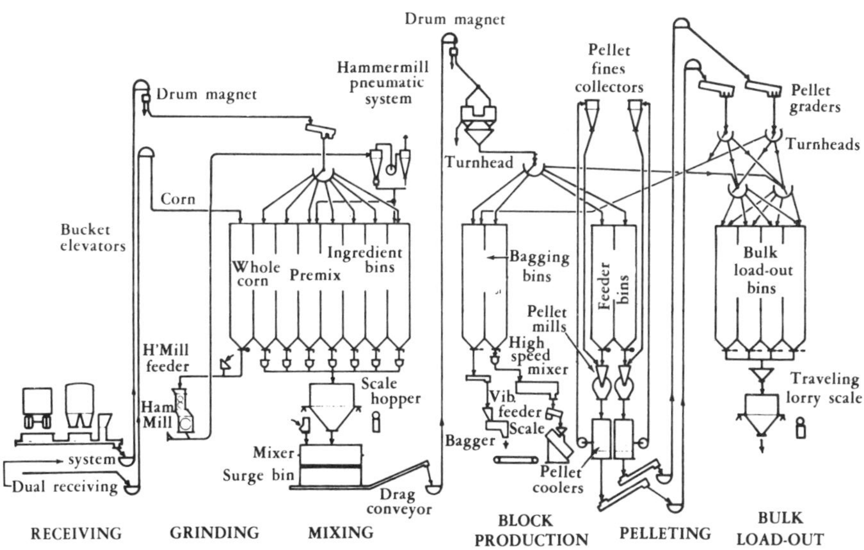

FIG. 39.1. FLOW DIAGRAM FOR MOOORMAN'S FEED PLANT, COLUMBUS, NEBRASKA, WHICH CAN MANUFACTURE 60,000 MT PER YEAR OF PREMIX FOR CONCENTRATE TYPE FEEDS

Twelve men operate the mill—one man each for mixing, pelleting, sewing and maintenance; two men for receiving, grinding and block press; three men in the warehouse. It has a head house above the bins containing collectors, turnheads, graders and sifter, spouting and related equipment.

Many small feed mills being constructed today are one-story prefabricated structures. They have many built-in features and can be expanded or altered with a minimum of effort.

The following list of equipment is often found in small hand-operated feed mills:

(1) Storage bins—
  Two outside storage bins for bulk-handling:
    One for soybean meal
    One for shelled corn
  Three inside storage bins for bulk-handling:
    One for oats
    One for milo
    One for finished feed
(2) One inside storage tank for fat
(3) Warehouse storage for bagged feed—
  Animal protein ingredients
  Alfalfa meal
  Calcium carbonate or ground limestone
  Phosphates
  Salt (NaCl)
  Trace mineral premix
  Vitamin, antibiotic and antioxidant premix
  Caged area for drugs and pesticides
(4) One hammermill with a good dust-collector
(5) One 2 to 5 ton vertical or horizontal feed mixer
(6) One weigh buggy, platform scales and portable screw-elevators.

FEED MANUFACTURE

Traditionally, American farmers have produced more than enough food for their fellow man. The average farmer now produces enough for himself and 25 others, yet with population continuing to expand at a phenomenal rate, production must be increased at least if we are to maintain present levels of consumption. The formula-feed industry must formulate, manufacture and distribute quality feeds to livestock and poultry farmers that will enable them to meet this challenge with quality products at the lowest possible price. This industry ranks among the top 15 United States manufacturing industries and is the largest serving the American farmer.

As the feed industry has decentralized, smaller plants have been built closer to customers. Ingredients and finished feeds now move shorter distances and trucks offer more convenience, timeliness, and flexibility.

Improved roads and better trucks have contributed much to the switch to truck transportation.

Dealer distribution was the accepted way to do business in the early 1900s. Major feed companies owned or controlled retail feed outlets; except in New England and the far west, the trend in recent years has been away from company-controlled retail feed stores. For the past 30 years, many dealers have handled several brands. This method of merchandising feed is now disappearing. Major feed companies offer a complete line of feeds that enables dealers to meet all customer needs. The trend resulting from decentralization of feed manufacturing has been the movement of more feed direct from the plant to the feeder.

Feed manufacturers with national distribution benefit from geographic ingredient diversification because supplies tend to balance and cancel out in the course of a year. Concentrates, in general, contain soybean meal, animal and fishery by-products, fat, minerals, and traces of antibiotics and other substances for disease and parasite prevention and growth promotion. In grain-deficit areas, such as New England and the Pacific Coast, a manufacturer typically sells a complete feed which is composed of grains and concentrates. In the last decade, many manufacturers of drugs and by-products have developed combinations of drugs and vitamins know as premixes, to which protein and grain must be added. Typical complete feeds contain about two-thirds grain, 25 to 35% concentrate, and 5 to 10% premix ingredients. Of course, the animals ultimately receive the equivalent of a complete feed.

Complete Feeds Are in Demand

However, the volume of the feed manufacturer depends on whether animals receive a complete feed or a concentrate plus grain. The sale of concentrates represents to the manufacturer only about one-third the volume of complete feeds. The remaining two-thirds—buying, milling and mixing of the grain—is performed by the feeder or the local grain dealer. The volume of concentrates has been growing more rapidly than total feed production. The American Feed Manufacturers Association estimates that 32% of the 1960 manufactured feed volume was in concentrate form.

While total tonnage has increased at the rate of only 2.1% annually, complete feed equivalents showed an annual growth of 4.5%. Milling and mixing of the feed grain components are important to feed manufacturers with fixed investments in milling capacity. However, in terms of margin dollars or value added, the grain portion is less important. It is in the concentrate ingredients that research and product differentiation are most important. The rapid growth of concentrates has had 2 effects:

First, concentrates have rarely retarded the growth of complete feeds. This has been a negative factor to the industry, as it sells 40% of its volume rather than 100%. Second, and more important, concentrates have expanded as the total feed market. Most of the growth of the concentrates has been in sales to feeders who previously used little or no manufactured feed rather than as a replacement of complete feeds. Use of concentrates varies by geographic area and is heaviest where the local grains are readily available. In grain-deficient areas, it is economical to manufacture complete feeds, as the grain must be imported in any case. Similarly, the use of concentrates varies by animal class.

Poultry feeds must be processed to be edible, and thus they consume great proportions of complete feeds. The 2 factors of geography and animal class combine with such other factors as labor, costs, nearness to markets, and alternate land uses, to determine where each animal class grows most economically.

The rates at which feed is converted into end products have changed significantly in recent years. Research by the feed industry, the United States Department of Agriculture and land-grant colleges has focused on conversion of the highest possible percentage of feed into meat. Gains in efficiency of feed utilization will be of smaller magnitude in the future than in the past. Research activity has moved from nutrition and breeding to environment of animals because environmental improvements are most fruitful. Present research is aimed at better operating efficiency and cost reduction for the feeder.

Premixing

Handling these active vitamins or drugs is usually accomplished through premixing. Basic factors of premixing consist of selection of a suitable carrier, and oil treatment of the carrier to reduce dustiness. Proper mixing requires the right kind of equipment to provide the proper length of time for a uniform blend. Good carriers for active ingredients are wheat middlings, finely-ground corn, corn-gluten meal, corn-gluten feed, a good grade of fishmeal or meat scraps, and even minerals, such as limestone. Look at the particle size. So-called 20- to 80-mesh means no particles larger than 20 or finer than 80. This keeps the very big and small particles out of the carrier. Next, treat it with soybean oil, corn oil, cottonseed oil or animal fat. Antioxidants, such as BHT (butylated hydroxytoluene), BHA (butylated hydroxyanisole) and santoquin are incorporated. Intimate contact of the inhibitor with the carrier prevents loss by oxidation of the oil-soluble vitamins such as A, D and E.

Vitamin A, per se, is very sensitive to destruction by oxidation. To prevent this, some companies produce a pellet containing Vitamin A.

Liquid Vitamin A palmitate, insoluble in water, is whipped up in a gelatin solution so that there is a very fine subdivision of these particles. It is then sprayed into cold oil. A cross section of a Vitamin A pellet under the microscope shows thousands of little cells within this pellet, each containing a droplet of Vitamin A palmitate. The Vitamin A is coated both inside and outside, and has good stability. The gelatin breaks down rather easily in the digestive tract, liberating the droplets of Vitamin A palmitate where they can be more readily absorbed.

Horizontal Ribbon Blender

An efficient piece of equipment in which to make a premix is a horizontal ribbon blender. The material not only goes around, but also goes back and forth lengthwise, a rolling action plus the back and forth motion. Ribbon tolerance should be as close as possible, without actual rubbing, to avoid dead spaces. The central axle should be covered, but the tops of the ribbons should be in sight. First, the mixer is charged, the oil treatment is given and well-distributed before any active ingredients are added. The formation of "oil balls" from fines from the carrier must be avoided.

If it is desired to find out how good a blend is, a laboratory analysis is made. Not all active ingredients blend to the same degree of uniformity or at the same rate. Some materials, such as riboflavin and coccidiostats, are electrostatic and blend very rapidly. A good blend with riboflavin may be made in 4 min, but it may take 6 min with calcium pantothenate.

The Food and Drug Administration allows on most of the drugs added to feeds ±20% of target, and it should be relatively easy to accomplish this with modern feed equipment. Large feed manufacturers covering some 57 plants nationwide had an average of ±7% of a fairly sensitive drug in 1 period.

Operational Requirements

Good manufacturing practices include weather-proof buildings with adequate space for all materials, designed to avoid contamination between different products and ingredients. They should provide adequate lighting and ventilation, as well as clean toilets and locker facilities.

Equipment should be of suitable design and size for efficient maintenance. Measuring, weighing and mixing should be as accurate as the products demand.

Personnel should have appropriate experience and education to ensure the safety and quality claimed by the products.

Ingredients should be identified as to origin, receipt and disposition. Master formula records should be maintained for all feeds. A complete batch formula, packaging methods, manufacturing instructions and any precautions needed to prepare the product should be on records retained for 2 years. Records include information on all critical steps, mixing, processing, inspection, packaging and warehousing.

Control measures include sampling, mixing time and any others needed to maintain product uniformity. Yields of each run of a product are checked against theoretical yields, and any significant variations immediately investigated. Packaging and labeling are controlled so products are not intermixed. Protection against deterioration after manufacture should be provided.

Laboratory Tests

Laboratory controls include the establishment of tests to be certain that products conform to specifications. Association of Official Agricultural Chemists (AOAC) analytical methods are used wherever possible. Bulk feed tags show the name and address of the first purchaser, and the date, product name and quantity shipped. These practices are of value, not only from an economic standpoint but from that of improved relations with customers, as well.

There should be detailed specifications on every ingredient, and these ingredients should be sampled and analyzed from time to time. Bags of feed must be kept closed and used on a first-in, first-out basis. Be aware of possible equipment residues, dust control and human error. Whether feed manufacturers purchase complete premixes or manufacture them from individual ingredients depends on individual considerations.

In the mechanics of mixing, partition devices promote mixing and breaking up of lumps. Grounding the equipment is necessary to prevent static charges building up on certain ingredients. Mixers should give efficient blending, complete discharge and dust tightness. Do not overblend.

Vertical Mixers.—The vertical mixer, the most common mixer used by the feed industry, enables employees to produce more complicated formulas than do other mixers. The horizontal mixer requires facilities to charge or discharge, and horsepower requirement per ton is higher. It is not advisable to add liquids to this mixer; liquids should be added in other equipment, transferred by a controlled feeder and introduced by ports or manifolds. Dry feed and liquid are moved the full length of the cylinder and thorough, agglomerate-free dispersion is achieved. Liquid addition is controlled by meters coordinated with dry-product flow.

Rotary Mixers.—Rotary mixers are between the vertical mixer and horizontal mixer with respect to horsepower per ton. The design of the baffles and shearing plates determines the quality of the mix. This mixer requires more cycle time than others. Liquids should not be added; extremely dense and critical formulations work ideally in these mixers.

Raw materials, as well as finished products, are mechanically handled. There are 4 types of mechanical-handling systems: (1) bags of material are unitized on conventional wooden pallets; (2) material is pelletized but the load can be moved from the pallet without hand handling; (3) the pullpack method eliminates pallets altogether and utilizes a corrugated or fibrous sheet; and (4) slip sheet transfer method. By unitizing products and handling with mechanical equipment, improved service is rendered to customers. Air is also utilized to move material.

Systems available for delivery of formula feeds to feeders are mechanical delivery, drag-chain paddle trucks and screw conveyors, and air pneumatic units. A conveyor-panel truck, transporting 1500 to 2000 lb per min, requires 25 to 45 min at the farm to unload.

Pellet Mills

Variables that affect production of pellets are the physical state of the product, the amount of applied pressure given the feed and the lubrication of the feed. A minimum amount of force and increase in feed temperature are used to extrude the feed through the die holes. The most desirable temperature is 160° to 180°F. A large percentage of animal feeds is pelleted through a 5/32-in.-die with an effective thickness of 2 in., requiring between 6000 and 8000 psi to extrude feed through the dies. An increase in pressure increases pellet hardness. When a pellet mill is shut down, die holes are cleaned out with a relatively inert ingredient. A pellet mill should never be run without feed.

Packaging involves holding-tanks, scales, sacking spouts, bag clamps, bags, bag shaker, bag closer, tagging and identification, and coding. Fifty pounds multiwall bags are used. Bag shakers and settlers are effective on multiwall bags. Tape closures are neat and attractive. Bags are tagged and coded for quality control. Required on the label are net weight, brand name, minimum protein and fat, maximum fiber content, list of ingredients and name and address of the manufacturer. Automatic packaging consists of high-speed scales, magazine-type feeders and automatic closing units.

Steaming may be used with rolling. Live steam is applied in a conditioner. A more uniform product with less fines can be produced in a roller mill with steam-conditioning. The product is preferred by many farmers but steamed grains are no more palatable or nutritious to poultry.

Automatic Weighing and Bagging

Raw materials are purchased and feed is produced, taxed and hauled on a weight basis. Weighing equipment that will accurately weigh and batch speedily either into bags, trucks or cars, is available. Material flow is the keystone of automatic batching scale. Batching systems are built around weight sensors. If there are many formulae, the punchcard method is available. The card controls selection, weight and other functions. A card reader, an integral part of the control panel, senses the holes in the card (80 vertical columns and 12 horizontal rows) for a total of 960 pieces of information. Once the card is inserted in the reader, its information is channeled to a digital indicator. This keeps the operator advised on what is happening.

Because of weight and measure regulations and public relations considerations, feed mills set scale weights so that no bag goes out underweight. Lifting is eliminated to increase production of the bagging personnel. A mechanical settler makes it possible to increase output. Automatic sewing eliminates some manpower. Finished bag weights are checked from time to time.

Eighteen 50-lb bags per minute can be sacked by a single worker. A bag-hanger comes out, picks up the sack and slides it underneath the opening. By electrical impulse, the feed drops ino the bag. A machine picks up a feed tag and sews it on the bag. All the worker has to do is to make sure that the machine is in adjustment. The product is dated as it goes out.

A typical feed plant has a capacity of about 25,000 tons in conjunction with a 2.5 million-bushel grain elevator and produces 25,000 to 30,000 tons feed per year. It is better to establish smaller, localized feed mills than a large installation. Most feed plants have a receiving elevator of 15 to 20 tons. Grain trucks carry from 20 to 34 tons of grain. Most feed grains come in by truck.

Approximatey 60 ingredients are used in a small mill. Soybean meal, cottonseed meal and alfalfa ingredients are used in large amounts, and storage space for 1.5 cars is essential. Formerly, a plant would have 5 days to unload a car. Now, it has 24 hr in which to get the car unloaded and an extra 24 hr for which no charge is made. After 72 hr, a charge of $17.50 per day is made for just standing on a siding. At 10 days, the cost is $38.00 per day. Consequently if severe weather conditions prevail, this may be very expensive. The purpose of demurrage is to free cars. Otherwise, companies might use railroad cars for storage, so it is necessary to have some penalty.

Storage affects the nutritive value of grains, and is dependent upon the moisture content, temperature, degree of maturity at which the grain is harvested, the manner of handling from the time harvested, the type and construction of storage bins and the length of storage period. Whole grains store better than ground grains. Deterioration of grains is accelerated by high temperatures and retarded by low temperatures.

Changes in Nutrients

Fats form free fatty acids which reduce palatability; in fact, free fatty acid content is a useful index of general deterioration. Thiamin, riboflavin, Vitamin E, carbohydrates and mineral constituents are, in general, unaffected during storage, although the availability of phosphorus is affected adversely. Ground grains become rancid more quickly than whole grains; when ground grains are mixed with good feed they make the entire mixture unsatisfactory. Heat and moisture encourage this deterioration, causing tallowy odors and flavors. Rancid oils and fats destroy certain vitamins. Storage produces changes in the physical and chemical properties of proteins. Cryptoxanthin content of corn, and carotene content of alfalfa, decrease with storage. Half the carotene is lost during a summer of storage. Exposure to light, air and moisture increases carotene loss. Sacking the meal in airproof and moistureproof sacks reduces loss, as does storage in airtight bins in which the air is replaced by nitrogen. Fish oils, stored under conditions of heat, light and moisture, rapidly oxidize and destroy Vitamin A. Fish liver oils should be purchased in small containers, but if purchased in large drums, it can be drawn off from the bottom through a faucet without allowing air to pass through the oil.

Corn completely cured and dried before going into storage does not change as much as if stored directly after harvest. Soybeans with higher moisture content are greater risks than beans with lower moisture content, although the storageability of higher-moisture beans can be extended at low temperatures. Grains that mold in storage give poorer feeding results than sound grains. Protein value of freshly-harvested ripe corn decreases rapidly during the first few wk of storage, but after 12 months, no significant decrease occurs.

Propionates in feeds minimize heating of beans, particularly enroute to their destination. Feed mixtures heat eventually when the moisture of hygroscopic ingredients in the feed goes above a critical level.

When the germ of the seed is destroyed, natural antioxidants, such as Vitamin E, disappear and the grain is difficult to store. Early freezing before the grain is dried down and matured destroys or damages the germ of the grain. In seed corn, for example, if the temperature falls below the moisture content of the corn, the germ can be damaged. Corn is considered mature at 35 to 38% moisture. A temperature of less than 32° destroys the germination of 35 to 38%-moisture corn. Heat from drying of grain can be harmful. The higher the moisture content of the grain, the faster the germ is destroyed by drying.

Grain stored for more than 1 year loses some of its germination. Mold damage may occur in the field or in storage. Sprouted grain has a dead germ when dried. Mold damage may be present in the germ or in the whole kernel. The germ of a seed is dead before it becomes invaded with mold. Dead germ grain forms pockets in storage, so it must be moved or aerated for sufficient air to be available, and it is cooled down. Such grain should not be blended with high-moisture corn. The moisture and temperature content of such corn must be watched very carefully; moisture rise of 0.5% in stored corn indicates the beginning of spoilage.

INSECT DAMAGE

Most of the damage done by insects to grain in storage is brought about by 4 species of insects: the granary weevil, the rice or black weevil, the lessor grain borer or Australian wheat weevil and the Angoumois grain moth. The primary pests are not ordinarily capable of a free existence outside the kernel, but live entirely within the kernel where they feed unseen and unsuspected. These pests cannot be removed by ordinary grain-cleaning machines. If the grain kernels are broken, secondary surface feeders take over. The latter can be largely removed by grain-cleaning operations.

Prevention of Infestation

To prevent infestation after harvest, store dry grain in weathertight, rodentproof bins, preferably of steel. Clean out all bins before loading with grain and spray the walls and floors of wooden bins and around the door frames of metal bins with pesticide. Clean and dispose of litter, waste grain and feeds that have accumulated around farm buildings. Apply effective powder or spray directly to the grain as it is binned, or fumigate promptly after binning. Inspect monthly and fumigate if an infestation is discovered. Exposure to freezing climate kills infestations in ear corn stored on the farm so that loss from insect attack is inconsequential if corn is to be used for food during the ensuing season. Information regarding the prevention of infestation or the treatment of infested grain is contained in publications obtainable from the United States Department of Agriculture.

Losses from Infestation

Insects in farm-stored grains cost American farmers millions of dollars per year. Some is destroyed by insects and the remainder made unfit for human consumption. Yet, the loss is preventable. Watch for insect build-up and fumigate if necessary. Fumigants are sold under various trade names, ingredients of which are shown on the label. Follow the manufacturer's recommended dosage and apply from the outside of the bin. About once per month, take probe samples from the center of the bin and inspect for insects.

Insect Life Cycle

The life cycle of these insects consists of 4 stages: adult, eggs, larvae and pupae. Most insects hatch from eggs. A typical insect, the confused flour beetle, has eggs only one-fifth in. long which hatch in about 10 days under summer conditions. The larvae or worms are very small when they first emerge from the eggs; as they eat and grow, they shed their skins and change from larvae to pupae which do not move about or feed. During this stage, the larvae are transformed into adult beetles with wings. During the pupae stage which lasts about 10 days, adult organs can be seen through the thin skin which is to be molted when the adult emerges. The larvae of most moths and of a few beetles form cocoons in which the pupal stage is spent. Adult insects mate and lay eggs. Adult beetles live and feed for 1 year, during which time they crawl and fly about. Adult moths feed very little, if at all, and live only a few days or wks.

Heating Effects from Insects

Grain-infesting insects can cause heating in small grains and shelled corn as a result of their feeding and other activities. This is called "dry grain heating," and may bring the temperature of the grain up to 108°F. The warm air from the heating grain rises to the surface of the grain mass, carrying moisture with it. At the surface, the warm air is cooled and some of the moisture is deposited on the suface of the grain where molds and fungi develop, causing caking and spoilage of the grain. Insect-caused heating develops in grains of less than 15% moisture content. Heating that occurs in grain with higher than 15% moisture is "wet-grain heating" and comes from the activities of yeasts, molds, fungi, and the respiration of the stored grain itself. In this type of heating, the temperature of the grain may rise as high as 144°F. Insects cannot survive under these conditions. Both dry-grain and wet-grain heating may develop simultaneously in different parts of the same bin. All feed mills become infested at one time or another. Realizing the importance of maintaining a clean plant, managers fumigate one or more times per year as a general practice. It is less costly to operate this way than to let the condition become so bad that they are forced to shut down the mill for a thorough cleaning and fumigation. Successful fumigation is assured only with a tight bin and a temperature above 65°F.

RODENT CONTAMINATION

Grain is also destroyed or contaminated by rats and mice, which each year destroy many millions of bushels of grain and feeds. They are also the source of diseases that are deadly to humans, pets, livestock and poultry. Control consists of: (1) not giving them shelter, (2) not feeding them, (3) killing them, (4) organized community rat control projects, and (5) following a year-round program. Remove places where rats and mice can hide, including piles of rubbish, stacks of old lumber and other materials and wooden floors close to the ground. Check doors and windows to prevent rodent entry and raise the floors of chicken coops and granaries 18 in. or more above the ground. Store feeds properly and keep trash in ratproof containers. Make granaries, corn cribs, feed rooms and storage rooms ratproof.

Control of Rodents

Rats and mice that are hungry are easily killed by poisoning and trapping. Poisons should be handled with great care and kept away from children, livestock and pets. They are only as effective as the bait with which they are mixed, which should be fresh and clean. Warfarin, a slow-acting poison, is an anticoagulant chemical which increases clotting time of the blood, causing death by internal bleeding, and must be consumed in small quantities over a period of several days. Adult poultry are very resistant to the chemical. Warfarin is a good follow-up rodenticide to use after infestations have been quickly reduced by other control methods. Many ready-mixed baits contain red squill. When using this poison, covered feeding stations which permit entry of rats and mice but exclude large animals, should be used. Fumigation with cyanide gas or gasoline exhaust fumes is effective in rat control in outside burrows, but should not be used indoors. Calcium cyanide dust is usually applied with a special dusting pump. Traps are used where poisons or fumigation would be unsafe. A properly-placed rat trap needs no bait. Mice are easily trapped in ordinary snap traps baited with peanut butter and rolled oats.

Community action is the most effective way to fight rats. Community campaigns should include not only farms but dumps, feed mills and other food storage areas in towns or villages. Rats raise 4 or more litters per year, averaging 6 to 10 in a litter. The presence of 1 pair of rats on the premises in the spring can mean 50 rats by fall. Rat control, therefore, must be a continuous job. Once their numbers are reduced in a community, they can be kept down at little cost by maintenance control. Modern rat poisons not only kill the rat but cause them to leave the building before dying. Precautions in new construction and simple repairs in old buildings aimed at the exclusion of rats should precede any control method. The cost of supporting only 1 rat has been estimated at from 50¢ to $2.00 per year. If rats are eating poultry feed, every 3 rats killed will save enough feed to take care of 2 hens. Cleanliness is the worst enemy of rats. Rats will not remain long in a building where all feed is kept in ratproof containers.

Fumigation

Closed recirculation is one of the 3 basic methods of fumigation. Others are the gravity-penetration method and the single-pass method. Besides outperforming the other methods, closed recirculation also eliminates fumigant losses caused by minor difficulties usually associated with the single-pass method. Closed recirculation is the most effective way to fumigate grain in flat storage buildings.

40

Feed Mill Sanitation

In the case of the feed industry, good housekeeping has been preached for many years. Keeping premises clean and insect-free demands eternal vigilance. The only insurance against contamination is to prevent the propagation of weevils and beetles or any sort of creature capable of contaminating feeds. If bugs live long enough to find a place of abode, contamination begins at once. There are accumulations of dusty materials in spouts, trunks and housings of machines. The primary cause for much of this is vapor condensed on the inner surface through which feed passes.

There are many methods by which adult insects can be prevented from instituting contamination. Sprays that are nonflammable and harmless to human beings are employed daily in guns and atomizers that are not expensive to kill insects on contact. Fumigants can be used with gratifying results within machines and equipment. Pulverized diatomaceous earth is used in the Southwestern United States to prevent insect damage to grain. It is nontoxic and is used at levels of 7.5 lb per ton.

THE NEED FOR INSPECTION

Inaugurating a program of sanitation consists of outlining the work to be done in simple, logical steps as follows: general cleanup and house cleaning measures; insect control; rodent control, and routine inspections which will reveal progress made in each of the above categories and will serve as periodic condition reports. The inspections must be thorough and complete, including both inside and outside and detailed enough to include suggested improvements in operations. Check the surroundings outside the plant for piles of accumulated trash, or spilled or souring grain. Have weeds been allowed to grow around the tanks and along the tracks, providing cover for rodents? Is a suitable place provided for burning trash? Rodents will not harbor under equipment that is on racks 18 in. above the ground. They do not feel secure enough to establish a quarter. Check windows to be certain that they are properly screened to prevent rodent entry. Screening should be of one-fourth in. hardware cloth to keep out both rats and mice. Eliminate standing pools of water due to poor drainage.

SANITARY PRECAUTIONS

Assign labeled coveralls and tools to specific areas in a plant and clean equipment to prevent buildup of pockets of Salmonella contamination. Prevent moisture from circulating through the plant. Separate raw materials and processed material areas by a solid, nonleaking wall or floor. Prevent air and dust circulation between raw material and processed material areas. Live animals and birds should not be permitted in processed material areas. There should be adequate showering, dressing and disinfecting facilities for employee use. Equipment used in the processed material area should be marked and used only in the area designated. Personnel assigned to raw material areas should not work in the processed material area, and vice versa. Processed products should be kept dry at all times. Bacteria require moisture to multiply, and thus finished products, or containers or areas where those are stored, should be kept dry. The footwear of all personnel should be scrubbed, using a disinfectant solution and a stiff brush before entering the area.

Key plant personnel should be trained as security/safety officers to be certain that sanitary guidelines are followed.

A partial breakthrough in insect control may be in sight. The structure of the long-elusive insect juvenile hormone has now been chemically synthesized and is being applied. Insect control may be achieved with as little as 1 g per acre. Since treated larvae do not mature or pupate, ecological suicide occurs in northern climates. It is not likely that insects will develop resistance against their own hormone systems.

PREPARATION OF COUNTRY ELEVATOR

In preparing the country elevator for receiving new wheat and other grains at harvest time, it is necessary to clean and spray the elevator bins to prevent infestation from beginning as soon as the grain has been placed in the bin. The entire wall surface should be brushed down to remove as much dust and loose grain as possible, and the bottom should be scraped clean. The bins should then be sprayed under high pressure to force the insecticide into crevices and holes. New grains arriving in an infested condition have picked up that infestation after leaving the field. If grain taken in July is to be housed for 30 days or longer in any type of storage, it should be fumigated. Fumigants are of 3 basic types: (1)

carbon tetrachloride, ethylene dichloride and ethylene dibromide combinations; (2) carbon tetrachloride, carbon bisulfide and ethylene dibromide combinations; and (3) carbon tetrachloride and carbon bisulfide combinations. Manufacturers supply dosages for different types of bins. Factors influencing dosages are temperature, type of grain, amount of dirt and moisture in grain, and most important of all, type of bin. Spraying the cars in which the grain is loaded is another means of holding down insect infestation. They should be sprayed the night before loading. At no time should materials be sprayed directly on the grain.

THE SALMONELLA PROBLEM

Public health agencies have alerted the poultry and feed industries to the Salmonella problem which causes food poisoning in humans. Over 650 different *Salmonella* species are known. The extent to which feed ingredients are contaminated is uncertain. In general, *Salmonella* comes from fecal contamination from rodents, wild birds, and probably insects. Most pathogens are easily destroyed by chemicals and heat. The usual heat-processing effectively destroys *Salmonellae* in meat and poultry by-products used as feed ingredients. Thus, ordinary precautions should prevent *Salmonella* from being a problem. Eggs are suspect, because poultry are susceptible to many diseases caused by *Salmonella*. *Salmonella typhimurium* causes paratyphoid infection in chickens and turkeys. *Salmonella* organisms are killed by heating up to 180°F (or 165°F for 20 min). Contaminated processed materials when mixed in feed, produce disease in livestock and poultry which can then spread to man. To prevent this, only authorized persons should be allowed in the plant.

SANITATION AND FEED MILL HOUSEKEEPING

Sanitation in the feed mill influences the health of the employee, the quality of the ingredients, the security of the mill, the health of the poultry eating the feed, the financial security of the feed mill, the poultry farm operation and the operating costs of the feed mill.

Mill Dust

Between 1973 and 1978, several explosions in feed mills and grain elevators were attributed to dust explosion. Mill dust often settles or becomes lodged in many parts of a feed mill or among the grains. A very high electrical charge develops, and moisture can ground the charged particles. Grounding can be accomplished by physical means. An electric spark can be produced, igniting the dust particles and causing an explosion. Therefore, dust-collectors should be used in the feed mill and "open" electric motors should never be used. The air in a feed mill should be continually scrubbed, and any time feed dust is observed, the air-filtering system should be improved.

The dust in the feed mill is principally fine particles of feed; however, some dust is soil dust which becomes attached to dry feed ingredients.

Dust settles most frequenty on electrical circuit boxes, in elevator shafts, on old unused mill machinery and in storage rooms occupied by used burlap feed bags.

Water Leaks

Water leaks can cause feed to spoil and also encourage mold growth. Wet feed heats and can cause a spontaneous fire.

Moldy Feed

When feed is spilled and allowed to accumulate in corners, in cracks or under equipment, it can collect moisture and mold. High-moisture feed ingredients, such as new corn, when ground and stored in closed bins, will mold and become warm.

High-fat ingredients, such as linseed meal, poultry by-product meal and high-fat fish meal, will mold if stored in closed bins.

When grains are allowed to stand in fields, molds will often develop. High-moisutre harvested grains will spoil and heat unless dried before storage.

Old Used Feed Bags

Often, old used burlap feed bags are soiled and contain high moisture and mold growth; they can also contain disease-producing organisms. When feed is stored in this type of container, it can become contaminated. A good policy is to clean and sterilize the used bags and never transport them from 1 farm to another.

Drugs and Toxic Sustances

Many drugs, antibiotics and trace minerals must be used in feeds in small amounts. Large amounts, being toxic, can cause death, poor feed efficiency and poor egg quality or production. These feed ingredients must be well-labeled and properly stored. The feed tag must show the amount present and the feeding program for all medicated feeds.

Old or Unsold Feeds and Ingredients

Old feed often becomes dry and unpalatable. This should not be fed to poultry. The fats and fat-soluble vitamins are destoyed or become rancid, reducing feed consumption and feed efficiency. All unused ingredients should be removed from the poultry feed operation. Much of this material is usable in hog rations or, if highly damaged, can be used as fertilizer of composting material.

Trash

Trash includes dirt, sand, metal fragments, plastics or anything else that could come from the field or machinery during harvest. Metal can be removed with screens and magnets. Air blasts are helpful in separating dirt, plastics and sand.

41

Poultry Farm Sanitation

Maintaining sanitation in poultry farm operations is a sound practice. A successful sanitation program is based on common sense and hard work. The use of federal and state inspectors encourages better housekeeping.

SUCCESSFUL SANITATION PROGRAMS

The cleaning of houses consists of the following: the removal of brooders, feeders, waterers, roosts and other equipment. Litter is also removed and the floor is swept clean. The house is washed down inside and out to eliminate dust and cobwebs that provide hiding places for organisms and reduce the biological load of the disinfectant. A disinfectant is applied with a power sprayer over every inch of the ceiling and walls, inside and out. A band of soil 6 ft wide around the exterior is sprayed with fuel oil. If leukosis has been a problem on the farm, another spray of disinfectant is applied before fresh litter is introduced. Litter is dusted with a suitable insecticide.

On the farm, isolation and sanitation are the best weapons of defense if they are properly planned and executed. Concrete floors should be used because it is impossible to get good disease control in houses with earth floors. They also reduce the amount of litter needed, and so less work is incurred in cleaning between broods. Brooding quarters must be restricted to certain personnel who are not permitted to visit other parts of the farm. No visitors should be allowed at any time. Screen houses to keep out wild birds. Provide a pan of disinfectant and a stiff brush to clean boots, a change of clothes, when possible, and a lock on the door to keep everyone out of the house but the caretaker and owner.

Baby chicks should be delivered in sanitary chick boxes (not previously used). On one farm a flock was so severely infected that loss from leukosis averaged over 20% by 20 wk of age; yet, when similar birds were placed on another farm under isolation, the breeder seldom had any losses.

Sanitation is cleanliness. It is not necessarily expensive, yet lack of it may prove very costly. Where installations are of sufficient size, the use of steam cleaners proves efficient and economical. The use of a chlorinated general cleaner for some general applications may be feasible. The presence of residual chlorine in the wash water tends to hold down bacterial growth. The domestic fowl assumes some importance as a carrier of disease. The virus of several strains of encephalomyelitis which infect horses and man find a natural reservoir in the hen. Many types of *Salmonella* are found in poultry, as well as in eggs. *Salmonella typhimurium*, the most common causing infection, is found more frequently in hens than in all other animals combined. Ornithosis (psittacosis) in domestic fowl assumes an increasing role in the epidemiology of the disease. The fowl mite causes an itchy dermatitis on the hands of the poultry attendants. Thus, public health officials have an increasing interest in sanitation of the poultry farm. Even a casual inspection of the available reports of mortality in commercial and farm flocks shows sufficient evidence that the loss of fowl by death often assumes serious proportions. With the concentration of the poultry population in particular areas, poultry mortality is sure to increase unless preventive measures are applied. Flock owners should be provided with essential instructions in the fields of bacteriology, immunology, parasitology, pathology, embryology, genetics, nutrition and poultry management. Spearheading it all is the necessity for an understanding of how bacteria live, multiply and travel from place to place, as well as what they are capable of accomplishing. It is obvious that carrying out the sanitation requirements is essential to the well-being of the poultry flock, and rests entirely in the hands of the owner and those he employs.

POOR SANITATION MAY CAUSE DISEASES

An example of the necessity for the application of sanitation is pullorum disease, starting with the hen which lays infected eggs from which are hatched infected chicks, resulting in an infected incubator and infected shipping boxes. Infected chicks infect other chicks in the brooder house which carry the infection to the feeding grounds and ranges. The infected young stock then become infected adults and the cycle is repeated. In the case of *Salmonella* infection, infected birds cause fecal contamination of feed, water troughs, yards and eggs. *Pasteurella multicoida* causes fowl cholera which may persist for several months, since organisms remain in soil and manure. Close contact by birds facilitates spread of infection. The organism causing fowl cholera has been isolated from mites taken from sick birds, as well as from house flies. Many other specific infections, including avian tuberculosis and pneumonia, *Spira-*

cheta gallinarium, pneumonia and *Staphylococcus aureus* demand control measures through cleaning and disinfection of houses and equipment for proper carcass disposal. Involved in these infections are soil, ranges, mites, flies, ticks, houses, brooders, incubators, chicks, eggs, fowl and humans. In each instance, sanitation plays an extremely important part in the disease-control program. Regular washing and sanitizing of housing equipment and utensils, immediate disposal of dead birds, eradication of rodents and insects and clean incubators and personnel are important. Proper cleaning compounds and an adequate scrubbing with suitable brushes are of great importance. Flies, roaches and ticks should be eradicated and rodent-control measures properly applied. A concrete cleaning platform, open or enclosed and properly drained, facilitates cleaning and sanitizing portable equipment in one central location. Isolation of diseased birds and immediate removal of dead fowl are important. Above all, the proper attitude toward sanitation on the part of the owners and employees is important to the health of the chickens. Thoroughly clean between flocks using a 400-lb pressure-spray rig, disinfectant and fumigation if possible. No feed is carried over between flocks; elimination of manure piles, bird-proofing buildings, locking doors, and rodent control are carried out; servicemen practice thorough disinfection between farms to avoid disease-carrying problems; personnel and equipment are sanitized thoroughly when leaving a farm and before going to another. Many of the same bacterial diseases and parasites that affect chickens also affect wild birds. Sparrows, cowbirds and starlings are the most likely carriers, but many other species are infected. By discontinuing ranges and building windowless and birdproof houses, the hazard of bringing in infection is reduced.

Poultrymen located in heavily populated suburban areas must accept certain responsibilities concerning their operations as they affect their neighbors. Avoid nuisance problems by keeping the farm neat and clean. If manure must be stored, see that it is well away from the neighbors' homes. Use discretion in the timing of cleanout operations. Haul manure during the early morning in cold weather, and not on hot muggy days. If you have a lagoon, watch overflow outlet. Lagoons should be located so there is no possibility that manure can contaminate the stream. Follow a sound fly- and rodent-control program, as well as local health laws in the disposal of dead birds and explain to neighbors how dead birds are disposed of.

Gamma radiation has potential value in destroying insects and bacteria in feedstuffs. Food and Drug Administration has approved the use of gamma radiation in processing wheat and wheat products. The use of radiation is expected to grow and include many food items and has become an important technique in the food industry in a decade. At present, the process is too costly for animal feeds, but it is anticipated that it will find limited use with specific items and become more economical as time goes on.

Dead Animal Disposal

Dead poultry, if not immediately disposed of, will decompose and be eaten and produce undesirable odors. Sometimes, flies will develop in the decaying carcass and produce toxins which are fatal to birds. The eating of dead birds may also initiate cannibalism, which can be very destructive.

Dead, crippled or sick birds should be destroyed and disposed of immediately. Several disposal methods are used:

Burying.—The dead or destroyed birds can be buried. This method places them in a pit and covers them with enough dirt to prevent wild animals from molesting the burial.

The use of burial containers in southern areas is practical when the temperature of the soil will support bacteriological development. Burial containers can be made from plywood and covered with soil, with an opening large enough to drop the birds through. About 24 in. of soil should cover the top; a 12-in. tile can be used for the opening. When filled, the tile can be removed and the hole filled with dirt, covering the birds. A container 4 ft × 4 ft × 6 ft high is sufficient for a 10,000-hen or 30,000-broiler operation. Such a burial-place must be well-drained and located near the poultry operation. If properly drained and covered, odors will not be observed as the bacteria decomposes the dead birds.

Burning—Use of Incinerator.—Although more expensive, an incinerator is a simple means of disposal, providing the escaping odors are not noticeable. The odors of burning feathers will produce environmental contamination and the operator subjected to legal costs. Odors can be somewhat reduced by the use of a tall smoke stack and by not over-filling or under-heating the unit.

Dead Animal Disposal Services.—Some producers of meat and bone scraps have services for collecting dead animals. Although such a service is principally for livestock, some incinerator operators will also collect dead poultry.

Man, Wild Birds, Stray or Wild Animals

Good feed efficiency cannot be obtained when poultry are subjected to diseases or fright from wild birds and stray or wild animals. Unnecessary visits or interruptions by outsiders can cause heavy losses. A good rule is to develop your poultry program before the new birds are obtained, then allow only planned visits or service workers to enter the poultry operation, such as by feed delivery trucks and veterinarians. Some supply and equipment companies make direct farm contacts and many have excellent services, supplies and equipment useful to the poultryman. However, as these representatives go from farm to farm, they can carry disease-producing organisms. They sometimes attempt to sell poultrymen new ideas which, if introduced into a well-planned, well-operating program could place both the poultryman and the birds under unnecessary stress. Changes should be initiated at the beginning of a starting period.

New Poultry Stock

New poultry stock should never be introduced into an existing operation without:

(1) Removing all birds from the building.
(2) Thoroughly cleaning the building and all the equipment.
(3) Keeping the facilities free of birds for 3 wks or longer before replacing stock.
(4) Keeping birds of the same age in each building.
(5) Never mixing stocks, even of the same age, because strain, variety and breed differences have different degrees of susceptibility to diseases. Growth, feathering and maturity may also be different.

Vaccination and Drug Programs

Follow the recommendations of your veterinarian, hatcheryman and feed manufacturer for disease prevention and control. There are no miracle treatments to eliminate diseases in poultry; the so-called "tip" treatments cannot be relied upon to correct or prevent a disease disaster. Chapter 43 discusses Disease and Parasite Control.

Odors

Many poultry odors are offensive and become absorbed into wood structures and are readily released when humid conditions develop.

A poultry farm produces many odors. These odors may be:

(1) Natural odors: All animals cast odors characteristic of the animal. They may come from fresh feces or by respiration.
(2) Filth: Broken eggs, leaking water equipment, improper manure removal all produce odors.

(3) Drainage: Water forming puddles or marshes around poultry buildings result in flies, mosquitoes and decayed soil organic matter.

Heating and Ventilation

The energy requirement for poultry maintenance is influenced by the temperature of the environment. Also, air composition influences productive capacity and health.

The ventilation of poultry houses involves the removal of water vapor, carbon dioxide and ammonia. Approximately 50% of water consumed daily is respired, carrying with it expanded heat and carbon dioxide. The following statistics emphasize the daily work required of a ventilation system per 1000 egg-production hens, assuming they consume 250 lb of a 17% protein-complete feed and 400 lb of water, laying at a rate of 60 to 70%.

Manure production	250	lb
[1]Water respired	200	lb
lb nitrogen excreted	3.125	lb
lb nitrogen volatilized	1.6	lb
lb fecal moisture	180	lb
lb fecal moisture volatilized	120	lb

The nitrogen excreted in the feces escapes as ammonia. The higher the moisture content of the environment, the greater the amount of escaping ammonia. When the ammonia becomes concentrated enough in the atmosphere to burn or irritate the poultrymen's eyes, it will also influence growth and feed efficiency of the chickens. Some poultrymen have lost entire flocks of broilers in fully-automated houses with forced ventilation when the electricity failed, the chief cause being nitrogen toxicity along with low oxygen (high carbon dioxide) levels.

The heat produced by poultry is enough to keep the environment satisfactory; however, in cold climates and bad weather, supplementary heat is desirable. In forced-ventilation houses, this supplementary heat can be obtained on days of sunshine from usage of hot-air solar units. These are economical and simple to construct. If supplementary heat is not provided, an increase in total feed or in energy content of the ration becomes necessary. This change in energy requirements has often been observed in poultry operations. Some nutritionists have made 2 dietary adjustments between northern and southern operations. The diets in the

---

[1]Each g water vapor respired required over 500 Cal. energy to transfer from 36°C to water vapor.

South use lower energy and higher protein, whereas those in the North use higher energy and lower protein. The biological action of poultry to consume feed until the energy needs are met makes it most important to keep the protein and energy contents in adjustment with the environmental temperature.

Water—Water Troughs

Water troughs must be kept clean. A broiler drinks over 1.5 lb water per lb feed consumed. The water must be clean, cool, and free from toxins and microbiological growth. When the water temperature is over 80°F (26.7°C), it is not palatable to poultry. Contaminated water is not readily consumed.

Filthy water containing feed particles and bacterial growth is often observed in feed troughs. Cleaning the water troughs daily improves the quality of the water and prevents the accumulation of microbiological growth. The usage of chlorinated water will control bacterial growth and thus help keep the troughs clean.

Floors

The 4 types of floors used in poultry houses are:

(1) Dirt
(2) Slat or wire
(3) Solid board
(4) Concrete

Concrete floors are the best from the standpoint of operation, but are very expensive. They are varmintproof, dry if properly drained, easy to clean, and do not appear to harbor pests or diseases. Solid board floors are ideal except that rats can gain entrance and destroy the birds. Dirt floors are used in most houses, and if properly constructed, give good performance. They are not easy to clean or keep free from pests, rats, worms and diseases, however. Slat and wire floors are installed in those operations using pits or mechanical cleaners. The major problem with these 2 types of floors is leg trouble and breast blisters. Rations for birds using either type of floor require additional calcium and phosphorus.

Lime is generally sprayed over dirt, concrete and wood floors before litter is added. Most diseases do not survive in strong alkaline conditions. Lime does not produce, or become, a toxic substance, but it does increase the loss of nitrogen from the manure. The nitrogen escapes as ammonia.

Litter

There are many types or kinds of poultry litter. Some of the most common are:

(1) Wood shavings or sawdust
(2) Sugar cane stalks
(3) Peanut hulls

Wood shavings, sawdust and sugar-cane stalks are the best litter material. When starting chicks or poults, a minimum of 3 in. should be used, with additional litter material added until an equivalent of 6 in. has been added.

When the chicks or poults are first started on the new litter material it will display poor handling qualities. The litter will not adsorb moisture during this period while the bacteria are becoming established. After an initial 3-wk period, the surface tension of the litter material changes and it readily absorbs moisture. The feces cast by broilers receiving a high-energy ration contain about 16 to 17% ash on the dry basis. This ash and nitrogen helps improve the handling qualities of litter.

Poultry eat their feces, and if the litter becomes contaminated with diseases such a Salmonella, Coccidia or worms, the young birds will become diseased or worm-infested.

Most broilers and egg producers reuse their litter—a practice known as "built-up" litter. The "built-up" litter is generally used for 1 year or longer. Attempts have also been made to feed the dried-poultry litter to broilers and egg producers. The results are highly variable and its usage is not warranted without some treatment, such as digestion. A careful study on the biological cycle in the poultry food chain suggests a microbiological cycle involving poultry feces before being consumed by poultry or hogs. Feeding the poultry-litter to ruminants would fulfill this microbiological step in the biological cycle. In this step, we find a destruction of poultry wastes, death of organisms common to poultry's digestive tract (Fig. 41.1).

Litter Treatment.—After the birds have been marketed, the litter is either removed and used as fertilizer, top-dressed with 2 to 3 in. of new litter material or given a microbiological treatment.

In operations where the old litter is "turned," the lumps broken up or removed, and smoothed and 2 to 3 in. of new litter added, there is a "build-up" of worms, insects and diseases. Some operators have reported that this type of management brings about development of disease-resistance in the birds.

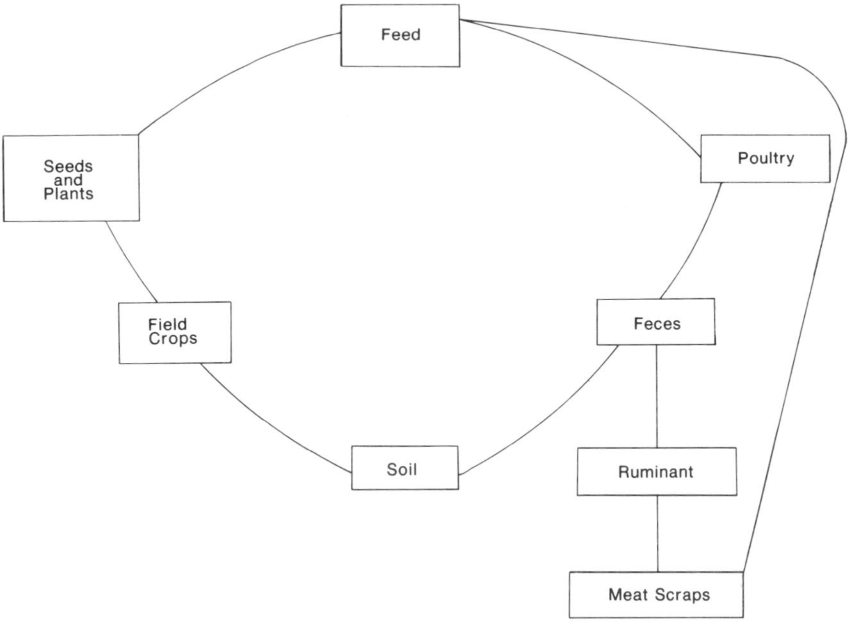

FIG. 41.1. BIOLOGICAL CYCLE OF POULTRY WASTES

The treatment involving microorganisms has proven to be helpful if properly performed. The process allows the water from washing the walls and equipment to be absorbed by the litter and additional water added to the old litter to provide enough moisture for excluding most of the air. After thoroughly mixing, it is put in long piles about 24 in. high. Within 48 to 72 hr, the microorganisms become active and heat is generated to the extent that the temperature reaches over 150°F. Sometimes a microbiological culture similar to that used in garbage digestion is added. This generally produces higher heat production, the heat reaching a temperature of 160°F. After about 14 days digestion, the temperature drops, the escaping ammonia is reduced and the litter becomes drier. The litter is respread over the floor and 3 in. of new litter are added. The feathers in the old litter are decomposed and the diseases and pests appear destroyed. The treatment offers, besides a digested-type litter, a "cycle-break" in the operation. Generally, about 21 days are required to complete this litter treatment.

The Cycle-break

Successful farmers have found that crop rotations are of value in fertility, and weed or pest control along with disease and erosion control.

Many years ago, Dr. Billings, of the University of Minnesota, developed a pasture-management program for turkeys which rotated field usage so as to provide a more disease-free environment. This management program was a type of cycle-break which allowed the fields to remain from turkey production long enough for the blackhead organism (Histomonas Meleagridis) to die before reusing the pasture for turkeys.

A cycle-break is most desirable, regardless of the drug program, for the successful production of confinement- or open-range rearing of broilers or poults.

42

Poultry Farm Management

A MANAGER

The poultryman is a manager and a manager's job is to secure maximum net income from the resoruces at his disposal and to combine the "correct amounts" of land, buildings, labor, equipment and technology. Technology means application of the most recent scientific information on nutrition, breeding, ventilation and equipment layout.

The manager is the farm's economist. Four sets of knowledge are necessary to him: (1) the appropriate economic principle to apply; (2) the correct analytical procedure to use in applying it; (3) the latest scientific technology (input-output data); and (4) correct predictions on costs paid and prices received.

It is not an easy or simple job. However, this is the only alternative to "guessing." The managerial function is a skill that is developed with practice. Economic principles can be learned; analytical procedures can be used until they come easily; and pertinent input-output data identified, filed and kept up-to-date to provide the latest information.

KEEN OBSERVATION NECESSARY

Poultry farming is, however, a worthwhile, satisfying occupation, requiring an alert, keen mind and a willingness to work. One must like poultry to succeed, for birds are very susceptible to the persons caring for them. The poultryman must observe his birds frequently and otherwise study his business. He must have "chicken sense" or good judgment in caring for and managing his flock. He should acquire knowledge and skill in conducting the business through training and experience. Training is offered in agricultural schools; he can obtain it also by an apprenticeship with a successful poultryman. The poultryman must also known how to deal with people, be courteous, prompt, tactful, honest and ambitious.

In a small operation, the poultryman can do most of the work himself or with his family. As the operation grows, he will require more help. A poultryman needs both cash and credit, depending on the size of the enterprise, the efficiency of the plant and the financial standing of the operator. If the premises are rented, less cash is needed. Credit for a poultry enterprise is often available through feed dealers or banks specializing in agricultural accounts. One method of starting a poultry business is to begin with a small plant and enlarge as earnings increase. There are risks involved. Because of the short life of the fowl, there is rapid depreciation of stock which requires frequent renewal. Each year, there are the hazards of brooding, rearing, and sometimes of incubation. Both young and old stock are susceptible to diseases and parasites. The use of incubators and brooders increase fire hazards.

PRODUCTION AND MARKETING PROBLEMS

A poultryman must understand both production and marketing problems, but there are special advantages to the poultry business. The egg is the most universally used of all animal products, except milk. It is essentially "liquid meat," being one of the most easily digested and assimilated forms of animal food. It is relatively easy to get into the poultry business financially and otherwise. There are the quick growth of the fowl, its early and heavy laying ability, and its natural vitality to resist disease in its favor. A foraging farm flock can get much of its living from feed which otherwise would be wasted.

Relatively little capital and land are required, and there is a quick turnover of capital. Raising poultry is a healthy vocation and does not require heavy manual labor as do most other kinds of farming. Many women and children on farms assist in the enterprise. Poultry are efficient transformers of raw materials and develop into high-priced finished product. About 100 lb feed and 50 gal. water are converted into approximately 25 lb eggs yearly, in addition to maintaining the hen. Producers who have established a reputation for high-quality eggs frequently receive several cents per doz higher than the market quotations.

A question often asked by poultrymen, is, "Is it possible to keep layers a second year?" Some think it is good business to carry the better pullets through the second year, but most studies show it is not a good practice. More money is made with an all-pullet flock. Old hens lay larger eggs but the interior and shell quality are poorer.

A popular starting time for egg-type chicks is the spring. More chicks are started in April than any other time. Years ago, poultrymen wanted to start their chicks in the spring and then get them out on the green grass as soon as possible. With modern rations, green grass is no longer necessary, and it is possible to raise chicks entirely in confinement.

Do layers lay less if the cockerels are in the pens? Experience shows that layers produce as well with cockerels as without. Of course, cockerels eat feed and produce no eggs. The male consumes the same amount of feed as the hen when she is laying at the rate of 50% because the male is larger.

To get the most from the feed (1) buy chicks bred to lay; (2) provide plenty of clean water; (3) provide light to eat and drink by; (4) provide plenty of clean air to breathe; (5) keep the temperature between 40° and 70°F; (6) supply all feed nutrients in the right amounts; and (7) provide enough feeders and waterers properly spaced. A flock of 500 birds divides into about 10 smaller groups of "peck orders." Each group must have its own waterers, feeders, etc., otherwise, fighting and poor production will result.

THE FEED SERVICEMAN

The business of the serviceman is to assure the steady growth of the birds and good feed conversion from starting to market. The last week before market is most critical; then, birds eat the most feed and build the most meat. At this point, there is a big investment and if anything happens to stress the birds, they slow down and lengthen their growing period; costs rise, and profits drop or may be wiped out That is why it is extra important to keep balanced nutrition and good health to the very end.

The serviceman is the key to profits. Every day he plans his work. Baby chicks arrive, a new building goes up, a disease problem crops up. The serviceman has to be on his toes and equipped for any emergency. Working closely with the grower, he aims to better his last record with a new flock. The production chart on one of his flocks looks favorable with daily feed consumption increasing and mortality low.

Keeping daily feed records is a bother, but it pays off. A drop in feed intake is the first sign of an undetected disease and can cause a higher rate of condemnation. At one farm, a serviceman noted the feed drop on the same day he saw the birds slipping and heard the sniffle and a few sneezes among the flock. That is the one unmistakable sound of trouble. He recognized a respiratory infection and took sample birds to the diagnostic laboratory. Medication was needed. Some growers prefer water medication; others medicate the feeds, but this is mostly prophylactic because sick birds do not eat well. Medication in the water works faster. The serviceman gave the right medication in the right amount, at the right time, and in the right way. Five weeks later, there was not a sign of the setback. Fine looking birds show that waxy, yellow shank, good feathering and bright eyes. The combination of good birds, good feed, good management and good cooperation is unbeatable.

The serviceman's company breeds only the very best stock that are sure to grow. The feed mill formulates high-quality feeds with the proper balance of nutrients from the best materials and best suppliers; these feeds are designed to speed sturdy, healthy growth with every safeguard included, and all the minerals and vitamins proved essential by long years of scientific research.

Today's feed mill has complete control with big-volume mixing and processing is handled by automation. With proven techniques micro-ingredients and drugs are blended and premixed before they are added to the feed. Carriers in the premix hold important trace elements. The exact amount of this premix is added after 25% of the batch, for which it is formulated, is in the big blender. The precise, controlled job of mixing and blending puts quality into every bag of feed. Sampling for analysis provides another quality control. When this feed goes into feeders, the serviceman can be sure it meets every specification of the formula. Feeds are analyzed for uniformity by mill control chemists or in outside laboratories. When flocks reach the market, the final proof of correct blending and mixing is evident.

The starting of a new flock is critical. A fresh, clean, well-ventilated house is needed. The serviceman makes a careful check. Temperature must be exactly correct for incoming chicks. A good start helps insure a good investment.

Disease Carriers

Wild birds carry disease. They enter the buildings through outside doors, holes, faulty window screens, etc. These openings must be watched carefully. Better wear protective boots. Older birds can carry typhoid and other diseases. The warning signs of disease are easy to hear and see but autopsy gives more clear-cut evidence. Servicemen are not trained pathologists. They depend on laboratories for a full diagnose. The serviceman's job is management. Disease detection and consultations on housing are very much his business. He makes available construction plans with vapor barriers and other insulating materials that are effective in temperature control and pay off in better birds.

Modern poultry growers are looking ahead. The 3 story house is a far cry from the old 1-story buildings. Seventy thousand birds may soon be housed in 1 enormous building. There may not be much excitement in good housekeeping, but attention and common sense are the foundations of all chicken-raising.

Complete Checks Made by Servicemen

Good servicemen do not miss a thing. What about the level of waterers and feeders? Get the type to avoid spilling and raise as the birds grow. How about temperature? A little on the high side? An extra fan may be needed. Nothing wrong possibly, but is is always wise to make sure. Possibly there is no sign of disease. Perhaps the most satisfaction is the spirit and and cooperation between men in all facets of the poultry industry. Servicemen, technical experts, and scores of scientists are helping growers. Members of a team, they join forces to combine their skills and abilities. When disease strikes, everyone organizes his force with action. Usually, servicemen work with local laboratories to identify the problems and diseases. The trained eye of the technical expert can be essential. Feed consumption going down? Sample feed. This helps in a complete diagnosis. Look at the birds; are many still lively but is the color of the feet and legs a little off? Or is there some heavy breathing and maybe a few birds are down? An inside look will tell more. The big job is quick action. A preliminary diagnosis, immediate medication to cut down mortality and halt or stop the disease may save them for market. Perhaps they have a low-grade infection which is not too serious. Immediately a program of treatment is initiated. Check further in the laboratory to confirm findings.

The servicemen see that those responsible live up to their contractual obligations: Among the benefits accrued are monetary settlements for failure to fulfill contracts, keeping suppliers on their toes since they know they are being checked, building a history of supplier performance and consistency, and allowing the feed manufacturer to reject lots of ingredients that are not usable in his mill.

FEED WASTAGE

Wastage of 1 lb feed per day per 100 chickens would hardly be noticeable in a cage- or floor-type operation. Yet, it adds up to the difference between profit and loss in some cases. How can this feed wastage be reduced? Consider the following:

Improperly adjusted feeders result in more wastage than anything else with floor birds. Proper height is where the lip of the feed trough is 1 in. above the level of the back of the average-sized chickens. Improper debeaking is a factor; scoop-billed birds must have deep feed or they waste a lot of feed just attempting to eat.

Rats and mice not only eat a lot of feed but they also ruin a lot more and spread disease. Control rodents the year around. Delivery men often don't take the time to place the feed carefully in the bins. Filling hoppers too full will result in birds spilling out feed. Feed more often if necessary.

DAY-OLD CHICKS OR STARTED PULLETS

Should day old chicks be purchased and raised into pullets or should "started pullets" be bought? There is no one answer to this; it varies from farm to farm and area to area. Some advantages and disadvantages of pullets are as follows:

Advantages

(1) Better use of time may be obtained by becoming a specialist in egg production; (2) better use of buildings and equipment may be made (other than for brooding and raising chicks); (3) no money tied up for long periods and money will be saved which would otherwise be invested in brooding facilities; (4) the risks of starting and growing chicks are cut; (5) the disease hazard if on an all-in, all-out program is reduced; (6) better pullets may be purchased at a cost less than the cost of raising them; and (7) the culls in the laying house may be reduced because of the purchase of Grade A pullets.

Disadvantages

(1) Disease hazard; (2) may increase costs (the grower has to make a profit, too); (3) it is not always possible to get started pullets of the strain desired; and (4) the feeding program or management that has been employed previously cannot be controlled.

In an area of mass production, efficiency of management is the price of survival. The change from open-range to confinement-rearing under crowded conditions has emphasized the importance of correct management practices. The trend is toward larger, more efficient production units, making use of labor-saving equipment. This requires larger capital investment. Eighty percent of the poultry meat and 50% of the eggs marketed in the United States are from large commercial flocks fed indoors and operated where no feed is produced.

FORCED MOLT

Force-molting at the end of the first laying year encourages a satisfactory lay rate of acceptable egg quality during the following year. There are a number of ways to cause molt. One is as follows: Along with light restriction, feed and water are removed for several days. Then the water is turned on and 3 lb grain per 100 birds are fed. In another 3 days, 6 lb grain are fed for 3 days. After 3 days, a limited amount of mash is offered and the next day, full mash for the remaining period.

Forced molting results in improved feed efficiency. The high consumption figure for the treated birds during the stress period is due to the fact that the hens cease to lay. The stress period continues up to the time the production of the treated birds equals that of the nonstressed birds. A somewhat higher egg production and lower feed conversion results in a substantial difference in egg income over feed costs in favor of the forced-molt group. Recent work indicates that if molting, or induced resting is to be effective, 2 or 3 rests during a 2-yr productive flock life are needed.

RECORD KEEPING

It is difficult to make money year after year without keeping meaningful records. Only by records can a poultryman tell what kind of a job he is doing. Is he wasting feed or breaking too many eggs? Is his help giving the birds necessary day-to-day attention? If records are not kept the reasons for poor performance cannot be pinpointed. Further, management cannot be improved because errors will not be determined.

Record forms provide spaces for recording the needed information; by keeping records and studying them to find ways to reduce costs, it is possible to increase income per dollar invested. Records should be kept on things that have to be done. Lack of meaningful goals and lack of attention to details assure failure of any operation that requires business management. Make sure goals are meaningful with respect to operation and then set out to reach them by following the old axiom, "If you take care of the little things, the big things will take care of themselves."

Types of Records

What records should be kept? The kind of records needed depends on the type of business, how it is operated and where it is located. A started-pullet grower keeps different records than an egg producer. With started pullets, your business requires records on daily growing periods, mortality during the growing period, percentage of saleable pullets produced from the number of pullets started, feed used per day, and total feed used per bird, dates the birds were debeaked, vaccination records including age at which vaccines were administered and the type and lot number of vaccines, drugs used, and when and how they were administered. This information spots areas of needed improvement.

On the other hand, the egg business requires records on the number of pullets housed, date housed, age at housing time, daily egg production, daily mortality, reason for mortality, culls, feed used daily, medication used and cost, and egg receipts. With this information, feed conversion, percentage hen-housed production, and all other data necessary to measure the success or failure of various management practices can be estimated.

Keeping a record of the feed used on a daily basis can present somewhat of a problem but it does not need to. Daily weighing of feed is most accurate, and this method should be used whenever possible. However, alternate methods provide usable information. If a motorized cart is used for feeding, mark several levels in the cart with a tape; then weigh the cart empty and record the cart weight as it is filled to the various levels. Each day records should be kept of the full or partially filled carts of feed put in the troughs. With simple multiplication, the pounds of feed used per day can be calculated. The cart should be weighed with each new batch of feed. If a poultry house contains automatic feeding equipment, the feed hopper should be marked to show various levels. Then the hopper should be filled by hand and the feed weighed. This will indicate the weight of feed the hopper contains when filled to the various levels. When the hopper is filled in the morning, subtract the weight of the remaining feed in the hopper from the weight of the hopper when it is full. This indicates how much feed your birds have used.

The small amount of time used in keeping records will be well spent. Management should set goals. Birds should lay 240 eggs or more in 12 months on 4.2 lb or less feed per 1 doz eggs. Not more than 1% of the flock per month should be lost. If the birds are using more feed to produce a dozen eggs, begin searching for the reason and take steps to correct the problems causing the decreased performance. By making comparisons and looking for ways to reduce feed costs, income per dollar invested can be increased.

Properly kept records will make business profitable. It gives an opportunity to size up weak points and make them strong points. Some poultry associations, hatcheries and servicemen provide record forms as a service. Records should be set up to give the kind of information needed to analyze the business. Eggs per hen housed, dozens of eggs to date, pounds of feed required per doz eggs and percentage livability are essential to know. A 28-day period (4 wk) is preferable to a monthly basis because months have different numbers of days. Egg production on a hen-day basis is different than that on a hen-housed basis. In figuring on a hen-housed basis, the number of eggs is divided by the number of hens placed in a laying house. On the hen-day basis, the number of eggs produced is divided by the average number of hens living during the laying period. Production figures are higher on a hen-day than a hen-housed basis, since the latter assumes no mortality while hen day takes mortality and culling into consideration.

Egg production averages about 70%. The birds start out with 6% rate of lay at 22 wk, which is increased at the rate of 12% per week until it is 84% at 29 wk. The peak is 86% at 30 wk. For the next 34 wk, it drops 0.5% per wk and for the next 13 wk at the rate of 1% per wk.

A budget for the layer shoud be prepared to determine needs and what profits are likely to be. To figure the budget, multiply the number of dozens of eggs the layers will lay times egg production costs. The figures should be those which apply on your own farm rather than standard figures. A goal should be set, whether hatching or market eggs are produced. If properly managed, either enterprise can be profitable.

CULLING

Culling provides an opportunity to save on feed costs. Close culling of pullets at housing time removes underdeveloped pullets. After the birds reach their peak of lay (95% of pullets are laying), removal of 2 to 4% of the birds not in lay will save on feed costs. When hens are kept 14 months, a large percentage of the birds go out of production at the tenth month of lay and are out of lay for 2 months. These birds will not be back in production long enough to pay their feed bill and should be sold. Any time production drops below 60%, there are 10 to 15% of the birds out of production.

DEBEAKING

Proper debeaking of pullets requires patience, and a knowledge of knowing how and when to debeak. Chick-debeaking, frequently done by the hatchery at 1 day of age, cuts water and feed intake for 48 hr. During the first 3 days of life, chicks eat very little, relying on their yolk for nourishment. Normal water and feed consumption are drastically reduced, causing severe stress. If they are debeaked on the first day, they have 2 days to recover while drawing nourishment from their yolks. When housing debeaked chicks, place the waterers on boards so that they can drink without lowering their heads; make sure the water level is within easy reach. Arrange the feed and water close to the source of warmth. Preheat the water to the same temperature as the brooding room. Feed mash for the first few days. During cold weather, be sure the brooding unit is preheated.

Debeaking places a stress on older pullets, primarily due to the drastically reduced water and feed consumption. Water consumption is cut by 75%. Put extra feeders and waterers in the laying house and make them readily available. Fill the hopper with mash so the bird's tender beak will not bang against the bottom. Do not feed pellets the first week.

For cage operations, make the feed deeper. Debeak during the coolest part of the day. If the temperature rises above 80°F, some birds will bleed heavily after the operation. Some poultrymen feed a tranquilizing drug for several days before debeaking. This helps keep the flock calm and minimizes the stress of handling and debeaking. Do not debeak when birds are sick. Debeaking cuts egg breakage and egg eating as well as feed wastage by 50%. It decreases blow-out, pickouts and cannibalism.

Because of differences in beak length and shape, there is no standard slice off the beak. Each bird is debeaked in relation to the length and shape of her beak. Some poultrymen debeak when birds are between 12 and 22 wk of age, 2 wk prior to housing time or at the first sign of trouble from picking. For pullets, cutting both upper and lower beaks produces the best results.

As for baby chicks, adjust the blade's heat so that it glows cherry red. Hold the bird's legs in one hand and the head in the other. Use your thumb and forefinger on the sides of the jaws to open the mouth. Place the upper beak over the bar so that the blade strikes a point two-thirds of the distance from the tip to the nostril. Incline the head so that the cut slants from this point backwards. Leave the top of the beak about one-eighth in. longer than the lower surface. Cut the lower beak squarely at the point where the fleshy tissue ends. After the cut, roll the cut surface against the hot blade to sear the wound and round the edges of the cuticle. Debeaking does not stop picking, but hampers the birds in their efforts to pick; because of this, it reduces cannibalism. Retouch a bird's beak whenever necessary, because the original debeaking job is almost never 100% effective.

FEEDERS

Feeders may be of several kinds: (1) open floor or yard, (2) flat boxes or egg cartons, (3) open box troughs, or (4) mechanical feeders.

Floor Feeders.—Early poultry producers often threw the grain portion of the feed on the ground or into the litter. This was thought to stimulate the birds. Today, many large-range feeding operations use spreaders to feed pellets to turkeys and developing pullets grown on open ranges. This method is cheap because the bulk feed trucks can often spread the pellets directly instead of unloading and reloading. However, it risks the spreading of disease and filth, which could bring about high drug costs and increased mortality.

Flat Boxes, Egg Cartons or Egg Case Dividers.—Feeding starting chicks and poults is often simplified by placing some starter feed on egg case dividers or cartons, as well as using box tops with about 1 in. edging. These feeders, after about 3 days, are replaced with regular starting feeders.

Open-box Troughs.—Open-box troughs are the type used throughout the poultry industry. They may be used in hand-feeding, self-feeding or mechanical feeding.

This type of feeder with its different modifications is desirable, providing the operator recognizes the physiological structure of the bird.

The birds eyes, as previously stated, are located on the side and slightly to the front and top of the head. This results in the bird pecking directly where it looks, resulting in a side-wise motion which throws feed out of the trough if the troughs are overfilled (Fig. 42.1). A box trough should never be filled over one-third, and the top of the trough should be at below floor level or level with the back. Feeders of this type are often placed at floor level because of construction cost.

An advantage of hanging feeders is that the box troughs can be placed high enough to force the birds to reach up for the feed. Birds which reach up seldom waste feed except in excessively overfilled feeders.

Mechanical Feeders.—Mechanical feeders use the open trough and have the advantage of being filled to the desired level. The movement of the feed also stimulates the birds to eat.

These are several designs of mechanical feeders and each allows control of offered amount and length of feeding time. They have the same disadvantage of all pen feeders, i.e., easy entry for rodents, and the feed can be eaten or destroyed. All automated poultry operations use mechanical feeders. Feeding requires more labor than any other poultry operation. Also, most of the expense in poultry is in the feeding operation and one of the greatest losses can be in feed wastage.

CONTROL OF SEXUAL MATURITY

Delaying sexual maturity of broiler-breeder pullets has consistently resulted in increased egg production, reduction in number of unsetttable eggs and increased average egg weight. Similar results have been obtained with egg production strains; however, the advantages are somewhat fewer.

Several methods have been used, but the daily weighing of the feed allotment using these methods has always increased labor. The more common methods are:

Restricting Energy.—The restriction of energy involves the use of a diet containing approximately two-thirds the energy required, and all other nutrients present in adequate amounts. The most effective restricted-energy diet uses ground oats, wheat bran and 17% dehydrated alfalfa meal. Sometimes oat hulls are used to replace part of the ground yellow corn. Only the mash form is used because the birds eat less of the bulky diet; therefore, this mash diet can be fed ad libitum. If pelleted, the amount fed is never greater than that consumed of the mash containing the regular formulation which is approximately 2700 Kcal of metabolizable energy per kg of diet. The restricted-energy diet contains around 1800 Kcal. The protein content is maintained at 15%.

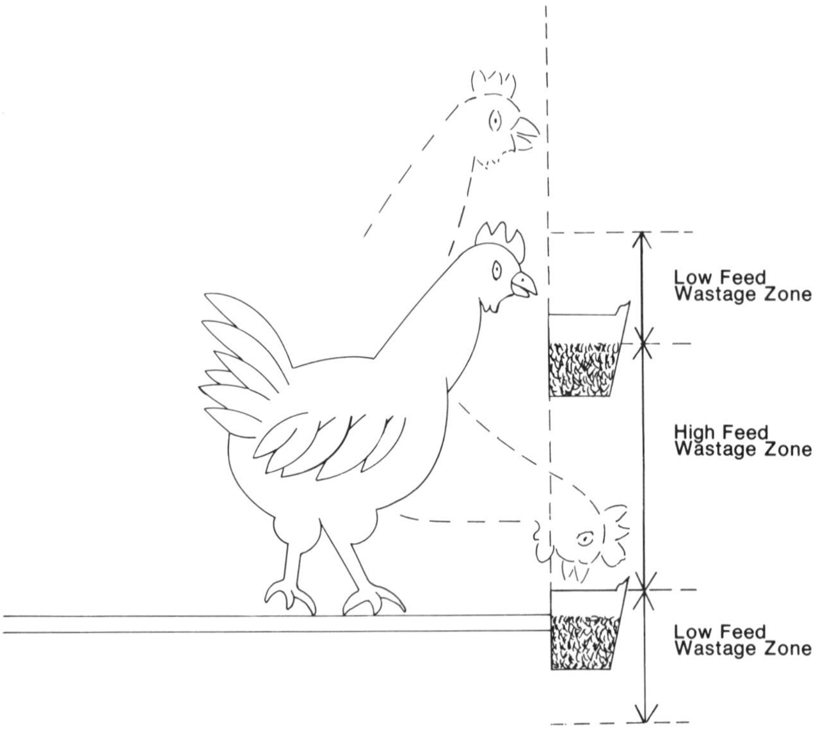

FIG. 42.1. RELATION OF HEIGHT OF FEEDER TO FEED WASTAGE

Restricting Light.—The restricted light program uses a 2700 Kcal/kg ration with 15% protein. This method uses a windowless house, and both lights and mechanical feeders are on time clocks. This method is better than total feed restriction because if properly handled, cannabalism is reduced and less litter is eaten.

Total Feed Restriction.—The total feed is restricted, and this program often gives poor results. When birds are without feed, they develop litter-eating and feather-picking (cannabalism) habits. Deficiencies of some vitamins and minerals often occur because their needs may be a function of body size rather than percentage of feed.

Low-protein or Low-lysine Uptake.—Low-protein diets contain 10% protein and 3100 Kcal/kg of metabolizable energy. The birds tend to become too fat and this program does not reduce maturity.

Low lysine gives results similar to low protein, in that maturity was not delayed appreciably nor egg production improved.

Skip-a-day Program.—Skip-a-day is often used. This method generally provides the birds with 65% of the expected feed desired every other day. A regular 2700 cal ration containing 15% protein is used. This program gives results similar to restricted-energy or restricted-feed.

Restricting Feeding.—Poultrymen often restrict the feed for heavy breed pullets to reduce cost and slow sexual development. Several investigators have reported a savings of 6 to 10% in feed without reducing egg production, egg weight or feed efficiency. Table 42.1 reports the results of a study showing the influence of social order and limited feeding performance. This study reduced the feed of layers 6 to 10% without influencing egg production.

LIGHTING

Different types of colored lights have been used with equal results in rearing pullets. To guard against having light of inadequate intensity during the laying period, test it with a light meter. A high incidence of cannibalism occurs in pullets with beaks raised under white light. Red, orange and green lights prevent cannibalism in White Leghorn cockerels. A pronounced eye enlargement occurs in birds placed under low intensity blue, green and red lights; it is evident at 12 wk of age and persists to 20 wk. Layers do not need more than 14 hr of light per day.

Pullets grown in darkness, and kept so during the laying cycle, averaged 59% production, whereas others that received a gradual increase of light during the laying period gave 73% production. Many experts recommend increase in light 15 min every other week up to 14 hr during the laying cycle. It is generally agreed that the length of daily light periods in the laying house should not be decreased. Using red light during the growing period and white light in the laying house produces excellent results.

The production of broilers in windowless houses and dry caves has been accomplished with only enough light for the attendant to feed, water and inspect the flock and facilities. Restriction of light is desirable when feed is restricted to reduce feather-picking and other forms of cannibalism.

WATERING

Frequency of watering is a factor in assuring proper water intake. Poultry drink small amounts at a time. Feeds are low in moisture, so birds need a constant supply of water in close proximity to the feed and with the proper allowance of drinking space. Cool drinking water is desirable in the summer rather than water from a tank that is continually heated by the hot sun. Birds on lush pasture may get almost enough water from this source alone to supply their needs and 1 watering per day is ample.

TABLE 42.1. SOCIAL ORDER AND LIMITED FEEDING PERFORMANCE[1]

Treatment	Hens/ Cage	Cage Density Sq In./ Bird	Feeder Space/ Bird, in.	HD Prod %	Av.	Egg Wt (g)	Av.	Feed/Dz (lb)	Av.
Full fed	2	64	4.0	74		58.4		3.46	
Full fed	5	51	3.2	72	74	59.2	58.8	3.50	3.49
Full fed	12	53	3.3	74		58.7		3.52	
Limited fed 94%	2	64	4.0	74		58.0		3.23	
Limited fed 94%	5	51	3.2	71	72	58.0	58.2	3.34	3.32
Limited fed 94%	12	53	3.3	71		58.6		3.41	
Limited fed 90%	2	64	4.0	73		59.0		3.15	
Limited fed 90%	5	51	3.2	72	72	57.9	58.2	3.15	3.17
Limited fed 90%	12	53	3.3	72		57.7		3.21	
Timed fed 94%	2	64	4.0	73		57.9		3.21	
Timed fed 94%	5	51	3.2	73	74	58.9	58.6	3.30	3.24
Timed fed 94%	12	53	3.3	76		58.9		3.21	

Source: Snetsinger and Zimmerman (1975).
[1]From 36 to 56 weeks of age.

TABLE 42.2. INFLUENCE OF TEMPERATURE ON DAILY WATER CONSUMPTION PER 100 PULLETS[1]

	21.1°C		32.2°C	
Week	Leghorns (Liters)	Meat Type (Liters)	Leghorns (Liters)	Meat Type (Liters)
4th	6.1	9.8	10.6	17.0
8th	9.2	13.3	15.9	23.0
12th	11.9	15.9	20.5	27.0
16th	14.2	18.8	24.5	31.0

Source: North (1972).
[1]Values for water consumption include evaporation from troughs and spillage.

Hens in laying cages drink more water than floor-managed hens. Also, the fecal moisture from caged hens is about 4% higher than from floor-managed hens.

Most poultry drink more water than is needed. A reduction of water allowing 15 min drinking time per hr during a 14 hr day has been used with success.

THE DUST PROBLEM

The dust problem in poultry operations is of great importance. In winter, poultry houses are low in humidity and high in dust. The litter, the feed, and the bird itself are sources of dust. Most dust consists of particles of fragmented feathers and tiny particles of dried skin. The type of feed is not as important as the type and age of the litter. Shavings, peat moss, sawdust and clay, in that order, are sources of dust. Reused litter is dustier than new material. There is a peak of dust concentration during midmorning and midafternoon and a gradual increase as birds age. Birds are able to tolerate a high-dust load, but it is generally believed that dust is a harmful factor. Dust has been studied from a bacteriological standpoint; it can act as a major carrier of disease-causing miroorganisms in the bird. It increases disease load in several ways: Organisms are actually transported on dust particles and irritation caused by the dust lowers natural resistance of the respiratory tissues to the organism. It may also set up an allergic reaction in the respiratory tract because it is mainly protein in nature. It increases the incidence of air-sac lesions in young turkeys. High-bird density and reduced ventilation produce a higher microbial contamination of the air.

To overcome the dust problem, several methods have been tried. Humidity has been increased with an air scrubber using a water spray. This reduced dust unfortunately, increased humidity, which resulted in wet litter. This is an even more serious problem. Pelleted feed that does not break into small fragments helps. However, most dust originates with the chicken itself. It is advised to maintain humidity as high as possible without causing wet litter. Increased ventilation helps but makes the house more difficult to heat. Improving the dust problem for the birds also makes conditions far more pleasant for the men who work with them.

Dust interferes with ventilation in that the hardware cloth over windows and ventilating grills becomes covered with it, as well as with lint, and closes off the flow of air. The structure of dust makes it difficult to remove with water or air. The screens or grills need to be removed once per year and scrubbed with a wire brush.

INFLUENCE OF ENVIRONMENT ON EGG QUALITY

Flocks that have been in production up to a year or longer produce poor shells, mottled yolks, watery whites and eggs with off-flavors during hot weather. Feed consumption drops in hot weather, resulting in a decrease of calcium intake. Also, the longer and higher the rate of lay, the poorer the interior egg quality. Eggs pick up odors and flavors from their environment in hot weather. Musty odors, vegetables and strong-smelling disinfectants are some of the most common causes of odors. Keeping the flock comfortable will help maintain production and good egg quality. This requires ventilation and increased watering space or time the water is on. Frequent removal of manure helps control flies and odor.

STRESS, BAD HABITS AND VICES

A caged life in a battery is unnatural for the hen; she can move freely but cannot supplement her feed. Caged birds are constantly in some stress—even if feeding and management are excellent. Little attention is paid to individual birds, although regular checkups are made. Diseases, behavioral abnormalties and vices are checked for, as well as consumption of the feed and drinking water.

Mechanization is only good if the bird is seen not as a machine but as a living being. Since most layers placed in the batteries are reared on litter, special attention should be paid to the prevention of coccidiosis and gastrointestinal parasites. Place birds into battery cages at 18 to 20 wk of age so the birds can get accustomed to the new housing before laying starts and can be wormed if required. Battery fatigue is less frequent in birds that have been in battery cages from an early age than in birds placed there when older.

The vicious habit of picking feathers, combs, wings, toes, tails, vents and other parts of the body is one of the serious problems in raising poultry. It is prevalent among chicks, growing pullets and mature birds, and leads to outbreaks of cannibalism if not promptly controlled.

TABLE 42.3. CHANGES IN SPACE REQUIREMENTS AND BODY WEIGHT FOR LAYING HENS

Year	Type Housing	Space per Bird sq ft	cu ft	Body Weight (lb per Bird)
19926	Floor	3	32	5
1946	Floor	3	24	4.5
1966–1967	Floor	1	8	4
1966–1967	Cage	0.5	4	3.7
1970	Cage	0.25	2	3.5

An English *Complete Poultry Book* published some 80 years ago refers to feather-picking as a pernicious habit and suggests the chopping block as the surest remedy. It also suggests a wire bit to prevent shutting of the beak firmly together so the birds are unable to grasp feathers. Dark nests tend to keep hens on the roosts until the vent has returned to its normal position. This takes away the temptation to start pecking.

Bad habits are usually attributed to faulty management, including overcrowding; overheating; underventilation; high humidity; too-bright light; insufficient nesting, eating and drinking space; presence of injured or sick birds; and housing together of pullets of different ages. Other factors alleged to cause cannibalism are: temperature too high in the brooding period, lack of fresh air, crowding, allowing birds to be without feed for considerable time, insufficient feeding and watering space and insufficient and poorly-designed nests. From the standpoint of nutrition, a deficiency of any essential nutrient results in a depraved appetite that could lead to cannibalism. Nutritional preventives suggested are: feeding of heavy whole oats, grass clippings, alfalfa hay, addition of more salt to the mash or to the drinking water, and reduction of the amount of scratch grains. The breed of chicken is also important. Most of the difficulty reported has been with White Leghorns which have a disposition and temperament quite different from the heavier breeds.

To a lesser extent, vices are also attributed to rations deficient in certain nutrients or containing too much of others. Too little protein, fiber, salt and bulk, and too much corn in the ration are said to favor the formation of oily, curly or rough feathers; whereas, oats, oat hulls and wheat bran in rations produce smooth and glossy feathers. It is assumed that birds are prone to pick at rough feathers rather than at smooth ones and that feather-picking is a prelude to cannibalism.

The complex nature of cannibalism and feather-picking is conceded by those who have made a study of the subject. Whereas the opinion prevails that nutrition is secondary to management as a cause of vices, it is recognized that inadequate rations produce syndromes which do not favor the contentment of the bird.

CANNIBALISM

Poultry of all kinds may become cannibalistic when confined. Breeding may influence slowness of feathering which, combined with faulty management, may encourage picking. When quills are picked and broken, the pigment spreads out under the skin, giving a bluish appearance. Prolapses of the oviduct may be a contributing factor. The oviduct becomes open to view and is picked by other layers. This causes pickouts. If the birds after laying an egg do not remain in the nest until the everted cloacal membrane is withdrawn, the other birds are attracted by the red membrane and begin picking it. Sometimes the oviduct may become loose and protrude from the vent when the bird lays an egg. The blood which accompanies the prolapse and the oviduct itself attracts other birds and the injured one is picked to death unless promptly isolated. The problem is most common during late winter or early spring. Internal rupture of the oviduct may result from severe inflammation, an attempt to pass large eggs, or from injury. The breaking up of the oviduct permits yolks or fully-formed eggs to pass into the body cavity. The yolks may rupture and the contents dry out, causing a yellow coating over the viscera. There is no treatment for rupture of the oviduct. Egg-bound troubles occur most frequently among pullets. They result from the attempt to pass large or double-yolked eggs through an oviduct that is too small for their passage. Birds appear listless and make frequent attempts to lay.

Research on social behavior of the domestic fowl, especially chickens, indicates the complexity of events. Social interactions differ in frequency and vigor as they function with age. Picking increases from hatching to about 10 wk, after which age data are not available. The amount of past social experience in groups accounts for differences between age groups. The more the social experience, particularly in changing social groups, the faster and more vigorous the organization processes. The organization of social groups within the pen is essentially linear. That is, alpha birds peck all the others; the beta birds peck all under them, and so on. The weight of birds and their status and social order usually bear an inverse relationship to each other. From a management standpoint, preventives of cannibalism include some of the following: provide more floor space, supply ample feed and watering space for rapid growth and egg production, feed scratch in the litter to layers morning and evening to keep the layers busy, and stir the litter. Crowding is relieved where pullets or stock are removed to the range as soon as age and weather permit. Cool-room brooding is helpful in combating cannibalism. Brooding rooms can be cooled rapidly if poultry buildings are designed to allow cross-ventilation.

PREVENTION OF PICKING

Devices are available for the prevention of picking among laying chickens. One type prevents the bird from seeing downward; the second type is a blinder which prevents the birds from seeing straight ahead. These devices do not affect adversely either feed consumption or egg production. Bits are used successfully to prevent turkeys from picking. The wire keeps beaks apart so they cannot be shut tightly and injure other turkeys by picking.

Debeaking is the most reliable method of controlling cannibalism. Debeaking may be done with only a slight drop or no drop in egg production. Debeaked birds have difficulty in cleaning up mash in hoppers, so at least ½ in. of mash should be kept in the hoppers. If scratch grain is fed, it should be fed on top of the mash in the hopper. About 6 months are required for the beak to grow back. Desnooding turkey poults is also practiced. Some growers cut off the snood at about 4 wk of age, but it can be done at 1 day old more easily and with no discomfort to the poults. Chickens may be debeaked at as early as 3 wk of age, and it actually improves the performance of the flock. Prompt, and sometimes drastic, action is necessary to avoid serious losses.

Cannibalism occurs more often among white than colored chicks and is manifested first by toe-picking, tail-picking or feather-pulling. This may be because blood shows more plainly through the quills of white feathers. Darkening the room and painting the windows with a ruby water-color paint are helpful in preventing cannibalism. Cannibalism usually disappears as soon as the chicks are given fresh grains. In the case of adult birds, cannibalism and kindred troubles develop as a habit, usually among birds in close confinement. Suggestions for preventing it include use of anti-picking devices before the birds begin to lay, and debeaking. Debeaking will stop the vicious habit for at least 3 months. Pickouts are usually encountered among pullets. Inspection of the flock for ringleaders and removing them from the flock, as well as removing injured birds from the pens, are helpful. Smearing the affected parts with quinine and red paint has been done. Some farm flocks experience little cannibalism; in others, it is a major source of mortality—as much as 30%, in some instances. Perhaps through selective breeding, some cannibalism may be eliminated.

PASTURAGE

If all variables, such as disease and parasites, could be eliminated, and only nutritional factors considered, range-reared replacement stock would consistently lay at a higher rate with a lower percentage of mortality than confinement-reared stock. The nutritive values obtained from the range come directly from the herbage consumed, from the soil, bugs and worms, and from irradiation by the sun. Palatable green herbage contains an abundant supply of such essentials as carotene, riboflavin, Vitamin K, pyridoxine, panthothenic acid, niacin, biotin, choline and others.

Today, poultry-raising is a year-round industry and a large proportion of meat birds, such as commercial broilers and laying birds, are now grown indoors until market age. Large commercial poultry enterprises require fast growth, high rate of egg production and efficient feed utilization, and these conditions are best met by high-energy rations. Forage does not fit into this picture. As has been pointed out, the digestive tract of the chicken is not adapted to the utilization of fibrous feeds. Legumes, such as Korean Lespedeza, ladino clover, crimson clover and Sudan grass have been proven especially valuable because they supply excellent grazing during the hot, dry summer months when grasses are usually dormant. Lespedeza reseeds itself even when grazed short, and once established, there is no need for reseeding.

Turkeys are, by nature, better adapted to grazing than chickens. They make the largest use of forage when the concentrate is fed in limited amounts. Turkeys reared on green forages are healthy and heavy, more so than those reared in confinement or under range conditions where good forage is limited. Poultry on good pasturage need as good mash and whole grain as do confinement-reared poultry. This is of particular interest to farmers who grow relatively few birds but have plenty of grass, clover, weeds and insects available. The pasture will not supplement a ration but will extend the feed. A choice green legume pasture containing 15% protein will contain enough of the other nutrients to make this pasturage a self-contained feed.

Good pasturage for chickens involves a number of factors, such as succulent green feeds, isolation and ample space. It is necessary to mow often during the late spring and summer to encourage continued growth and succulence of the grass, weeds and clover. A good range provides insects which make a valuable supplement for the green feeds. The range for chicks and growing pullets should be isolated from older birds carrying diseases and parasites. Attempts to raise chickens and grow pullets on a range frequented, or recently contaminated, by older chickens is a most frequent cause for the failure of the farm poultry raiser to grow first-class pullets which will live and lay. A well isolated, clean range with plenty of herbage is the keystone to successful and economical poultry-raising on pasture.

Fresh, immature grass is the finest kind of supplementary feed that can be produced, being particularly high in many of the essential vitamins and minerals. The same grass at a later stage is less valuable, and in case it is dry and stringy, may be positively detrimental. When grass is mowed frequently, the stubble is short, stiff and stands up straight. Droppings from the birds work down into this stubble and are soon out of sight. Mowing keeps young grass growing, and the result is a fresh, high-potency immature grass for the birds to eat. Frequent clipping of the immature grasses increases the acreage yields of protein 40 to 60% over that recovered by the usual haymaking procedure.

Alfalfa is grown for pasturage quite generally in the United States, except where the soil is too acid; such soil requires tremendous applications of lime and very careful innoculation of the seed in order to get a good stand. Alfalfa does have some unfavorable features. The crop grows well, so a dense growth is obtained, allowing little or no sunlight penetration of the soil. Chickens flatten down the alfalfa and droppings accumulate on this matted mass so that the birds are in contact with contamination at all times. Very much the same conditions occur where clover is used. However, ladino makes a taller and ranker growth than ordinary white clover and is excellent for pasturage.

PASTURES

Before development of our modern drug and mechanization program, it was necessary to use pastures. Blackhead nearly destroyed turkey production in some states until a pasture rotation program was developed. Many pastures served only as clean range land and provided very little food because of trampling down of the green plants. Soybean was found to be a good pasture crop because of its strong central stalk. Most grasses proved to be desirable only when the flocks were small and the range shelters moved often.

Pasturage has proven of value when growing small flocks less than 500, on range using range shelters holding about 125 birds. Data obtained with small flocks are not applicable to our large commercial flocks of 10,000 to 100,000 hens.

The successful usage of a pasture depends on the management program. The ranging of the birds from the range shelter is dependent upon the location of the feeders. When the range shelters have been moved to a new area, the feeders should be near the range shelters the first day and them moved farther away each day. This daily moving of feeders will result in the birds grazing at comfortable distances and also will result in a cleaner grazing of the field. If feeders are not moved daily, the birds will continue to eat at the shelter or return to it often. Waterers can be kept near the shelters. Fences are not needed to keep the birds near the range shelter or to get them to return at night if feed is provided in the hoppers. Roaming or "far" ranging birds are generally hungry birds.

There are many range crops, but the major use of pasturage is to provide some feed, plus a clean, disease-free "floor" and a place to exercise. Grasses constitute more floor value than feed value because once a bird walks on grass, it becomes, and remains, flat. Crops such as alfalfa, soybeans and cereal grains give some pasture, s well as a clean floor.

TURKEYING-DOWN SORGHUMS

In the fall, after the sorghum grain has matured, the practice of letting turkeys harvest the crop has proven effective. Wastage is small and the turkeys do a uniform job of harvesting.

FUTURE TRENDS IN POULTRY MANAGEMENT

Several practices used in management of confined chickens will be challenged in the future by both legal and social groups. Organizations concerned about the cruelty to animals are developing effective practices on humane treatment and environment contamination. Some of these are being practiced today in the British Isles.

(1) Overcrowding.—Some cage operations do not provide enough space for the hen to move freely. A minimum space for hens will be specified. Overcrowding results in high mortality, as well as broken bones and bruises.

(2) Debeaking.—The debeaking of chickens to reduce cannibalism will probably be prohibited. Light control, special feeds and breeding for less nervousness in birds will help solve the problem of cannibalism.

(3) Liquid Manure Pits.—The flooding of manure pits with water will be eliminated. The odors and possible disease condition will bring about effective manure-management programs.

(4) Force Feeding.—The force feeding of geese and squabs will be given careful evaluation. This method can cause injury or death.

43

Disease Control

Many diseases gain entrance into poultry through feed, water, air and litter. Throughout history, there has been a tendency to blame feed as the carrier of most diseases; however, there is little proof to document such a theory. It is difficult to disassociate contamination of feed from that of litter, equipment, air and water. Nevertheless, where nutrition and sanitation have been controlled as much as possible, results have suggested that there may be some transmission of pathogens by way of the feed under certain circumstances. When birds are together in large-scale confinement production, diseases spread rapidly and losses are sometimes devastating. New diseases and complications of old ones plague the industry.

COMPETITION BETWEEN MAN AND ORGANISMS

There is constant strife between man and many other organisms for the same raw material. The battle of survival will be won by whichever can take advantage of situations. These other organisms are highly-efficient competitors for things humans need and want. Although man has been successful so far, rodents, insects, plant diseases, nematodes, and weeds have not given up the battle. More than 3000 species of insects, as many plant diseases and uncounted numbers of other competitors, fight to attain what humans need. Losses to these pests amount to between 8 and 15 billion dollars per year, one quarter of our annual production, in direct defiance of the best protective measures that we can put up against them. It is necessary to plant an extra 88 million acres for them to harvest and another 32 million acres to make up for past harvest losses. What would happen if all our pesticides were prohibited? Cereal, forage, and farm crops might be reduced from 10 to 25% and the quality of the products would diminish. The same would be true of domestic animals and animal products.

The greatest concern over the use of pesticides has been the relationship between their use and human health, through exposure in one way or another to the toxic material. As a result of some unfavorable publicity, the World Health Organization, the United States Public Health Service and the Food Protection Committee of the National Research Council concluded that: (1) the use of pesticides is recommended consistent with the rules and regulations promulgated under existing laws; and (2) careless and unauthorized use of pesticidal chemicals might pose a potential hazard which would require further careful observations, although at the moment there is no cause for alarm. It is clear that wildlife may be subjected to hazards which are not problems to humans or domestic animals.

Complete Diagnosis

When the poultryman suspects disease, several specimens that are typical of the whole flock— not just a single bird—should be sent to a diagnostic laboratory. In many instances, help can be obtained from the highly-specialized laboratories at the universities and drug manufacturers to back up the local diagnostic laboratory. Such laboratories have equipment and skills to make the differential diagnosis. A complete laboratory diagnosis of a disease problem is sometimes necessary for adequate protection of the investment in the huge flocks of today. A complete diagnosis means more than an examination of a specimen of tissue and litter. Sometimes it is necessary for the poultry specialists to visit the flock and make first-hand observations of environmental factors.

Modern poultry management is depending more on science and less on intuition. Drugs are essential to the control of disease. They are profit tools when used judiciously along with good management practices. New drugs developed to meet the changing needs of the poultry industry are potent, versatile, and have a wide range of margin of safety. More birds go to market, weight gains are better and the profit picture is improved when these drugs are used along with good management. The effectiveness of the partnership between science and agriculture is exemplified by the poultry industry in the United States, including 3.5 billion broilers per year, 140,000,000 turkeys, astonomical numbers of pigs and cattle, and production efficiency that is the envy of the world. Continued research gives assurance of continuing success.

Battery System

Battery systems may include individual or multiple cages and large confined slatted or wired-mesh floored open-sided structures. The slatted-floor battery system was developed for turkeys to control blackhead and coccidiosis. However, sanitation was often overlooked and manure and wasted feed was allowed to accumulate. This resulted in heavy fly population. Since flies carry worms and pathogens, this management program did not give the poultryman freedom from giving needed attention to the sanitation program.

Battery Diarrhea

Some hens in laying cages develop diarrhea which seems to spread to hens nearby. The condition appears to develop in only 1 hen and then spreads slowly to adjoining cages. Some operators have expressed a theory associating this to hot-air pockets or improper ventilation of the house. When a hen shows this condition, the best remedy is to remove her from the cage. These hens go out-of-production and lose weight if not returned to floor management.

Happiness

Hens in cages do not show signs of nervousness or unhappiness. Many of them sing throughout the day.

Battery Fatigue

Many diseases are transmitted in batteries, and there are typical battery diseases which are often the result of restricted exercise. While the background of battery fatigue is still unknown, symptoms occur in birds that have been in production for about 6 to 8 wk, paralysis being especially characteristic. Spontaneous fractures of the legs occur due to a decalcification of the bones. An increase in the calcium or Vitamin D levels in the feed has only limited preventive value.

Dosages of phosphorus have some positive effect. Battery fatigue is limited to only 5 to 10% of the birds but it is still a serious economic handicap. The birds recover quickly if they are placed on the ground.

Studies on loss of specific gravity in men making prolonged space flights indicated that there was excessive loss of potassium. Since battery-housed hens developed weak bone structures, there could be a change in potassium metabolism and other minerals, such as calcium, phosphorus and magnesium. Apparently, confinement of animals has a heavy bearing on mineral metabolism.

Other Conditions Associated with Batteries

A high percentage of the mortality in battery cages is due to inflammation of the kidneys (nephritis). Wetness of droppings occurs because it is believed the birds drink too much water because of boredom. However, a hormonal abnormality may possibly be involved. Fatty degeneration of the liver is associated with excessive feed intake, especially when the feed is rich in energy and poor in protein. Reduction in exercise may cause an energy surplus, fatty degeneration being the result. Fatty degeneration causes severe changes, especially in the liver. Again the birds can be cured by placing them on the ground for some time or taking them out of production. Extra choline, methionine, Vitamin E and Vitamin B_{12} have been proposed, but have no beneficial effect.

Stress

Turkey leg problems are often caused by stress. Stress conditions, such as wet litter in the brooder, may cause a leg problem which leads to crippling later in the growing period. Also, the presence of wet cold litter and the feeding of a diet containing excess Vitamin D can result in excess excretion of zinc and magnesium. This leads to leg weakness.

Overcrowding is another cause of leg problems. Poultry should be given enough space to allow them to walk and not be forced to crowd around the feeders and waterers.

Another form of stress is bad air—i.e., high in ammonia, carbon dioxide, water vapor and dust. Ammonia can damage the lung tissue and reduce oxygen uptake. Also, dust can carry fungi which readily grow in lung cavities. Aspergillosis, a fungus, is generally obtained from contaminated feed litter and water and gains entrance into the lungs by breathing. During damp weather, the loss can be high from this disease.

Coccidiosis

One disease that threatened to bring the developing poultry business to a grinding halt was coccidiosis. Because of a 1-celled parasite, millions and millions of chicks died each year. Geneticists tried unsuccessfully to breed coccidiostat-resistant birds; management experts developed systems involving frequent changes of litter. Nutritionists tried to develop feeds that would protect the birds; milk products, iodized feeds, vinegar, sulfur in various forms were tried; but lacking basic knowledge of the disease, were not successful. Coccidiosis still remains a major problem of the poultry industry.

Coccidiosis once accounted for mortality rates as high as 40%, but its impact has been greatly diminished by effective coccidiostats. There are 9 species of chicken coccidia, of which *E. tenella* and *E. necatrix* cause the highest mortality. Seven species of coccidia have been identified in turkeys (Table 43.1).

It was recognized early that there were 2 classifications of coccidiosis—cecal and intestinal. Identifying these was a first step in defining the disease process. The universities, experiment stations and the drug industry began studying the disease. After years of study, parasitologists have identified about 20 species of coccidia in poultry, as well as in other animals. Coccidia look similar under the microscope but differ in size and economic importance. Pathologists can identify lesions caused by the organisms in specific areas of the bird's body. The ceca or the pouches may be enlarged or filled with blood and cheesy exudates. Lesions may occur in the digestive tract causing damage in many areas, and also in the upper digestive tract of turkeys. Lesions may be found, also, in the duodenal loop and the upper part of the small intestine. Coccidiosis reduces absorption of nutrients from the intestinal tract. Zinc was found to be poorly absorbed in diseased birds.

Infection Cycle.—Coccidiosis is a disease that starts with the oocyst—the egg-like structure of the parasite. Moisture and warmth stimulate their development. Oocysts mature and become infective in the litter of infected hens. Once within the bird's digestive tract, the parasites multiply rapidly and invade and damage the tissue of the intestinal wall. On the sixth or seventh day after infection, many oocysts are passed into the litter. One coccidium is capable of producing 10,000 parasites during its lifetime. It is obvious, then, how a sick bird can start a cycle of infections and cause heavy losses in the flock. Quail disease, often confused in the field with coccidiosis, causes bloody droppings and affects the cecum and lower intestines in a manner similar to coccidiosis. However, quail disease does not respond to commonly-used coccidiostats.

Scientists found that damage to the intestinal wall slowed growth and lowered feed efficiency, even when the disease did not kill. They found that previous infections acted like vaccinations. Protecting the bird against every infection is impossible. Chickens can develop immunity to 1 species of coccidiosis and still be susceptible to infection by 8 others. Thus, coccidiosis of chickens is, in effect, 9 diseases, not just one. Community protection can be of brief duration, so control of the disease is attempted through programs of moderate exposure. Without drugs, treatment is impractical, but knowledge of the mechanisms of immunity is essential to the successful use of drugs.

Some of the most devastating outbreaks of coccidiosis have occurred in clean, new houses. Starting in a coccidia-free environment with birds having had no exposure, there is no immunity to the various strains of the infective organisms. Oocysts may be carried on the feet of wild birds, caretaker's shoes or equipment. Just a few occysts can produce as many as 500,000 parasites. One infected bird may harbor as may as 65 million oocysts. Thus, infections can spread rapidly.

Control of Coccidiosis.—When sulfonamides began to show remarkable control of infectious diseases in human medicine, poultry special-

TABLE 43.1. SPECIES OF COCCIDIA FOUND IN POULTRY

Characters		*E. tenella*	*E. mitis*	*E. acervulina*	*E. maxima*
Oocyst	Size in μ	14.2–31.2 × 9.5–24.8; av. 22.9 × 19.1	10.8–19.8 × 9.6–16.8; av. 15.8 × 13.4	11.7–22.7 × 9.2–17.5; av. 16.4 × 13	26.9–35.4 × 20.9–25.1; av. 31.8 × 22.7
	Shape	Broad ovoid	Subspherical	Egg-shaped	Egg-shaped
	Sporulation time	48 hr[1] (18 hr)	48 hr (18 hr)	21 hr (17 hr)	48 hr (30 hr)
	Prepatent period	6 days	4–5 days	4 days	5–6 days
Region of intestine most heavily parasitized		Schizonts and oocysts in ceca	Anterior small intestine	Anterior small intestine	Middle, anterior, and posterior small intestine
Position of parasites in tissue		Second generation schizonts subepithelial	Schizonts generally above and gametocytes generally below nuclei of epithelial cells	Above nuclei of epithelial cells	Schizonts above nuclei. Gametocytes deep in epithelium
Macroscopic lesions		Hemorrhagic ceca	None	Intestinal mucosa streaked with transverse whitish opacities composed of oocysts Catarrhal exudate	Exudate or flecks of blood on mucosa; intestinal wall thick
Degree of pathogenicity		++++	+	++	++

TABLE 43.1. *(Continued)*

	Characters	*E. necatrix*	*E. praecox*	*E. hagani*	*E. brunetti*
Oocyst	Size in μ	12.1–28.9 × 10.8–23.8; av. 19.7 × 16.7	19.8–24.7 × 15.7–19.8; av. 21.3 × 17.1	15.8–20.9 × 14.3–19.5; av. 19.1 × 17.6	13.8–33.7 × 12.4–26.2; av. 23.4 × 19.7
	Shape	Oblong ovoidal	Ovoidal	Broadly ovoid	Egg-shaped
	Sporulation time	48 hr (18 hr)	48 hr	36 hr (18 hr)	24–48 hr
	Prepatent period	6 days	4 days	6 days	5 days
Region of intestine most heavily parasitized		Schizonts in small intestine, oocysts in ceca	Upper third of small intestine	Anterior half of small intestine	First generation schizonts throughout small intestine, later stages in ileum and posteriorly
Position of parasites in tissue		Similar to *E. tenella*	Below nuclei of epithelial cells	?	Oocysts in all parts of cecal mucosa
Macroscopic lesions		Whitish opacities and hemorrhage and exudate in small intestine	None, except mucous cast of intestine	None	Catarrhal enteritis with blood-tinged exudate
Degree of pathogenicity		++++	+	+	++++

Source: Biester and Schwarte (1965).
[1]Sporulation times in parentheses are minimal at 29°C.

ists were quick to try these new drugs on birds. Sulfaquinoxaline enabled poultrymen to reduce coccidiosis mortality effectively, even when birds were grown in crowded conditions. Continuous feeding of sulfaquinoxaline made it possible for poultrymen to keep large flocks with less than 2% mortality attributed to coccidiosis. This was a dramatic achievement.

In 1954, a new drug, Nicarb, from Merck Laboratories, virtually eliminated coccidiosis as a serious problem in chickens. It was so effective that poultry raisers obtained high levels of performance, even though some began to neglect important management details. The poultry industry learned the importance of management details the hard way because neglect led to outbreaks of disease. Even the best of drugs cannot substitute for good management. Adequate feeding space, good equipment, proper litter and close supervision of the flock are more essential today than ever before, in spite of the increased effectiveness of modern drugs. While high temperature and humidity slow down the bird's feed intake, they speed up the development of sporulated oocysts. So, when it is hot and humid, it is necessary to check to be sure the coccidiostat intake is up to required levels.

Different coccidiostats work in different ways. One type of action is metabolite antagonism, wherein the drug substitutes itself for the Vitamin B_1 needed for coccidia to survive. Good management practices are: reduce moisture and keep humidity down, keep the house clean, prevent contamination from outside sources, feed a preventive level of satisfactory coccidiostat, and at the first sign of a disease, call the serviceman. A drop in feed consumption, loose and bloody droppings, sickly appearance of birds, and coughing and sneezing are nature's warning signals.

DISEASES OF TURKEYS

Blue Comb

Blue Comb disease is a cause for concern to the turkey industry, and is more prevalent in areas which practice year-round turkey production. It is a disease of turkeys of all ages and breeds, and both sexes. It appears suddenly and affects almost 100% of the flock. Turkeys go off feed and have reduced water intake. Body temperature falls 5° or 6°. Turkeys lose weight, become emaciated, and have loose, watery droppings which contain mucous threads and urates. Heads and necks tend to become a dark blue color. Laying birds lose production and produce chalky egg shells. Usually, flock mortality will average from 5 to 25%. The younger the poults, the higher the mortality. The course of the disease runs 10 to 14 days. Major losses lie in decreased feed conversion and weight losses, with increased litter, medication and labor costs. Lower grades and yields at the processing plant increase the cost of production in flocks experiencing a moderate case of the disease.

Method of Spreading Disease.—Turkeys which have recovered from a case of Blue Comb have immunity against a second attack, but continue to eliminate the infective agent in their droppings. Fecal contamination of the drinking water makes for very rapid spread of the disease. Chickens, pheasants and sea gulls are not carriers of the agent. However, any free-flying birds which obtain food from turkey yards and immediately fly to a neighboring flock would be mechanical carriers of infective fecal material. Canadian workers have identified the infective agent from the intestinal tracts of turkeys as a bacterium of the genus, *Vibrio*. The lack of a convenient test for the presence of the infective agent is holding back control of the disease.

Control of Disease.—Antibiotics, feed additives, immune serum, drugs, and chemicals of all types have been used, but have been ineffective in preventing the disease. A strict sanitation and isolation program is recommended. A crude vaccine from suspensions of infected turkey intestinal tracts is promising but far from perfected.

Blackhead (Histomoniasis)

Blackhead, common to turkeys, is a disease caused by a protozoan parasite. It has been credited with the near-obliteration of the turkey industry from the eastern and midwestern sections of the United States.

The organism causing blackhead *(Histomonas meleagridis)* is harbored by the common poultry cecal worm. This parasite generally does not injure chickens and can live undetected in poultry worms. A general recommendation is not to maintain turkeys and chickens on the same farm.

Blackhead is often classed as a "filth-born" disease, and prevention is recommended because of the ineffectiveness of most drugs.

INFECTION WITH PARASITES

Parasites damage the animal body by living on tissues and substances that feed these tissues. Infections develop gradually. In the long run, damage becomes so great that profitable management is impossible.

Parasites cause: (1) damage to the intestinal wall and thus impede absorption of nutrients; (2) loss of nutrients intended for the host; (3) occlusions to blood and lymphatic vessels; (4) wounds by which other pathogenic organisms enter the body; and (5) harmful substances to develop in the body. Once present, parasites multiply and are easily transmitted by footwear, clothing, car tires and packing materials. To limit damage, hygienic methods, good management, specific control measures and identification of the disorder are needed.

One method of parasite prevention is good nutrition, a factor of the animal's general health. Certain substances, such as Vitamin A, have very positive effects on the resistance of the body to parasites because they have protective values for epithelial tissue. The mucous membrane of the gastrointestinal tract secretes protective values for epithelial tissue. The mucous membrane of the gastrointestinal tract secretes protective substances which attack infectious organisms and parasites. This function is disturbed if Vitamin A is deficient, particularly in the case of coccidiosis in poultry.

Internal parasites in poultry are tremendously prolific—1 large female round worm can lay as may as 5000 eggs per day. Thus, a worm population explosion can take place in the intestines of a bird, and a light infection of a few birds can develop quickly into a major problem throughout the flock. By the time they are detected, worms have already done damage. Effective worm-control calls for action against several stages in the worms' development. Purge-wormers expel large round worms and cecal worms but are ineffective against immature worms or capillaria worms. Continuous in-feed antibiotics, such as Hygro Mix, used at proper levels, halt worm-egg production, destroy mature worms and prevent a large share of immature worms from reaching the egg-laying stage. They are effective against large round worms, cecal worms and capillaria worms, and offer a means of controlling the worm population.

Avian Hysteria

Avian hysteria is a condition of complex causes affecting all strains of chickens. Suspected causes are insufficient protein intake, poor ventilation with dust or ammonia, insufficient floor space, irritation or poorly debeaked birds, and warm weather which decreases feed intake. It is common in high-producing birds with large egg size. Affected birds lose all normal social behavior and sense of direction, mill around, hide, and fly in every direction, making unusual crying and squawking sounds. They go into a molt and egg production drops. Genetics may be involved. Attempts to reproduce the disease have failed. It is not contagious. Crowding plays some role in triggering the outbreak. Many drugs, feeds and management practices have been tried with varying degrees of success. Increasing the protein level, adding cultured yeast or adding 100 g niacin per ton feed has also been advocated. Tranquilizers have been used. Hysteria stems from malfunction or degeneration of the central nervous system, either the brain or spinal cord.

The spraying of poultry houses and poultry equipment in the presence of poultry leads to rapid storage of pesticide residues in poultry tissue and eggs. As much time as possible should be allowed after spraying before broilers or layers are housed. The outer layers of grain seeds retain the largest amount of pesticide residue. The greatest degree of contamination of animal products with farm-used insecticides and pesticides comes about through the feeding of forages. Grains and grain products are less implicated and commercial feeds implicated the least. Feedstuffs lend themselves to contamination by insecticides through various means unless proper precautions are taken. The occurrence of these insecticides in feeds leads to their presence in foods of animal origin. Consequently, it is necessary to control the presence of insecticides in feeds.

Feed manufacturers should take precautions to prevent contaminated materials from entering the plant and should attempt to retain identification of ingredients and feed additives as they are delivered in order that any trouble which may develop may be traced to the ingredients. In the interest of a wholesome and economical supply of food of animal origin, the feed and poultry industries recognize the need to hold pesticide residues to a minimum. They subscribe to the policy of exercising all reasonable precautions, but at the same time, demand a realistic approach toward the establishment of safe tolerances for all feed products. Chlorinated hydrocarbons have been used widely for less than 20 years and have been directed primarily toward the control of various insect pests. Much additional research must be done on the question of carryover of pesticide residues in products so extensively used in commercial feeds—oilmeals, meat and cereal by-products, and distillers' and brewers' by-products.

Chlorinated Hydrocarbon Insecticides

When poultry ingest feeds contaminated with chlorinated hydrocarbon insecticides, the compounds deposit in the tissues, and hence in meat products prepared from the animal. This is particularly true of meat high in fat, since this insecticide is fat-soluble and so tends to become concentrated in fatty tissues. Some of the insecticides are not only retained in the fatty tissues but are actually converted into more toxic forms before they are stored. This is true of Aldrin and Heptachlor. Nonfatty tissues contain very low concentrations of the chemical, although they are not completely free from contamination. DDT appears in eggs even when relatively low levels of insecticides occur in the diet.

Feedstuffs become contaminated with chlorinated hydrocarbon insecticides in a number of different ways: absorption from the soil, application of the chemicals during growth and use of water contaminated with the compounds in the preparation of the feed. The insecticide is most persistent in sandy soils. Uptake and deposition from the soil have been demonstrated for numerous feed crops. Direct application of these chemicals to control insects is probably the greatest source of contamination.

Mode of Contamination

Alfalfa, wheat and other major components of poultry feeds, including corn, oats, oilmeals and meat meals may be contaminated. Bodies of water may become contaminated with insecticides through direct application for control of various pests or through run-off water which has picked up the chemicals from contaminated plants or soil. Probably DDT is the most widespread contaminant, but appears at a maximum concentration of less than 1 ppm. A survey showed that more than 50% of the feeds tested contained BHC, Aldrin, DDT, Endrin, Chlordane and Dieldrin. The average level of contamination was generally quite low, although individual feeds contained enough insecticide to appear in animal products. Thus, the presence of insecticide residues in feedstuffs must be of concern to the feed and poultry industries, for these materials, if consumed by animals, are deposited in animal tissues or appear in animal products.

Feedstuffs may become contaminated from a variety of sources, including the soil, spray, or dusting treatments. When insecticides are used properly, the magnitude of contamination should be quite low. Agricultural chemicals are necessary for growing food to feed the present world population and will become more essential as the population increases. Rachel Carson's book, *Silent Spring*, has virtue because it emphasizes that those who use agricultural chemicals should be aware of the problem and exercise reasonable care in their use. Excessive residues are undesirable and illegal. Label precautions, careful use levels and proper withdrawal periods are designed to leave food products free from harmful residues. The Food and Drug Administration is charged with the responsibility of the necessary care to keep our food supply safe; but needless to say, there is no such thing as absolute safety. Nearly everything we do involves a calculated risk.

Position of Pesticide Residue Committee

The Pesticide Residue Committee of the National Academy of Sciences, National Research Council, is cognizant of the advances that have been made through discovery, manufacture and application of new chemicals for the control of pests of all types and of the necessity of their use in the health, nutrition and economy of the nation. The Committee believes that the valuable properties of these chemicals can be utilized without exposing humans, domestic animals, fish or wildlife to undue risk. It has recommended that the concept of "no residue" and "zero tolerance" as employed in the registration and regulation of pesticides is scientifically and administratively untenable and should be abandoned. It recommends an expanded research program on the persistence of pesticides in the total environment and on the toxicology, pharmacology and biochemistry of pesticides that would improve the reliability and precision of animal studies and their relevance to man.

Pesticide Transfer to Eggs

Hens exposed to pesticide sprays will, in a short time, transfer pesticides to the egg. DDT concentrates in the yolk of the egg and, to a much lesser degree, in the albumen. The selection of the right pesticide for a given purpose, used in the minimum amount on a carefully chosen carrier, and applied in such a manner as to minimize uptake of pesticide, will go far toward solving the pesticide residue problem. If birds have been exposed to high pesticide residue levels, they should be removed to uncontaminated feedlots as soon as possible, to eliminate the ingested residue before marketing time.

Control of Flies

A female fly lays up to 3000 eggs in her 3- to 20-day life in poultry manure or in any moist organic material. Offspring grow in about 7 days, to the egg, larvae, pupal and adult stages. Control flies by:

(1) Removing droppings every 7 days to break the life cycle (effective but costly).
(2) Provide ventilation to dry and cone droppings.
(3) Prevent rain and drinking water from spilling into dropping pits.
(4) Apply dry sawdust to wet pit areas.
(5) Clean and adjust waterers regularly.
(6) Water 30 min every 2 hr.
(7) Use mash feed, not crumbles.
(8) Reduce salt in feed during hot weather.
(9) Remove dead birds and other dead animals.
(10) Mow grass and weeds near houses.
(11) Store manure under water or use lagoons.
(12) Darken houses to prevent larval growth.
(13) Screen buildings.

Fly-control requires, in addition, chemical sprays or baits, i.e., Diasinion, Cygaon, Vapolna and Ronnell. Don't use lindane, DDT or other chlorinated hydrocarbons that leave residues in eggs and meat.

Medicated Feeds

The standards of state and federal agencies promote honesty and fair dealing. There are several kinds:

(1) "Identity" tells what a feed is or what a poultryman can expect to receive when he selects the feed by its usual name.
(2) "Quality" sets minimum standards and specifications for color, odor, flavor, nutritional value, cleanliness, texture, tenderness, and freedom from defects.
(3) "Fill" is how full the container must be to avoid deception and to meet the net weight or volume claim stated on the package.
(4) "Microbial growth" may produce food poisoning.

Federal Laws Controlling Pesticides

Laws and regulations have been set up for the protection of human and animal food supplies. Two federal laws control pesticides in interstate commerce and are designed to give protection to the public. They are the Federal Food, Drug, and Cosmetic Act as amended by the Miller Bill and the Federal Insecticide, Fungicide, and Rodenticide Bill. The first provides for pretesting and the establishment of safe tolerances before new material intended for use on food crops can be put on the market. The second provides that new materials shall not be registered for sale in interstate commerce until complete specifications have been filed and adequate data have been presented to establish their usefulness for the purposes claimed, as well as their safety when used as prescribed. These 2 pieces of legislation are in the public interest and make it necessary for manufacturers to spend many hundreds of thousands of dollars before a new product can be marketed. In providing for the safety of the consuming public, safety factors from 10- to 100-fold have been built into recommendations. These controlling laws exist for a purpose, even though some of the tolerances represent legal, rather than scientific biological, limitations. Their purpose is to protect humans. More than 200 basic chemicals are now produced and offered for sale on our farms.

The Miller bill was passed in 1954, and is now known as the Pesticide Amendment of the Food, Drug, and Cosmetic Act. It gives the Food and Drug Administration authority to establish and enforce tolerances for pesticides and chemicals in raw agricultural products. Since 1954, more than 2000 such tolerances or exceptions from tolerances have been established for a wide variety of pesticide chemicals. Special attention has been given to the chlorinated hydrocarbon group of chemicals, among which DDT and it homologues are, by far, the most widely used. Because they have the unhappy faculty of being transmitted through the biological system of the animal and stored in the fatty tissue, feeds treated with these pesticides clearly are a significant source of residue. Feeds containing pesticide residues cause some transfer of these to the tissues of turkeys and hens, especially in the fat. Dusting birds and spraying poultry houses also lead to transfer of chlorinated hydrocarbons to the flesh of poultry. Keeping down the amount of pesticides used and allowing as much time after spraying as possible before housing the birds, are helpful in reducing transfer of pesticides.

Food and Drug Inspection

Compliance is achieved by the inspection of mills. Official sampling of animal tissues for excessive residues of drugs is done at processing plants. It is not difficult to comply if sound business practices are followed in all operations and the company is obviously doing its best to follow the rules.

All drug substances (including antibiotics) and additives are listed by the Food and Drug Administration. Those which are "old" drugs do not require a N.D.A. (new drug application); those which are "new" drugs require either a N.D.A. or an Antibiotic Form 10. Receiving records, perpetual inventory records, and mix sheets which demonstrate that each additive is being properly used, are filed. Each drug report shows name of substance, amount, manufacturer's lot number, your lot number (if any) and receiving date. Drugs are stored separately in a locked room under the supervision of a responsible person and are properly labeled. Materials from broken or open bags are stored in covered drums with the appropriate label on the side of each drum. Spilled material is cleaned up immediately. A perpetual inventory card for each lot of drug, feed additive and microingredient showing lot number and each withdrawal is supplied and checked against mixing sheets and physical inventory each day.

An accurate mixing sheet is provided for each feed, preferably showing lot number for each drug substance in each mix. Personnel trained to handle drug substances and microingredients accurately and safely are supplied with appropriate safety equipment.

Mixing of Drugs

Mixing procedures are clearly written out. The addition of each ingredient is checked before charging the mixer, to assure that it is the correct ingredient, in the right amount, and that it is put in the mixer or mixing line. Drug substances, micronutrients and other small quantity items are added when the mixer is partially (not over 75%) charged, or in such manner on a continuous operation as to assure adequate dispersion. Mixing time after fully charging the mixer is controlled to assure complete mixing. Adequate evidence is available showing that mixing times and methods are adequate for different types of mixes in order to establish minimum and maximum mixing times and procedures.

Mixing and conveying equipment is grounded to avoid electrostatic charges. Equipment is designed to avoid contamination of feeds from medicated feeds. This includes mixers, conveyors, bins, bulk tanks, bulk trucks and used bags. Equipment is cleaned or flushed out with a common ingredient after each run of medicated feed or delivery of medicated feed. Such cleanout material is set aside in labeled containers for use as an ingredient in the next run of feed containing the same drug substance and ingredients.

Food and Drug Clearance

To be mentioned in the drug additive compendium, full Food and Drug Administration clearance is first required for legality and for claims to be made on feed tags. Even though the drug company has this number, the feed manufacturer must get his own new drug application, providing the drug is classified as new. A request must be made to the drug company to refer to the new drug application number.

The following example summarizes what needs to be done. There are several new coccidiostats. One coccidiostat may be in use; its desired use is at low levels along with an antibiotic. Even though the coccidiostat has been cleared, an antibiotic producer must get clearance for this double combination. The Food and Drug Administration may insist on tests to show the effects of the additive on each part or component of the combination.

Coccidiostats are used in feeds if a reasonable degree of immunity has not already been established. No 1 coccidiostat remains effective for very long because of drug resistance. Drugs are withdrawn from the feed for a period of time before meat birds are marketed.

Regulation of the United States Drug Industry

The United States drug industry comprises some thousand companies, yet its total sales are less than those of any one of a dozen other single companies in other fields. In recent years, there have been attacks on the American pharmaceutical industry; its motives, profits, ethics, and even product quality have been challenged. As a result, federal regulations for obtaining rights to market new drugs have been put into effect, resulting in costs of millions of dollars.

It is costly to discover and develop new drugs. It appears that the drug industry has been somewhat like a whipping boy. Through patents, the industry is given the right to a period of exclusive use of the drugs it discovers through research. The drug industry must be given the opportunity to complete on the basis of merit and grow and prosper on the basis of performance. Of course, drug research could be centralized under the government. The free enterprise system has outperformed governmental research. The drug industry can look with pride on its many years of performance.

Poison-treated or Mercury-treated Seed Grains

The United States Food and Drug Administration prohibits the storage, sale, and use of poison-treated and mercury-treated seed grains. Such grains will be seized by the government and violators may face fine and imprisonment. One lone kernel of treated grain contaminates the entire lot. Destroy all leftover treated seeds. *Do not mix any* treated seed or grain with other feedstuffs.

44

Turkey Feeds—Feeding

Turkeys are natives of the New World and were recorded in history when Cortez landed in Mexico. They were here when the Pilgrims landed at Plymouth Rock and were served for their first Thanksgiving feast in the New World. In the early part of the twentieth century, Naragansetts, White Hollands, Bourbon Reds and Standard Bronze turkeys were kept on small farms, and virtually no improvement in breeding was made. Between 1930 and 1940, great improvements were made in the Broad Breasted blood lines. Breeders are now developing pedigreed stock by strain and cross-breeding, even 3- and 4-way crosses, so that the quality of breeders is far better than previously. Today's Broad Breasted turkey strains are possible only through artificial insemination. Turkeys with a conformation and meat yield that satisfy the processor and consumer have neither the agility nor desire to mate as often and successfully as the narrow-breasted turkeys.

Only a few years ago, the turkey industry was a 4-month, dressed-bird market for Thanksgiving, Christmas and New Year. Today, turkeys are processed year-round and can be served any day of the year.

Broad Breasted Bronze birds are produced because of their large, mature body size and the greater profit per head in comparison with smaller birds. Broad White turkeys have the advantage of white pin feathers, giving an improved carcass appearance.

Increased use of bulk-handling equipment, mechanical feeding and watering, pole-shed housing and other labor-saving devices, have been responsible for the efficient manner in which turkeys are raised today. These devices permit the producer to grow larger flocks. In recent years, many producers have concentrated on the production of turkey broilers, most of which are reared in confinement. Increased production demands more breeder hen and more hatchery production. The dynamic growth of the turkey industry has doubled per capita consumption during the past 20 years.

The early production of turkeys was centered in Virginia, Tennessee and Kentucky, with limited production in a number of midwestern states. In general, grain farmers often raise turkeys and chickens.

Turkey producers in Virginia, Tennessee and Kentucky drove their turkeys to railroad terminals where the birds were loaded into live poultry railroad cars and shipped to the New York live poultry market. These live poultry cars originated in Chattanooga, Tennessee. Poultry Extension Director, Chadwell, of the University of Tennessee, generally directed the operation and accompanied the shipment to the New York sales location. These shipments were discontinued during World War II.

The driving of turkeys to market was a great occasion. Since the turkeys were fat and heavy, their feet could become sore during the drive; so the farmers generally drove them through a narrow walkway leading over a shallow trough of pine tar. When the feet were covered with tar, a pile of sand was reached, and the sand with the tar formed a "black-top" padding under the feet. This prevented the formation of sores during the long walk, which often was as much as 30 mi. The local term for this process was "tarring-down." The disease commonly known as blackhead nearly eliminated the production of turkeys in the United States. A range rotation program was developed by Dr. A. B. Billings of the University of Minnesota which provided a control of the losses from blackhead. Minnesota is the leading turkey producing state in the United States. In 1976, Minnesota produced 24,370,000 turkeys.

Turkey production became established in California during the period following the decline in production in the eastern part of the United States. Today, it ranks second in the nation, producing over 17,506,000 turkeys in 1976.

The development of drugs specifically for diseases in turkeys has reestablished turkey production in many states. Both confinement- and range-rearing methods are used.

Turkey production in the United States has increased rapidly. In 1966, there were 115 million turkeys; in 1976, a record number of 140 million turkeys was produced. A primary reason for this phenomenal increase is that turkeys are now produced at a cost which provides high quality meats for the consumer at low cost. Feed manufacturers wasted no time in translating research findings into improved turkey feeds which resulted in faster and more economical returns to the commercial turkey grower, thus enabling him to provide meat at a reasonable cost to the consumer. Prior to 1948, most turkey rations contained high-fiber, low-energy ingredients, such as wheat bran, alfalfa meal and oats. More recent research has shown that many of the problems starting poults encountered were because such rations are not palatable to young turkeys. Starveouts and uneven rates of growth, and poor feed efficiency result. Earlier difficulties in getting poults to eat were largely eliminated by the use of high-energy rations. These rations also markedly improved the efficiency of feed utilization. Thus, improvements in turkey nutrition over the years have emanated from research on energy and protein levels.

MARKETING OF TURKEYS

In the United States, the Thanksgiving market is the largest, but there is also a very substantial one for Christmas.

The average yearly consumption of turkey per person in the United States has risen from 5.9 lb in 1957, to 8 lb in 1967. Freezing the whole turkey and turkey parts encourages the consumer to buy turkey meat the year round, and institutions are making wide use of turkey rolls. Years ago, 32 wk were required to market turkeys; today, they are ready in 24 to 25 wk with an average weight of 26 to 27 lb for the tom turkeys and 18 lb for the hens.Turkeys weighing 70 lb have been produced, but these are not common. In their natural state, turkeys lay eggs only 6 or 7 months of the year. By keeping the gobblers in windowless houses and varying the lighting, it is possible to stimulate egg production 1 or 2 months early and to have egg production the year round (not from the same birds).

RESEARCH ON TURKEY NUTRITION

Antibiotics and UGF have been important fields of research with turkeys. For the future, more precise information is needed concerning the exact balance of energy, protein, amino acids, vitamins, minerals, antibiotics, enzymes and hormones required. More information is also needed on the digestibility of feedstuffs and on stress, disease, vaccination and moving birds. As science learns more of the potentialities through optimum methods of breeding, management and feeding, the turkey will become one of the best converters of feeds into highly-nutritious foods.

Manganese and choline are involved in the prevention of perosis in turkey poults. Riboflavin, pantothenic acid, niacin, Vitamin B_{12}, Vitamin A, Vitamin D and choline are now added routinely to turkey feeds by feed manufacturers. Antibiotics improve growth when fed to young turkeys. Growing turkeys have one of the highest protein requirements of any class of poultry or livestock. There is a theory among certain turkey growers that turkeys need rations high in fiber during the growing period after the first 8 wk of life. Such a theory is without foundation in the light of experimental evidence that indicates it is not good practice to use feeds high in fiber during the growing period. Excellent turkeys are raised on feeds that are low in fiber and high in energy. Vitamin A is one of the critical vitamins in the practical feeding of turkey poults and a source of this vitamin should be added to the feed by the feed manufacturer either in the form of pure Vitamin A, Vitamin A from fortified fish oil, stabilized Vitamin A in the form of a dry Vitamin A concentrate or carotene (Provitamin A). Vitamin A is necessary for growth and for maintenance of the inner linings of the digestive tract, respiratory, reproductive and other similar surfaces in the body of the poult. Most feed manufacturers meet this requirement by using high-grade, dehydrated alfalfa leaf meal, and supplementing with one of the forms of Vitamin A.

Vitamin D_3, or activated animal sterol, is the best form of Vitamin D to use for feeding turkey poults. It is utilized more efficiently and assists in the utilization of phosphorus to a greater extent than Vitamin D from fish oils. Vitamin D is not required if poults have access to sunshine, but feed manufacturers cannot take the risk of eliminating Vitamin D in the hope that birds will be grown in sunshine.

Brewers' dried yeast is effective in preventing hock disease. The yeasts have always been an excellent source of UGF, and when available, should be added to turkey poult, pheasant and duck rations.

Imprinting

One of the best study methods used in determining eating and breeding habits of wild turkeys is the imprinting method, actually the attendant spending the daytime with the young poults, away only while the birds are nesting for the night and returning before the birds are ready to leave for the open fields. The young turkeys both lead and follow the attendant, who becomes their adopted mother. The attendant can watch and observe the birds' eating habits.

Nutritive Requirements of Turkeys

The turkey's requirement for Vitamins A and D, niacin and choline, is substantially higher than the chick's. The other vitamin requirements are similar, except Vitamin B_{12}, which is 3 times higher for the chick. Vitamins routinely added to turkey rations include A, D, E, K, riboflavin, pantothenic acid, niacin, B_{12} and choline. In recent years, there has been evidence of a biotin deficiency in turkeys fed corn-soybean type diets. At least, biotin has been helpful in correcting the trouble. Researchers have not, as yet, identified all the variables involved in this problem. Many turkeys have been raised without biotin supplementation with no biotin deficiency signs. Possibilities suggested as causative agents include usage of certain drugs, improperly heated soybean meal, consumption of wild seeds, molds, genetic changes and nutritional imbalance. Biotin is added to the ration to prevent dermatitis on foot pads.

Margins of safety are necessary for certain of the vitamins. This is partly because certain vitamins have a beneficial effect at higher than requirement levels under disease or nutritional stress conditions. The vitamin fortifications for turkey rations have changed as the cost of synthetic vitamins has decreased. As a result, there is no need to risk borderline vitamin deficiencies during periods of stress or depressed feed intake.

Nutrition of Breeding Turkeys

Insufficient research has been done on the nutrition of breeding turkeys. For high hatchability of fertile eggs, turkeys require very high levels of Vitamin E. Fresh grass juice produces a marked increase in the hatchability of fertile eggs, but apparently, this is not due to its Vitamin E content. Turkey breeders require more pantothenic acid than is needed for breeder chickens. Vitamin A is very important, particularly for its carry-over values to the young poult at hatching time. Improvements in the efficiency of feed utilization have had much to do with keeping turkey producers in business. There are slight variations in fertility as a result of nutrition in turkey breeders.

An all-vegetable diet results in a slight reduction in numbers of copulations and fertility by single males. A ration low in Vitamin A reduces sperm development. A Vitamin E deficiency affects sperm development and activity, and feed restriction gives a pronounced decrease in fertility and semen volume in males. Excessively fat toms are relatively inactive and clumsy in mating. Sexual maturity in turkey toms varies from 20 to 38 wk, with the average around 30 wk. Artificial lights affect sexual activity and sperm development. Mating precedes the onset of egg production so it is important that toms be subjected to artificial light ahead of the hens. Frequency of mating is usually highest just prior to egg production and declines as egg production reaches its peak. The highest degree of fertility is observed during January, February and March. Sudden changes in temperature have a definite effect on fertility. Preferential mating is another factor that affects fertility of single tom matings as well as large flock mating. Under good management, a season's fertility rate of 85% can be expected.

After the toms and hens have been selected, they are placed on special feeds providing them all the nutritional needs of breeder birds. The feed should have additional levels of essential vitamins and trace minerals to supply any deficiencies that have occurred during the growing period. The grain is reduced as much as possible to prevent accumulation of fat on the birds prior to going into the breeding pens. If grain is fed, it should be good quality oats. Green feeds should be made available when possible during the interim between the time birds are selected and blood-tested and the time they go into the breeding pens. If early egg production is expected, toms are placed under artificial lights at least 3 wk prior to placing them in the pen with the hens. This stimulates the less active toms and it insures greater activity at the time mating begins. In the breeding pen, use 1 tom for every 20 hens. Spare toms should be kept on the breeder diet and under lights when in the resting pens.

Breeders should be fed a complete breeder pellet with all grain feeding eliminated. This provides uniform feed intake. Do not mix toms from 1 group with toms from another. Keep each group separate during the breeding season as a precaution against fighting. This permits rotation and a spare group to give an opportunity to rest the birds. The breeder feed should be fed 30 days before egg production is expected. Protect the birds against rain and cold. Sudden drops in temperatures do more toward reducing fertility than the actual cold and rain. Remove broody birds from the nest each night and place in separate pens on the breeder feed. Hatching eggs should be gathered 3 times per day during extremely cold weather. The eggs should be allowed to cool before being placed in cases. If the eggs are to be held, they should be turned to prevent damage to the germ. Low fertility is a baffling and costly problem in turkey breeding operations. Both fertility and hatchability tend to decrease rather rapidly as the breeding season progresses, particularly in out-of-season hatching. Age and staleness of sperm influence egg hatchability. More frequent matings might result in high hatchability of turkeys. Fertility is improved by artificial insemination, either alone or in with natural matings. The late season decline of fertility of naturally-mated females is reduced by artificial insemination during the last 6 wk of their breeding seasons. Turkeys produce semen of about 50% volume (0.33 ml) and twice the sperm concentration (8,400,000 per cu mm) as chickens. Increasing the interval between successive inseminations from 2 to 4 or 6 wk results in lower fertility, hatchability and egg production. There is no significance in either degree or duration of fertility and hatchability between groups of females inseminated with a single dose of semen of different concentrations. Since turkeys produce relatively small amounts of semen, these results suggest that more efficient use of the semen might be made by utilizing smaller doses.

TURKEY RATIONS

Energy accounts for a large part of the cost of the turkey ration. For many years, turkeys were fed low-energy rations. As much as 30% of a ration would consist of wheat bran, wheat middlings and alfalfa. As feed grade fats became available economically, 2% or more fat was added to produce medium- to high-energy rations. The common practice is to start broods on rations containing 1200 to 1300 Cal. metabolizable energy per lb. This is increased during the growing period with the birds being finished on rations containing 1300 to 1400 Cal. per lb. The cost per Cal. is not the only factor to be considered, as the time required to obtain market weights and the carcass finish of the bird must be considered, as well. Energy from fat is more efficiently metabolized than that from carbohydrates. The explanation is that there is a ceiling on the energy needed to synthesize tissue fat from carbohydrates. As the level of added fat in the turkey ration increases, the feed efficiency also increases.

Years ago, turkeys were fed somewhat as follows: poults were fed wheat bread soaked in milk, squeeze-dried and mixed with egg and sting nettle chopped fine. In the morning, they were fed lettuce, and at night, sting nettle with bread soaked in milk, and a little chopped onion. They were given all the skim and sour milk they would drink. In the 1930s it was discovered that blackhead was due to a microorganism carried by chickens and transmitted to turkeys. It rarely caused disease in chickens but was disastrous when outbreaks occurred in turkeys. Thus, turkey growers began to raise turkeys in confinement. This greatly reduced mortality due to blackhead but with it came many nutritional problems. Confined turkeys no longer had access to lush, green pastures, and many of them grew poorly. A large percentage became crippled.

The quantitative nutritional requirements of turkeys are higher than those of any other farm animal. They need approximately 1200 to 1400 Cal. of metabolizable energy per lb of diet and, in addition, 13 vitamins, 13 inorganic elements and 13 amino acids. Turkeys also need UGF present in such materials as soybean meal, distillers' solubles, brewers' yeast and fish products. Because turkeys regulate feed consumption primarily to meet energy requirements, less feed is required to rear a turkey when diets of high-energy value are used as compared to low-energy feeds. Starting and growing rations should contain enough energy for maximum growth with no excessive amounts to be deposited as fat. Finishing diets should contain wider energy-to-protein ratios. When energy is in excess of that needed for growth, the extra energy is converted to fat.

Efficiency of Feed Conversion

Turkey hens and toms can be produced experimentally with an efficiency of feed conversion of 1.85 to 2.25 lb of feed per lb of turkey, respectively. Great advances have been made in the last several decades in turkey production as a result of the advances in turkey nutrition and practical application of these findings in the field. Still, much of the knowledge was obtained by trial and error, and more precise knowledge is necessary. Using present methods of feeding, at least 20% of the protein fed to turkeys is wasted. Until recently, the protein and amino acid requirements were expressed as a percentage of the ration, without any consideration being given to the energy level or the feed intake. The importance of relating the protein and energy requirements of the ration is obvious. The energy level regulates the amount of feed consumed. When rations containing different energy densities are fed to turkeys, there is no difference in the daily energy intake per turkey. Birds fed a low-calorie ration consume approximately twice as much feed as birds fed high-calorie rations. The environmental temperature, sex and strain of birds also affect protein requirements, and must be considered in formulating turkey rations. Prolonged hot or cold weather causes a decrease or increase in energy intake of 15 to 25% from the normal intake at moderate temperature.

The National Research Council requirement of poultry lists the protein requirement of starting poults at 25% of the ration and growing turkeys of 8 to 16 wk of age at 20%. These recommendations apply to rations containing 1110 Cal. of metabolizable energy per lb for the starting poults and 1180 Cal. for the older birds. If high levels of added fat are used in the ration, higher levels of protein are necessary for optimum growth.

In high-protein corn or milo soy-type rations, the first limiting amino acid is methionine, the seond lysine. In low-protein rations, lysine is the first limiting and methionine is second. In low-protein rations, much of the protein is derived from grains which are poor sources of lysine. It is cheaper at today's prices to add synthetic methionine than to increase protein. A multiple amino-acid deficiency occurs when the protein in the starter ration is reduced to 18% or lower.

Special Requirements

Differential feeding cuts turkey costs. Turkey toms need a higher protein content in their finisher ration than hens to achieve maximum economy and weight gains. Feed costs can be significantly reduced by adopting a different dietary regime for both sexes after birds are 8 wk old. The best economic response from toms is obtained on a finisher feed containing 21% protein. The optimum return from hens is obtained on a 18% protein ration. This is logical, because toms grow 30 to 35% faster than hens. Rearing the sexes separately on different feeds saves at least $.05 per bird. Turkeys require more protein than chicken broilers, particularly during their early starting-growing period. Soybean protein is adequate in most respects for good growth, but somewhat deficient in some amino acids compared with animal proteins. High quality fishmeal is frequently added during the starting and growing periods of the poults. Since corn supplies more energy than either oats or barley, producers make maximum use of corn in turkey fattening and finishing and in breeder-hen diets.

Growth Rate Control

Growth rate of turkeys can be limited by restricting feed consumption to 70% of the amount consumed by full-fed birds, or feeding high-fiber oat hulls. Restricted birds weigh 1 to 2 lb less than full-fed birds. Restricted birds at the start of egg production start to lay about 5 to 8 days later than the controls. Fertility, hatchability and egg weight are not diminished by the growth restriction. In fact, restricted birds show a significant improvement in average fertility. Growth restriction of turkeys by means of restricting the nutrient intake leads to a decrease in the number of eggs laid. Therefore, it appears wise to feed young breeder turkeys conventional starter-grower and finisher rations that have been proved by past experience to be satisfactory.

Diet Protein Level Adjustment

If both methionine and lysine are available for feeding, the protein level could be reduced to approximately 20% by adding synthetic amino acids. This would give a diet relatively higher in energy and efficiency than a comparable 28%-protein diet. However, synthetic lysine, so far, is not available at sufficiently low cost for animal feeding, so it must be supplied from natural sources. For practical purposes, a carefully selected protein mixture at an approximate 24% level, properly supplemented with methionine, presents the best possibility for top poult performance. Protein quality must be put ahead of protein quantity in the formulation of turkey diets. This is imperative in poultry rations in which the lysine requirement is high and increased energy is necessary to improve efficiency. It is evident that increased efficiency in turkey diets should come from increased dietary energy levels properly balanced with all other nutrients. A 900-Cal., 24%-protein diet of high quality (especially in lysine content) and supplemented with DL-methionine, seems to be the best approach. The use of fat and 50% protein soybean meal are necessary under present conditions. As more knowledge is gained about the poult's nutritive requirements, perhaps 24%-protein diets at 1000 Cal. will become a reality.

TABLE 44.1. GROWTH RATE AND FEED CONSUMPTION OF BROAD-BREASTED BRONZE AND WHITE TOMS

Age in Weeks	Live Weight Avg (lb)	Live Weight Gain for Period (lb)	Feed Required Total Cumulative (lb)	Feed Required Per lb of Turkey to Date (lb)
1	0.27	0.15	0.2	0.75
2	0.55	0.28	0.6	1.1
3	1.00	0.45	1.2	1.2
4	1.50	0.50	2.1	1.4
5	2.1	0.6	3.4	1.6
6	2.8	0.7	4.8	1.7
7	3.7	0.9	6.7	1.8
8	4.7	1.0	8.8	1.9
9	5.8	1.1	12.1	2.1
10	6.9	1.1	16.2	2.3
11	8.0	1.1	20.6	2.6
12	9.2	1.2	25.0	2.7
13	10.5	1.3	29.5	2.8
14	11.8	1.3	34.3	2.9
15	13.1	1.3	39.0	3.0
16	14.6	1.5	43.8	3.0
17	16.1	1.5	48.5	3.0
18	17.6	1.5	52.9	3.0
19	19.1	1.5	57.5	3.0
20	20.6	1.5	62.5	3.0
21	22.2	1.6	67.7	3.1
22	23.8	1.6	74.0	3.1
23	25.4	1.6	80.5	3.2
24	27.0	1.6	87.3	3.2
25	28.5	1.5	94.0	3.3
26	30.0	1.5	100.0	3.3
27	31.3	1.3	107.0	3.4
28	32.5	1.2	116.0	3.6

TABLE 44.2. THE EFFECT OF A CHANGE IN PROTEIN LEVEL IN A POULT DIET ON THE ENERGY LEVEL AND THE RELATIVE AMOUNT OF METHIONINE AND LYSINE

% Corn	% SBOM	The Diet Contains Prod. Energy Cal. per lb	% Protein	Methionine % of Protein	Lysine % of Protein
42	33	815	28	1.98	5.51
52	23	873	24	2.07	5.43
62	13	931	20	2.19	5.30

Diet Adjustment in Hot Weather

During extremely hot weather, it may be necessary to feed as much as 20% or more oats in place of corn to keep birds eating at full capacity. With cooler weather, corn should be fed as the chief grain for lower cost. Several additives may be included in a turkey mash. High levels of antibiotics or nitrofurans have been used to advantage during the first 2 wk of life. Generally speaking, the cost of these drugs becomes prohibitive if they are fed beyond the first few wk. Control of certain types of intestinal or respiratory disorders is effected by feeding these drugs for a period not to exceed 5 to 7 days. Organic arsenical drugs improve weight gains and feed efficiency when added to feed. A tranquilizer in the mash reduces blood pressure in turkeys, particularly between the eighth and sixteenth wk of age, to prevent aortic rupture. Antioxidants and coccidiostats are used in turkey feeds, and histostat and hepzide (20%) are used for blackhead control. These drugs are added according to the manufacturer's directions to avoid toxicity or excessive cost of medication.

For Broad Breasted Bronze or large white turkeys between 8 and 20 wk of age, the minimum calcium requirement for maximum weight gains is 1.6% of the diet, and the minimum phosphorus requirement is 0.5 to 0.6% of the diet. Male and female poults differ widely in their requirements. This may mean that each sex could be grown better separately and fed balanced feeds to fit their specific nutritive requirements, if maximum growth and maximum efficiency or feed conversion are to be obtained.

Feeding Methods

Methods of feeding turkeys were adapted from those for chickens. The mash and grain, or pellets and grain, systems which keep the feed before the turkeys all the time, are the most popular methods. Increased production and the lower selling price of turkeys have created greater interest in methods of feeding turkeys which secure lower production costs. The term, "concentrate" or "free choice," feeding of turkeys is applied to 2 different feeding methods. In 1 method , the high-protein concentrate is fed and supplemented with whole grains kept separate in the same hopper. In the other, the high-protein concentrate is mixed with ground grains to approximate the composition of commercially-prepared mash.

Hen turkeys mature at an earlier age than tom turkeys. It is most economical to begin feeding a finishing diet high in energy and low in protein to hens at about 16 wk of age and to market them as soon as they show a satisfactory degree of finish. Tom turkeys are still growing rapidly at 18 wk of age and are fed growing diets until they are approximately 21 wk, before giving them the high-energy, finishing diets. By reducing the protein and increasing the energy content of the feed approximately 3 wk before marketing, toms are finished at an earlier age and lower weight than they would ordinarily achieve at maturity. Common vegetable and animal sources of protein are combined to provide the essential amino acids in approximately the ratio required by turkeys, and higher use is made of synthetic methionine, lysine and other amino acids when the need arises. Both growth and efficiency of feed utilization are markedly improved by the adjustment.

Curled Toe Paralysis

"Curled toe" paralysis is a condition where poults sit on their hocks with their toes curled inward, and is a symptom of a deficiency of riboflavin. Since only small quantities of this vitamin are found in ingedients normally used in turkey rations, riboflavin should be added to the feed to ensure that a riboflavin deficiency does not occur. Pantothenic acid is required for the prevention of dermatitis, or sores at the corners of the mouth, on the eyelids and on the tops of the feet. It is required for growth and maximum utilization of feed. Pantothenic acid deficiency has occurred in the field in a few instances.

Conditions from Niacin, Choline, and Vitamin B_{12} Dificiency

Niacin deficiency symptoms in turkey poults are similar to those described for chicks; namely, perosis, poor feathering, poor growth and inflammation of the tongue. The addition of niacin to a turkey feed is especially important with high-energy rations, composed largely of corn and soybean meal. Wheat by-products are good sources of niacin, and if they are removed in an atttempt to increase energy content, niacin should be added to the feed. Choline is required for the prevention of perosis or slipped tendon in turkey poults, as well as for growth and good feed utilization. Soybean meal is an excellent source of choline, but choline chloride is usually added as a feed supplement, too. Growing poults have a sufficient amount of Vitamin B_{12} carried over from the mother through the egg. The poults respond to Vitamin B_{12} supplementation of the ration when antibiotics are fed to produce maximum growth.

Conditions from Biotin Deficiency

Perosis or leg weakness in turkeys occurs from a biotin deficiency. The biotin requirement of the turkey is approximately 2.5% as high as that of the chick. The young poult consumes approximately 20 g of feed per day during the first 2 wk of life. The starter diet must contain 0.25 mg of biotin per kg in order to provide 6 mcg biotin per poult per day. Biotin that exists in natural materials occurs in both bound and free form, but much of the bound biotin is unavailable. Synthesis of biotin in the intestinal tract supplies a variable and undependable amount of biotin for the nutrition of poults.

Losses from Leg Weakness

Leg weakness and hock problems in large tom turkeys cost the turkey industry millions of dollars per year. Leg weakness is not due to any single factor, but results from a wide variety of different stresses, some of which are pathological, others nutritional, still others genetic; some are due to poor management. Leg weakness in turkeys may result from different diseases, 3 of which are nutritional and 2 pathological. The gross symptoms of these diseases are similar. At first, the turkeys have a stilted, unsteady gait with evidence of leg weakness. Soon, 1 or both hock joints become enlarged and swollen, whereupon the legs may bow or take on a knock-kneed appearance. In some diseases, it is noted that the Achilles tendon slips from its chondyl and typical perosis and enlarged hock disorders occur. Certain strains of turkeys are more susceptible to leg weakness than others. Weaknesses may be caused by breeding strains of turkeys with such a high ratio of meat to bone that its skeletal structure is no longer capable of supporting its weight. Poor footings on slat or wire floors, with no litter, may cause leg weakness at times when better flooring would produce normal birds. Thus, breeding, management, disease prevention and nutrition are all important in prevention of leg weakness in turkeys.

Good management consists in preventing stresses, hock disease and nutritional deficiency. Although brewers' dried yeast is relatively expensive compared to some other feedstuffs, prevention of leg weakness may justify its use.

Many turkey toms injure their hocks by flying from roosts. Open range-managed turkeys should not use roosts because turkeys frighten easily at night.

Antibiotics

Potentiated antibiotics are used in starter rations. This reduces mortality and enables poults to put on weight early when the gains are least costly. Antibiotics protect against paratyphoid, PPLO infection, blue comb, hexamataxiasis and synovitis; antibiotics also overcome infections during stress periods, including vaccination, moving, handling and weather extremes. Automation of feeding and watering is common. Clay-type floors are covered with wood shavings. There are 3 broods per year. There is a cleanup of houses and equipment between each brood. As many as 2 million birds are grown each year on some turkey ranches which market over 90% of all poults started.

The feeding of antibiotics has been one of the more important advancements in turkey production. Antibiotics increase the rate of growth and improve feed efficiency. Aureomycin, terramycin, bacitracin and penicillin are the more commonly-used antibiotics. The growth-promoting level is approximately 10 g per ton of feed.

Losses from Crooked Breast

Crooked breast is of some economic loss in turkey production. There are several factors and combinations of factors which cause the deformity; heredity appears to be one. Thus, the careful selection of breeding stock seems to be of special importance. Feeding seems to have a more or less direct bearing on crooked breast production. Protein, green feeds, phosphorus and calcium in proper amounts are especially important for growing birds. Carbohydrates are probably a secondary consideration during the finishing period. An ample and convenient water supply at all times is important.

Mineral Supplements

It is common practice to supplement turkey feeds with calcium, phosphorus and zinc, as well as iron, copper and iodine. Poults are frequently slow learning to eat. Outbreaks of rickets have been traced to inadequate intake of rations normally adequate in calcium and phosphorus. A zinc deficiency can also occur with practical rations. Symptoms are enlarged hocks and poor feathering.

Selenium Deficiency

A recent discovery is that a selenium deficiency may also be a practical problem with turkeys. Poor growth, gizzard myopathy and high mortality occur if the deficiency is severe. Most soils east of the Mississippi River are deficient in selenium. The selenium requirements of poults are estimated to be 130 to 150 mcg per lb feed. Turkey producers use selenium-rich feed ingredients, such as wheat mill by-products, standard middlings and mill-run wheat from wheat grown in selenium-rich areas.

45

Game, Cage and Aviary Bird Feeds and Feeding

The development of nutrition standards for feeding and management of cage and aviary birds has been slow and difficult. Research on feeds often involves artificial feeding methods which frequently frighten the bird. Success in feeding depends on establishing voluntary feed intake, triggered by the manner in which the food is offered.

When studying feeds and feeding of birds, the nutritionist must recognize several types and classifications of birds. Three classifications are made of birds according to their eating habits.

(1) Seed-eaters or hard bills.
(2) Insect-eaters or soft bills.
(3) Fruit- and nectar-eaters.

Another important division of birds is based on condition at hatching. The 2 types at hatching are:

(1) Nidicolous chicks—young which hatch at an early stage of development and are generally naked, helpless and blind.
(2) Nidifugous chicks—young which have a longer incubation period, are covered with down when they leave the shell and are able to run and pick up food for themselves. Also, small tail and wing feathers can often be seen protruding from under the skin when the chicks are first hatched. Birds of this type are easy to start and to study.

Birds kept in captivity for many years, have been used for food, dress, sports and fundamental research.

Although birds appear to be highly selective of their food, they will eat, in most instances, the feed available. Many birds change their diets as they migrate or as feed is provided throughout the season.

The nutritional requirements of caged birds are similar to those of the quail or pheasant, except for fiber and iodine. Caged birds require high-energy food and supplemental iodine. Most seeds are very low in iodine and the hulls are chiefly fiber. Many cases of iodine deficiency have been observed in budgerigars throughout the United States. Some surveys have reported as much as 85% visual thyroid trouble.

Some birds that eat fruit and drink nectar may require Vitamin C. The enzyme involved in the synthesis of Vitamin C in the red-vented bulbul is not present in functioning amounts. The response to Vitamin C in the fruit- and nectar-eating birds may be a dependency need which has arisen from consuming such a high level of Vitamin C that the body's mechanism for synthesizing this vitamin is nearly absent or inactive. We know that over-uptake of calcium by pullets results in an undeveloped parathyroid-control program. When man consumes excessive Vitamin C, he will readily develop a deficiency of this vitamin if the dietary level is reduced to a normal uptake. Also, those birds which secrete "milk" seem to have a higher requirement for iodine. Some have indicated that birds which secrete milk and regurgitate not only require supplemental iodine but the water-soluble vitamins and trace elements known to be required by quail.

All seed-eating birds require grit, but it must not become contaminated with feces.

Water is very important. The bird releases over 40% of maintenance heat by evaporative cooling. Water is critical for birds, such as sparrows and finches, that live on dry seeds. Canaries cannot survive over 48 hr without water, whereas budgerigars have the ability to survive several weeks or months without water. Although this is reported, the best practice is to have clean water available at all times.

FEEDING OF PARAKEETS (Pudgies)

Parakeets (pudgies) are among the most common household pets. They are considered by geneticists to be rather weak, fragile and sensitive. They, like many song birds or pet birds, seem to prefer a special seed for most of their food. Solutions to dietary problems with whole seed-eating birds are difficult because of the instinct of the bird. The lack of knowledge regarding food supplements acceptable by the bird encourages the owners (fanciers) to adhere to the regular diet.

The parakeet's preferred seed is millet; therefore, an understanding of its production, composition and storage must be recognized and understood. The average composition of millet grain is presented in Table 45.1.

Protein

The protein content of millets varies from 10 to 16%, and since the protein requirement of most growing game birds is 16 to 18%, the higher

TABLE 45.1. MILLET GRAIN COMPOSITION [ALL ANALYSIS (AVERAGE)]

D.M. (%)	89.8 (87.9–98)
Crude protein (%)	13.5 (10–16)
Ether extract (%)	4.5 (2.3–5.8)
Crude fiber (%)	9 (3–11.1)
Ash (%)	3.9 (1.6–6.7)
Nitrogen free extract (%)	69.1 (65.6–7.64)
Calcium (%)	0.06 (0.01–0.13)
Phosphorus (%)	0.31 (0.20–0.39)
Copper (mg/lb)	10.9
Potassium (%)	0.48
Magnesium (%)	0.18
Iron (%)	0.005 (0.004–0.005)
Manganese (mg/lb)	14.7 (4.1–42.1)
Thiamin (mg/lb)	3.3 (2.7–3.7)
Riboflavin (mg/lb)	0.8 (0.6–1.2)
Pantothenic acid (mg/lb)	3.7 (1–7.9)
Niacin (mg/lb)	2.65 (1.8–3.6)

Source: Natl. Res. Counc. Pub. *585* (1958).

protein millets provide only the minimum protein needs of the growing bird. Also, the protein in millet seeds is like other seed proteins in being deficient in one or more amino acids. Most seeds are deficient in lysine.

Vitamins

Millet seeds do not contain Vitamin A, D_3 or B_{12}. They are deficient in riboflavin, pantothenic acid and choline. These voids or deficiencies make it necessary that a premix or supplement be fed.

Minerals

A deficiency of calcium in millet seed exists, and egg shells will be weak unless supplemental calcium is provided. The manganese content varies from 4.1 to 42.1 mg per lb, so on low-manganese soils, the seeds will be deficient in this mineral. The trace element content of seeds often reflects the soil analysis. Due to many years of cropping, soils will become depleted or not provide enough of the trace elements for the plant to accumulate minerals in the seeds. A soil could provide adequate amounts for plant growth, yet not provide for the extra accumulation necessary for game birds. The trace elements which influence birds most rapidly depleted from soils are manganese, zinc and copper.

When soils become depleted in a required trace element, the trace element nearest this element in the periodic table may be present, taken up by the plants and deposited in the seeds. Often, these substitute elements are toxic, hence resulting in poor growth or reproduction of the birds.

Molds

Seeds must be mature and dry when harvested and stored or mold spores will develop and grow. Since some molds form toxins (aflatoxins) which are very harmful to game birds, it is extremely important that only clean, bright and sound seeds be fed. Also, game birds must always have a clean, dry cage floor. The floor requires regular cleaning so that the birds will not eat contaminated or decayed seed.

The symptoms resulting from mold toxins often resemble nutritional deficiency symptoms. Table 45.2 presents information identifying nutritional deficiencies and symptoms similar to those from mold toxins.

TABLE 45.2. DEFICIENCY SYMPTOMS

Symptoms, Deficiency or Toxicity	Aflatoxins[1]	Vitamin B_{12} (Cobone)	DDT or Polychlorinated Hydrocarbons	Vitamin D	Vitamin K
Rachitic or soft bones	Yes		Reported	Yes	
Blood clotting	Yes		Reported		Yes
Reproduction	Yes	Yes	Yes	Yes	
Livability	Yes	Yes	Yes	Yes	Yes
Congenital malformations (birth defects)	Yes	Yes	Yes	Yes	

Source: Patrick (1960).

[1]Nonspecific diseases may be recognized by symptoms similar to aflatoxin toxicity. Control seems to vary from use of drugs or sanitation cycle-breaking management. When animals are subjected to toxins, poor nutrition, improper management or sanitation, the nonspecific disease seems to develop.

THE DIET

Many zoological gardens have developed complete and adequate feed and feeding programs for each species of bird or animal. The Philadelphia Zoological Garden diet for cage birds, such as parakeets, is presented in Table 45.3. Diet O can be mixed and stored for 2 to 4 wk in a cool, dry place, but the fresh vegetables, eggs or meat portions should be prepared daily.

Diet O contains approximately 24% protein, and when fed with the fresh supplements, will provide about 18 to 20% protein.

FEEDS FOR PHEASANTS, BOBWHITE, QUAIL AND COTURNIX (JAPANESE) QUAIL

Pheasants and bobwhite quail are raised primarily for private hunting preserves, while the coturnix (Japanese) quail are raised primarily for

TABLE 45.3. CAGE BIRD DIET USED BY PHILADELPHIA ZOOLOGICAL GARDENS

Diet 0[1]	Percentage of	Feeding Mixture %
Ground yellow corn-9	15	
Ground whole oats-11	15	
Ground whole barley-11	10	
Ground rolled oats-11	10	
Brewers' yeast-45	10	
Cottonseed meal-45	10	
Soybean meal-50	10	
Dried skim milk-32	10	
Alfalfa leaf meal-17	5	
Oyster shell flour	2	
Iodized salt	1	
A-D feeding oil	2	
Diet 0		50
Ground raw carrots-1		20
Ground cooked meat-20		20
Ground raw cooked meat-18		7
Ground cooked eggs and shells-12		3
A-D feeding oil		
		100
Provide supplements of lettuce, apples or other fruits (free choice).		

Source: Altman and Dittmer (1968).
[1]Approximately 24% protein, the figure after each ingredient indicates protein content.

research purposes. Also, many pheasants are mass-produced for sale as a ready-to-cook item in supermarkets or through private sales to fine restaurants.

Many feed manufacturers use a similar ration for pheasants and all quail. These feeds have many specifications which are similar to turkey rations (Table 45.4). The following ranges of protein and energy are: starter feeds 26–29% protein, growing feeds 16 to 19% protein, and breeder feeds 18 to 20% protein with metabolizable energy of 1100 to 1200 Cal per lb. Rations and premixes for pheasants and quail are presented in Tables 45.5 and 45.6. Some investigators recommend larger amounts of choline, niacin, pantothenic and riboflavin for pheasants, and quail than for turkeys. A ration high in fiber (lower energy) generally results in slightly slower growth but fewer leg deformities. Additional methionine is added to each ration because of the rapid feather growth.

FEEDING PHEASANTS AND QUAIL

Pheasants are nervous and have a highly-developed social order. To assure adequate feed and water for pheasants, three or more feeders and

TABLE 45.4. TURKEY-PHEASANT BREEDER RATION REQUIREMENTS

Protein (%)	16–18
Metabolizable energy (calories/lb)	1250–1350
Calcium (%)	2.2–2.4
Phosphorus (%)	0.75–0.80
Available phosphorus (%)[1]	0.60
Zinc (mg/lb)	60
Manganese (mg/lb)	60
Vitamin A[2] (units/lb)	5000
Vitamin D_3[2] (units/lb)	1000
Riboflavin (mg/lb)	3.5
d-Pantothenic acid (mg/lb)	8–10
Niacin (mg/lb)	30–34
Choline (mg/lb)	900–1000
Vitamin E (units/ton—added)	10,000–20,000
Vitamin B_{12} (mg/ton—added)	10–12

Source: Patrick (1965).
[1]Available phosphorus for growing birds is defined as that from inorganic sources and animal products plus 30% of that supplied by plant materials. Available phosphorus for layers and breeders is defined as that from inorganic sources and animal products plus 50% of that supplied by plant materials.
[2]Dry stabilized vitamins A and D_3 (added).

TABLE 45.5. GAME BIRD PREMIX (USE 10 LB PER T RATION)

Manganese sulfate	4 oz
Zinc carbonate	4 oz
Antioxidant (Ethoxyquin)	4 oz
Zinc B_a citracin or Aureomycin or Terramycin	20 g
DL-Methionine or hydroxyanalogue	24 oz
Stabilized Vitamin A	20,000,000 U.S.P. Unit
Vitamin D_3	1,500,000 I.U.
D1-Alpha-tocohperol	5,000 I.U.
Choline chloride	320 g
Riboflavin	5 g
Niacin	40 g
Calcium pantothenate	12 g
Menadione sodium sulfate	2 g
Cobone	6 mg
Corn meal or soybean meal to make total of 10 lb	

Source: Patrick (1965).

TABLE 45.6. RATIONS FOR PHEASANT AND QUAIL

Feeding Period Form of Ration Ingredients	Starter 0–4 Weeks Crumbes lb	Grower 4–9 Weeks ⅛ in. Pellets lb	Breeder ⅛ in. Pellets lb
Ground No. 2 yellow corn	784	782	882
Oats—pulverized	100	200	200
Wheat midds	100	200	200
Fat (stabilized)	20	20	
Soybean meal—49% protein	600	500	300
Fish meal—60% protein	100		
Poultry meat scraps 55% protein	100	100	200
Whey, dried	50	50	50
Brewers' yeast or dist. solubles	50	50	50
Alfalfa meal 17% protein	60	60	60
Dicalcium phosphate	20	20	40
Salt (NaCl + iodine)	8	8	8
Vitamin T.M. + antibiotic premix	10	10	10

Source: Patrick (1965).

waterers should be provided. Also, some form of "escape-room" must be provided or cannabalism will occur.

Management of pheasants and quail is similar to that of turkeys, except for the development of their social order. Each grows fast and is an efficient feed converter. The disease problems of each are similar.

Young wild pheasants and quail depend on dew as a source both of water and minerals.

FEEDING PIGEONS

Pigeons must be provided a well-balanced diet, grit and clean water. They may survive on inferior feed for a time, but as the body becomes depleted in essential nutrients, they become weak, show deficiency symptoms, and pathological conditions appear.

There are several special ready-mixed feeds available, but some producers feed a high-energy duck-starting ration. When supplemented with iodine, the latter gives good results.

The feeding of a grain mixture is common practice; however, the grains must be of excellent quality. Yellow corn, field peas, kaffin corn or milo, millet, rice, oats, barley and hemp are all good grains for pigeons. Soybeans are not always good and wheat must be used with great caution. Clean, dry, well-handled hard winter wheat is often used to supply not over 20% of the total grain mixture. Damaged wheat often causes digestive disturbances.

Water

The pigeon needs water for drinking and bathing. The bath water should be separate. Pigeons should be allowed to bathe 3 times per wk, and after completing their bath, the facility should be cleaned and removed from the cage. Some pigeon fanciers have reported that bathing stimulated reproduction.

Drinking water must be clean, fresh and uncontaminated. A pigeon will die without water, whereas some caged birds may live for a week or longer without it.

Grit, a mineral mixture, is a very important feed for pigeons, especially when they are given a diet high in whole grains which are generally high in fiber and deficient in iodine and most minerals. The majority of birds in the wild obtain minerals from dew and field stone.

Grit contains:

40% medium-crushed oyster shell
35% limestone or granite grit
10% medium-sized hardwood charcoal

5%	ground bone
5%	ground limestone
4%	salt (iodinated)
1%	Venetian red

Common grain or seed mixtures are prepared for pigeons. One commonly used is a cafeteria-feeding mixture:

Corn	39.5%
Peas	22.7%
Wheat	19.8%
Milo	18.0%

A self-feeder needs 4 to 5 compartments. This use of compartments eliminates the most serious criticism of the self-feeder, which is the scattering of grain by the birds in search of some select morsels.

Cafeteria feeding is generally considered unadvisable for any kind of air-performing bird. With food before them at all times, these birds are sluggish in performance, slow to trap when desired, and obedience to the breeder's wish is nullified. Many pigeon producers prefer feeding twice daily; however, each operator has his reasons for the feeding method employed.

Pellets

High-energy duck-starting feed pellets with iodine have been used with success, although some producers feel that a pigeon feed without peas leads to disaster. A pigeon program should be better by feeding pellets rather than individual seeds.

Force Feeding Squabs

Force feeding is a very old practice and a proven method of finishing squabs. The labor requirement is great but the carcass finish is excellent.

The young squabs, after 10 to 12 days of age, are put in batteries and fed as follows: Ordinary pigeon feed or finishing duck feed is soaked in water for 4 hr and the squab given enough to fill the crop 3 times per day. A small funnel is often used to intubate the pigeon. Squabs seem to have a high capacity for utilizing feed if it is placed in the crop.

Pigeon "Milk"

The pigeon, dove and African Sand Grouse are the only birds which feed their young on a substanced called "milk." A few birds feed their young through a process called regurgitation. The process of regurgitation is feeding of a partially-digested food.

Pigeon milk is not a secretion, but a creation of both the male and female. It is a strange phenomenon of nature, a cream-colored material that appears like a curd in the crop.

Each pigeon crop has 2 lobes which develop in the parents' crops, beginning on the eighth day of incubation, for formation of the milk. On the eighteenth day of incubation, the cells begin to develop fat globules and break down (desquamate). The white masses of dead cells become the pigeon milk and the process of formation of the milk lasts for about 7 days. By the seventh day, the milk is nearly gone and a high grain milk mixture is present. This grain milk mixture first appears on the third day.

Pigeon milk has the following composition:

Water	72.4%
Crude protein	16.0%
Fat	10.0%
Ash	1.6%
Carbohydrates	None detected

PEAFOWL

Peafowl are ornamental birds—colored blue, white or green. They originated in Northern India centuries ago. They are often found on large estates as decorative features. The male matures in 3 years and the hen in 2 years. The care and feeding of peafowl is similar to that of turkeys and pheasants. These birds are excellent foragers and pick up most of their food requirements from the "wild" if allowed to range in open and wooded areas.

GUINEA FOWL

The guinea fowl is a close relative of the pheasant. Its feathers are marked dark grey with small white spots and the head and neck are bare. Guinea fowl eggs and meat are delicacies. These fowl are difficult to raise, and their continual harsh crying makes them unpopular, except as "watchdogs." They are commonly fed regular chicken or turkey feeds. They prefer fence rows or underbrush to buildings as a place to live, but will seek shelter in poultry houses during winter storms.

AVIARY BIRD FEEDS

There are many types or species of birds in zoos or public parks, and each requires special diets or feeding methods. The rations or diets in Tables 45.7, 45.8 and 45.9 are being used by some zoos. For more information on the feeding of birds in zoos, one should contact such leading zoos as:

(1) New York Zoological Park
(2) Philadelphia Zoological Garden
(3) Lincoln Park Zoological Garden of Chicago
(4) San Diego Zoological Garden

TABLE 45.7. NEW YORK ZOOLOGICAL PARK'S FEED FOR HUMMINGBIRDS

Formula A	
Millin's food	4 tsp
honey	5 tsp
condensed milk	1 tsp
beef extract	5 ml
vitamins (multiple)	4 drops
warm water	1 qt
Formula B	
honey	5 tsp
beef extract	5 ml
vitamins (multiple)	4 drops
warm water	1 qt

Source: Data assembled from Altman and Dittmer (1968) and Petrak (1976).

TABLE 45.8. SPECIAL ZOO BIRD DIETS FOR OMNIVORES

Grain Mixture	Grams
Rolled oats	20
Whole wheat meal	20
Soybean oil meal	20
Peanut oil meal	10
Yellow corn meal	10
Dry buttermilk	5
Dried brewers' yeast	5
Alfalfa leaf meal	5
Oyster shell flour	2
Iodized salt	2
Cod liver oil	1
	Parts
Fix grain mixture	9
Cooked ground horsemeat or boneless chuck	1
	10 parts

Source: Ratcliffe (1940).

An excellent reference on zoological diets is summarized in Biological Handbooks—Metabolism, 1968; compiled and edited by P.L. Altman and D.S. Dittimer, Federation of American Societies for Experimental Medicine, 9650 Rockville Pike, Bethesda, Maryland, 20014.

TABLE 45.9. SPECIAL ZOO BIRD DIET FOR CARNIVORES

Feed a mixture of 60% ground horsemeat or boneless chuck, 10% ground fresh vegetables and 30% of the following mixture:

	%
Soybean oil meal	30
Peanut oil meal	25
Dry buttermilk	25
Oyster shell flour	6
Alfalfa leaf meal	4
Dried brewers' yeast	4
Cod liver oil	4
Iodized salt	2

Source: Ratcliffe (1940).

TABLE 45.10. DIET FOR PARROTS, MACAWS AND COCKATOOS

	(g)
Sunflower seed	10
Hemp seed	10
Canary seed	20
Large millet seed	10
White millet seed	5
Peanuts	10

Source: Data assembled from Altman and Dittmer (1968) and Petrak (1976).

TABLE 45.11. CANARIES SEED MIXTURE

	(%)
Canary seed	47.0
Rape seed	27.7
Red proso millet	5.6
White proso millet	5.4
Yellow proso millet	14.3

Source: Data assembled from Altman and Ditmer (1968) and Petrak (1976).

46

Duck and Geese Feeds—Feeding

Raising ducks is either a specialized, large-scale operation or a secondary operation on general farms. Except for the Muscovy, all economic breeds of ducks originated from the Wild Mallard. The Runner, commonly called the Indian Runner, is an egg-laying class, and much smaller than breeds of the meat type. The Pekin duck is best for commercial farms where ducks are the principal source of income. Young ducks, forced for rapid growth and marketed at 7 to 8 wk of age, are called "green" or "junior" ducks, and weigh from 4.5 to 6 lb each. The demand for ducks is not as general as for chickens and the demand for duck eggs is even more limited. Ducks lay a good-sized white egg, considerably larger than the chicken egg. Some cooks prefer duck eggs to chicken eggs for cooking. The egg albumen is very firm and the yolk well-pigmented.

LOCATION OF INDUSTRY

Duck farming on a large scale has developed as a special business on Long Island and in other areas within easy shipping distances of large centers of population. Pekin ducks stand confinement well, are easily brooded, and less subject to disease than chickens. Artificial methods of hatching and rearing, and labor-saving machinery, have been used very successfully.

PRODUCTION

The White Pekin duck lays about 120 eggs during the hatching season. The Indian Runner duck is grown mainly for eggs, many producing over 200 per year. When based on weight of eggs, the Indian Runner duck is considered an excellent producer of food.

Ducks and geese are among the most rapidly-growing and efficient producers of human food. Ducks attain a weight of 5.5 to 6.5 lb in 7 to 8 wk. Ducks require up to half again as much feed for rearing and egg production as do chickens. Growth to 4 wk of age is improved by approximately 30% when pelleted rations are used instead of crumbled mashes. Starter rations for ducklings are made up in one-eighth-in. pellets; for ducks over 2 wk of age, three-sixteenth-in. pellets. Ducklings do not require water for swimming, but they need clean water for drinking. They achieve maximum growth on rations varying widely in energy content. Rations containing 16% or more protein have produced normal growth with energy levels ranging from 950 to 1350 Cal. metabolizable energy per lb ration. With rations containing 16 to 18% protein, higher energy levels produce ducks with higher carcass-fat content. As the energy content of the diet is increased over ranges from 800 to 1250 Cal M.E. per lb diet, efficiency of feed utilization is improved.

Three rations are often used for production of "green" or table ducks. These are (1) starting — up to 2 wk of age, (2) growing — 2 wk to 5 wk and (3) finishing.

The starting ration is high-protein, high-energy, vitamin-fortified and high-density. Special pellet binders are often used to make high-density pellets. The duck's digestive system seems capable of digesting food as fast as supplied; therefore, the high-density ration is important. Before use of the high-density ration, it required 12 wk to grow a table duck; but when the high-density pellets were developed, growing time was reduced to 8 wk, and hence a great saving in feed and labor was achieved (Tables 46.1 and 46.2).

Since the duck grows very fast, the big management and nutrition problem is to develop normal legs to support the body. Some starting ducklings grow from 40 g at 1 day of age to 640 g at 14 days — a 16-fold increase in weight. The characteristic bone problem is called "bowed-legs." Added niacin and choline play a very important part in helping alleviate this bone abnormality.

Breeding-ducks respond well to regular chicken-egg rations. They do not need high energy or excessive calcium. A level of 2.2 to 2.6% calcium is adequate.

GEESE

Geese-raising has developed rapidly in recent years, mainly for use in the weeding of cotton, sugar cane, strawberries, onions, garlic and other crops commonly hoed. Even unruly Johnson and Bermuda grasses can be controlled. Concentration of geese in such areas makes processing and marketing possible. Geese are spring layers, starting about the middle of February. Peak production is reached by April and continues into May. Highest production occurs in the second or third layer years, gradually lessening in succeeding years. Geese respond to light-stimulation outside of the normal reproduction period. Commercial goslings grow so rapidly and in such concentrations that pasture is hard to keep up. They are permitted to have clean, bare runs, and exercise on well-drained, sandy soil. Geese are largely herbivorous, and once started, can grow to maturity on grasses alone, if these are young and tender. Goslings can be reared in confinement, if fed a properly balanced diet, but they are much more easily managed and economically reared on pasture with grain added. Breeding-geese relish pasture, which greatly reduces carry-over costs. Geese cannot be grown satisfactorily on dried-out or mature pasture.

Geese are reared in small numbers on many farms for their meat, feathers and ability to weed crops. They are frequently part of a scavenging program because they can live mainly on weeds, except for the finishing period. Goslings are usually available only in the spring of the year, and the best market for geese is at Thanksgiving and Christmas. The soft, curved body feathers are a valuable by-product of market geese because they produce warm quilts and soft pillows. Geese forced to rely on weeds or pasture for nourishment are usually emaciated, so it is desirable to have a special finishing period to give them a proper degree of finish for market.

Geese can be sold in the fall or processed in early life and sold as "green" or "junior" geese. Young geese are good gainers to the age of about 8 wk, showing a very fast growth rate, decidedly greater than other kinds of poultry. After this fast growth, a plateau is reached. Starting at about 20 wk of age, geese again begin gaining rapidly during the fattening period. Geese are produced in 2 ways—by relying completely upon forage for nutrients or feeding complete feeds. In the former case, gains are comparatively slow; with a complete feeding program, gains are extremely rapid. Geese go into a molt at 10 wk of age and are not completely feathered out until 16 wk. Geese weigh 7 lb at 6 wk of age, using 2.25 lb feed per lb weight. At 9 wk, geese weigh 9.9 lb, using 3 lb feed per lb, and are comparatively free of pin-feathers. Geese receiving a low-energy feed do not grow as fast as those receiving a high-energy feed. Goslings fed a high-fat ration are heavy and show excellent efficiency of feed utilization.

The most frequent objection to geese for table use is the fatiness of the bird; but with "green" geese, overfattening is not a problem. From a nutrition standpoint, as stated, dietary addition of the vitamin, niacin, corrects a troublesome leg weakness problem. Good chick-starter rations supplemented with niacin produce good gains in goslings. Riboflavin tends to be deficient in practical-type starter diets, but other B-vitamins in corn-soybean meal type rations are adequate. Although the protein requirement of goslings is 20%, 24% produces better feathers and growth. Breeding-geese have been fed chicken-breeder supplement plus free-

TABLE 46.1. PRACTICAL FEED FORMULAS FOR DUCKS AND GEESE (1000 LB MIX)

	Duck Feeds (Pellet Form Only)				Goose Feeds (Pellet Form Only)		
% Protein	Starter 18%	Finisher 16%	Breeder Developer 18%	Breeder 18%	Starter 24%	Grower Finisher 16%	Breeder 18%
Time to Feed/Weeks	0–2	3–8	Feed as a Mainte-nance Ration After 9th Week	Start Feeding 4 Weeks Before Collecting Hatching Eggs	0–6	7–18	Start Feeding 4 Weeks Before Collecting Hatching Eggs
Ingredients	lb	lb	lb	lb	lb	lb	lb
Corn, yellow, medium grind	485	615	320	490	395	600	480
Alfalfa meal, dehyd., 17% protein	30	25	15	100	30	25	100
Soybean meal, solvent, 45% protein	145	170	—	130	320	160	140
Wheat middlings, standard	100	50	175	50	100	100	50
Oats or barley, ground	100	50	—	50	40	—	50
Corn distillers' dried solubles	40	—	—	25	20	—	25
Grain screenings	—	—	400	—	—	—	—
Molasses, cane	—	25	50	—	—	50	—
Meat and bone scraps, 50% protein	20	20	15	20	20	20	20
Whey, dried	20	10	—	30	20	10	30
Fishmeal, 60% protein	25	—	5	40	25	5	40
Salt	5	5	5	5	5	5	5
Dicalcium phosphate	15	15	5	10	10	10	15
Limestone, ground	10	10	5	45	10	10	40
Vitamin-trace mineral premix	2.5 or 5	2.5 or 5	2.5 or 5	2.5 or 5	2.5.or 5	2.5 or 5	2.5.or 5
TOTAL	1,000	1,000	1,000	1,000	1,000	1,000	1,000

TABLE 46.1. *(Continued)*

Approximate Analysis		Duck				Goose		
		Starter	Finisher	Breeder Developer	Breeder	Starter	Grower Finisher	Breeder
Crude protein	%	18.0	16.1	13.4	18.1	24.2	16.3	18.4
Crude fat	%	3.4	3.1	4.2	3.2	2.8	3.1	3.2
Crude fiber	%	5.2	4.3	6.3	5.8	5.3	4.0	5.8
Productive energy	Cal. per lb	872.0	919.0	592.0	829.0	784.0	907.0	818.0
Calcium	%	1.0	0.9	0.7	2.6	1.1	1.0	2.4
Phosphorus	%	0.9	0.7	0.6	0.7	0.8	0.7	0.8
Salt (added)	%	0.5	0.5	0.5	0.5	0.5	0.5	0.5
Manganese (added)	mg per lb	28.0	28.0	28.0	28.0	28.0	28.0	28.0
Vitamin A	IU per lb	4,985.0	4,680.0	3,216.0	12,373.0	4,843.0	4,658.0	12,350.0
Vitamin D_3	ICU per lb	700.0	700.0	700.0	500.0	700.0	700.0	500.0
Vitamin E	IU per lb	5.5	4.3	3.8	12.2	5.1	4.6	12.4
Vitamin B_{12}	mcg per lb	4.9	4.1	3.7	6.2	5.0	4.1	6.2
Riboflavin	mg per lb	2.9	2.5	2.0	3.1	2.9	2.5	3.1
Niacin	mg per lb	23.9	20.1	20.7	21.9	23.7	23.0	21.9
Pantothenic acid	mg per lb	7.1	6.6	5.3	7.3	7.5	7.2	7.3
Choline	mg per lb	667.0	551.0	377.0	605.0	784.0	552.0	613.0

TABLE 46.2. GROWTH RATE AND FEED CONSUMPTION (PER 100 DUCKLINGS)

Age Weeks	Live Weight lb	Gain During Week lb	During Week lb	Total to Date lb	Per lb Duckling to Date lb	Per lb Gain During Week lb
At hatching	12	—	—	—	—	—
2	110	98	170	170	1.54	1.73
3	204	94	187	357	1.75	2.00
4	294	90	270	627	2.13	3.00
5	380	86	300	927	2.43	3.48
6	480	80	330	1257	2.60	4.12
7	525	65	335	1592	3.03	5.10
8	590	65	350	1942	3.29	5.38
9	646	56	340	2282	3.53	6.07
10	680	46	325	2607	3.83	7.00

Source: Schaible 1972.

choice of corn and oats for many years. Given a choice, breeder-geese select 20 to 25% of a 27% protein mash, 30 to 25% of oats and 35 to 60% of corn.

Fattening Geese

Although geese are fast-growing and most efficient in feed conversion, practically free of disease and excellent as foragers, they are poor reproducers. This is a deterrent to large-scale production. Goslings do not need feed until they are from 36 to 48 hr old. Green grass makes up most of their feed and only a very limited amount of grain is used. Fresh, clean drinking water should be supplied. After 2 or 3 wk, if the goslings have plenty of young grass, they will usually not need any other feed. If the pasture is good, most goslings are raised from the time they are 2 to 3 wk old to fattening without any additional feed. Whole grains are not fed until the goslings are well-feathered. In a few areas, young geese are fattened in large numbers by buyers who make this a specialty. Corn-on-the-cob and plenty of water are kept before the geese at all times, and they eat the leaves of the cornstalks for roughage. Some farmers pen-fatten their geese in flocks of 20 to 25 and feed 3 times daily a moist, sloppy mash made of one-third shorts and two-thirds cornmeal. Some roughage or vegetables are also provided. An increase in weight from 4 to 6 lb can be obtained by this method of feeding.

Stuffing Geese with Noodles.—Another method which produces a much better fattened goose but involves considerably more work is to stuff geese with noodles for 3 to 4 wk. From 8 to 10 weeks, geese are confined to a pen. The feeder holds a goose between his legs and stuffs it with noodles, beginning with 3 to 5 noodles 3 times daily and gradually increasing to 6 or 7 noodles 5 times daily. Noodles are made of scalded corn meal, ground oats, ground barley or wheat flour of about equal parts of each. A small amount of salt is added and the material is put through a sausage stuffer, cutting the product into pieces 3 in. long. They are boiled 10 to 15 min, dipped in cold water and then flour-dusted to keep them from sticking together. Just before they are fed, hot water is poured over them to warm them and make them slippery. The noodles are put into the mouth one at a time and worked down by using the hands on the outside of the neck.

These feeds—starting, growing and breeding—are used on large commercial geese operations (Table 46.3). A young goose can develop normally on a chick-growing ration and on pasture. Many receive only pasture with limited grain. If given too much grain, they will not eat as much pasture.

TABLE 46.3. FEEDING PROGRAMS FOR GEESE

	Starter lb	Grower on Pasture lb	Breeder lb
Ground yellow corn	1014	1105	604
Oats, pulverized	100	100	100
Soybean oil meal, 44%, solvent	600	500	490
Meat and bone scrap, 50%	100	100	300
Fish solubles, condensed[1]	40	20	30
Alfalfa meal, dehydrated	50	—	150
Dried whey[2]	30	20	60
Bone meal, steamed	40	40	60
Limestone (calcium carbonate)	10	—	—
Salt, iodized	10	10	30
Manganese sulfate (65%)	0.50	0.50	1.0
Vitamin A (5000 IU/g)	1.0	1.0	4.0
Vitamin D (1500 ICU/g)	1.0	1.0	4.0
Vitamin mixture "2-4-9"[3]	1.0	1.0	2.0
Vitamin B_{12} (6 mg/lb)	1.0	1.0	2.0
Choline chloride (25% product)	1.0	—	—
Antibiotic supplement[4]	1.0	1.0	3.0
Niacin (grams of pure vitamin)	40 g	—	—
Total, lb	2000.5	2000.5	2000.0
Calculated analysis:			
Crude protein, %	22.3	20.2	27.2
Crude fat, %	3.2	3.3	3.7
Crude fiber, %	4.0	3.0	5.0
Calcium, %	1.49	1.12	2.56
Phosphorus, %	0.88	0.78	1.43
Productive energy, calories per lb	860.0	210.0	725.0

Source: Schaible (1972).

[1]Fish solubles may be replaced with an equal amount of fishmeal.

[2]Dried whey may be replaced by dried skim milk, dried buttermilk, distillers' dried solubles or dried yeast products.

[3]Vitamin mixture "2-4-9" refers to a mixture containing 2 g riboflavin, 4 g pantothenic acid and 9 g niacin per lb. Commercial supplements with this vitamin potency are readily available.

[4]The antibiotic supplement should contain one of the following per lb of supplement: Penicillin, 4 g; Aureomycin, 10 g, Terramycin, 10 g; or bacitracin, 10 g.

Feeding directions:

1. Starting period (0—4 weeks): Use starter mash. Goslings may be placed on pasture, but feed no supplemental grain.
2. Growing period (after 4 weeks): With geese on good pasture, the grower mash should be fed in a free choice manner with whole corn and oats. With poor or no pasture, use starter mash in place of grower, and allow free choice with corn and oats.
3. Breeder geese: Feed breeder supplement in free choice manner with whole corn and oats.

Feed conversion is good and growth is very rapid (Table 46.4).

TABLE 46.4. AVERAGE WEIGHT OF YOUNG GEESE PER WEEK IN POUNDS

	Toulouse		Emden					
Week	Males and Females	Males	Males and Females	Males	Emden × Toulouse	Toulouse × Emden	African × Emden	All Crosses
0	0.25	0.26	0.25	0.25	0.23	0.23	0.24	0.24
1	0.65	0.69	0.49	0.49	0.41	0.43	0.43	0.43
2	1.4	1.4	1.5	1.6	1.4	1.4	1.4	1.4
3	2.9	2.8	2.9	3.0	2.7	2.6	2.6	2.6
4	3.8	4.0	3.7	3.7	3.5	3.7	3.7	3.6
5	5.2	5.5	4.8	4.9	4.7	4.9	4.9	4.8
6	6.5	7.0	6.1	6.3	6.1	6.3	6.4	6.3
7	7.6	8.1	7.2	7.6	7.3	7.2	7.8	7.5
8	8.3	8.9	8.1	8.5	8.1	7.9	8.5	8.2
9	8.4	9.1	8.4	8.7	8.5	7.9	8.8	8.5
10	8.7	9.3	8.6	9.0	9.1	7.8	9.1	8.9
11	—	9.8	—	9.6	—	—	—	—
12	—	10.1	—	10.1	—	—	—	—
13	—	11.3	—	11.2	—	—	—	—
14	—	10.9	—	10.8	—	—	—	—
15	—	11.5	—	11.6	—	—	—	—
16	—	11.5	—	11.9	—	—	—	—

47

Commonly Reported Nutritional Disturbances

Poultry are raised throughout the United States, and each year, many new poultry enterprises are started and conversely, many established businesses are discontinued. Whether as a practical means of making a living, as an investment, an avocation or the fulfillment of a dream these businesses provide opportunities, and each year, a new "crop" of poultry-production students begin training "on the job." Some have schooling, whereas others are learning by experience. Many problems will confront the beginner but also, a wide-open free exchange of knowledge will be available to him.

The free discussion of poultry problems has created some confusion. Problems related to nutrition most often "broadcasted" are reported in this chapter.

EGGS

Soft Shelled Eggs

There are many dietary factors influencing egg-shell quality.

Mineral and Vitamin Deficiency.—If the diet is deficient in calcium or Vitamin D or both, egg shells will be soft. If the deficiency is only in Vitamin D, small calcium crystals will appear on the surface of the egg. The brown pigment in turkey eggs and normally brown-shelled eggs will appear bleached.

Parathyroid Gland.—If the parathyroid gland is not secreting adequate amounts of its hormone, the shell will be thin, or even absent. This is not a nutrition problem.

Number of Eggs in the Clutch.—After a hen has laid approximately 18 eggs in a clutch, she will begin laying eggs with weak shells. Twenty-four to 30 eggs per clutch may have as much as 18% "cracks."

Age of Hen.—Hens which have been kept for the second laying year produce larger eggs, yet the amount of calcium mobilized or processed for egg-shell formation remains the same as for the first year. This means the amount of calcium available for the larger egg is the same as was available for the smaller egg; hence, the shell on the larger egg is thinner.

Pathological.—Egg shells can be thinner and show an abnormal shape with ridges when the hens have the disease, New Castle.

Drugs.—Some drugs used to control coccidiosis will cause abnormal egg shells. The drug, Nicarb, has been reported to influence egg-shell formation.

Small Eggs

The diet has a very limited influence on egg size. Eggs are graded into extra large, large, medium, small and pullet or "peewees." These sizes are clearly stated on the cartons.

When pullets begin to lay, the eggs are very small, i.e., peewees, for the first 3 to 4 wk of production; then most of the eggs are large and medium. Some flocks of "mini" hens lay largely medium eggs, whereas the standard hen lays larger eggs.

Watery Whites

"Watery" whites generally suggest that the eggs were not cooled properly or became overheated before gathering. Old and stored eggs may be "watery."

Generally, the geneticist produces a strain of birds which lays a very high percentage of eggs with firm whites. The watery white contains the same amount of water and protein as the thick white. The difference lies in the colloidal structure of the water protein micelle.

Color of Egg Yolk

Normal egg yolks are colored by a carotenoid pigment called xanthophyll. The rate of lay, amount of xanthophyll in the ration, quality of dietary fat, diseases and genetic makeup of the hen all influence the degree of pigmentation. Xanthophyll is not a required nutrient, and so far, scientists have not been able to show any nutritional value of the substance except for eye-appeal.

When hens are fed excess cottonseed meal, the egg yolks are often colored by the pigment, gossypol. Most cottonseed meals are relatively free of gossypol; however, that produced by the hydraulic method is generally not satisfactory for egg production.

Cholesterol in Eggs

Eggs are among the richest sources of cholesterol. This is one of the most common sterols in animal tissues and occurs in the free form or in combination in ester formation with fatty acids. Cholesterol is present only in animal tissues or animal eggs. It is readily synthesized from acetate in animal cells, the first step being the formation of acetyl CoA and the acetoactyl CoA.

Many compounds essential for life are derived from cholesterol.

(1) Cholic Acid
(2) 7-dehydrocholesterol Vitamin D_3
(3) Cortisone
(4) Progesterone
(5) Corticosterone
(6) Testosterone
(7) Estriol

The use of eggs as a food for man has decreased in recent years because of 2 major factors:

(1) Dietary habit changes
(2) Labeling of high-cholesterol foods, such as eggs.

This labeling has resulted in consumer resistance because of the association of cholesterol with atherosclerosis and heart disease.

Pathological studies have often, but not always, associated heart disease with cholesterol level in the body. Many studies have shown that the serum cholesterol levels respond to changes in dietary intake of fatty acids. Also, the serum level of cholesterol is lowered by ingesting a diet with a high ratio of polyunsaturated to fatty acids. Although circumstantial evidence suggests cholesterol as the causative agent of atherosclerosis and heart trouble, the only positive position one should take is that of possible vascular involvement. The extent to which diet is related to the cause of heart disease is not known nor the dietary factors implicated.

A study of the embryological development of the blood circulatory system shows that it was derived from the mesoderm. In fact, all connective tissues are derived from the mesoderm, and the connective tissue most often thought to be involved with crystal formation is bone. If properly treated, it is possible to get crystals to form in most connective tissues. The crystal formation is not necessarily dependent upon the concentration of a substance capable of forming a crystal but upon some outside force, such as a bruise or perhaps a foreign chemical, to serve as the nucleus for crystallization. This means that crystals form (grow), not by being in a supersaturated environment but in a desirable biocolloidal or biophysical chemical environment conducive to crystal growth.

In summary, as far as is known from available data, eggs remain the most complete food with a cholesterol content that is not detrimental.

DEFICIENCIES AND DISEASES

Dermatitis

Three types of dermatitis are observed: (1) pantothenic acid deficiency, (2) biotin deficiency and (3) niacin deficiency.

These deficiencies are seldom observed under farm conditions. A pantothenic acid deficiency results in dermatitis around the mouth, and nasal openings and between toes. It also causes scaly shanks. Exudates collect around the corners of the mouth.

A biotin deficiency results in deep fissures around the toe joints, foot pads and beak joints.

A niacin-type dermatitis may involve most of the skin on the body and the throat appears red. A slight dermatitis appears on the feet and legs.

A type of scales appears on the legs due to mites, observed only in older birds. Sores can be found on turkeys and chickens reared on open wooded ranges, due to chiggers and ticks.

Early Turkey Poult Mortality

Early turkey mortality is divided into 2 groups: One to 3 days and 6 to 10 days. Death of poults within 72 hr is often due to a breeding-ration deficiency in some vitamin, such as riboflavin. Mortality, starting at 6 days and ending at about 10 days, is generally due to starvation. If a poult has not eaten by the third day, it probably will never eat. When poults do not eat early, the lining of the gizzard sloughs off and no eating stimulus is displayed. The starving poults will cry and become dehydrated. They generally die before the tenth day of life.

There are several ways to prevent starvation in young poults:

(1) Dip beaks in water, then in feed, when placing day-old poults under brooders.
(2) Purchase poults close by so they can be given feed and water within 36 hr of hatching.
(3) Place some newly-hatched White Leghorn cockerels with poults for 3 to 5 days. One or 2 cockerels per hundred poults will suffice.
(4) Put bright-colored marbles, or dyed feed particles, or shine a spotlight on the feed.
(5) Overfill the feed troughs the first 2 days and keep the poults comfortable.

Note—Once starvation occurs and the gizzard lining sloughs off, the birds cannot be saved.

Bumblefoot—Infected Footpad

This condition is often observed in White Leghorn laying hens. The footpad becomes swollen and deep cracks are observed with a hard core in the center. Microbiological cultures will show signs of the infection. So far, this is not prevented by diet and is seldom observed except in light breeds. The feeding of high levels of Vitamin A has been used without success.

Dark Brown-colored Bones in "Fresh" Dressed Broilers

The appearance of dark brown-colored bones in "fresh" broilers displayed in meat sales units is frequently observed. These dressed broilers have probably been frozen during holding or shipping. The nutritional value has not been changed, only appearance. This discoloration is due to the rupture of the immature red blood cells brought about by freezing.

Excess Visceral Fat

Excessive visceral fat is caused by feeding very high energy rations. When high levels of fat are included in the ration, the chicken's body may not metabolize the fatty acids to produce fat characteristic of the species. Sometimes, this fat results in a type of "greasy fat" which filters through the skin, causing a greasy effect. The best control of excess fat is by keeping the free fat low and including some oats or barley as a replacement of 5 to 10% of the ground yellow corn.

Perosis or Hock Disease

Perosis or hock disease is generally a nutritional problem. It may be the result of a breeding-ration or a starting-ration deficiency in any one or more of the recognized antiperosis factors. The antiperosis factors are choline, biotin, manganese and zinc. Also calcium, magnesium and iron phosphates can reduce the availability of manganese. Calcium and phytic acid reduce the availability of zinc. Perosis must be prevented — it cannot be corrected. Birds with perosis should be destroyed.

Abnormality of Embryos

A choline and manganese deficiency in breeding rations will cause bone deformities in developing embryos.

If the breeding ration contains excess selenium, the embryo will be abnormal — an eye, beak or appendage may be missing. Vitamin B_{12} and Vitamin A deficiencies have been reported to result in bone deformities.

Soft Bones or Rickets

Rickets can be caused by a deficiency of Vitamin D, calcium and phosphorus. Also, mold toxins have been found to cause soft bones.

The general symptoms found in ricketic chicks are low bone ash, soft beaks, crooked breasts, and "beads" on ribs. Chickens will remain in a sitting position or walk as if in pain.

Hemorrhages Under Skin (Blue Spots)

Both live and dressed birds often show dark blue areas under the skin. These are most frequently observed in the breast and hock-joint areas. This condition can be caused by a Vitamin K deficiency, bruising, mold toxins or pathological conditions. Birds showing hemorrhages should not be used for food.

FEATHERS

Neck Molting

Bird often show neck molting. The feather tract holding the feathers around the neck seems to release its feathers. The causes are common "colds," parasites, need for water and feed and end of egg production.

Feathers on most birds are arranged in definite tracts. Some of these tracts are paired. There are recognizable differences between the sexes in the appearance of feathers on the neck, back, saddle and tail tracts. Each tract or pair of tracts sheds its feathers separately from other tracts.

The feather is a protein structure which is high in sulfur amino acids. Therefore, when poultry are growing feathers, the diet should be high in the sulfur amino acids. The major mineral present in feathers is silicon; however, birds which perform long flights have apatite crystals in their primary and tail feathers.

Feather Pigmentation

Amino acids, such as lysine and methionine, and some trace minerals, influence pigment formation in feathers.

Pin Feathers

When feathers first appear under the skin, they are called pin feathers. The geneticist selects birds for rapid feathering. A desirable meat bird is one which has its feathers developed when the bird is fleshed enough for market. Good nutrition helps feather growth but rapid feathering is a quality controlled by genetic makeup.

POULTRY HOUSE MAINTENANCE

Wet Litter

Wet litter can be caused by: (1) poor ventilation, (2) leaking water equipment, (3) improper soil drainage, (4) leaking roof, (5) genetic makeup of the bird, (6) excess salt or magnesium in the ration and (7) high levels of molasses or milk in the ration.

Maintaining both a dry, suitable litter and moisture in the poultry house has always been the No. 1 problem in poultry management.

Dust in the Poultry House

Dust is often observed in dry poultry houses. In very dry houses, the litter contributes to the dusty condition. However, most of the dust samples collected and analyzed show that dust is high in protein, being derived from feathers and skin scales of the chicken.

48

Measurements

Two systems of measurements are used in nutrition: (1) The English or customary system and (2) the metric system. The English system was developed in England around the thirteenth century, and the metric system was created by a group of French scientists in the 1790s. The official name of the latter is International System of Units. The term, metric, comes from the basic unit of length in the system, the meter.

The scientists who created the metric system designed it to fit their needs for units in measurements of length, temperature, weight, volume and time. This system was developed to be logical and exact, yet flexible enough for use by the layperson.

The metric system is simple to use for 3 reasons:

(1) Metric units increase or decrease in size by tens.

Prefix	Abbreviation	Increase or Decrease in Units
mega	m	1,000,000 (one million)
kilo	k	1,000 (one thousand)
hecto	h	100 (one hundred)
deka	da	10 (ten)
deci	d	0.1 (one-tenth)
centi	c	0.01 (one-hundredth)
milli	m	0.001 (one-thousandth)
micro	m	0.000001 (one-millionth)

(2) The metric system has only 7 basic units that make up all the measurements, whereas the English system has 20 basic units for the common measurements.

(3) Conversion. Sometimes, it is desirable or necessary to convert from weight to capacity in liquids. This can be accomplished by deter-

mining the weight and specific gravity at 16°C. At this temperature, 1 cubic centimeter (cc) of water weighs 1 gram (g).

METRIC MEASURING UNITS

(a) The Meter (Linear Measurement)

Linear Measure

10 millimeters	=	1 centimeter
10 centimeters	=	1 decimeter
10 decimeters	=	1 meter
10 meters	=	1 decameter
10 decameters	=	1 hectometer
10 hectometers	=	1 kilometer
10 kilometers	=	1 myriameter

Conversion:

When You Know Length and Distance	Multiply By	To Find
inches (in.)	25	millimeters
feet (ft)	30	centimeters
yards (yd)	0.9	meters
miles (mi)	1.6	kilometers
millimeters (mm)	0.04	inches
centimeters (cm)	0.4	inches
meters (m)	1.1	yards
kilometers (km)	0.6	miles

(b) The Kilogram (Mass or Weights)

Weights

10 milligrams	=	1 centigram
10 centigrams	=	1 decigram
10 decigrams	=	1 gram
10 grams	=	1 decagram

10 decagrams	=	1 hectogram
10 hectograms	=	1 kilogram or kilo
10 kilograms	=	1 myriagram
10 myriagrams	=	1 quintal
10 quintals or 1000 kilos	=	1 metric ton

Conversion:

When You Know Weight and Mass	Multiply By	To Find
ounces (oz)	28	grams
pounds (lb)	0.45	kilograms
short tons	0.9	metric tons
grams (g)	0.035	ounces
kilograms (kg)	2.2	pounds
metric tons (t)	1.1	short tons

(c) Time (Second)

60 seconds	=	1 minute
60 minutes	=	1 hour
24 hours	=	1 day
365 days	=	1 year
10 years	=	1 decade
10 decades	=	1 century
10 centuries	=	1 millennium

(d) Temperature (Celsius or Centigrade Scale)

Celsius = 0°C water freezes

Celsius = 100°C water boils

Normal Body Temperature = 37°C

To Convert °C to °F

$(°C \times 9/5) + 32 = °F$

To Convert °F to °C

(°F − 32) 5/9 = °C

(e) Capacity (Liter)

Measure of Capacity

10 milliliters	=	1 centiliter
10 centiliters	=	1 deciliter
10 deciliters	=	1 liter
10 liters	=	1 decaliter
10 decaliters	=	1 hectoliter
10 hectoliters	=	1 kiloliter

When You Know Volume and Capacity (Liquid)	Multiply By	To Find
fluid ounces (fl oz)	30	milliliters
pints (pt), U.S.	0.47	liters
pints (pt), imperial	0.568	liters
quarts (qt), U.S.	0.95	liters
quarts (qt), imperial	1.137	liters
gallons (gal.), U.S.	3.8	liters
gallons (gal.), imperial	4.546	liters
milliliters (ml)	0.034	fluid ounces
liters	2.1	pints, U.S.
liters	1.76	pints, imperial
liters	1.06	quarts, U.S.
liters	0.88	quarts, imperial
liters	0.26	gallons, U.S.
liters	0.22	gallons, imperial

(f) Area (Square Meter)

Square Measure (Area)

100 square millimeters	=	1 square centimeter
100 square centimeters	=	1 square decimeter
100 square decimeters	=	1 square meter

100 square meters	=	1 square decameter (are)
100 square decameters	=	1 square hectometer
100 square hectometers	=	1 square kilometer
10,000 square meters	=	1 hectare

When You Know Surface or Area	Multiply By	To Find
square inches (sq in.)	6.5	square centimeters
square feet (sq ft)	0.09	square meters
square yards (sq yd)	0.8	square meters
square miles (sq mi)	2.6	square kilometers
acres	0.4	hectares
square centimeters (cm^2)	0.16	square inches
square meters (m^2)	1.2	square yards
square kilometers (km^2)	0.4	square miles
hectares (ha)	2.5	acres

(g) Volume (Cubic Measurements)

Cubic Measure (Volume)

1000 cubic millimeters	=	1 cubic centimeter
1000 cubic centimeters	=	1 cubic decimeter
1000 cubic decimeters	=	1 cubic meter (stere)

Frequently Used
Conversion Table for Weights and Measures

1 acre	=	0.4047 hectares
1 foot	=	0.3048 meters
1 foot, cubic	=	0.0283 cubic meters
1 foot, square	=	0.0929 square meters
1 inch	=	2.54 centimeters
1 inch	=	0.0254 meters
1 mile	=	1.6093 kilometers
1 ounce	=	28.3495 grams (avdp.)
1 pint, dry	=	0.5506 liters

1 pint, liquid	=	0.4732 liters
1 pound (avdp.)	=	0.4536 kilograms
1 rod	=	5.0292 meters
1 yard	=	0.9144 meters
1 yard, cubic	=	0.7646 cubic meters
1 yard, square	=	0.8361 square meters
1 centimeter	=	0.3937 inches
1 gram	=	0.0353 ounces (avdp.)
1 hectare	=	2.4710 acres
1 kilogram	=	2.2046 pounds (avdp.)
1 kilometer	=	0.6214 miles
1 liter	=	1.8162 dry pints
1 liter	=	2.1134 liquid pints
1 meter	=	3.2808 feet
1 meter	=	39.37 inches
1 meter	=	0.1988 rods
1 meter	=	1.0936 yards
1 meter, cubic	=	35.3144 cubic feet
1 meter, cubic	=	1.3079 cubic yards
1 meter, square	=	10.7639 square feet
1 meter, square	=	1.1960 square yards

Vitamin and Trace Mineral Conversion

One of the most common conversions is parts per million (ppm) to grams per ton.

1 English short ton	=	2000 pounds
1 pound	=	453 grams
1 ton = 2000 × 453 gram	=	906,000 grams

Therefore, for practical use, when vitamins and trace elements are used at 1 ppm is equivalent to one gram per ton (exactly 0.906 grams).

Feed ingredient standard units of weights and measures used in the United States are shown in Table 48.1.

TABLE 48.1. WEIGHTS AND MEASURES OF FEEDS IN UNITED STATES[1]

Commodity	Unit	Approximate U.S. lb	Net Weight Metric kg
Barley	bu	48	21.8
Corn			
ear husked	bu	70	31.8
shelled	bu	56	25.4
meal	bu	50	22.7
Cottonseed	bu	32	14.5
Millet seed	bu	48 to 50	21.8 to 22.7
Molasses	gal.	11.70	5.3
Oats	bu	32	14.5
Peanuts (unshelled)	bu	17 to 25	7.7 to 11.3
Peas (dry)	bu	60	27.2
Rice (rough)	bu	45	20.4
Sorghum	bu	56	25.4
Soybeans	bu	60	27.2
Wheat	bu	60	27.2

Source: Agric. Statistics (1977).
[1]To convert from Avoirdupois pounds to kilograms, multiply by 0.4536; to metric tons, multiply by 0.0004536.

LABELING REQUIREMENTS

Government at all levels has become a dominating factor in the feed business. Originally, the states provided a measure of regulation and control. In the past decade, the United States Food and Drug Administration has overtaken the states in regulation, and recently, local governments have been taking direct action in the field of pollution. The Food and Drug Administration concerns itself principally with feed additive combinations, levels and sanitation in handling food and feedstuffs. It is desirable to have available appropriate subsections of Title 21 Code of Federal Regulation and the Feed Additive Compendium. No mill is too small to be exempt from Food and Drug Administration scrutiny. Common sense application of these regulations and the use of reasonable care and intelligence in the storage and handling of feed additives will create a more responsible industry.

Feed laws and others concerned with poisons, remedies, fertilizers and liming materials are enforced by the states. The law is basically a labeling law. There is no provision for questioning the efficacy of the feed. The state law is predicated on the theory that the feeder has freedom of choice, and so long as he is provided with the necessary information on the label, it is up to him to read the label and make the choice in purchasing it. Although they are licenses, basically they are label laws. There is no question of value of the product. The feed law requires basic information on the label, stating the percentages of protein, fat and fiber, as well as minerals. In the case of salt, there is a minimum and maximum guarantee. Vitamins are guaranteed when reference is made on the label. If a statement is made that this product is enriched with vitamins, a vitamin guarantee is necessary. Medication guarantees are also required. Control officials want only necessary information on the label. It is up to educators to help consumers make wise decisions.

STATE LICENSING LAWS

Licensing is strictly the privilege of doing business and can be revoked. When an application for a license is made, the label is reviewed to make sure it contains the necessary information. Then, the license is issued. Officials prefer conciseness, and in most cases, labeling is done to their satisfaction. From then on, they are involved in quality control.

Minerals, vitamins and medications are checked to be sure that the products on the market contain what they claim. Medications are in the feed for therapeutic or preventive purposes or for growth promotion. If the efficacy of medications is questioned, reference is made to the Federal Food and Drug Administration levels of feed additives. The Feed Additive Compendium is used in reviewing labels to ensure that the levels of medication said to be in the feed have been recognized for the stated value and purposes—prevention of coccidiosis, growth promotion, feed efficiency. Another publication used by the feed control officials is the official Feed Control Officials publication which contains definitions of standard ingredients used in the manufacture of feeds. Officials are bound by law to use the terminology of this association for each specific feed ingredient.

Here is one example. There are several different kinds of alfalfa. If the label uses the term "sun-cured" alfalfa, this means that the aerial portion of the alfalfa plant, which is reasonably free of other crop plants or weeds and molds, has been sun-cured and finely-ground. If it is chopped instead of finely ground, it must be designated as "chopped alfalfa" or "chopped alfalfa hay." On some classes of alfalfa, there are protein and fiber limitations. Alfalfa leaf meal is defined as leaves that are separated from the alfalfa plant, dried and finely ground. It must contain not less than 20% crude protein and not more than 18% crude fiber.

Each and every ingredient in the feed must be on the label and named in descending order of amount in the feed. Percentages are no longer required. Ingredient terms are universal throughout the United States.

The feed production committee of the American Feed Manufacturers Association is promoting uniform feed laws for all states. Some states, to prevent violations, use the same approach as the Federal Food and Drug Administration. Other states use "random sampling." They sample 5000 to 8000 samples of feed and base their enforcement upon the analytical results of the samples they have drawn. This is quite expensive. Another means is "selective sampling"; samples are taken only when a feed is believed to be in violation of the law. Some do this by a system of mill inspections and examination of feedstuffs, mixing procedures and labels, without drawing samples.

Suppose a label called for a coccidiostat in the feed but it had none? The feed manufacturer should be questioned as to how many chickens are being fed and how many farms are using his feed. Then he should be asked who would pay for the losses if a flock were infected with coccidiosis and the feeders discovered the omission of a coccidiostat in his feed. His reaction would be to correct the situation at once. This system of enforcement makes most feedmen readily alert to labeling. The law provides that control officials have the authority to inspect records.

Product names cannot be misleading. A groing mash cannot be termed a laying mash or vice versa. Misbranding is considered a violation. Suppose Vitamin D_2 was put in a feed for chickens instead of D_3? The source of vitamins, either animal sterol or plant sterol, must be declared.

Control laboratories analyze feed samples, and the results are used to enforce the law. The law states that if a state standard does not exist, any prevailing feed standard, such as a federal standard, can be used for enforcing the law.

UNITED STATES FOOD AND DRUG REGULATIONS

In 1895, Connecticut passed a law containing provisions for regulation of food sales for man and animals. Massachusetts passed the first feed law in 1896. By 1920, 38 states had some form of regulation in effect. Today, all states, except Nevada, have laws regulating the sale of feeds. Justification for such laws is the inability of the buyer to judge the quality of ground ingredients, by products or mixtures by visual inspection. The buyer needs assistance in determining value, and protection from those who sell material of poor or varying quality. The fact that laws were passed to control the sale of ingredients and formula feeds implies that everything was not always as it should be. State and Federal Government publications continue to advise the farmer to beware of feeds containing low-grade ingredients.

The United States Food and Drug Act was passed in 1906, to control feeds shipped in interstate commerce. In 1938, the Food, Drug and Cosmetic Act added certain requirements for labeling of feeds. An additive amendment restricted the use of estrogenic compounds and arsenicals that were not proven safe. Considerable confusion arose regarding compliance procedures. For the first time, feed manufacture was brought under Federal Government scrutiny. Inspection of feed plants and methods for storing and handling additives were suggested to feed manufacturers. Better housekeeping and production methods had to be followed in order to comply. Design, engineering and layout of new feed mills have also been influenced by passage of the Delaney Amendment.

There is a word of warning from the Food and Drug Administration on feeds prescribed by veterinarians. Where new drug substances are involved, they cannot be used legally, even on the prescription of a veterinarian, unless the prescription is the subject of a new drug application. Nor can old drugs be mixed, unless they comply. Feedmen should be cautious about mixing anything for a farmer unless the farmer knows what it is; otherwise, the feedmen are increasing their liability.

THE ASSOCIATION OF AMERICAN FEED CONTROL OFFICIALS

The Association of American Feed Control Officials (AFCO) was organized in 1938. It adopted uniform standards of feed analysis. In 1937, AFCO collaborated with AFMA in preparing a model feed law. Uniform laws throughout all states would make the manufacture and marketing of feed much easier for firms selling in two or more states. Through the joint efforts of industry and state control officials, farmers have been provided with product information. Uniform feed laws and tonnage reporting systems would move the feed industry forward on the road to progress.

Is a poultryman who mixes his own feed a "feed manufacturer"? More than half of the feed in the United States is manufactured by someone other than a feed manufacturer. Should the farmer abide by control laws written to protect the farmer as a buyer of feed? The so-called feed industry is basically a part of the framework of agriculture and food economics. Agriculture has already changed tremendously in recent years. Most of the predictions are already realities. Farming as a mode of living has passed. Poultry production is a matter of converting crops into human foods; it is food production. Poultry feed is not formulated and manufactured for consumers but for producers of meat and eggs for consumers. The feed industry is a part of agriculture and will shift with future agricultural changes. Business, government and consumers are teammates in the production of food.

Glossary

Part 1—Glossary

Absorption Absorption pertains to the uptake of nutrients, water, or other substances by stomach or intestinal walls following digestion of food. For example, glucose, a simple sugar, is absorbed without change but starches must be broken down into sugars before absorption can take place. In food processing, absorption may also refer to uptake of other substances by foods, such as absorption of fats by foods during cooking in deep fat, or absorption of water by cereals during cooking.

Adipose Adipose is commonly used in describing the part of the body where fat is stored, which is adipose tissue.

Adrenal The adrenal glands are ductless glands near the upper end of the kidneys. Their secretions are essential for the maintenance of life.

Adsorption Adsorption pertains to the binding or attraction of other molecules upon surfaces of particulates which possess unneutralized fields of force or free valences. Each atom at the surface has free valences or attractive forces which are capable of attracting one and only one molecule from a solution or gas. The greater the surface of the adsorbing agent the greater is the adsorption.

Amino acid Organic compound of carbon, hydrogen, oxygen, and nitrogen. Each amino acid molecule contains one or more amino groups (-NH2) and at least one carboxyl group (-COOH). In addition, some amino acids (cystine and methionine) contain sulfur. Many amino acids linked together in some definite pattern form a molecule of protein.

Antibiotics Chemical substances, "against life," which are produced by certain living cells, such as bacteria, yeasts, and molds that are antagonistic or damaging to other living cells, such as disease-producing bacteria. Antibiotics may kill living cells or prevent them from growing and multiplying. Penicillin is an example of an antibiotic that inhibits or destroys certain bacteria that cause disease in a man.

Antibody Specific substances produced in the body to react against disease-producing or other foreign materials in the blood stream. Some antibodies remain in the blood stream for many years and help to give a person immunity to certain diseases.

Antioxidant A substance capable of chemically protecting other substances against oxidation.

Ascorbic acid The chemical name for Vitamin C.

Available A nutrient is available to the body when it is in the form that can be absorbed from the digestive tract and then used for its intended function in the body.

Avidin A protein material that can combine with biotin, causing the vitamin to be unavailable to the body. Cooking renders avidin inactive.

Bacteria Very small, one-celled organisms visible only under a microscope and widely distributed in the air, water, soil, and animal and plant tissues. They have some useful functions, such as in decaying dead matter and in fermentation of fruit or vegetable juices—as in the making of sauerkraut. Many bacteria produce disease or cause harmful spoilage of foods.

Basal metabolism The energy produced by an individual during physical, digestive, and emotional rest, measured directly by the heat evolved and indirectly by the oxygen consumed and carbon dioxide given off.

Bile A thick green or yellow fluid formed in the liver, collected in the gall bladder, and emptied into the intestinal tract at intervals, particularly during the digestion of fats. It is a complex mixture containing salts of bile acids, which aid in digestion of fats, and cholesterol and other substances from different body sources. It carries cholesterol from the liver into the intestine for excretion or for reuse in digestion as needed.

Biochemistry The chemistry of living things, plant, and animal.

Biological Pertaining to the science of life.

Biological value The biological value of a food protein is the efficiency with which that protein furnishes the proper proportions and amounts of the amino acids needed, at the time of synthesis of body proteins, by man or animals eating these foods. The more nearly a protein supplies the tissues with the necessary proportions and amounts of these amino acids, the higher is its biological value.

Biosynthesis The coming together of chemical building units to form new materials in the living plant or animal.

Biotin One of the ten recognized vitamins of the B complex. It is widely distributed in foods. It is needed in the diet of certain laboratory animals, such as the chick, rat, and mouse. A deficiency in man has never been observed under usual conditions.

Bran The outer coarse coat of grains.

Calcification Process by which organic tissue becomes hardened by a deposit of calcium salts.

Caloric Pertaining to heat or energy; used in reference to the caloric value of a food, it means the heat or energy that can be obtained as muscular work and heat when the body uses or metabolizes that food.

Calorie The unit by which heat is measured. It is defined in terms of the amount of heat required to raise the temperature of a specified amount of water 1°C. A small calorie (written with a small c) is the amount of heat required to raise one gram of water from 14.5°C to 15.5°C. A large Calorie (capital C) is the amount of heat required to raise 1,000 gm (1 kg) of water from 14.5° to 15.5°C. The large Calorie is used exclusively in expressing the caloric value of foods and the caloric needs of humans and animals. (Calorie and caloric, as adjectives, usually are not capitalized.)

Calorimeter An instrument for measuring the heat exchange and the energy in any system.

Carbohydrate An important group of organic substances that contain carbon, hydrogen, and oxygen. The hydrogen and oxygen are present in the same proportion as in water (H_2O), and there is one molecule of water for every carbon. Starch and sugar are carbohydrates.

Carcass The part of an animal's body that is used for meat. After an animal is slaughtered and the entrails, head, feet, and lower parts of legs and hide are removed, the carcass remains. It is hung and chilled and later is cut up into steaks, roasts, etc.

Carotene A yellow compound of carbon and hydrogen that occurs in plants and is a form of vitamin A. Alpha, beta, and gamma carotenes may be converted into vitamin A in the body.

Cartilage A special form of white connective tissue that is attached to the ends of bones that are either divided into joints or united by joints. It is more flexible but not so strong as bone. Cartilage is the first substance to form in growing bone; then calcium and phosphorus are deposited in the cartilage, thus changing it to bone.

Catalyst A substance that speeds up the rate of a chemical reaction but is not itself used up in the reaction.

Centigrade (C) A thermometer scale in which water freezes at 0°C and boils at 100°C. To change to degrees Fahrenheit, multiply degrees centigrade by 9/5 and add 32. Also CELSIUS (C.), the preferred form approved by the Ninth General Conference on Weights and Measures in 1948.

Chemical additives Substances added to foods to improve their flavor, color, texture, or keeping quality.

Chlorophyll The green coloring matter present in growing plants, which under stimulus of light is active in the manufacture of carbohydrates from carbon dioxide and water.

Cholesterol The most common member of the group of sterols; it is present in many foods and also can be made within the body.

Coagulation The change from a fluid state to a thickened jelly, curd, or clot.

Collagen A protein that forms the chief constituent of the connective tissue, cartilage, tendon, bone, and skin. Collagen is changed to gelatin by the action of water and heat.

Conduction The transmission of heat or electricity through an object, or from one object to another in direct contact with it, without the motion of the conducting bodies. Conduction is different from convection, which is the transmission of heated particles, and different from radiation, which is the transmission of heat effected through the atmosphere.

Decalcification The withdrawal of calcium from the bones where it has been deposited. It may be caused by an inadequate supply of calcium in the diet so that calcium has to be taken from the bones to help meet the body's needs. It may be caused also by an imbalance in some of the hormone activity in the body.

Deficiency disease A disease resulting from an inadequate dietary intake of something required nutritionally; most commonly refers to diseases resulting from dietary deficiencies of vitamins or trace elements.

Dehydrated foods Products from which most of the water has been removed in order to improve their stability during storage.

Dietetics The application of the science of nutrition to the feeding of individuals and groups of people.

Digestion In physiology, the breaking down of foods into simpler components in the alimentary or digestive tract. Foods may be digested by natural body enzymes in the stomach and intestines, or be broken down similarly by the chemist using chemicals and prepared enzymes, heat or microorganisms. Proteins are digested to peptides and amino acids, fats to fatty acids and glycerol, carbohydrates to dextrins and sugars. Buttermilk is an example of a food partially digested by natural enzymes and microorganisms in the milk.

Edema Swelling of a part or the entire body due to the presence of an excess of water. Edema is most noticeable at the end of the day around the ankles, which increase in size. When the condition becomes severe, an impression remains for a few minutes where a finger was pressed against the skin.

Emulsify To make into an emulsion. When small drops of one liquid are finely dispersed (distributed) in another liquid, an emulsion is formed. The drops may be held in suspension by an emulsifying agent, which surrounds each drop and makes a coating around it.

Endocrine Secreting internally, or into the blood stream, as endocrine glands, or glands of internal secretion.

Endosperm The starchy portion within the kernel of wheat, corn, or other cereal, from which refined flour is reduced after the germ and fibrous outer layers are removed.

Energy Capacity to perform work.

Enzymatic Related to that class of complex organic substances called enzymes, such as amylase and pepsin, that accelerate (catalyze) specific chemical reactions in plants and animals, as in digestion of food.

Enzyme One of a class of substances formed in living cells. It speeds up chemical reactions but is not changed during the process.

Epidemic A disease is epidemic when many people in a region are attacked at the same time or when the disease is spreading rapidly.

Epithelial Refers to those cells that form the outer layer of the skin, those that line all the portions of the body that have contact with the external air (such as the eyes, ears, nose, throat, and lungs), and those that are specialized for secretion as the liver, kidneys, and urinary and reproductive tracts.

Evisceration Of poultry and fish, refers to removal of the intestinal tract and other organs from the carcasses.

Excretion The product of digestion and metabolism that are discharged from the body—feces from the intestinal tract and urine from the kidneys.

Factor In nutrition, any chemical substance found in foods. A factor might be a vitamin, mineral, or any other nutrient or nonnutrient. Usually it has some effect on growth or reproduction of animals. A factor may be "identified" or remain "unidentified." In arithmetic, factor is a value or ratio expressing the relationship between two items, such as liters of oxygen and the equivalent amount of oxygen in grams.

Fahrenheit (F) A thermometer scale that marks the freezing point of water at 32°F and the boiling point at 212°F.

Fat A glyceryl ester of fatty acids. Fats generally are substances of plant and animal origin. Fat may be in solid form, as butter, margarine, lard, or other shortening, or in liquid form, as the vegetable oils.

Fat-soluble Refers generally to substances that cannot be dissolved in water but can be dissolved in fats and oils, or in fat solvents. The fat-soluble vitamins are Vitamins A, D, E, and K.

Fatty acid Organic compound of carbon, hydrogen and oxygen with carboxyl group which combines with glycerol to make a fat.

Fermentation The chemical changes brought about by the activity of enzyme systems of microorganisms. For example, yeast contains enzymes that produce carbon dioxide and alcohol from sugar. In bread-making, it is carbon dioxide that causes the dough to rise. The souring of milk is lactic fermentation, some of the milk sugar being converted by the action of lactic acid bacteria into lactic acid. Lactic fermentation also is utilized in making pickles.

Fortify To add one or more nutrients to a food so that it contains more of the nutrients than were originally present before processing. Milk often is fortified with Vitamin D.

Fractionation A term used by the chemist when he takes natural materials apart in the laboratory by physical means. He does this for various reasons—usually to isolate or purify some specific compound present in feeds or foods.

Gastric Pertaining to the stomach.

Gene The particle in the cell that carries hereditary characteristics. Genes control the development of the body and its specific functions.

Genetic Pertaining to heredity.

Germ The part of a cereal seed that grows and produces new plants.

Gestation Pregnancy.

Glandular Adjective of gland. A gland is an organ that makes and discharges a chemical substance that is used elsewhere in the body or eliminated.

Gluten An elastic substance that gives adhesiveness to dough. It is formed when the proteins in flour, especially those in wheat flour, absorb water. Gluten assists in giving shape to the cooked product as it coagulates when heated.

Gonadotropins Hormones from embryonic sex glands.

Growth-regulating substances Chemicals that in extremely small amounts will affect the rate or type of growth of cells, tissues, and organs and sometimes referred to as hormone-like substances.

Hard (water) Water containing soluble salts of calcium and magnesium and sometimes iron. Hardness caused by bicarbonate salts of these metals is known as temporary hardness, because boiling expels the carbon dioxide and converts the bicarbonate to the insoluble carbonate, forming incrustation on the walls and bottom of the container. Hardness from chlorides and sulfates of calcium and magnesium is not affected by boiling.

Heat-labile Changeable by heat; unstable to heat.

Hemicellulose A complex carbohydrate that occurs widely in plants as a structural part of their cell walls. It differs chemically from cellulose by being subject to hydrolysis with dilute mineral acids.

Hemoglobin A protein in the blood that contains iron and carries oxygen from the lungs to the tissues.

Hemolytic Refers to the destruction or breakdown of the red blood cells.

Hemorrhage Loss of blood.

Homogenized Breaking up a liquid of two or more components into small droplets of the same size. Homogenized milk has been treated mechanically to break the fat into such small globules that it will not rise to the top as cream.

Hormone A chemical substance that is produced in an organ called an endocrine gland and is transported by the blood or other body fluids to other cells. A hormone greatly influences the functions of some specific organ and of the body as a whole. Thyroxin is a hormone secreted by the thyroid gland, and insulin is a hormone secreted by the pancreas.

Hyper A prefix meaning above, beyond, in excess of the normal or average.

Hypertension An abnormally high blood pressure.

Hyperthyroidism Overactivity of the thyroid gland so that it secretes abnormal amounts of the hormone thyroxin.

Hypertrophied Increased in size of an organ, independent of natural growth.

Hypervitaminosis The undesirable effects produced by taking an excess of a vitamin concentrate or pure vitamin.

Hypo A prefix denoting a deficiency or lack or less than the normal or desirable amount.

Inactivate To suspend or terminate certain biological activities, such as by heat, irradiation, or other forms of energy.

Inedible Not fit for food, such as poisonous nuts and plants. Tough skins, seeds, and decayed spots of fruits and vegetable, and bones of meat are considered inedible parts because they are not suitable for human consumption.

Inert Relatively inactive.

Ingest To eat or take in through the mouth.

Inorganic Describes chemical compounds that do not contain carbon.

Intake Substances or amounts of substances which are taken in by the body, e.g., the intake of food.

Intestinal juices The digestive juices secreted by the intestinal walls (in contrast to gastric juices secreted by the stomach walls) and pancreatic juices secreted by the pancreas. The intestinal juices contain enzymes which complete the final stages of digestion of protein, fat and carbohydrate.

Intestinal tract The entire intestines, both small and large.

Irradiation Treatment with ultraviolet rays from sunlight or an artificial source; treatment with X-rays or other radioactive agent.

IU The abbreviation for International Units, the measure of the potency of a vitamin.

Labile Easily destroyed.

Lactic acid A compound formed in the chemical metabolic processes which accompany muscular activity; also a substance formed by the fermentation of lactose, the sugar in milk.

Lactose A sugar that occurs in milk. It is a white, crystalline sugar that is less soluble and less sweet than ordinary cane sugar (sucrose).

Linoleic acid One of the digestion products from certain fats, which is essential to body tissues. See *polyunsaturated fatty acids.*

Lipids A broad term for fats and fat-like substances; characterized by the presence of one or more fatty acids. Lipids include fats, cholesterol, lecithins, phospholipids, and similar substances, which do not mix readily with water.

Lymph A yellowish liquid that contains corpuscles and is present in the lymphatic vessels or channels of the body.

Malformation A deformity; an abnormal development or formation of a part of the body.

Matrix The intercellular framework of a tissue

Maturation The process of coming to full development, maturity, or adulthood.

Meat The flesh of animals used as food. Meat contains proteins, fats, minerals, vitamins, and water. The most common meats used as food are beef, veal, lamb, and pork. Poultry and fish are sometimes included with the foods classified as meat.

Metabolic Refers to metabolism.

Metabolism The sum of the chemical changes that go on in the body as food is made into body tissues, energy is produced, and body tissue is broken down. There are two parts to body metabolism: anabolism—is constructive and includes building, maintaining, and repairing tissue; catabolism—is destructive and includes changing or breaking down tissue or other materials in the body into simple substances for producing energy and for excretion.

Microbiological Pertaining to microorganisms—that is, microscopic plants or animals. Refers usually to a method by which certain microorganisms are used to determine the amounts of a particular nutrient, like a vitamin or an amino acid, in a food. Such assays are possible because these microorganisms must have these vitamins and amino acids in order to grow. These determinations are called microbiological assays or analyses.

Microorganisms Very small living beings. Bacteria, yeasts, and molds are microorganisms found in foods.

Miscible Capable of being easily mixed. For example, corn syrups are miscible in water; fats cannot be mixed with water but are partly miscible in alcohol and are completely miscible in ether. If the end product is a solution, the products are soluble.

Moisture content The amount of water in a substance.

Molecule A chemical combination of two or more atoms that form a specific substance. For example, the combination of an atom of sodium and an atom of chlorine makes a molecule of sodium chloride, or table salt. This is a comparatively simple molecule. There are also large, complex molecules, such as hemoglobin. Proteins and starches are examples of even larger and very complex molecules containing many atoms.

Morbidity Has the same meaning as sickness. The term is used to indicate the extent of illness in a population as contrasted to mortality or deaths.

Mucosa The mucous membrane that lines the passages and cavities of the body, as in the gastrointestinal tract.

Musculature The muscular apparatus of the body or any part of it.

Neural Refers to nerves or nervous tissue.

Nitrogen A chemical element essential to life. Plants can use some nitrogen compounds direct from the soil and nitrogen-fixing bacteria can use nitrogen directly from the air, but animals must have their nitrogen supplied by foods containing protein.

Nitrogenous A substance containing nitrogen is referred to as nitrogenous. Proteins contain nitrogen, as do the chemical components of proteins—amino acids. Protein-decomposition products containing nitrogen are called nitrogenous extractives. They are found in well-ripened meat and contribute to the flavor of meat.

NRC An abbreviation for National Research Council; usually used when referring to the recommended dietary allowances.

Nutrient A chemical compound with specific functions in the nourishment of the body, such as typtophan, an amino acid; thiamin, a vitamin; or calcium, a mineral. The body obtains about 50 different nutrients from food.

Nutritionist A professionally trained person who applies the science of nutrition and related subjects in research, teaching, or advisory services.

Obese Excessive overweight due to the presence of a surplus of body fat.

Organic Describes a large group of chemical compounds that contain carbon.

Ossification The process of forming bone. Cartilage is made into bone by the process of ossification. The minerals, calcium and phosphorus are deposited in the cartilage, changing it into bone.

Ossify To change into bone or bony substance.

Oxidation The removal of electrons, in the most general sense; may also mean the combining with oxygen or the removal of hydrogen.

Oxidative Refers to the processes of oxidation.

Oxygen One of the most plentiful chemical elements. Oxygen makes up 65% of the human body. When other chemical substances combine with oxygen, energy is released.

Palatability The quality characteristics (such as color, flavor, and texture) of a food product that make an impression on the organs of touch, taste, smell, or sight and have significance in determining the acceptablity of the food product to the user.

Pancreas A glandular organ extending across the upper abdomen close to the liver. It secretes into the intestinal tract digestive juices containing enzymes to act upon protein, fat, and carbohydrate. It also secretes directly into the blood the hormone insulin, which is essential for one stage in the oxidation of sugar to carbon dioxide and water. Adjective: *pancreatic.*

Parathyroid Four small glands in the neck situated beside the thyroid gland. They secrete the hormone that affects calcium metabolism.

Parts per million A way of expressing amounts, especially of trace minerals in diets of foods. Examples of how small a part per million is: It is equal to 1 lb in 500 tons, 1 in. in about 16 miles, or 1 cent in $10,000.

Pathology The branch of medicine that deals with the special nature of disease.

Peroxidase An enzyme that speeds up oxidative changes in certain plant constituents.

Phospholipids Fat-like substances containing phosphorus and nitrogen, along with fatty acids and cholesterol. Phospholipids are abundant in brain and nerve tissues, in egg yolk, and in some plant sterols, as in soybean lecithin.

Phosphorus A mineral element necessary for the formation of bone and essential in the blood and soft tissues of the body.

Physiological Refers to the science of physiology, which deals with functions of living organisms or their parts.

Pigment Any of the coloring materials in the cells and tissues of plants and animals. In fruit and vegetables, the green pigment is chlorophyll; orange to red pigments are cartenoids; red to blue colors are anthocyanins; light-yellow pigments are flavones and flavonols. In meat, the chief pigment producing the pink or red color is myoglobin.

Pituitary gland A gland, in the lower part of the brain, which produces a number of hormones. These hormones regulate the growth of all body tissues and regulate the development and action of other endocrine glands such as the thyroid, pancreas, and adrenal glands.

Plasma The colorless fluid portion of the blood in which the cells are suspended.

Polyunsaturated fatty acids A class of fatty acids that have more than 1 unsaturated linkage in the chain, each lacking 2 hydrogens. Saturated fatty acids have all the hydrogens the carbon chain can hold. Monounsaturated fatty acids have only one unsaturated linkage. Although there are many kinds of polyunsaturated fatty acids, linoleic appears to be the only one which the body cannot synthesize and so must receive from food sources.

Pregnancy The condition of having a developing embryo in the body; the state of being with child.

Progesterone A sex hormone secreted by the corpus luteum in the ovary.

Protein One of a group of complex organic compounds that contain nitrogen, carbon, hydrogen, and oxygen and are essential for life and growth. They are formed by various combinations of different amino acids.

Provitamin Any of a number of substances, called carotenes, that occur in nature and can be converted into Vitamin A in the body.

Putrefaction The decomposition of proteins by microorganisms under anaerobic conditions, resulting in the production of incompletely oxidized compounds, some of which are foul-smelling.

Pyridoxine One of the B Vitamins, commonly designated as Vitamin B_6. Strictly speaking, Vitamin B_6, includes a group of three vitamins of nutritional interest—pyridoxine, pyridoxamine and pyridoxal.

Radical In chemistry, a group of elements joined in a set formation, which appears as a unit in a series of compounds of participates without decomposition in chemical reactions. Examples are the glycerol radical in fats, the carboxyl group in organic acids and the phenyl radical (benzene ring) in certain amino acids. The amino acids themselves act as larger radicals in making up proteins.

Radioactive Giving off atomic energy in the form of radiations, such as in alpha, beta, or gamma rays.

Radioisotopes One of a broad class of elements capable of becoming radioactive and giving off atomic energy which is detectable with a Geiger counter. Some radioisotopes occur naturally; others are produced artificially. The word is synonymous with radioactive elements and includes tracer elements.

Rancid Having a disagreeable odor or flavor; usually used to describe foods with a high content of fat when oxidation or hydrolysis of unsaturated fatty acids has occurred. Noun: *rancidity.*

Reducing (chemical action) The taking up of oxygen from air and from other materials.

Rehydration Soaking or cooking or using other procedures to make dehydrated foods take up the water they lost during drying.

Ruminant Any of a group of hoofed mammals including oxen, sheep, goats, antelopes, giraffes, deer, and camels, that have a special stomach compartment, called the rumen, which is designed to aid in the digestion of grasses and other plants that are indigestible to humans.

Salmonella A large genus of bacteria, some of which are associated with food poisoning. Certain *Salmonella* species are sometimes found in raw and dried eggs and in poultry products. They can be destroyed by sufficient heating.

Salt One of a class of compounds formed when the hydrogen atom of an acid radical is replaced by a metal or metal-like radical. The most common salt is sodium chloride, the sodium salt of hydrochloric acid. Other metal or metal-like salts in food may include phosphorus, calcium, potassium, sodium, magesium, sulfur, manganese, iron, cobalt, zinc, and other metals. They may be present as chlorides, sulfates, phosphates, lactates, citrates, or in combination with proteins, as in calcium caseinate.

Secretory The formation of a secretion. Thus, the salivary glands of the mouth secrete saliva and are secretory glands.

Serum The colorless fluid portion of the blood that separates when blood clots.

Solid A substance that does not perceptibly flow.

Soluble Capable of being dissolved or going into solution.

Solution A uniform liquid mixture containing a solvent and a solute. Salt dissolved in water is a solution—water is the *solvent*; salt is the *solute*.

Specific gravity The relation of the weight of a definite volume of a substance to the weight of an equal volume of water.

Specific heat Heat or thermal capacity of a substance in relation to that of water.

Spoilage Organisms Bacteria, yeasts, and molds that cause food to spoil. They exist everywhere—in air, soil, water, food, plants, and animals.

Sterile Free from living microorganisms. A food container may be made sterile by heating it in a pressure cooker where the temperature will reach 240°F at 10 lb pressure or by using dry heat at a high temperature in an oven for designated periods of time.

Sterilize Describes the process by which microorganisms, such as bacteria, yeasts, and molds are destroyed. In food preparation, sterilization is usually done by the use of high temperatures.

Sterol A complex steroid alcohol that can be dissolved in ether and other fat-solvents. Sterols are found in plant and animal tissues. Cholesterol is a sterol that helps the body to use fat and is used also in the body for making bile salts, hormones, and Vitamin D.

Stomach A sac-like organ into which the food passes upon being swallowed. The early stages of digestion, particularly digestion of proteins, occur in the stomach before the partly digested food passes on to the intestines.

Stress Intense strain. In medicine, any circumstances great enough to disrupt the normal, steady functioning of the body.

Subclinical disease A disease, usualy mild, that has no definite symptoms or signs that can be recognized by the usual visual or clinical means.

Syndrome A medical term meaning a group of symptoms that occur together.

Synthesis Combining two or more sustances to form a new material.

Therapeutic Refers to curing a disease.

Therapy The medical treatment of disease.

Thermal Refers to heat.

Thyroid The gland in the neck that secretes the hormone thyroxin.

Toxicity The quality of a substance that makes it poisonous or toxic; sometimes refers to the degree of severity of the poison or the possibility of being poisonous.

Tracer element A radioactive element used in biological and other research to trace the fate of a substance or follow stages in a chemical reaction, such as the pathway of metabolism of a nutrient or growth formations in plants or animals. Radioactive elements that have proved useful for tracer work in nutrition research are carbon 14, calcium 45, cobalt 60, strontium 90, and phosphorus 32. Carbon 14 is used widely in studies of fat and sugar metabolism and cholesterol formation.

Trauma Physically, wound or injury; psychologically, an emotional shock. Adjective: *Traumatic.*

Unsaturated fatty acid A fatty acid that has a double bond between 2 carbon atoms at 1 or more places in the carbon chain. Hydrogen can be added at the site of the double bond. An unsaturated fat is one that contains an unsaturated fatty acid. A saturated fatty acid has no double bonds.

Urea An end product of protein metabolism excreted in the urine.

Uric acid A chemical compound that contains nitrogen and is present in small amounts in the urine.

Urinary Refers to the urine, which is formed by the kidney and excreted.

Vapor phase The part of any substance that exists as a gas.

Vertebrates Animals with backbones or spinal cords (so-called "higher" animals in the animal scale). All mammals, birds, reptiles, and fish are vertebrates.

Vitamin One of a group of substances that are essential for life and growth.

Water-extract Whatever can be removed or dissolved out of a substance with water. A substance like sugar is completely soluble in water, whereas when yeast is shaken up with water only a small portion of it goes into solution. What remains is insoluble and does not pass into the water-extract.

Wheat germ The heart or kernel of the wheat; the embryo, from which a new plant may develop.

Work An accomplishment of body activity measurable in work units, such as foot-pounds.

Yeast One-celled fungi widely distributed in nature. Some convert sugar in fruit juices to alcohol. Some are used to produce carbon dioxide for a leavening agent in bread-making.

Glossary

Part 2—Feed and Processing Definitions

Commercial Feed

Commercial feed All materials which are distributed for use as feed or for mixing in feed for animals other than man.

Commercial formula feed (Manufactured, commercial, branded, formula, prepared). A formula feed processed to the specifications of the manufacturer.

Customer formula feed (Custom, custom feed, custom mixed) A mixture of commercial feeds and/or materials, each batch of which mixture is mixed according to the specific instructions of the final purchaser.

Feed ingredient Each of the constituent materials making up a commercial feed.

Formula feed Two or more ingredients proportioned, mixed, and processed according to specification.

Feed Form Classifications

Agglomerated feeds A compacted or extruded form of individual ingredients or a mixture of nonprocessed and/or individually processed ingredients or a combination of both.

Crumbles PELLET FEEDS reduced to granular form.

Cubes LARGE PELLETS

Large pellets PELLETS with minimum cross section dimensions of over 32/64 in.

Medium pellets PELLETS with minimum cross section dimensions of over 13/64 in. and maximum cross section dimensions of 32/64 in.
Pellet feeds AGGLOMERATED FEEDS formed by extruding individual ingredients or mixtures by compaction and forcing feed through die openings by any mechanical process.
Small pellets PELLETS with maximum cross section diameter of 13/64 in.
Textured feed Refers to pellets and/or crumbles in this text.

Feed Classifications

Complete feed A complete balanced ration for fowl, animal, or fish. Complete feeds are capable of sustaining life, growth, and/or production without any additional feed being consumed except for water.
Concentrates A fortified *supplement* feed which requires the addition of other feed ingredients to form a complete balanced ration for animal, fowl, or fish. Concentrate feeds are capable of sustaining life, but generally require the addition of other feed ingredients to obtain or sustain maximum production.
Mash feeds A mixture of nonprocessed and/or individually processed ingredients or combinations of both.
Meal feeds A ground ingredient, or ingredient otherwise reduced in particle size, of sometimes referred to as combinations of ground ingredients.
Scratch feeds A mixture consisting entirely of whole or individually prepared grains or a combination of both.
Supplements (Feeding) A feed ingredient or mixture which requires the addition of other feed ingredients to form a complete balanced ration for animal, fowl, or fish. Supplement feed may be capable of sustaining life, growth, and production, but generally only when other feed requirements are met by additional feed ingredients.

Feed Processing Terms

Aspirating Separating materals by use of air.
Cleaning Removing foreign material by such methods as scalping, aspirating, magnetic separation, or other methods.
Drying Removing moisture from ingredients by thermal means.
Grading Separating (frequently applied).
Magnetic separating Removing ferrous material by magnets.
Scalping Removing larger material by passing product through a screen.
Separating Classifying ingredient particles by size, shape and/or density.

Particle Reduction

Clipping Removing ends of whole grain.

Cooking Using heat in the presence of moisture to alter chemical and physical characteristics.

Cracking The reducing of particle size by a combined breakup and crushing action.

Crimping A rolling process using corrugated rolls. Usually entails steaming or conditioning before rolling.

Cutting The reducing of ingredient particle size by severing, using edges or knives.

Dehulling The removing of the outer covering from intact grains or seeds.

Dressing The processing of breaking or screening off of lumps from ingredients or finished feed to obtain uniformity of texture.

Flaking A rolling process using smooth rolls (usually entails steaming or conditioning before rolling).

Grinding The reduction of particle size by impact, shearing, or attrition.

Mixing Agitating feeds or ingredient until the dispersion reaches the specific degree of uniformity.

Premixing The preliminary blending of ingredients used at low levels with or without diluents (carriers).

Proportioning Measuring amounts of various ingredients for a mix by gravimetric or volumetric means.

Rolling Changing the shape and/or size of ingredients by rolling or compressing between rolls.

Scalping A form of dressing done by screening.

Steaming Treating ingredients with steam to alter physical and/or chemical properties.

Tempering or Conditioning Achieving predetermined uniform moisture and/or temperature of ingredients.

Pelleting

Conditioning Achieving predetermined moisture and/or temperature of ingredients or mixture of ingredients prior to pelleting.

Cooling Reducing temperature of pellets by air movement.

Crumbling Reducing pellets to granular form.

Pelleting Agglomerating individual ingredients or mixtures by compacting and forcing through die openings.

Sizing Classification of pellets or crumbles by particle size.

Special Ingredient Classifications

Additive Feed additive refers to any substance which becomes a component of or affects the characteristics of a feed or food if such substance is not generally recognized among experts qualified by scientific training and experience to evaluate its safety, as having been adequately shown through scientific procedures to be safe under the conditions of its intended use (as defined by Food and Drug Administration).

Microingredients Added vitamins, trace minerals, antibiotics, drugs, and other materials used in minute amounts.

Premix A combination of one or more microingredients with diluents.

Special Process

Blocking Agglomerating individual ingredients or mixtures into a large mass by means of reciprocating piston machinery.

Expanding Agglomerating mash feeds by gelatinizing the cereal portion and then extruding and expanding to form expanded pellets.

Kibbling (1) Cracking or crushing of a baked dough to form kibbled feed; or (2) extrusion of cooked feed mixtures and subsequent cooling and drying to form kibbled feed.

Wafering Agglomerating feeds of fibrous basic ingredients, by compressing into a finished form usually having a diameter or cross section measurement greater than its length.

Glossary

Part 3—Feed Terms

Ammoniated (Ammoniating) Combined with or impregnated with ammonia or an ammonium compound.
Antibiotics Drugs usually synthesized by a living microorganism and in proper concentration capable of inhibiting the growth of or destroying other microorganisms.
Aspirated (Aspirating) Having removed chaff, dust, or other light materials by use of air.

Bagasse The crushed remains of sugar cane after removal of the juice.
Balanced feed A feed having all known required nutrients in proper amount and proportion based upon recommendations of recognized authorities in the field of animal nutrition, such as the National Research Council. The species for which it is intended and conditions of growth and development should be specified.
Beans Seeds of leguminous plants, especially of the general *Phaseolus, Dalichos,* and *Vigna.*
Biscuits Shaped and baked dough.
Blending Mingling or combining two or more ingedients or feed. It does not imply a uniformity of dispersion.
Blocked (Blocking) Having agglomerated individual ingredients or mixtures into a large mass.
Blocks Agglomerated feed compressed into a solid mass cohesive enough to hold its form and usually weighing over 2 lb.
Blood Red vascular fluid of animals.

Bolted (Bolting) Separated by means of a bolting cloth, as bran from flour.
Bone Skeletal parts of vertebrates.
Bran Pericarp of grain.
Buttermilk Liquid residue from churning cream.
By-product Product produced in addition to the principal product.

Cake The mass resulting from the pressing of seeds, meat, or fish in order to remove oils, fats, or other liquids.
Calcined (Calcining) Treated at high temperature in presence of air.
Cannery waste Residue suitable for feeding from a canning process.
Carcass meat trimmings Clean flesh obtained from slaughtered animals. It is limited to striate, skeletal, and cardiac muscles, but may include the accompanying and overlaying fat and the portions of skin, sinew, nerve, and blood vessels which normally accompany the flesh.
Carcass residue, mammals Residues from animal tissues exclusive of hair, hoofs, horns, and contents of the digestive tract.
Carrier An edible material to which ingredients are added to facilitate the uniform incorporation of the latter into feeds.
Chaff Glumes, husks, or other seed covering combined with other plant parts separated from seed in threshing or processing.
Charcoal Dark-colored porous form of carbon made from the organic parts of vegetable or animal substances by their incomplete combustion.
Chipped (Chipping) Cut or broken into fragments, also meaning prepared into small, thin slices.
Chopped (Chopping) Reduced in particle size by cutting with knives or other sharp instruments.
Cleaned (Cleaning) Having removed material by such methods as scalping, aspirating, magnetic separation, or by other method.
Cleanings Chaff, weed seeds, dust, and other foreign matter removed from cereal grains.
Clipped (Clipping) Having removed the end of whole grain.
Coagulated (Coagulating) Curdled, clotted, or congealed, usually by the action of a coagulant.
Cobs The fibrous inner portion of the ear of maize from which the kernels have been removed.
Commercial feed See Uniform State Feed Bill, Section 3(d).
Complete feed A nutritionally adequate feed for animals other than man; is compounded by specific formula to be fed as the sole ration and is capable of maintaining life and/or promoting production without any additional substance being consumed except water.
Concentrate A feed used with another to improve the nutritive balance

of the total and intended to be further diluted and mixed to produce a supplement or a complete feed.

Condensed (Condensing) Reduced to denser form by removal of moisture.

Conditioned (Conditioning) Having achieved predetermined moisture characteristics and/or temperature of ingredients or a mixture of ingredients prior to further processing.

Cooked (Cooking) Heated in the presence of moisture to alter chemical and/or physical characteristics or to sterilize.

Cooled (Cooling) Temperature reduced by air movement, usually accompanied by a simultaneous drying action.

Cracked (Cracking) Having reduced particle size by a combined breaking and crushing action.

Cracklings Residue after removal of fat from adipose tissue or skin of animals by dry heat.

Crimped (Crimping) Rolled by use of corrugated rollers. It may entail tempering or conditioning and cooling.

Crumbled (Crumbling) Having reduced pellets to granular form.

Crumbles Pelleted feed reduced to granular form.

Crushed (Crushing) See *rolled (rolling).*

Cubes See *pellets.*

Cubes, range See *pellets* and *range cubes.*

Cull Material rejected as inferior in the process of grading or separating.

Culture Nutrient medium inoculated with specific microorganisms which may be in a live or dormant condition.

Cultured (Culturing) Having biological multiplied or produced material in a nutrient medium.

Cut (Cutting) See *chopped (chopping).*

D-activated (D-activating) Describes plant or animal sterol fractions which have been Vitamin D-activated by ultraviolet light or by other means.

Defluorinated (Defluorinating) Having fluorine removed.

Dehulled (Dehulling) Having removed the outer covering from grains or other seeds.

Dehydrated (Dehydrating) Having been freed of moisture by thermal means.

Diet Feed ingredient or mixture of ingredients including water that is consumed by animals.

Digested (Digesting) Subjected to prolonged heat and moisture, or to chemicals or enzymes with a resultant change or decomposition of the physical or chemical nature.

Diluent An edible substance used to mix with and reduce the concentration of nutrients and/or additives to make them more acceptable to animals, safer to use, or more capable of being mixed uniformly in a feed. (It may also be a carrier.)

Distillation solubles Stillage filtrate.

Dressed (Dressing) Made uniform in texture by breaking or screening of lumps from feed and the application of liquids.

Dried (Drying) Having removed water or other liquid from materials.

Drugs Substances (1) intended or represented for the cure, mitigation, treatment of prevention of disease of man or other animals, or (2) intended to affect the structure of any function of the body of man or other animal. (Under study by special committee).

Dry-rendered (Dry-rendering) Describes residues of animal tissues cooked in open steam-jacketed vessels until the water has evaporated. Fat is removed by draining and pressing the solid residue.

Dust Fine, dry, pulverized particles of matter usually resulting from the cleaning or grinding of grain.

Ears Entire fruiting heads of *Zea mays* including the cob and grain.

Egg albumin Whites of eggs of poultry.

Emulsifier A substance capable of holding in suspension two or more liquids which do not normally dissolve in each other.

Endosperm Starchy portion of seed.

Evaporated (Evaporating) Reduced to a denser form; concentrated as by evaporation or distillation.

Expanded (Expanding) Subjected to moisture, pressure, and temperature to gelatinize the starch portion. When extruded, its volume is increased, due to abrupt reduction in pressure.

Extracted, mechanical Having removed fat or oil from materials by organic solvents. Similar terms:expeller extracted, hydraulic extracted, "old process."

Extracted, solvent Having removed fat or oil from materials by organic solvents. Similar term: "new process."

Fat A substance consisting chiefly of glycerides of fatty acids insoluble in water but soluble in certain organic solvents.

Fatty acids Aliphatic monobasic acids containing only the elements carbon, hydrogen, and oxygen.

Feather The light, horny epidermal outgrowths that form the external covering of birds.

Feed(s) Edible material(s) which are consumed by animals and contribute energy and/or nutrients to the animal's diet. (Usually refers to animals rather than man.)

Feed grade Suitable for animal consumption.

Feed mixture See *formula feed.*

Feedstuff (See *feed(s)*)

Fermented (Fermenting) Acted upon by yeasts, molds, or bacteria in a controlled aerobic or anaerobic process in the manufacture of such products as alcohols, acids, vitamins of the B-complex group, and antibiotics.

Fines (in *crumbles* or *pellets*) Any material which will pass through a screen whose openings are immediately smaller than the specified minimum crumble size or pellet diameter.

Flaked (Flaking) See *rolled.*

Flakes An ingredient rolled or cut into flat pieces with or without prior steam conditioning.

Flour Soft, finely ground and bolted meal obtained from the milling of cereal grains, other seeds or products. It consists essentially of the starch and gluten of the endosperm.

Fodder The green or cured plant, containing all the ears or seed heads, if any, grown primarily for forage. (It has been applied more specifically to corn and sorghum.)

Food(s) When used in reference to animals, is synonymous with feed(s). See *feed(s).*

Formula feed Two or more ingedients proportioned, mixed, and processed according to specifications.

Free choice A feeding system by which animals are given unlimited access to the separate components or groups of components constituting the diet.

Fused (Fusing) Melted by heat.

Gelatinized (Gelatinizing) Having had the starch granules completely ruptured; brought about by a combination of moisture, heat, and pressure, and in some instances by mechanical shear.

Germ The embryo found in seeds and frequently separated from the starch endosperm during milling.

Gluten The tough, viscid, nitrogenous substance remaining when flour or wheat or other grain is washed to remove the starch.

Gossypol A phenolic pigment in cottonseed that is toxic to some animals.

Grain Seed from cereal plants.

Grease Animal fats with a titer below 104°F.

Grits Coarsely ground grain from which the bran and germ have been removed, usually screened to uniform particle size.

Groats Grain from which the hulls have been removed.

Ground grinding Reduced in particle size by impact, shearing, or attrition.

Hay The aerial portion of grass or herbage especially cut and cured for animal feeding.

Heads The seed or grain-containing portions of a plant.

Heat-processed (Heat-processing) Subjected to a method of preparation involving the use of elevated temperatures, with or without pressure.

Heat-rendered (Heat rendering) Melted, extracted, or clarified through use of heat. Usually, water and fat are removed.

Homogenized (Homogenizing) Fat particles broken down to evenly distributed small globules and allowed to be emulsified for long periods of time.

Hulls Outer covering of grain or other seed.

Husks Leaves enveloping an ear of maize; or the outer coverings of kernels or seeds, especially when dry and membranous.

Hydrolyzed (Hydrolyzing) Complex molecules having been split to simpler units by chemical reaction with water, usually by catalysis.

Ingredient Any component in a mixture.

Irradiated (Irradiating) Treated, prepared or altered by exposure to a specific radiation.

Juice The aqueous substance obtainable from biological tissue by pressing or filtering with or without addition of water.

Kernel A whole grain. For other species, dehulled seed.

Kibbled (Kibbling) Describes cracked or crushed, baked dough, or extruded feed that has been cooked prior to or during the extrusion process.

Lard Rendered fat of swine.

Leaves Lateral outgrowth of stems of plants.

Lecithin A specific phospholipid. The principal constituent of crude phosphatides derived from oil-bearing seeds.

Liver The hepatic gland.

Malt Sprouted and steamed whole grain from which the radicle has been removed.

Malted (Malting) Converted into malt or treated with malt or malt extract.

Mash An ingredient which has been ground or otherwise reduced in particle size.

Microingredients Vitamins, minerals, antibiotics, drugs and other materials normally required in low concentration.

Middlings A by-product of flour milling comprising several grades of granular particles containing different proportions of endosperm, bran, germ, each of which contains different levels of crude fiber.

Milk Total lactal secretion from the mammary gland.

Mill by-product A secondary product obtained in addition to the principal product in milling practice.

Mill run The state in which a material comes from the mill, ungraded and usually uninspected.

Mixing Combining by agitation two or more materials to a specific degree of dispersion.

Molasses The thick, viscous by-product resulting from refined sugar production or the concentrated, partially dehydrated juices from fruits.

Offal, poultry The eviscerated portion of slaughtered poultry such as heads, feet, undeveloped eggs, and intestines, exclusive of feathers.

Oil Liquid fat.

Pearled (Pearling) Having reduced dehulled grains by machine brushing into small, smooth particles.

Peel See *skin.*

Pellets Agglomerated feed formed by compacting and forcing through die openings by a mechanical process. Similar terms: pelleted feed, hard pellet.

Pellets, soft Pellets containing a large percentage of liquids requiring immediate dusting and cooling.

Pelleted (Pelleting) Having agglomerated feed by compacting and forcing through die openings.

Pomace The solid residue remaining after the extraction of juice or other products from fruits.

Precipitated (Precipitating) Separated from suspension or solution as a result of some chemical or physical change brought about by a chemical reagent, by cold, or by other means.

Premix A uniform mixture of one or more microingredients with diluent and/or carrier. Premixes are used to facilitate uniform dispersion of microingredients in a larger mix.

Premixing The preliminary mixing of ingredients with diluents and/or carriers.

Pressed (Pressing) Compacted or molded by pressure; also meaning having fat, oil, or juices extracted under pressure.

Presswater The aqueous extract of fish or meat free from the fats and/or oils. Presswater is the result of hydraulic pressing of the fish or meat followed by separation of the oil either by centrifuging or other means.

Product A substance produced from one or more other substances as a result of chemical or physical change.

Protein Any of a large class of naturally occurring complex combinations of amino acids.

Pulp The solid residue remaining after extraction of juices from the sugar beet root, other roots, or fruits. (See *pomace*.)

Pulverized See *ground (grinding).*

Range cake See *cake.*

Range cubes Large pellets designed to be fed on the ground. Similar term: range wafer.

Ration The amount of the total feed or any specified portion of it which is provided to 1 animal over a 24-hr period.

Residue Part remaining after the removal of a portion of its original constituents.

Rolled (Rolling) Having changed the shape and/or size of particles by compressing between rollers. It may entail tempering or conditioning.

Rumen contents Contents of the first two compartments of the stomach of a ruminant.

Scalped (Scalping) Having removed larger material by screening.

Scratch Whole, cracked or coarsely cut grain. Similar term: scratch grain, scratch feed.

Screened (Screening) Having separated various sized particles by passing over and/or through screens.

Self-fed A feeding system where animals have continuous free access to some or all components of a ration, either individually or as mixtures.

Separating Classification of particle by size, shape, and/or density.

Separating, magnetic Removing ferrous material by magnetic attraction.

Shells The hard, fibrous, calcareous coverings of a plant or animal product, i.e., nut, egg, or oyster.

Shorts Fine particles of bran, germ, flour, or offal from the tail of the mill from commercial flour milling.

Sizing See *Screening.*

Skin Outer coverings of fruits or seeds, as the rinds, husks, or peels. May also apply to dermal tissue of animals.

Solubles Liquid containing dissolved substances obtained from processing animal or plant material. They may contain some fine suspended solids.

Spent Exhausted of active or effective properties, i.e., absorbing activity.

Stalk(s) The main stem of an herbaceous plant often with its dependent parts, as leaves, twigs, and fruit.

Starch A white, granular glucose polymer of plant origin.

Steamed (Steaming) Having treated ingredients with steam to alter physical and/or chemical properties. Similar terms: steam cooked, steam rendered, tanked.

Steep-extracting (Steep-extracted) Soaked in water or other liquid (as in the wet milling of corn) to remove soluble materials.

Steepwater Water containing soluble materials extracted by steep-extraction, i.e., by soaking in water or other liquid (as in the wet milling of corn).

Stems The coarse, aerial parts of plants which serve as supporting structures for leaves, buds, and fruit.

Sterol A class of solid cyclic alcohols widely distributed in animal and plant lipids.

Stick See *stickwater, presswater.*

Stickwater, fish The aqueous extract of fish free from the oil. Stickwater contains the aqueous cell solutions of the fish and any water used in processing.

Stickwater, meat The aqueous extract of meat free fom the fat. Meat stickwater is the result of the wet rendering of meat products and contains the aqueous cell solution and the soluble glue proteins and the water condensed from steam used in wet rendering.

Stillage The mash from fermentation of grains after removal of alcohol by distillation.

Straw The plant residue remaining after separation of the seeds in threshing; it includes chaff.

Stover The stalks and leaves of corn after the ears, or of sorghum after the heads have been harvested.

Supplement A feed used with another to improve the nutrient balance at the total and intended to be: (1) feed undiluted as a supplement to other rations; or, (2) offered free-choice with other parts of the ration separately available; or (3) further diluted and mixed to produce a complete feed.

Syrup Concentrated juice of a fruit or plant.

Tallow See *carcass residue.*

Tempered (Tempering) See *conditioned (conditioning).*

Titer A property of fat determined by the solidification point of the fatty acids liberated by hydrolysis.

Trace minerals Mineral nutrients required by animals in small amounts only (measurable in milligrams per pound or smaller units).

Unsaponifiable matter Ether soluble material extractable after complete reaction with strong alkali.

Vines Any plants whose stems require support, or lie on the ground.

Viscera All the organs in the great cavity of the body, excluding contents of the intestinal tract.

Vitamins Organic compounds that function as parts of enzyme systems essential for the transformation of energy and the regulation of body metabolism.

Wafered (Wafering) Having agglomerated a feed of a fibrous nature by compressing into a form usually having a diameter or cross section measurement greater than its length.

Wafers A form of agglomerated feed based on fibrous ingredients in which the finished form usually has a diameter or cross section measurement greater than its length.

Waste Damaged, defective, or superfluous edible material produced during or left over from a manufacturing or industrial process.

Wet-rendered (Wet rendering) Cooked with steam under pressure in closed tanks.

Whey The watery part of milk separated from the curd.

Whey solids The solids of whey (proteins, fats, lactose, ash, and lactic acid).

Whole pressed (Whole pressing) Having pressed the entire seed to remove oil.

Wort The liquid portion of malted grain. It is a solution of malt sugar and other water soluble extracts from malted mash.

Glossary

Part 4—U.S. Food and Drug Administration Definitions

(1) A "complete feed" is an article intended to be adminstered as the sole ration to an animal.

(2) A "feed additive supplement" is an article for the diet of an animal which contains one or more food additives, and is intended to be:

- (i) further diluted and mixed to produce a complete feed; or
- (ii) fed undiluted as a supplement to other rations; or
- (iii) offered free choice with other parts of the ration separately available.

A "feed additive supplement" is safe for the animal and will not produce unsafe residues in the edible products from food-producing animals if fed according to directions.

(3) A "feed additive concentrate" is an article intended to be further diluted to produce a complete feed or a feed additive supplement and is not suitable for offering free choice without dilution. It contains, among other things, one or more additives in amounts, in a suitable feed base, such that from 100 to 1000 lb of concentrate must be diluted to produce 1 ton of a complete feed. A "feed additive concentrate" is unsafe if fed free choice or as a supplement, because of danger to the health of the animal or because of the production of residues in the edible products from food-producing animals in excess of safe levels.

(4) A "feed additive premix" is an article that must be diluted for safe use in a feed additive concentrate, a feed additive supplement, or a complete feed. It contains, among other things, one or more additives in high concentration in a suitable feed base such that up to 100 lb must be

diluted to produce 1 ton of a complete feed. A "feed additive premix" contains additives at levels for which safety to the animal has not been demonstrated and/or which may result, when fed undiluted, in residues in the edible products from food-producing animals in excess of the safe levels.

(5) In feeding chickens:

(i) "Broiler chickens" are chickens raised for meat purposes only.

(ii) "Replacement chickens" are chickens being raised for the purpose of egg production.

(iii) "Laying chickens" are chickens producing eggs for food.

(iv) "Breeding chickens" are chickens producing eggs used for hatching.

(6) "Taste" describes the collective sensations of smell, taste, texture, and a multitude of other factors that can contribute to the stimulation that occurs with ingestion of food.

(7) Poultry includes all domesticated birds, such as chickens, ducks, geese, guineas, ostriches, peafowl, pigeons, swans, turkeys, pheasants, quail, partridges, and grouse. Many persons, however, use the terms poultry and chickens interchangeably.

Glossary

Part 5—Terms For Structure and Composition of Foods

Cellular Tissues

Collenchyma Cells with conspicuous thickenings at the angles; walls of cellulose or a modification, not lignified.

Cork Protective tissue of tabular cells in radial rows with suberin as the characteristic constituent of the walls.

Cortex The layer in stems between the epiderm or cork tissue and the central cylinder with the bundle tissues.

Cuticle The membranous coating often covering the epiderm, consisting of cutin, a substance related to lignin.

Emergences Multicellular outgrowths with tissues derived from subepidermal as well as epidermal layers.

Endocarp The innermost layer of the pericarp, often consisting of a dense woody zone.

Epicarp The outer epiderm of the pericarp (fruit coat).

Epiderm The outer and inner tissue layer of organs as, for example, the spermoderm, also the upper and lower tissue layers of leaves.

Fibro-vascular bundles Compound strands of elongated elements consisting of (1) phloem with sieve tubes, cambiform cells and parenchyma, (2) xylem with pitted, reticulated, spiral and annular vessels, and (3) sheath of vast fibers.

Hairs or Trichomes Unicellular or multicellular outgrowths of the epiderm.

Palisade layer A layer of elongated cells arranged vertically to the surface of the organ.

Parenchyma Thin-walled, isodiametric or moderately elongated cells, with walls usually of cellulose, with or without intercellular spaces at the angles. In *spongy parenchyma,* the intercellular spaces are conspicuous and the contour of the cells is much distorted.

Sclerenchyma Cells with lignified, often strongly pitted walls. Includes stone cells, fibers, and a great variety of other forms.

Stomata The epidermal openings that facilitate assimilation (photosynthesis), respiration, and transpiration.

Water stomata Openings at the ends of nerves through which water is discharged.

Cell Contents

Aleurone grains Protein bodies, present in certain oil seeds, consisting of ground substance containing one or more of the following: (1) crystalloids or protein crystals, (2) globoids believed to be phosphates with organic salts and (3) calcium oxalate crystals. Tannins, fats, fatty oils, waxes, volatile or essential oils, and resins do not occur as definite organized bodies, but are recognized by their solubilities or chemical reactions. Calcium oxalate occurs often as (1) single monoclinic crystals, (2) crystal rosettes, (3) raphides or needle-shaped crystals, and (4) crystal sand. Silica occurs as rounded bodies in the stegmata of palms and some other plants.

Inulin Is present in roots and rhizomes. It forms sphaerocrystals with alcohol.

Latex Milky secretions in tubes in various leaves, stems, roots, and fruits.

Mucilage A general term for gelatinizing substances such as gums and pectins.

Protoplasm The living matter of cells. Among the forms are (1) cytoplasm, which is granular or stringy; (2) cell nucleus, a round or oval body brought out clearly by special stains; and (3) plastids or chromatophores, including chlorophyll grains or chloroplasts, the green bodies of leaves, leucoplasts or starch formers, and chromoplasts, the yellow, orange, or red colored bodies of various organs.

Starch Starch has come to be regarded as a mixture of amylose and amylopectin, the latter containing H_3PO_4 in its molecule.

Sucrose Is the only soluble carbohydrate usually present in crystalline form, and this only when deposited by evaporation.

Glossary

Part 6—Poultry Terms

Assimilation Digestion, absorption, transporting, and metabolism of nutrients. (Transformation of food into animal tissue.)

Digestible nutrients The difference between the nutrients consumed and those excreted. In the case of the chicken, which excretes the undigested food and the urinary products together, correction is made for the quantity of uric acid present.

Essential amino acids Those required in the diet, because the animal is either unable to synthesize them or unable to synthesize them at a sufficient rate to meet needs for growth or production.

Feedstuff Any material included in a diet because of nutritional properties.

Maintenance requirements That portion of the energy of the food used in the life processes of the animal for keeping the animal warm and for movement of the body.

Mammals Mammals are distinguished by the presence of hair and mammary glands. Birds are distinguished by their covering of feathers.

Metabolism A term applied to all chemical processes occurring within the body contributing to the transformation of matter and energy involved in growth, muscular activity, heat production, and the maintenance of vital functions.

Metabolizable energy The energy of the food eaten less the energy of the excrement derived from it, both fecal and urinary, and, in the case of ruminants, in gases produced by fermentation.

Net energy The metabolizable energy minus that used in utilizing it. It may be used for maintenance, for the production of fat, flesh, eggs or other animal products, or body movement or work.

Nutrient Any food constituent, or group of food constituents of the same general chemical composition that aids in the support of animal life. These are protein, fat, carbohydrates, vitamins, and minerals. Oxygen and water are sometimes included.

Nutrition The sum of the processes by which an animal or plant absorbs or takes in and utilizes food substances.

Ornithology Ornithology is the study of birds not classified as poultry.

Poultry Used to designate those species of birds which render man an economic service and reproduce freely under his care. It includes the following: chickens, turkeys, ducks, geese, guineas, pigeons, pheasants, ostriches, pea fowl, and swans.

Poultry concentrate Contains protein, fat, minerals, vitamins, and a very small amount of carbohydrate.

Poultry science Poultry Science is the study of principles and practices, involved in the production and marketing of poultry and poultry products. It includes breeding, incubation, brooding, housing, feeding, curing of diseases, marketing, and management.

Productive energy The net energy as measured by the energy stored up as fat and protein in a growing or fattening animal.

Protein quality Ability of a protein supplement to meet the protein needs of an animal. Is a measure of the distribution (kind and amount) of the required amino acids and their availability.

Principal References

1. GENERAL—TEXTS AND JOURNALS

Journals

INST. NUTR. 1928–1978. Journal of Nutrition. Vol. 1–60. Institute of Nutrition, Bethesda, Md.

POULT. SCI. ASSOC. 1978. Poultry Science, Vol. 1–58. Poultry Science Association, Champaign, Ill.

Texts

ALTMAN, P.L. and DITTMER, D.S. 1968. Metabolism. Fed. Proc. Fed. Am. Soc. Exp. Biol., Washington, D.C.

ASSOC. AM. FEED CONTROL OFF. 1978. Official Publication. Assoc. of Am. Feed Control Officials, College Station, Tex.

BRESTER, H.E. and SCHWARTE, L.H. 1965. Diseases of Poultry, 5th Edition. Iowa State Univ. Press, Ames.

BROOKER, D.B. and BAKKER-ARKEMA, F.W. 1974. Drying Cereal Grains. AVI Publishing Co., Westport, Conn.

CRAMPTON, E.W. and HARRIS, L.E. 1969. Applied Animal Nutrition, 2nd Edition. W. H. Freeman and Co., San Francisco.

CULLISON, A. 1975. Feeds and Feeding. Reston Publishing Co., Reston, Va.

GRIST, D.H. 1959. Rice, 3rd Edition. Longmans, Green and Co., London.

INGLETT, G.E. 1970. Corn: Culture, Processing, Products. AVI Publishing Co., Westport, Conn.

KUTSKY, R.J. 1973. Handbook of Vitamins and Hormones. Van Nostrand Reinhold Co., New York.

LEONARD, W.H. and MARTIN, J.H. 1963. Cereal Crops. Macmillan Co., New York.

LEVI, W.M. 1969. The Pigeon. Levi Publishing Co., Sumpter, S.C.

MATZ, S.A. 1969. Cereal Science. AVI Publishing Co., Westport, Conn.

NATL. RES. COUNC. 1971. Nutrient Requirements of Poultry No. 1, Nutrient Requirements of Domestic Animals No. 1, 7th Revised Edition. National Research Council, Washington, D.C.

NORTH, M.O. 1978. Commercial Chicken Production Manual, 2nd Edition. AVI Publishing Co., Westport, Conn.

SCHAIBLE, P.J. 1970. Poultry: Feeds and Nutrition. AVI Publishing Company, Westport, Conn.

SINHA, R.N. and MUIR, W.E. 1973. Grain Storage: Part of a System. AVI Publishing Co., Westport, Conn.

SMITH, A.K. and CIRCLE, S.J. 1978. Soybeans; Chemistry and Technology Vol. 1, Revised 2nd Printing. AVI Publishing Co., Westport, Conn.

U.N. DEP. ECON. SOC. AFFAIRS STAT. OFF. 1976. Statistical Yearbook 1976. U.N. Dep. Econ. Soc. Affairs Stat. Off., New York.

UNDERWOOD, E.J. 1962. Trace Elements in Human and Animal Nutrition, Revised Edition. Academic Press, New York.

U.S. DEP. AGRIC. 1977. Agricultural Statistics 1977. U.S. Dep. Agric., U.S. Govt. Print. Off., Washington, D.C.

WALL, J.S. and ROSS, W.M. 1970. Sorghum Production and Utilization. AVI Publishing Co., Westport, Conn.

2. AMINO ACIDS

ADAMS, R.L., ANDREWS, F.N., ROGLER, J.C. and CARRICK, C.W. 1962. The sulfur amino acid requirement of the chick from four to eight weeks of age as affected by temperature. Poult. Sci. *41*, 1801–1806.

ADKINS, J.S., HARPER, A.E. and SUNDE, M.L. 1962. The L-arginine requirement of the laying pullet. Poult. Sci. *41*, 657–663.

ADKINS, J.S., MILLER, E.C., BIRD, H.R., ELVEHJEM, C.A. and SUNDE, M.L. 1958. An estimate of the threonine requirement of laying hens. Poult. Sci. *37*, 1362–1367.

ALMQUIST, H.J. 1952. Amino acid requirements of chickens and turkeys—a review. Poult. Sci. *31*, 966–981.

ALMQUIST, H.J. 1965. Nutritional values of methionine isomers and hydroxy analog. Proc. Nutrition Counc., Amer. Feed Manufacturers Assoc. 25th Ann. Meet., May 10, Chicago.

BOOMGAARDT, J. and BAKER, D.H. 1973. The lysine requirement of growing chicks fed sesame meal-gelatin diets at three protein levels. Poult. Sci. *52*, 586–591.

BOOMGAARDT, J. and BAKER, D.H. 1973. Effect of age on the lysine and sulfur amino acid requirement of growing chicks. Poult. Sci. *52*, 592–597.

CAREW, L.B. and HILL, F.W. 1961. Effect of methionine deficiency on the utilization of energy by the chick. J. Nutr. *74*, 185–190.

CARLOTTI, R.J. 1971. Manifestations of metabolic bursts during development of the cockerel. Unpublished Ph.D. Dissertation. West Virginia Univ., Morgantown.

CARLSON, C.W. and GUENTHER, E. 1969. Response of laying hens fed typical corn-soy diets to supplements of methionine and lysine. Poult. Sci. *48*, 137–143.

DEAN, W.F. and SCOTT, H.M. 1965. The development of an amino acid reference diet for the early growth of chicks. Poult. Sci. *44,* 803–808.

FISHER, H., GRIMINGER, P. and LUTZ, H. 1960A. The amino acid requirement of laying hens. Poult. Sci. *39,* 173–175.

FISHER, H., JOHNSON, D., JR. and LEVEILLE, G.A. 1957. The phenylalanine and tyrosine requirement of the growing chick with special reference to the utilization of the D-isomer of phenylalanine. J. Nutr. *62,* 349–355.

FITZSIMMONS, R.C. and WAIBLE, P.E. 1962. Determination of the limiting amino acids in corn-soybean oil meal diets for young turkeys. Poult. Sci. *41,* 260–268.

GAZO, M., GERGELYIOVA, V. and GROM, A. 1970. Lysine requirement of fattening ducklings. Sitzungsber. Dtsch. Akad. Landwirtschaftswiss. Berlin *19* (3) 59–63.

GRABER, C. and BAKER, D.H. 1973. The essential nature of glycine and proline for growing chicks. Poult. Sci. *52,* 892–896.

HARMS, R.H. and DAMRON, B.L. 1969. Protein and sulfur amino acid requirement of the laying hen as influenced by dietary formulation. Poult. Sci. *48,* 144–149.

HEYWANG, B.W., VAVICH, M.G. and REID, B.L. 1963. Supplemental methionine in a sixteen percent protein diet for laying chickens. Poult. Sci. *42,* 245–249.

HILL, D.C. and OLSEN, E.M. 1963. Effect of starvation and a nonprotein diet on blood plasma amino acids, and observations on the detection of amino acids limiting growth of chicks fed purified diets. J. Nutr. *79,* 303–316.

HURWITZ, S. and BORNSTEIN, S. 1973. The protein and amino acid requirements of laying hens: Suggested models for calculation. Poult. Sci. *52,* 1124–1134.

JENSEN, L.S., FALEN, L. and SCHUMAIER, G.W. 1974. Requirement of White Leghorn laying and breeding hens for methionine as influenced by stage of production cycle and inorganic sulfate. Poult. Sci. *53,* 535–544.

KELLY, M. and SCOTT, H.M. 1968. Plasma lysine titers in the chick in relation to sources of lysine and mode of administration. J. Nutr. *94,* 326–330.

KIM, S.M. and MCGINNIS, J. 1972. Methionine requirement of white leghorn pullets. Poult. Sci. *51,* 1735–1740.

KUMMERO, V.E., JONES, J.E. and LOADHOLT, C.B. 1971. Lysine and total sulfur amino acid requirements of turkey poults, one day to three weeks. Poult. Sci. *50,* 752–758.

LEONG, K.C., SUNDE, M.L., BIRD, H.R. and ELVEHJEM, C.A. 1959. Interrelationships among dietary energy, protein and amino acids for chickens. Poult. Sci. *38,* 1267–1285.

MARCH, B.E. and BIELY, J. 1972. The effects of protein level and amino acid balance in wheat-based laying rations. Poult. Sci. *51,* 547–557.

MULLER, R.D. and BALLOUN, S.L. 1973. Further studies with methionine additions and leucine imbalance in corn-soybean meal laying hen diets. Poult. Sci. *52,* 2068. (Abstract)

MULLER, R.D. and BALLOUN, S.L. 1974. Response to methionine supplementation of leghorn hens fed low-protein corn-soybean meal diets. Poult. Sci. *53*, 1463–1475.

NESHEIM, M.C. and HUTT, F.B. 1962. Genetic differences among White Leghorn chicks in requirements of arginine. Science *137*, 691–692.

REID, B.L. and WEBER, C.W. 1973. Dietary protein and sulfur amino acid levels for laying hens during heat stress. Poult. Sci. *52*, 1335–1343.

ROBERSON, R.H. and FRANCIS, D.W. 1966. The lysine requirement of white Chinese goslings. Poult. Sci. *45*, 324–329.

ROSENBERG, H.R., BALDINI, J.T. and TOLLEFSON, C.I. 1957. Histidine requirement of the growing chick. Poult. Sci. *36*, 1381–1382.

SAVAGE, J.E. and O'DELL, B.L. 1960. Arginine requirement of the chick and the arginine sparing value of related compounds. J. Nutr. *70*, 129–134.

SCOTT, M.L., HOLM, E.R. and REYNOLDS, R.E. 1963. Studies on the protein and methionine requirements of young bobwhite quail and young ring-necked pheasants. Poult. Sci. *42*, 676–680.

SVACHA, A.J., WEBER, C.W. and REID, B.L. 1969. Lysine, methionine and glycine requirements of growing Coturnix quail. Poult. Sci. *48*, 1881. (Abstract)

SVACHA, A., WEBER, C.W. and REID, B.L. 1970. Lysine, methionine and glycine requirements of Japanese quail to five weeks of age. Poult. Sci. *49*, 54–59.

TUTTLE, W.L. and BALLOUN, S.L. 1974. Lysine requirements of starting and growing turkeys. Poult. Sci. *53*, 1698–1704.

WAIBEL, P.E. 1959. Methionine and lysine in rations for turkey poults under various dietary conditions. Poult. Sci. *38*, 712–721.

WAIBEL, P.E. 1968. Amino acids and protein for growing turkeys. Minn. Nutr. Conf. Proc. (personal communication)

WARNICK, R.E. and ANDERSON, J.O. 1973. Essential amino acid levels for starting turkey poults. Poult. Sci. *52*, 445–452.

WATERHOUSE, H.N. and SCOTT, H.M. 1961. Glycine need of the chick fed casein diets and the glycine, arginine, methionine and creatine interrelationships. J. Nutr. *73*, 266–272.

WILBURN, D.R. and FULER, H.L. 1974. The effect of methionine and lysine levels on the arginine requirement of the chick. Poult. Sci. *54*, 248–256.

3. ANTIBIOTICS AND ARSENIC COMPOUNDS

AL-TIMIMI, A.A. and SULLIVAN, T.W. 1972. Safety and toxicity of dietary organic arsenicals relative to performance of young turkeys. 1. Arsanilic acid and sodium arsanilate. Poult. Sci. *51*, 111–116.

BARE, L.N., WISEMAN, R.F. and ABBOTT, O.J. 1964. Effect of dietary antibiotics on coliform bacteria and lactobacilli in the intestinal tract of uric acid-fed chicks. J. Bacteriol. *87*, 329–331.

EYSSEN, H. and DESOMER, P. 1963. The mode of action of antibiotics in stimulating growth of chicks. J. Exp. Med. *117*, 127–138.

FOOD AGRIC. ORGAN. U.N. 1968. Specifications for the indentity and purity of food additives and their toxicological evaluation: some antibiotics. 12th Rept. Joint FAO/WHO Expert Comm. Food Additives. FAO, Rome.

HEYWANG, B.W. 1959. The effect of arsanilic acid and low levels of antibiotics on laying chickens during hot weather. Poult. Sci. *38*, 854–858.

HUHTANEN, C.M. and PENSACK, J.M. 1965. The role of Streptococcus faecalis in the antibiotic growth effect in chickens. Poult. Sci. *44*, 830–834.

MARCH, B.E, AKINWANDE, A. and SOONG, R. 1972. The effect of feeding antibiotics for different periods on growth rate, feed conversion and metabolizability of dietary energy in growing chickens. Poult. Sci. *51*, 1409–1414.

RICHMOND, M.H. 1972. Some environmental consequences of the use of antibiotics. J. Appl. Bacteriol. *35*, 155–176.

SULLIVAN, T.W. and AL-TIMIMI, A.A. 1972A. Safety and toxicity of dietary organic arsenicals relative to the performance of young turkeys. 2. Carbarsone. Poult. Sci. *51*, 1498–1501.

SULLIVAN, T.W. and AL-TIMIMI, A.A. 1972B. Safety and toxicity of dietary organic arsenicals relative to the performance of young turkeys. 3. Nitarsone. Poult. Sci. *51*, 1582–1586.

SULLIVAN, T.W. and BIRD, H.R. 1957. Effect of quantity and source of dietary nitrogen on the utilization of the hydroxy analogues of methionine and glycine by chicks. J. Nutr. *62*, 143–150.

SWANN REPT. 1968. Report of the Joint Committee on the Use of Antibiotics in Animal Husbandry and Veterinary Medicine (Swann Report). H.M. Stationary Office, London.

WICKER, S.L., ISGRIGG, W.N., TRAMMELL, J.H. and DAVIS, R.B. 1977. The control and prevention of necrotic enteritis in broilers with zinc bacitracin. Poult. Sci. *56*, 1229–1231.

4. BIOCHEMICAL

NEEDHAM, J. 1931. Chemical Embryology, Vol. 1, 2, and 3. Macmillan Co., New York.

PANE, L.F. 1919. Distribution of mortality during the period of incubation. J. Am. Assoc. Inst. Invest. Poult. Husb. *6*, 9–12.

WAGNER, A.F. and FOLKERS, K. 1964. Vitamins and Coenzymes. Wiley (Interscience Div.) New York.

WEST, E.S., TODD, W.R., MASON, H.S. and VAN BRUGGEN, J.T. 1966. Text Book of Biochemistry, 4th Edition. Macmillan Co., New York.

5. BIOTIN

ANDERSON, J.O. and WARNICK, R.E. 1970. Studies of the need for supplemental biotin in chick rations. Poult. Sci. *49*, 569–578.

COUCH, J.R., CRAVENS, W.W., ELVEHJEM, C.A. and HALPIN, J.G. 1948. Relation of biotin to congenital deformities in the chick. Anat. Rec. *100,* 29–48.

JENSEN, L.S. and MARTINSON, R. 1969. Requirement of turkey poults for biotin and effect of deficiency on incidence of leg weakness in developing turkeys. Poult. Sci. *48,* 222–230.

PATRICK, H., BOUCHER, R.V., DUTCHER, R.A. and KNANDEL, H.C. 1942. The nutritional significance of biotin in chick and poult nutrition. Poult Sci. *21,* 476. (Abstract)

PAYNE, C.G., GILCHRIST, P., PEARSON, J.A. and HEMSLEY, L.A. 1974. Involvement of biotin in the fatty liver and kidney syndrome of broilers. Br. Poult. Sci. *15,* 489–498.

WAIBEL, P.E., KRISTA, L.M., ARNOLD, R.L., BLAYLOCK, L.G. and NEAGLE, L.H. 1969. Effect of supplementary biotin on performance of turkeys fed corn-soybean meal type diets. Poult. Sci. *48,* 1979–1985.

6. CALCIUM (AND PHOSPHORUS)

AITKEN, J.R., LINDBLAD, G.S. and HUNSAKER, W.G. 1958. The calcium and phosphorus requirements of goslings. Poult. Sci. *37,* 1180. (Abstract)

CONSUEGRA, P.F. and ANDERSON, D.L. 1967. Studies on the dietary calcium and phosphorus requirements of immature Coturnix quail. Poult. Sci. *46,* 1247. (Abstract)

DAVIDSON, J. and BOYNE, A.W. 1970. The calcium and phosphorus requirements of laying hens. Br. Poult. Sci. *11,* 231–240.

DEAN, W.F., SCOTT, M.L., YOUNG, R.J. and ASH, W.J. 1967. Calcium requirement of ducklings. Poult. Sci. *46,* 1496–1499.

DEWITT, J.B., NESTLER, R.B. and DERBY, J.V., JR. 1949. Calcium and phosphorus requirements of breeding bobwhite quail. J. Nutr. *39,* 567–577.

FERGUSON, T.M., SEWELL, C.E. and ATKINSON, R.L. 1974. Phosphorous levels in the turkey breeder diet. Poult. Sci. *53,* 1627–1629.

FRITZ, J.C., ROBERTS, T., BOEHNE, J.W. and HOVE, E.L. 1969. Factors affecting the chick's requirement for phosphorus. Poult. Sci. *48,* 307–320.

O'DELL, B.L., BIRD, B.W., RUGGLES, D.L. and SAVAGE, J.E. 1966. Composition of aortic tissue from copper-deficient chicks. J. Nutr. *88,* 9–14.

ROLAND, D.A., SR. and HARMS, R.H. 1976. The influence of feeding diets containing different calcium-phosphorus ratios on the laying hen. Poult. Sci. *55,* 637–641.

SCOTT, M.L., HOLM, E.R. and REYNOLDS, R.E. 1958. The calcium, phosphorus and vitamin D requirements of young pheasants. Poult Sci. *37,* 1419–1425.

SIMKISS, K. 1961. Calcium metabolism and avian reproduction. Biol. Rev. *36,* 321–367.

SULLIVAN, T.W. 1960. An estimate of the phosphorus requirement of Broad Breasted Bronze turkeys, 8–20 weeks of age. Poult. Sci. *39,* 1321–1327.

SUMMERS, J.D., GRANDHI, R. and LEISON, S. 1976. Calcium and phosphorus requirements of the laying hen. Poult. Sci. *55,* 402–413.

SUNDE, M.L. and BIRD, H.R. 1956. A critical need of phosphorus for the young pheasant. Poult. Sci. *35,* 424–430.

WAIBEL, P.E., JOHNSON, E.L. and PILKEY, A.M. 1961. Successful turkey growth with reduced calcium and phosphorus levels. Poult. Sci. *40,* 256–258.

WALDROUP, P.W., AMMERMAN, C.B. and HARMS, R.H. 1963. The relationship of phosphorus, calcium and vitamin D_3 in the diet of broiler-type chicks. Poult. Sci. *42,* 982–989.

WALDROUP, P.W., MAXEY, J.F. and LUTHER, L.W. 1974. Studies on the calcium and phosphorus requirements of caged turkey breeder hens. Poult. Sci. *53,* 886–888.

7. CHOLINE

LATSHAW, J.D. and JENSEN, L.S. 1971. Choline level and its effect on egg weight in the Japanese quail. Poult. Sci. *50,* 790–794.

LATSHAW, J.D. and JENSEN, L.S. 1972. Choline deficiency and synthesis of choline from precursors in mature Japanese quail. J. Nutr. *102,* 749–755.

NESHEIM, M.C., NORVELL, M.J., CEBALLOS, E. and LEACH, R.M., JR. 1971. The effect of choline supplementation of diets for growing pullets and laying hens. Poult. Sci. *50,* 820–831.

SERFIN, J.A. 1974. Studies on the riboflavin, niacin, pantothenic acid and choline requirements of young Bobwhite quail. Poult. Sci. *53,* 1522–1532.

SINGSEN, E.P., MATTERSON, L.D., KOZEFF, A. and DECKER, L. 1950. The choline requirements of turkey breeders. Poult. Sci. *29,* 780. (Abstract)

VON VOGT, H. 1970. Choline requirement of quail. Arch. Gefluegelkd. *34,* 41–44.

WELCH, B.E. and COUCH, J.R. 1955. Homocystine, vitamin B_{12}, choline and methionine in the nutrition of the laying fowl. Poult. Sci. *34,* 41–44.

WEST, J.W., CARRICK, C.W., HAUGE, S.M. and MERTZ, E.T. 1951. The relationship of choline and cystine to the methionine requirement of young chickens. Poult. Sci. *30,* 880–885.

8. ENERGY

DEGROOTE, G. 1974. A comparison of a new net energy system with the metabolizable energy system in broiler diet formulation, performance and profitability. Br. Poult. Sci. *15,* 75–95.

EWAN, R.C. 1976. Utilization of energy of feed ingredients by young pigs. Distill. Feed Res. Counc. Proc. *31,* 16–21.

FULLER, H.L. 1975. Limitations in the use of metabolizable energy values. Proc. Ga. Nutr. School. (personal communication)

HURWITZ, S., SKLAN, D. and BARTOV, I. 1978. New formal approaches to the determination of energy and amino acid requirements of chicks. Poult. Sci. *57,* 197–206.

JENSEN, L.S. 1974. Calorie efficiency of fats versus carbohydrates for poultry and their use in formulation of rations. Proc. Cornell Nutr. Conf., 1957, Ithaca, N.Y.

SIBBALD, I.R. 1975. The level of feed intake on metabolizable energy values measured with adult roosters. Poult. Sci. *54,* 1990–1997.

SIBBALD, I.R. 1976A. A bioassay for true metabolizable energy in feedstuffs. Poult. Sci. *55,* 303–308.

SIBBALD, I.R. 1976B. The true metabolizable energy value of several feedstuffs measured with roosters, laying hens, turkeys and broiler hens. Poult. Sci. *55,* 1459–1463.

9. FEEDING METHODS

COUCH, J.R. 1960. Restricted feeding broiler type replacement stock. Poult. Sci. *39,* 339–346.

GOWE, R.S., JOHNSON, A.S., CRAWFORD, R.D., DOWNS, J.H., HILL, A.T., MOUNTAIN, W.F., PELLETIER, J.R. and STRAIN, J.H. 1960. Restricted versus full-feeding during the growing period for egg production stock. Br. Poult. Sci. *1,* 37–56.

PROUDFOOT, F.G. and GOWE, R.S. 1974. The influence of an increased photoperiod, a modified natural day length and feed restriction during the rearing period on the performance of five S.C.W.L. genotypes. Poult. Sci. *53,* 518–528.

SINGSEN, E.P., MATTERSON, L.D., TLUSTOHOWEZ, J. and POTTER, L.M. 1958. The effect of controlled feeding, energy intake and type of diet on the performance of heavy type laying hens. Conn. Agric. Exp. Stn. Bull. *346.*

SNETSINGER, D. and ZIMMERMAN, R. 1974. Limited feeding of egg strain layers. Feedstuffs *48* (4) 70–93.

SNETSINGER, D. and ZIMMERMAN, R. 1975. Limited feeding of layers. Feedstuffs *47* (4) 17–18, 92.

TOUCHBURN, S.P., NABER, E.C. and CHAMBERLIN, V.D. 1968. Effect of growth restriction on reproductive performance of turkeys. Poult. Sci. *47,* 547–556.

WILSON, H.R., FRY, J.L., HARMS, R.H. and ARRINGTON, L.R. 1967. Performance of hens molted by various methods. Poult. Sci. *46*, 1406–1412.

10. FOLACIN

KRATZER, F.H., DAVIS, P.N. and ABBOTT, U.P. 1956. The folic acid requirement of turkey breeder hens. Poult. Sci. *35,* 711–716.

MARCH, B.E. and BIELY, J. 1955. Fat studies in poultry. 3. Folic acid and fat tolerance in the chick. Poult. Sci. *34,* 39–44.

MARCH, B.E. and BIELY, J. 1956. Folic acid supplementation of high-protein, high-fat diets. Poult. Sci. *35,* 550–551.

SAXENA, H.C., BEARSE, G.E., MCCLARY, C.F., BLAYLOCK, L.G. and BERG, L.R. 1954. Deficiency of folic acid in rations containing natural feedstuffs. Poult. Sci. *33,* 815–820.

SCHWEIGERT, B.S., GERMAN, H.L., PEARSON, P.B. and SHERWOOD, R.M. 1947. Effect of the pteroylglutamic acid intake on the performance of turkeys and chickens. J. Nutr. *35,* 89–102.

YOUNG, R.J., NORRIS, L.C. and HEUSER, G.F. 1955. The chick's requirement for folic acid in the utilization of choline and its precursors betaine and methylaminoethanol. J. Nutr. *55,* 353–362.

11. IODINE

CREEK, R.D., PARKER, H.E., HAUGE, S.M., ANDREWS, F.N. and CARRICK, C.W. 1957. The iodine requirements of young chickens. Poult. Sci. *36,* 1360–1365.

ROGLER, J.C., PARKER, H.E., ANDREWS, F.N. and CARRICK, C.W. 1961. The iodine requirements of the breeding hen. Poult. Sci. *40,* 1546–1562.

SCOTT, M.L., VAN TIENHOVEN, A., HOLM, E.R. and REYNOLDS, R.E. 1960. Studies on the sodium, chlorine and iodine requirements of young pheasants and quail. J. Nutr. *71,* 282–288.

12. IRON AND COPPER

AL-UBAIDI, Y.Y. and SULLIVAN, T.W. 1963. Studies on the requirements and interaction of copper and iron in Broad Breasted Bronze turkeys to 4 weeks of age. Poult. Sci. *42,* 718–725.

CARLTON, W.W. and HENDERSON, W. 1964. Studies on the copper requirements of growing chicks. Avian Dis. *8,* 227–234.

DAVIS, P.N., NORRIS, L.C. and KRATZER, F.H. 1968. Iron utilization and metabolism in the chick. J. Nutr. *94,* 407–417.

FRITZ, J.C., BOCHNE, W. and ROBERTS, T. 1969. A study of the chick's requirements for dietary iron. Poult. Sci. *48,* 1810. (Abstract)

HILL, C.H. and MATRONE, G. 1961. Studies on copper and iron deficiencies in growing chicks. J. Nutr. *73,* 425–431.

KUBENA, L.F., MAY, J.D., DEATON, J.W., REECE, F.N. and LOTT, B.D. 1972. Effect of dietary iron and copper on mortality of chicks at two temperatures. Poult. Sci. *51,* 1470–1471.

13. FATS—FATTY ACIDS

BALNAVE, D. 1970. Essential fatty acids in poultry nutrition. World's Poult. Sci. J. *26,* 442–460.

BALNAVE, D. and WEATHERUP, S.T.C. 1974. The necessity of supplementing laying hen diets with linoleic acid. Br. Poult. Sci. *15,* 325–331.

CALVERT, C.C. 1965. The performance of female Japanese quail on linoleic acid deficient diets. Poult. Sci. *44,* 1358. (Abstract)

FULLER, H.L. 1973. Effects of Processing on the Nutritional Value of Feeds. Natl. Acad. Sci., Washington, D.C.

HEYWANG, B.W. and VAVICH, M.G. 1962. Energy level of a sixteen percent protein diet for layers in a semi-arid, subtropical climate. Poult. Sci. *41,* 1389–1393.

KETOLA, H.G., YOUNG, R.J. and NESHEIM, M.C. 1973. Linoleic acid requirement of turkey poults. Poult. Sci. *52,* 597–603.

MENGE, H. 1970A. Comparative requirements of linoleic acid for male and female chicks. Poult. Sci. *49,* 178–183.

MENGE, H. 1970B. Further studies on the linoleic acid requirement of the hen using purified and practical type diets. Poult. Sci. *49,* 1027–1030.

14. MAGNESIUM

ALMQUIST, H.J. 1942. Magnesium requirement of the chick. Proc. Soc. Exp. Biol. Med. *49,* 544–545.

GARDNER, E.E., ROGLER, J.C. and PARKER, H.E. 1960. Magnesium requirement of the chick. Poult. Sci. *39,* 1111–1115.

HAJJ, R.N. and SELL, J.L. 1969. Magnesium requirement of the laying hen for reproduction. J. Nutr. *97,* 441–448.

SELL, J.L., HAJJ, R., COX, A. and GUENTER, W. 1967. Effect of magnesium deficiency in the hen on egg production and hatchability. Br. Poult. Sci. *8,* 55–60.

SULLIVAN, T.W. 1964. Studies on the dietary requirement and interaction of magnesium with antibiotics in turkeys to 4 weeks of age. Poult. Sci. *43,* 401–405.

VOHRA, P. 1972. Magnesium requirement for survival and growth of Japanese quail *(Coturnix coturnix japonica).* Poult Sci. *51,* 2103–2105.

15. MANAGEMENT

FULLER, H.L. and CHANEY, L.W. 1974. Effect of delayed maturity of White Leghorn chickens on subsequent production. Poult. Sci. *53,* 1348–1355.

FULLER, H.L., KIRKLAND, W.M. and CHANEY, L.W. 1973. Methods of delaying sexual maturity of pullets. 2. Restricted energy consumption. Poult. Sci. *52,* 228–236.

HARMS, R.H. and WALDROUP, P.W. 1962. Protein restriction. Another method of delaying sexual maturity in pullets. Feedstuffs. *34* (8) 70.

KIRKLAND, W.M. and FULLER, H.L. 1971. Methods of delaying sexual maturity of pullets. 1. Water restriction. Poult. Sci. *50,* 1761–1767.

NATL. RES. COUNC. 1977. Nutrient requirements of domestic animal. 1. Nutrient Requirements of Poultry. Seventh Revised Edition, NAS, Washington, D.C.

WAGNER, O.P., THOMAS, O.P. and GRABER, G. 1978. An adaptive growth response of chicks fed rye. Poult. Sci. *57,* 230–234.

16. MANGANESE

COX, A.C. and BALLOUN, S.L. 1969. Manganese requirements of laying hens as related to diet calcium. Poult. Sci. *48,* 745–747.

GALLUP, W.D. and NORRIS, L.C. 1939A. The amount of manganese required to prevent perosis in the chick. Poult. Sci. *18,* 76–82.

GALLUP, W.D. and NORRIS, L.C. 1939B. The effect of a deficiency of manganese in the diet of the hen. Poult. Sci. *18,* 83–88.

WATSON, L.T., AMMERMAN, C.B., MILLER, S.M. and HARMS, R.H. 1971. Biological availability to chicks of manganese from different inorganic sources. Poult. Sci. *50,* 1693–1700.

17. MANUFACTURING AND CALCULATIONS

HUTTON, R.F. and MCALEXANDER, R.H. 1957. A simplified feed-mix model. J. Farm Econ. *39* (Aug.) 714–730.

NIR, I., KAFIR, I. and COHEN, R. 1978. Effect of pelleting on the Vitamin K activity of various Vitamin K_3 preparations in chicks. Poult. Sci. *57,* 206–208.

SIBBALD, J.R. 1977. Effects of steam pelleting on the true metabolizable energy values of poultry diets. Poult. Sci. *56,* 1686–1688.

18. MANURE

HALLORAN, H.R. 1965. Effect of molasses on poultry droppings. Anim. Nutr. Health, April, 21–24.

MOORE, B.W., PATRICK, H., JOHNSON, J.R. and HYRE, H.M. 1964. Composition and production of poultry manure. W. Va. Univ. Agric. Exp. Stn. (Morgantown) Tech. Bull. *496T.*

PATRICK, H. 1955. Influence of protein source on consumption and excretion of water and excreta voided by broiler chicks. Poult. Sci. *34,* 155–157.

19. NIACIN

ADAMS, R.L. and CARRICK, C.W. 1967. A study of the niacin requirement of the laying hen. Poult. Sci. *46,* 712–718.

BATTIG, M.J., HILL, E.G., CANFIELD, T.H. and SLOAN, H.J. 1953. Prevention of perosis in goslings by nicotinic acid. Poult. Sci. *32,* 550–551.

GROPP, J., SCHULZ, V., SAUERWEIN, G. and TIEWS, J. 1972. The nicotinic acid activity of nicotinic acid nitrile and estimation of the niacine requirement of Japanese quail. Int. J. Vitam. Nutr. Res. *42,* 205–213.

HEGSTEAD, D.M. 1946. Nutritional studies with the duck. III. Niacin deficiency. J. Nutr. *32,* 467–472.

PATTERSON, E.B., HUNT, J.R., VOHRA, P., BLAYLOCK, L.G. and MCGINNIS, J. 1956. The niacin and tryptophan requirements of chicks. Poult. Sci. *35,* 499–504.

SCOTT, M.L., HOLM, E.R. and REYNOLDS, R.E. 1959. Studies on the niacin, riboflavin, choline, manganese and zinc requirements of young ringnecked pheasants for growth, feathering, and prevention of leg disorders. Poult. Sci. *38,* 1344–1350.

SUNDE, M.L. and BIRD, H.R. 1957. The niacin requirement of the young ringneck pheasant. Poult. Sci. *36,* 34–42.

20. PANTOTHENIC ACID

BALLOUN, S.L. and PHILLIPS, R.E. 1957. Interaction effects of vitamin B_{12} and pantothenic acid in breeder hen diets on hatchability, chick growth and livability. Poult. Sci. *36,* 929–934.

FOX, M.R.S., HUDSON, G.Q. and HINTZ, M.E. 1966. Pantothenic acid requirement of young Japanese quail. Fed. Proc. *25,* 721. (Abstract)

GILLIS, M.B., HEUSER, G.F. and NORRIS, L.C. 1948. Pantothenic acid in the nutrition of the hen. J. Nutr. *35,* 351–363.

HEGSTED, D.M. and PERRY, R.L. 1948. Nutritional studies with the duck. J. Nutr. *35,* 411–417.

KRATZER, F.H., DAVIS, P.N., MARSHALL, B.J. and WILLIAMS, D.E. 1955. The pantothenic acid requirement of turkey hens. Poult. Sci. *34,* 68–72.

SCOTT, M.L., HOLM, E.R. and REYNOLDS, R.E. 1964. Studies on the pantothenic acid and unidentified factor requirements of young ringnecked pheasants and Bobwhite quail. Poult. Sci. *43,* 1534–1539.

SERAFIN, J.A. 1974. Studies on the ribloflavin, niacin, pantothenic acid and choline requirements of young Bobwhite quail. Poult. Sci. *53,* 1522–1532.

21. POTASSIUM

BEN-DOR, B. 1941. Requirement of potassium by the chick. Proc. Soc. Exp. Biol. Med. *46,* 341–343.

GILLIS, M.B. 1948. Potassium requirement of the chick. J. Nutr. *36,* 351–357.

LEACH, R.M. 1974. Studies on the potassium requirement of the laying hen. J. Nutr. *104,* 684–686.

SULLIVAN, T.W. 1963. Studies on the potassium requirement of turkeys to 4 weeks of age. Poult. Sci. *42,* 1072–1075.

22. PROTEIN

ANDREWS, T.L., HARMS, R.H. and WILSON, H.R. 1973. Protein requirement of the Bobwhite chick. Poult. Sci. *52,* 2199–2201.

BALLOUN, S.L. and SPEERS, G.M. 1969. Protein requirements of laying hens as affected by strain. Poult. Sci. *48,* 1175–1188.

BLAIR, R. and YOUNG, R.J. 1974. Egg production responses of Coturnix quail to dietary additions of nitrogen in the form of amino acids, diammonium citrate and intact protein. Poult. Sci. *53,* 391–400.

BRAGG, D.B. and AKINWANDE, A.I. 1973. The nutritional value of wheat protein for early growth of layer and broiler breeder pullets. Poult. Sci. *52,* 1646–1651.

COUCH, J.R. and ROYTON, J.K. 1974. Amino acids and protein in broiler nutrition. Poult. Sci. *53,* 750–755.

JONES, D.B. 1931. Factors for converting percentages of nitrogen in foods and feeds into percentages of protein. U.S. Dep. Agric. Circ. *183.*

LEPORE, P.D. and MARKS, H.L. 1971. Growth rate inheritance in Japanese quail. 5. Protein and energy requirements of lines selected under different nutritional environments. Poult. Sci. *50,* 1335–1341.

MARCH, B.E. and BIELY, J. 1972. The effects of protein level and amino acid balance in wheat-based laying rations. Poult. Sci. *51,* 547–557.

MERRITT, E.S. and AITKEN, J.R. 1961. Raising geese. Can. Dep. Agric. Publ. *848.*

MINEAR, L.R., MILLER, D.L. and BALLOUN, S.L. 1972. Protein requirements of turkey breeder hens. Poult. Sci. *51,* 2040–2043.

NIVAS, S.C. and SUNDE, M.L. 1969. Protein requirements of layers per day and phase feeding. Poult. Sci. *48,* 1672–1678.

NORRIS, L.C., ELMORE, L.J., RINGROSE, R.C. and BUMP, G. 1936. The protein requirement of ringnecked pheasant chicks. Poult. Sci. *15,* 454–459.

PATRICK, H. 1960. Unpublished data. W. Va. Agric. Exp. Sta., Morgantown.

PONTON, D.K. 1972. Nutritional and biochemical aspects of lysine metabolism in the fasting chick. Unpublished Ph.D. Dissertation. West Virginia Univ., Morgantown.

ROBERSON, R.H. and FRANCIS, D.W. 1963A. The effect of protein level, iodinated casein and supplemental methionine on the performance of White Chinese geese. Poult. Sci. *42,* 863–867.

ROBERSON, R.H. and FRANCIS, D.W. 1963B. The effect of energy and protein levels of the ration on the performance of White Chinese geese. Poult. Sci. *42,* 867–871.

ROBERSON, R.H. and FRANCIS, D.W. 1963C. The effect of protein and energy levels of the ration on carcass characteristics of White Chinese geese. Poult. Sci. *42,* 872–875.

STOCKLAND, W.L. and BLAYLOCK, L.G. 1974. The influence of temperature on the protein requirement of cage reared replacement pullets. Poult. Sci. *53,* 1174–1187.

THAYER, R.H., HUBBELL, G.E., KASBOHM, J.A., MORRISON, R.D. and NELSON, E.C. 1974. Daily protein intake requirement of laying hens. Poult. Sci. *53,* 354–364.

WEBER, C.W. and REID, B.L. 1967. Protein requirements of Coturnix quail to five weeks of age. Poult. Sci. *46,* 1190–1194.

23. PYRIDOXINE

FULLER, H.L. and DUNAHOO, W.S. 1959. The effect of various drug additives on the vitamin B_6 requirement of chicks. Poult. Sci. *38,* 1150–1154.

FULLER, H.L., FIELD, R.C., RONCALLI-AMICI, R., DUNAHOO, W.S. and EDWARDS, H.M. 1961. The vitamin B_6 requirement of breeder hens. Poult. Sci. *40,* 249–253.

FULLER, H.L. and KIFER, P.E. 1959. The vitamin B_6 requirement of chicks. Poult. Sci. *38*, 255–260.

GRIES, C.L. and SCOTT, M.L. 1972. The pathology of pyridoxine deficiency in chicks. J. Nutr. *102,* 1259–1267.

HOGAN, A.G., RICHARDSON, L.R., PATRICK, H., O'DELL, D.L. and KEMPSTER, H.L. 1941. Vitamin B_6 and chick nutrition. Poult. Sci. *20,* 180–183.

SULLIVAN, T.W., HEIL, H.M. and ARMINTROUT, M.E. 1967. Dietary thiamine and pyridoxine requirements of young turkeys. Poult. Sci. *46,* 1560–1564.

WALDROUP, P.W., MAXEY, J.F., LUTHER, L.W., JONES, B.D. and NESHEIM, M.L. 1976. Factors affecting the response of turkeys to biotin and pyridoxine supplementation. Univ. Arkansas, (Fayetteville) Exp. Stn. Bull. *805.*

24. RIBOFLAVIN

BOLTON, W. 1947. The riboflavin requirement of the White Wyandotte Chick. J. Agric. Sci. *34,* 198–206.

BOUCHER, R.V., HILL, F.W., PATRICK, H. and KNANDEL, H.C. 1941. The riboflavin requirement of turkeys for hatchability. Poult. Sci. *20,* 456. (Abstract)

BOUCHER, R.V., PATRICK, H. and KNANDEL, H.C. 1942. The riboflavin requirement of turkeys for hatchability and growth. Poult. Sci. *21*, 466. (Abstract)

HEUSER, G.F., WILGUS, H.S. and NORRIS, L.C. 1938. The quantitative vitamin G requirement of chicks. Poult. Sci. *17,* 105–108.

PATRICK, H., DARROW, M.I. and MORGAN, C.L. 1944. The role of riboflavin in turkey poult nutrition. Poult. Sci. *23,* 146–148.

SCOTT, M.L., HOLM, E.R. and REYNOLDS, R.E. 1959. Studies on the niacin, riboflavin, choline, manganese and zinc requirements of young ring-necked pheasants for growth, feathering, and prevention of leg disorders. Poult. Sci. *38*, 1344–1350.

SERAFIN, J.A. 1974. Studies on the riboflavin, niacin, pantothenic acid and choline requirements of young Bobwhite quail. Poult. Sci. *53,* 1522–1532.

WAIBEL, P.E., and CANFIELD, T.H. 1957. Studies on riboflavin needs of goslings. (unpublished)

WYATT, R.D., TUNG, H.T., DONALDSON, W.E. and HAMILTON, P.B. 1973. A new description of riboflavin deficiency syndrome in chickens. Poult. Sci. *52,* 237–244.

25. SELENIUM

GRIES, C.L. and SCOTT, M.L. 1972. Pathology of selenium deficiency in the chick. J. Nutr. *102,* 1287–1296.

MOXON, A.L. 1937. Alkali disease or selenium poisoning. S. D. Agric. Exp. Stn. Bull. *311.*

NATL. RES. COUNC. COMM. ANIM. NUTR. SUBCOMM. SELENIUM. 1971. Selenium in Nutrition. Natl. Acad. Sci., Washington, D.C.

POLEY, W.E. and MOXON, A.L. 1938. Tolerance levels of seleniferous grains in laying rations. Poulty Sci. *17,* 72–76.

ROSENFELD, I. and BEATH, O.A. 1964. Selenium: Geobotany, Biochemistry, Toxicity, and Nutrition. Academic Press, New York, London.

SCOTT, M.L. 1966. Studies on the interrelationship of selenium, vitamin E and sulfur amino acids in a nutritional myopathy of the chick. Ann. N.Y. Acad. Sci. *138*, 82–89.

SCOTT, M.L., OLSON, G., KROOK, L. and BROWN, W.R. 1967. Selenium-responsive myopathies of myocardium and of smooth muscle in the young poult. J. Nutr. *91,* 573–583.

THOMPSON, J.N. and SCOTT, M.L. 1969. Role of selenium in the nutrition of the chick. J. Nutr. *97,* 335–342.

TULLY, W.C. and FRANKE, K.W. 1935. A new toxicant occurring naturally in certain samples of plant foodstuffs. VI. A study of the effect of affected grains on growing chicks. Poult. Sci. *14,* 280–284.

WALTER, E.D. and JENSEN, L.S. 1963. Effectiveness of selenium and non-effectiveness of sulfur amino acids in preventing muscular dystrophy in the turkey poult. J. Nutr. *80,* 327–331.

26. SILICA

CARLISLE, E.M. 1970A. Silicon, an essential element for chickens. Science *169,* 279.

CARLISLE, E.M. 1970B. Silicon, a possible factor in bone calcification. Science *167,* 279.

FRIGERT, S. 1958. Silica in chicken feathers. J. Invest. Dermatol. *31*, 95.

FROMAN, S.A. and SAUER, F. 1962. Some changes in the urine of sheep fed a hay high in silica. Can. J. Animal Sci. *42,* 9–14.

KURNICK, A.A. and REID, B.L. 1960. Poultry nutrition studies with bentonite. Feedstuffs Dec. 24.

QUISENBERRY, J.H. 1967. Some physiological responses to dietary clay supplements. Clay-Mineral Soc. Ann. Meet., Aug. 18, Denver.

QUISENBERRY, J.H. and BRADLEY, J.W. 1964. Sodium bentonite feeding experiment, progress report. Feedstuffs Nov. 24.

27. SULFUR

EMERY, R.S., SMITH, C.K. and HUFFMAN, C.F. 1957. Utilization of inorganic sulfate by rumen micro-organisms. 1. Incorporation of inorganic sulfate into amino acids. Appl. Microbiol. *5,* 360–362.

KELLOGG, W.W., CADLE, R.D., ALLEN, E.R., LAZRUS, A.L. and MARTELL, E.A. 1972. The sulfur cycle. Science *175,* 587–595.

MACHLIN, L.J., DENTON, C.A., PEARSON, P.E. and BIRD, H.R. 1952. Synthesis of cystine from inorganic sulfate by the hen. Fed Proc. *11,* 448–449.

MARTIN, W.G. and PATRICK, H. 1960. Radionuclide mineral studies. 2. The effect of certain dietary constituents, including protein and vitamin B_{12} on the metabolism of inorganic S^{35} sulfate in the chick. Poult. Sci. *39,* 1501–1540.

MUTH, O.H. and OLDSFIELD, J.E. 1969. Symposium: Sulfur in Nutrition. AVI Publishing Co., Westport, Conn.

28. SODIUM AND CHLORINE

BURNS, C.H., CRAVENS, W.W. and PHILLIPS, P.H. 1953. The sodium and potassium requirements of the chick and their interrelationship. J. Nutr. *50,* 317–329.

DEAN, W.F. 1972. Recent findings in duck nutrition. Proc. Cornell Nutr. Conf. (personal communication)

KUMPOST, H.E. and SULLIVAN, T.W. 1966. Minimum sodium requirement and interaction of potassium and sodium in the diet of young turkeys. Poult. Sci. *45,* 1334–1339.

LEACH, R.M., JR. and NESHEIM, M.C. 1963. Studies on chloride deficiency in chicks. J. Nutr. *81,* 193–199.

MCWARD, G.W. and SCOTT, H.M. 1961. Sodium requirement of the young chick fed purified diets. Poult. Sci. *40,* 1026–1029.

NESHEIM, M.C., LEACH, R.M., JR., ZEIGLER, T.R. and SERAFIN, J.A. 1964. Interrelationships between dietary levels of sodium, chlorine and potassium. J. Nutr. *84,* 361–366.

SCOTT, M.L., VAN TEINHOVEN, A., HOLM, E.R. and REYNOLDS, M.E. 1960. Studies on the sodium, chlorine, and iodine requirements of young pheasants and quail. J. Nutr. *71,* 282–288.

29. THIAMIN

GRIES, C.L. and SCOTT, M.L. 1972. The pathology of thiamin, riboflavin, pantothenic acid and niacin deficiencies in the chick. J. Nutr. *102,* 1269–1285.

ROBENALT, R.C. 1960. The thiamine requirement of young turkey poults. Poult. Sci. *39,* 354–360.

THORNTON, P.A. 1960. Thiamine requirement of growing chicks as influenced by breed differences. Poult. Sci. *39,* 440–444.

THORNTON, P.A. and SHUTZE, J.V. 1960. The influence of dietary energy level, energy source, and breed on the thiamine requirement of chicks. Poult. Sci. *39,* 192–199.

30. TOXICITY: INORGANIC ELEMENTS

ARTHUR, D., MOTZOK, I. and BRANION, H.D. 1958. Interaction of dietary copper and molybdenum in rations fed to poultry. Poult. Sci. *37,* 1181. (Abstract)

BERG, L.R. 1963. Evidence of vanadium toxicity resulting from the use of certain commercial phosphorus supplements in chick rations. Poult. Sci. *42,* 766–769.

BERG, L., BEARSE, G.E. and MERRILL, L.H. 1963. Vanadium toxicity in laying hens. Poult. Sci. *42,* 1407–1411.

BERG, L.R. and MARTINSON, R.D. 1972. Effect of diet compostition on the toxicity of zinc for the chick. Poult. Sci. *51,* 1690–1695.

CHICCO, C.F., AMMERMAN, C.B., VAN WALLEGHEM, P.A., WALDROUP, P.W. and HARMS, R.H. 1967. Effects of varying dietary rations of magnesium, calcium, and phosphorus in growing chicks. Poult. Sci. *46,* 368–373.

DAMRON, B.L., SIMPSON, C.F. and HARMS, R.H. 1969. The effects of feeding various levels of lead on the performance of broilers. Poult. Sci. *48,* 1507–1509.

DAVIES, R.E., REID, B.L., KURNICK, A.A. and COUCH, J.R. 1960. The effect of sulfate on molybdenum toxicity in the chick. J. Nutr. *70,* 193–198.

DOBERENZ, A.R., KURNICK, A.A., HULETT, B.J. and REID, B.L. 1965. Bromide and fluoride toxicities in the chick. Poult. Sci. *44,* 1500–1504.

HARPER, J.A. and ARSCOTT, G.H. 1962. Salt as a stress factor in relation to pendulous crop and aortic rupture in turkeys. Poult. Sci. *41,* 497–499.

HATHCOCK, J.N., HILL, C.H. and MATRONE, G. 1964. Vanadium toxicity and distribution in chicks and rats. J. Nutr. *82,* 106–110.

HILL, C.H. 1974. Influence of high levels of minerals on the susceptibility of chicks to Salmonella gallinarum. J. Nutr. *104,* 1221–1226.

HILL, C.H., MATRONE, G., PAYNE, W.L. and BARBER, C.W. 1963. In vivo interactions of cadmium with copper, zinc, and iron. J. Nutr. *80,* 227–235.

HILL, C.H., STARCHER, B. and MATRONE, G. 1964. Mercury and silver interrelationships with copper. J. Nutr. *83,* 107–110.

JENSEN, L.S. 1975A. Precipitation of selenium deficiency by high dietary levels of copper and zinc. Proc. Soc. Exp. Biol. Med. *149*, 113–116.

JENSEN, L.S. 1975B. Modification of a selenium toxicity in chicks by dietary silver and copper. J. Nutr. *105,* 769–775.

JOHNSON, D., JR., MEHRING, A.L., JR., SAVINS, F.X. and TITUS, H.W. 1962. The tolerance of growing chickens for dietary zinc. Poult. Sci. *41,* 311–317.

KRISTA, L.M., CARLSON, C.W. and OLSON, O.E. 1961. Some effects of saline waters on chicks, laying hens, poults and ducklings. Poult. Sci. *40,* 938–944.

LEPORE, P.D. and MILLER, R.F. 1965. Embryonic viability as influenced by excess molybdenum in chicken breeder diets. Proc. Soc. Exp. Biol. Med. *118,* 155–157.

MEHRING, A.L., JR., BRUMBAUGH, J.H., SUTHERLAND, A.J. and TITUS, H.W. 1960. The tolerance of growing chickens for dietary copper. Poult. Sci. *23,* 149–151.

MOXON, A.L. and WILSON, W.O. 1944. Selenium-arsenic antagonism in poultry. Poult. Sci. *23,* 149–151.

NESHEIM, M.C., LEACH, R.M., JR., ZEIGLER, T.R. and SERAFIN, J.A. 1964. Interrelationships between dietary levels of sodium, chlorine, and potassium. J. Nutr. *84,* 361–366.

NUGARA, D. and EDWARDS, H.M., JR. 1963. Influence of dietary Ca and P levels on the Mg requirement of the chick. J. Nutr. *80,* 181–184.

PARKHURST, C.R. and THAXTON, P. 1973. Toxicity of mercury in young chicks. 1. Effect on growth and mortality. Poult. Sci. *52,* 273–276.

POUPOULIS, C. and JENSEN, L.S. 1976. Effect of high dietary copper on gizzard integrity of the chick. Poult. Sci. *55,* 113–121.

ROBERSON, R.H. and SCHAIBLE, P.J. 1960. The tolerance of growing chicks for high levels of different forms of zinc. Poult. Sci. *39,* 893–896.

SELL, J.L. and ROBERTS, W.K. 1963. Effects of dietary nitrate on the chick; growth, liver vitamin A stores and thyroid weight. J. Nutr. *79,* 171–178.

STORER, N.L. and NELSON, T.S. 1968. The effect of various aluminum compounds on chick performance. Poult. Sci. *47,* 244–247.

SUNDE, M.L. 1964. The use of the turkey poult as a test animal for nitrite toxicity. Poult. Sci. *43,* 1368. (Abstract)

SUPPLEE, W.C. 1961. Production of zinc deficiency in turkey poults by dietary cadmium. Poult. Sci. *40,* 827–828.

TURK, J.L., JR. and KRATZER, F.H. 1960. The effects of cobalt in the diet of the chick. Poult Sci. *39,* 1302. (Abstract)

VOHRA, P. and KRATZER, F.H. 1968. Zinc, copper and manganese toxicities in turkey poults and their alleviation by EDTA. Poult. Sci. *47,* 699–704.

WEBER, C.W., DOBERENZ, A.R., WYCKOFF, R.W.G. and REID, B.L. 1968. Strontium metabolism in chicks. Poult. Sci. *47,* 1318–1323.

31. UNIDENTIFIED GROWTH FACTORS

BERG, L.R. and LAWRENCE, W.W. 1971. Cottonseed meal, dehydrated grass and ascorbic acid as dietary factors preventing toxicity of vanadium for the chick. Poult. Sci. *50,* 1399–1404.

BLAIR, R., SCOTT, M.L. and YOUNG, R.J. 1972. Unidentified factor activities in whole soybeans required for optimum growth of Coturnix quail. J. Nutr. *102,* 1529–1541.

CHANG, C.H., FALEN, L. and JENSEN, L.S. 1974. Effect of maternal diet on the response of Japanese quail to purified and practical diets. Poult. Sci. *53,* 265–272.

CHANG, C.H. and WAIBEL, P.E. 1970. Efficacy of zinc bacitracin and sources of unidentified growth factors with corn-soybean meal type diet for turkey poults. Poult. Sci. *49,* 733–743.

DIXON, T. and COUCH, J.R. 1970. Distillers dried solubles as a source of unidentified growth factors required by the chick and poult. Poult. Sci. *49,* 393–401.

JENSEN, L.S. and CHANG, C.H. 1976. Fractionation studies on a factor in linseed meal protecting against selenosis in chicks. Poult. Sci. *55,* 594–599.

TOUCHBURN, S.P., SILVA, R.D.M. and NABER, E.C. 1974. Further evidence for an unidentified nutrient in fish solubles which improves hatchability and progeny growth of turkeys and Japanese quail. Poult. Sci. *53,* 1745–1758.

WALDROUP, P.W. and RUTHERFORD, H.O. 1971. Acceptability of corn dried steep liquor concentrate for laying hens and turkeys. Poult. Sci. *50,* 1863–1867.

32. VITAMIN A

BERGER, H. and GEBHARDT, G. 1969. Relations between vitamin A and protein supplies of growing ducks. 4. Effect of different amounts of protein on the requirement for vitamin A. Jahrb. Tierenahrung Futterung *7,* 399–409.

CASTANO, F.F., BOUCHER, R.V. and CALLENBACH, E.W. 1951. Utilization by the chick of vitamin A from different sources. J. Nutrition *45,* 131–141.

CHAVEZ, R., CREGER, C.R. and COUCH, J.R. 1963. Vitamin A requirements of growing turkeys. Poult. Sci. *42,* 1259–1260. (Abstract)

COLES, B., BIELY, J. and MARCH, B.E. 1970. Vitamin A deficiency and Eimeria acervulina infection in the chick. Poult. Sci. *49,* 1295–1301.

COUCH, J.R., CREGER, C.R. and CHAVES, R. 1971. The vitamin A requirement for growing turkeys. Br. Poult. Sci. *12,* 367–371.

GURCAY, R., BOUCHER, R.V. and CALLENBACH, E.W. 1950. Utilization of vitamin A by turkey poults. J. Nutrition *44,* 565–577.

HILL, F.W., SCOTT, M.L., NORRIS, L.C. and HEUSER, G.R. 1961. Reinvestigation of the vitamin A requirements of laying and breeding hens and their progeny. Poult. Sci. *40,* 1245–1254.

JENSEN, L.S. 1965. Vitamin A requirement of breeding turkeys. Poult. Sci. *44,* 1609–1610.

MARUSICH, W. and BAUERNFEIND, J.C. 1963. The biological activity of beta-carotene in poultry and rats. Poult. Sci. *42,* 949–957.

PARRISH, D.B., ZIMMERMAN, R.A., SANFORD, P.E. and HUNG, E. 1963. Utilization of alfalfa carotene and vitamin A by growing chicks. J. Nutr. *79,* 9–17.

STOEWSAND, G.S. and SCOTT, M.L. 1961. The vitamin A requirement of breeding turkeys and their progeny. Poult. Sci. *40,* 1255–1262.

THORNTON, P.A. and WHITTET, W.A. 1962. The influence of dietary energy level, energy source, breed, and sex on vitamin A requirement in the chick. Poult. Sci. *41,* 32–36.

33. VITAMIN B_{12}

BALDINI, J.T., ROBERTS, R.E. and KIRKPATRICK, C.M. 1953. Antibiotic and vitamin B_{12} supplements as related to crude protein level of Bobwhite quail diets. Poult. Sci. *32,* 563–567.

CHIN, D., ANDERSON, J.B., MILLER, R.F., NORRIS, L.C. and HEUSER, G.F. 1958. The vitamin B_{12} requirement of White Leghorn hens. Poult. Sci. *37,* 335–343.

DAVIS, R.L. and BRIGGS, G.M. 1951. The vitamin B_{12} requirement of the chick. Poult. Sci. *30,* 628–629.

JOHNSON, E.L. 1955. Turkeys require vitamin B_{12} and choline. Poult. Sci. *34,* 1013–1016.

LOOI, S.H. and RENNER, R. 1974. Effect of feeding "carbohydrate-free" diets on the chicks' requirement for vitamin B_{12}. J. Nutr. *104,* 394–399.

OTT, W.H. 1951. Further studies of the activity of crystalline vitamin B_{12} for chick growth. Poult. Sci. *30,* 86–91.

RYS, R. and KORELESKI, J. 1974. The effect of dietary propionic acid on the requirement of chicks for vitamin B_{12}. Br. J. Nutr. *31,* 143–146.

WIESE, A.C., PETERSEN, C.F., DAHLSTROM, R.V. and LAMPMAN, C.E. 1952. Effect of vitamin B_{12} intake of hens upon carry-over in chicks. Poult. Sci. *31,* 851–854.

34. VITAMIN D

ASSOC. OFF. AGRIC. CHEM. 1955. Official and Tentative Methods of Analysis, 8th Edition. AOAC, Washington, D.C.

BIELY, J. and MARCH, B.E. 1967. Calcium and vitamin D in broiler rations. Poult. Sci. *46,* 223–232.

BRAGG, D.B., FLOYD, J. and STEPHENSON, E.L. 1971. Factors affecting the transfer of Ca^{45} from the hens' diet to the eggshell. Poult. Sci. *50,* 167–173.

BRITTON, W.M. and WATT, R.D. 1978. Effects of dietary aflatoxins and Vitamin D_3 metabolism in chicks. Poult. Sci. *57,* 163–165.

GARLICH, J.D. and WYATT, R.D. 1971. Effects of increased vitamin D_3 on calcium retention and eggshell calcification. Poult. Sci. *50,* 950–956.

PATRICK, H. and SCHWEITZER, G.K. 1955. Usage of Ca^{45} in determination of vitamin D depletion and response time in chicks. Poult. Sci. *34,* 832–833.

SCOTT, M.L., HOLM, E.R. and REYNOLDS, R.E. 1958. The calcium, phosphorus, and vitamin D requirements of young pheasants. Poult. Sci. *37,* 1419–1425.

SHUE, G.M. 1967. Vitamin D requirements of young Japanese quail. Fed. Proc. *26,* 697. (Abstract)

SINGSEN, E.P., MATTERSON, L.D. and SCOTT, H.M. 1947. Phosphorus in poultry nutrition. III. The relationship between the source of vitamin D and the utilization of cereal phosphorus by the poult. J. Nutr. *33,* 13–26.

WALDROUP, P.W., AMMERMAN, C.B. and HARMS, R.H. 1963. The relation of phosphorus, calcium, and vitamin D_3 in the diet of broiler-type chicks. Poult. Sci. *42,* 982–989.

WALDROUP, P.W., STEARNS, J.E., AMMERMAN, C.B. and HARMS, R.H. 1965. Studies on the vitamin D_3 requirement of the broiler chick. Poult. Sci. *44,* 543–548.

YANG, H.S., WAIBEL, P.E. and BRENES, J. 1973. Evaluation of vitamin D_3 supplements by biological assay using the turkey. J. Nutr. *103,* 1187–1194.

35. VITAMIN E

BARTOV, I. and BORNSTEIN, S. 1972. Nutritional factors affecting the occurrence of experimental encephalomalacia in chicks. Poult. Sci. *51,* 868–876.

BARTOV, I., BUDOWSKI, T. and BORNSTEIN, S. 1965. The relation between alpha-tocopherol content of the breeder diet and that of the newly hatched chick. Poult. Sci. *44,* 1489–1494.

COMBS, G.F. and SCOTT, M.L. 1974. Dietary requirements for vitamin E and selenium measured at the cellular level in the chick. J. Nutr. *104,* 1292–1296.

JENSEN, L.S. 1968. Vitamin E and essential fatty acids in avian reproduction. Fed. Proc. Fed. Am. Soc. Exp. Biol. *27,* 914–919.

JENSEN, L.S. and MCGINNIS, J. 1957. Studies on the vitamin E requirement of turkeys for reproduction. Poult. Sci. *36,* 1344–1350.

MACHLIN, J.L. and GORDON, G.S. 1962. Etiology of exudative diathesis, encephalomalacia, and muscular degeneration in the chicken. Poult. Sci. *41,* 473–477.

TENGERDY, R.T. and NOCKELS, C.F. 1973. The effect of vitamin E on egg production, hatchability and humoral immune response of chickens. Poult. Sci. *52,* 778–783.

36. VITAMIN K

CHARLES, O.W. and HUSTON, T.M. 1972. The biological activity of vitamin K materials following storage and pelleting. Poult. Sci. *51,* 1421–1427.

NELSON, T.S. and NORRIS, L.C. 1961A. Studies on the vitamin K requirement of the chick. Poult. Sci. *40,* 392–395.

NELSON, T.S. and NORRIS, L.C. 1961B. Studies on the vitamin K requirement of the chick. J. Nutr. *73,* 135–142.

37. WATER

ADAMS, A.W. 1973. Consequences of depriving laying hens of water a short time. Poult. Sci. *52,* 1221–1223.

DUNSON, W.A. and BUSS, E.G. 1968. Abnormal water balance in a mutant strain of chickens. Science *161,* 167.

ENOS, H.L., MORENZ, R.E. and WHITTET, W.A. 1967. Shade materials, growth and water consumption of turkeys. Poult. Sci. *46,* 1412–1421.

HALLER, R.W. and SUNDE, M.L. 1966. The effects of withholding water on the body temperature of poults. Poult. Sci. *45,* 991–997.

HEYWANG, B.W. 1941. The water consumption of hens. Poult. Sci. *20,* 184–187.

KELLERUP, S.U., PARKER, J.W. and ARSCOTT, G.H. 1965. Effect of restricted water consumption on broiler chickens. Poult. Sci. *44,* 78–83.

MULKEY, G.J. and HUSTON, T.M. 1967. The tolerance of different ages of domestic fowl to body water loss. Poult. Sci. *46,* 1565–1569.

PARKER, J.T., BOONE, M.A. and KNECHTGES, J.F. 1972. The effect of ambient temperature upon body temperature, feed consumption and water consumption using two varieties of turkeys. Poult. Sci. *51,* 659–664.

PATRICK, H. 1955. The influence of protein source on consumption and excretion of water and excreta voided by broiler chicks. Poult. Sci. *34,* 155–157.

PATRICK, H. and FERRISE, A. 1962. The water requirements of broilers. Poult. Sci. *41,* 1363–1367.

38. XANTHOPHYLL

COUCH, J.R. and FARR, F.M. 1971. Evaluation of beta-apo-8′-caratenal. Br. Poult. Sci. *12,* 87–93.

FRY, J.L. and HARMS, R.H. 1975. Yolk color, candled egg grade and xanthophyll availability from dietary natural pigmenting ingredients. Poult. Sci. *54,* 1094–1101.

MADIEDO, G. and SUNDE, M.L. 1964. The effect of algae, dried lake weed, alfalfa and ethoxyquin on yolk color. Poult. Sci. *44,* 1056–1061.

RUFF, M.D. and BRITTON, W.M. 1976. Reduced yolk color in layers with coccidiosis. Poult. Sci. *55,* 1712–1716.

SUNDE, M.L. 1962. The effect of different levels of vitamin A, beta-apo-8′-caratenal and alfalfa on yolk color. Poult. Sci. *41,* 532–541.

39. ZINC

BLAMBERG, D.L., BLACKWOOD, U.B., SUPPLE, W.C. and COMBS, G.F. 1960. Effect of zinc deficiency in hens on hatchability and embryonic development. Proc. Soc. Exp. Biol. Med. *104,* 217–220.

DESAI, I.D. and SCOTT, M.L. 1965. Mode of action of selenium in relation to biological activity of tocopherols. Arch. Biochem. Biophys. *110,* 309–315.

EDWARDS, H.M., JR., DUNAHOO, W.S. and FULLER, H.L. 1959. Zinc requirement studies with practical rations. Poult. Sci. *38,* 436–439.

KRATZER, F.H., BHORA, P., ALLRED, J.B. and DAVIS, P.N. 1958. Effect of zinc upon growth and incidence of perosis in turkey poults. Proc. Soc. Exp. Biol. Med. *98,* 205–207.

LEASE, J.G., BARNETT, B.D., LEASE, E.J. and TURK, D.E. 1960. The biological unavailability to the chick of zinc in a sesame meal ration. J. Nutr. *72,* 66–70.

O'DELL, B.L. and SAVAGE, J.E. 1960. Effect of phytic acid on zinc availability. Proc. Soc. Exp. Biol. Med. *103,* 304.

ROBERSON, R.H. and SCHAIBLE, P.J. 1958. The zinc requirement of the chick. Poult. Sci. *37,* 1321–1323.

SULLIVAN, T.W. 1961. The zinc requirement of Broad Breasted Bronze poults. Poult. Sci. *40,* 334–340.

SUNDE, M.L. 1972. Zinc requirement for normal feathering of commerical Leghorn-type pullets. Poult. Sci. *51,* 1316–1322.

TURK, D.E. and STEPHENS, J.T. 1967. Gastro-intestinal tract disease and zinc absorption: cecal coccidiosis. Poult. Sci. *46,* 938.

VOHRA, P. and KRATZER, F.H. 1964. Influence of various chelating agents on the availability of zinc. J. Nutr. *82,* 249–256.

ZIEGLER, T.R., LEACH, R.M., JR., NORRIS, L.C. and SCOTT, M.L. 1961. Zinc requirement of the chick: factors affecting requirement. Poult. Sci. *40,* 1584–1593.

40. MISCELLANEOUS

ALTMAN, P.L. and DITTMER, D.S. 1968. Metabolism. Fed. Amer. Soc. Exp. Biol., Bethesda, Md.

BIESTER, H.E. and SCHWARTE, L.G. 1965. Diseases of Poultry, 5th Edition. Iowa State Univ. Press, Ames.

BISHOP, C. and TAYLOR, T.G. 1963. Studies on the vitamin content of bird seeds. Vet. Rec. *75,* 688–691.

BLACKMORE, D.K. 1963. The incidence and aetiology of thyroid dysplasia in budgerigars (Melopsittacus undulatus). Vet. Rec. *75,* 1068–1072.

HILLERMAN, F.W., KRATZER, F.H. and WILSON, W.O. 1953. Food passage through chickens and turkeys and some regulating factors. Poult Sci. *32,* 332.

LEVI, W.M. 1951. The Pigeon. The R.L. Gryan Co., Columbia, S.C.

NATL. RES. COUNC. 1958. Composition of cereal grains and forages. Comm. on Feed Composition, Agric. Brd. NAS Pub. *585.*

O'DELL, B.L., WOODS, W.D., LAERDAL, A., JEFFAY, A.M. and SAVAGE, J.E. 1960. Distribution of the major nitrogenous compounds and amino acids in chicken urine. Poult. Sci. *39,* 426–432.

PATRICK, H. 1965. Comparative Animal Nutrition Supplement. W. Va. Univ. Coll. Agric. and Forestry, Morgantown, W. Va.

PETRAK, M.L. 1969. Diseases of Cage and Aviary Birds. Lea and Febeger, Philadelphia.

RATCLIFFE, H.L. 1940. Diets for a zoological garden. Zoologica *25,* 463–472.

STAT. YEARBK. 1976. U.N., Dep. Econ. and Soc. Aff., Stat. Off., U.N., N.J.

WILLIAMS, D.M. 1973. The effects of recycling dehydrated poultry wastes through feed on several economic traits and on excreta composition in poultry. Unpublished M.S. Thesis. West Virginia Univ., Morgantown.

Index

Other AVI Books

COMMERCIAL CHICKEN PRODUCTION MANUAL
North *Second Edition*

EGG SCIENCE AND TECHNOLOGY
Stadelman and Cotterill *2nd Edition*

ENCYCLOPEDIA OF FOOD TECHNOLOGY
Peterson and Johnson *Vol. 3*

EVALUATIONS OF PROTEINS FOR HUMANS
Bodwell

FOOD ANALYSIS: THEORY AND PRACTICE
Pomeranz and Meloan *Revised Edition*

FOOD COLLOIDS
Graham

FOOD PRODUCTS FORMULARY: MEATS, POULTRY, FISH AND SHELLFISH
Komarik and Tressler *Vol. 1*

FOOD QUALITY ASSURANCE
Gould

NUTRITIONAL QUALITY INDEX OF FOODS
Hansen et al.

PHOSPHATES IN FOOD PROCESSING
deMan and Melnchyn

POULTRY PRODUCTS TECHNOLOGY
Mourteny *2nd Edition*

PROGRESS IN HUMAN NUTRITION
Margen and Ogar *Vol. 1 and 2*

PROTEIN RESOURCES AND TECHNOLOGY
Milner et al.

QUALITY CONTROL FOR THE FOOD INDUSTRY
Kramer and Twigg *3rd Edition* *Vols. 1 and 2*

SOURCE BOOK FOR FOOD SCIENTISTS
Ockerman